Recent Advances in Theoretical and Computational Modeling of Composite Materials and Structures

Recent Advances in Theoretical and Computational Modeling of Composite Materials and Structures

Editors

Francesco Tornabene
Rossana Dimitri

MDPI • Basel • Beijing • Wuhan • Barcelona • Belgrade • Manchester • Tokyo • Cluj • Tianjin

Editors
Francesco Tornabene
University of Salento
Italy

Rossana Dimitri
University of Salento
Italy

Editorial Office
MDPI
St. Alban-Anlage 66
4052 Basel, Switzerland

This is a reprint of articles from the Special Issue published online in the open access journal *Applied Sciences* (ISSN 2076-3417) (available at: https://www.mdpi.com/journal/applsci/special_issues/Theoretical_and_Computational_Modeling).

For citation purposes, cite each article independently as indicated on the article page online and as indicated below:

LastName, A.A.; LastName, B.B.; LastName, C.C. Article Title. *Journal Name* **Year**, *Volume Number*, Page Range.

ISBN 978-3-0365-4261-4 (Hbk)
ISBN 978-3-0365-4262-1 (PDF)

© 2022 by the authors. Articles in this book are Open Access and distributed under the Creative Commons Attribution (CC BY) license, which allows users to download, copy and build upon published articles, as long as the author and publisher are properly credited, which ensures maximum dissemination and a wider impact of our publications.

The book as a whole is distributed by MDPI under the terms and conditions of the Creative Commons license CC BY-NC-ND.

Contents

About the Editors . . . . **vii**

Francesco Tornabene and Rossana Dimitri
Special Issue on Recent Advances in Theoretical and Computational Modeling of Composite Materials and Structures
Reprinted from: *Appl. Sci.* **2022**, *12*, 4715, doi:10.3390/app12094715 . . . . **1**

Lijun Zhao, Tiesheng Dou, Bingqing Cheng, Shifa Xia, Jinxin Yang, Qi Zhang, Meng Li and Xiulin Li
Theoretical Study and Application of the Reinforcement of Prestressed Concrete Cylinder Pipes with External Prestressed Steel Strands
Reprinted from: *Appl. Sci.* **2019**, *9*, 5532, doi:10.3390/app9245532 . . . . **5**

Liang Zhang, Shengjie Dong, Jiangtao Du, Yi-Lin Lu, Hui Zhao and Liefeng Feng
First-Principles Forecast of Gapless Half-Metallic and Spin-Gapless Semiconducting Materials: Case Study of Inverse Ti_2CoSi-Based Compounds
Reprinted from: *Appl. Sci.* **2020**, *10*, 782, doi:10.3390/app10030782 . . . . **23**

Farshid Allahkarami, Hasan Tohidi, Rossana Dimitri and Francesco Tornabene
Dynamic Stability of Bi-Directional Functionally Graded Porous Cylindrical Shells Embedded in an Elastic Foundation
Reprinted from: *Appl. Sci.* **2020**, *10*, 1345, doi:10.3390/app10041345 . . . . **35**

Cheng Shen, Shixun Fan, Xianliang Jiang, Ruoyu Tan and Dapeng Fan
Dynamics Modeling and Theoretical Study of the Two-Axis Four-Gimbal Coarse–Fine Composite UAV Electro-Optical Pod
Reprinted from: *Appl. Sci.* **2020**, *10*, 1923, doi:10.3390/app10061923 . . . . **65**

Jian Chen, Xiaolei Bi, Juan Liu and Zhengcai Fu
Damage Investigation of Carbon-Fiber-Reinforced Plastic Laminates with Fasteners Subjected to Lightning Current Components C and D
Reprinted from: *Appl. Sci.* **2020**, *10*, 2147, doi:10.3390/app10062147 . . . . **85**

Hong-Xia Jing, Xiao-Ting He, Da-Wei Du, Dan-Dan Peng and Jun-Yi Sun
Vibration Analysis of Piezoelectric Cantilever Beams with Bimodular Functionally-Graded Properties
Reprinted from: *Appl. Sci.* **2020**, *10*, 5557, doi:10.3390/app10165557 . . . . **101**

Mohammad Sadegh Nematollahi, Hossein Mohammadi, Rossana Dimitri and Francesco Tornabene
Nonlinear Vibration of Functionally Graded Graphene Nanoplatelets Polymer Nanocomposite Sandwich Beams
Reprinted from: *Appl. Sci.* **2020**, *10*, 5669, doi:10.3390/app10165669 . . . . **121**

Dmitry Gritsenko and Roberto Paoli
Theoretical Analysis of Fractional Viscoelastic Flow in Circular Pipes: Parametric Study
Reprinted from: *Appl. Sci.* **2020**, *10*, 9080, doi:10.3390/app10249080 . . . . **141**

Dmitry Gritsenko and Roberto Paoli
Theoretical Analysis of Fractional Viscoelastic Flow in Circular Pipes: General Solutions
Reprinted from: *Appl. Sci.* **2020**, *10*, 9093, doi:10.3390/app10249093 . . . . **155**

Ahmed Amine Daikh, Mohammed Sid Ahmed Houari, Behrouz Karami, Mohamed A. Eltaher, Rossana Dimitri and Francesco Tornabene
Buckling Analysis of CNTRC Curved Sandwich Nanobeams in Thermal Environment
Reprinted from: *Appl. Sci.* **2021**, *11*, 3250, doi:10.3390/app11073250 . . . . **177**

Mehran Safarpour, Ali Forooghi, Rossana Dimitri and Francesco Tornabene
Theoretical and Numerical Solution for the Bending and Frequency Response of Graphene Reinforced Nanocomposite Rectangular Plates
Reprinted from: *Appl. Sci.* **2021**, *11*, 6331, doi:10.3390/app11146331 . . . . **197**

Faraz Kiarasi, Masoud Babaei, Kamran Asemi, Rossana Dimitri and Francesco Tornabene
Three-Dimensional Buckling Analysis of Functionally Graded Saturated Porous Rectangular Plates under Combined Loading Conditions
Reprinted from: *Appl. Sci.* **2021**, *11*, 10434, doi:10.3390/app112110434 . . . . **223**

Salvatore Brischetto and Roberto Torre
3D Stress Analysis of Multilayered Functionally Graded Plates and Shells under Moisture Conditions
Reprinted from: *Appl. Sci.* **2022**, *12*, 512, doi:10.3390/app12010512 . . . . **245**

About the Editors

Francesco Tornabene

Francesco Tornabene is an Associate Professor at the School of Engineering, Department of Innovation Engineering, University of Salento. He was born on 13 January 1978 in Bologna, where he received his high school degree at Liceo Classico San Luigi, in 1997. In 2001, he achieved a National Patent Bologna (Italy) for the Industrial Invention: Friction Clutch for High Performance Vehicles Question BO2001A00442. From the University of Bologna, he received an Alma Mater Studiorum, a M.Sc. degree in Mechanical Engineering (Curriculum in Structural Mechanics) on 23/07/2003, with a thesis entitled: "Dynamic Behavior of Cylindrical Shells: Formulation and Solution". In December 2003, he was admitted to the PhD course in Structural Mechanics at the University of Bologna; in 2004, from the University of Bologna, he received a thesis in memory of Carlo Felice Jodi; in 2007, he received his Ph.D. degree in Structural Mechanics at the University of Bologna, with a thesis entitled "Modeling and Solution of Shell Structures Made of Anisotropic Materials". From 2007 to 2009, he received a research fellowship from the University of Bologna, working on the "Unified Formulation of Shell Structures Made of Anisotropic Materials. Numerical Analysis Using the Generalized Differential Quadrature Method and the Finite Element Method". From 2011 to 2012, he became a junior researcher within the research program entitled "Advanced Numerical Schemes for Anisotropic Materials"; from 2012 to 2018, he was an Assistant Professor and Lecturer at the Alma Mater Studiorum - University of Bologna; since 2018 he has been an Aggregate Professor in Structural Mechanics and Lecturer at the University of Salento, Department of Innovation Engineering (Lecce). For a long time, his scientific interests have included include Structural Mechanics, Solid Mechanics, Innovative and Smart Materials, Computational Mechanics and Numerical Techniques, Damage and Fracture Mechanics. He is the author of more than 280 scientific publications and collaborates with many national or international researchers and professors all over the world, as visible from his scientific production. He is the author of 11 books (including "Meccanica delle Strutture a Guscio in Materiale Composito. Il metodo Generalizzato di Quadratura Differenziale" (2012); "Mechanics of Laminated Composite Doubly-Curved Shell Structures. The Generalized Differential Quadrature Method and the Strong Formulation Finite Element Method" (2014); "Laminated Composite Doubly-Curved Shell Structures I. Differential Geometry. Higher-Order Structural Theories" (2016); "Laminated Composite Doubly-Curved Shell Structures II. Differential and Integral Quadrature. Strong Formulation Finite Element Method" (2016), "Anisotropic Doubly-Curved Shells. Higher-Order Strong and Weak Formulations for Arbitrarily Shaped Shell Structures" (2018), among many others). He is a member of the Editorial Board for 44 international journals. He is also Editor-in-Chief for two international journals: *Curved and Layered Structures*, and *Journal of Composites Science*; furthermore, he is an Associate Editor for seven international journals. In the last few years, he has received various important awards, including "Highly Cited Researcher by Clarivate Analytics" (years 2018, 2019, 2020), "Ambassador of Bologna Award for the organization of 21st International Conference on Composite Structures ICCS21, 4–7 September 2018, Bologna, Italy" (2019), "Member of the European Academy of Sciences" (since 2018). He collaborates as reviewer with more than 260 prestigious international journals in the structural mechanics field. Since 2012, his teaching activities have included Dynamics of Structures; Computational Mechanics; Plates and Shells; Theory of Structures; and Structural Mechanics. He is habilitated as Associate Professor and Full Professor in the area 08/B2 (Structural Mechanics) and as Associate Professor in Area 09/A1 (Aeronautical and Aerospace Engineering and Naval Architecture).

Rossana Dimitri

Rossana Dimitri is an Associate Professor at the School of Engineering, Department of Innovation Engineering, University of Salento, Lecce, Italy (since 2019). From the University of Salento, she received an M. Sc. degree in "Materials Engineering" in 2004, a Ph.D. degree in "Materials and Structural Engineering" in 2009, and a Ph.D. degree in "Industrial and Mechanical Engineering" in 2013. In 2005, from the University of Salento, she received the "Best M. Sc. Thesis Price 2003–2004" in memory of Eng. Gabriele De Angelis; in 2013, she was awarded by the Italian Group for Computational Mechanics (GIMC) for the Italian selection of the 2013 ECCOMAS PhD Award. Her current interests include Structural Mechanics, Solid Mechanics, Damage and Fracture Mechanics, Contact Mechanics, Isogeometric Analysis, High-Performing Computational Methods, Consulting in Applied Technologies, and Technology Transfer. During 2010 and 2011, she received a research fellowship by ENEA Research Centre of Brindisi (UTTMATB-COMP) for the development and the characterization of some thermoplastic composites for thermal solar panels and adhesively bonded turbine blades under severe environmental conditions. During 2011 and 2012, she was a visiting scientist with a fellowship at the Institut für Kontinuumsmechanik Gottfried Wilhelm Leibniz Universität Hannover to study interfacial problems with isogeometric approaches. From 2013 to 2016, she was a researcher at the University of Salento, within the ERC starting research grant "INTERFACES" on "Computational mechanical modelling of structural interfaces based on isogeometric approaches". From 2013 to 2019, she has collaborated as researcher RTD-B with the University of Bologna and the Texas A&M University for a comparative assessment of some advanced numerical collocation methods with lower computational cost for fracturing problems and structural modelling of composite plates and shells, made by isotropic, orthotropic, and anisotropic materials. She is the author of 112 scientific publications, and she collaborates with many national or international researchers and professors worldwide, as visible from her scientific production. She also collaborates with different prestigious international journals in the structural mechanics field, as reviewer, member of the editorial board, and guest editor for different Special Issues.

Editorial

Special Issue on Recent Advances in Theoretical and Computational Modeling of Composite Materials and Structures

Francesco Tornabene * and Rossana Dimitri *

Department of Innovation Engineering, Università del Salento, Via per Monteroni, 73100 Lecce, Italy
* Correspondence: francesco.tornabene@unisalento.it (F.T.); rossana.dimitri@unisalento.it (R.D.)

1. Introduction

The advancement in manufacturing technology and scientific research have improved the development of enhanced composite materials with tailored properties depending on their design requirements in many engineering fields, as well as in thermal and energy management. Some representative examples of advanced materials in many smart applications and complex structures rely on laminated composites, functionally graded materials (FGMs), and carbon-based constituents, primarily carbon nanotubes (CNTs), and graphene sheets or nanoplatelets, because of their remarkable mechanical properties, electrical conductivity, and high permeability. For such materials, experimental tests usually require a large economical effort because of the complex nature of each constituent, together with many environmental, geometrical, and/or mechanical uncertainties in nonconventional specimens. At the same time, the theoretical and/or computational approaches represent a valid alternative for the design of complex manufacts with more flexibility. In such a context, the development of advanced theoretical and computational models for composite materials and structures is a subject of active research, as explored here for a large variety of structural aspects, involving static, dynamic, buckling, and damage/fracturing problems at different scales.

2. Enhanced Theoretical and Computational Models

In a context where an increased theoretical/computational demand is required to solve solid mechanics problems, this Special Issue has collected 13 papers regarding the application of high-performing computational strategies and enhanced theoretical formulations to solve different linear/nonlinear problems, even from a multiphysical perspective. To this end, classical and nonclassical theories have been proposed together with multiscale approaches, homogenization techniques, and different fracturing models.

The first paper, authored by S. Brischetto and R. Torre [1], proposes a steady-state hygro-thermomechanical stress analysis of single-layered and multilayered plates and shells with FGMs under different moisture conditions and introduces a novel exact solution as a valid benchmark for moisture diffusion problems in composite materials. Different environmental conditions (primarily temperature and moisture) of structural components can significantly affect their internal stress distributions and overall response during their service life, with possible premature damage and failure mechanisms. Among advanced composite materials, FGMs represent heterogeneous materials with enhanced stiffness properties, hardness, thermal conductivity, moisture diffusivity, and corrosion resistance due to the combination of two of more different phases, primarily metallic and ceramic phases, as shown in more common examples [2]. In this setting, several recent works in literature focused on structures embedding FGMs, even with possible defects and porosities, and developed innovative analytical and numerical models combined with different higher-order assumptions to handle uncoupled or coupled multiphysical problems [3–8]. For multiscale electromechanical applications (i.e., sensors, actuators, and energy conversion devices), piezoelectric materials with functionally graded properties (FGPMs) are

Citation: Tornabene, F.; Dimitri, R. Special Issue on Recent Advances in Theoretical and Computational Modeling of Composite Materials and Structures. *Appl. Sci.* **2022**, *12*, 4715. https://doi.org/10.3390/app12094715

Received: 28 April 2022
Accepted: 5 May 2022
Published: 7 May 2022

Publisher's Note: MDPI stays neutral with regard to jurisdictional claims in published maps and institutional affiliations.

Copyright: © 2022 by the authors. Licensee MDPI, Basel, Switzerland. This article is an open access article distributed under the terms and conditions of the Creative Commons Attribution (CC BY) license (https:// creativecommons.org/licenses/by/ 4.0/).

increasingly attracting the interest of many researchers [9–12]. This interest is mainly related to their capability to produce large displacements while minimizing the internal stress concentration, creep fatigue proneness, and interfacial failures, with improved reliability and life cycle in many intelligent devices, generally in the form of flexible cantilever elements [13]. As detailed in the work by H.X. Jing et al. [13], indeed, purely FGMs and FGPMs can exert bimodular effects to a certain degree, which can modify the mechanical behavior of structures, with a further influence on design applications of electromechanical devices based on piezoelectric effects.

Nowadays, with the advancement of nanotechnology, CNTs and graphene sheets represent two valid alternatives of structural reinforcement due to their outstanding properties. This has led to extensive research on the behavior of sandwich structures reinforced with nanocomposites [14–18]. Among different reinforcement possibilities, graphene nanoplatelets (GPLs) provide uniform reinforced assembly, as well as the easiest manufacturing process. In the work by M.S. Nematollahi [17], for example, a higher-order laminated beam theory is applied to include the shear and rotation effects on a thick GPL-reinforced sandwich beam, where the nonlinear governing equations of the problem are solved in a straightforward manner by means of the multiple timescale method. The sensitivity of the vibration response to the total amount of GPLs is explored by the authors, together with the possible effect related to the power-law parameter, structural geometry, and environmental conditions. Unlike traditional engineering structural problems, the design of micro-electromechanical systems (MEMS) usually involves microstructures, novel materials, and extreme operating conditions, where multisource uncertainties usually exist. In such a context, the work by M. Safarpour et al. [18] determines a general thermo-elasticity solution to treat both the static and frequency problems of functionally graded, GPL-reinforced composite plate structures under different boundary conditions and embedding foundations, as typically applied in many lightweight mechanical and biomedical components, as well as in membranes and flexible wearable sensors and actuators. Another kind of carbon-based reinforcement relies on CNTs, in lieu of conventional fibers, for which different molecular dynamic simulations have been successfully performed in the literature to exploit the elastic properties of polymer–CNT composites embedded in polymeric matrices [19,20]. Among sandwich CNT-based nanostructural applications, the work authored by A.A Daikh et al. [21] provides a mathematical continuum model to investigate the buckling behavior of cross-ply, single-walled, CNT-reinforced curved beams in thermal environment, based on a novel quasi-3D higher-order shear deformation theory and nonlocal strain gradient method accounting for any possible nanoscale size effect. An efficient numerical model based on a fractional calculus is, instead, established by D. Gritsenko and R. Paoli [22,23] to study the viscoelastic flow in circular pipes, for different geometrical radii, fractional orders, and elastic moduli ratios, compared to classical models. This mathematical tool allows for a significant improvement of predictive power for numerous practical applications from heat conduction to anomalous diffusion and viscoelastic properties of fluids and solids.

The extensive use of composite materials and structures in many engineering applications with complex microstructures and manufacturing processes requires a thorough attention to their mechanical performances, such as the structural deflection damage and load capacity [24,25], as well as the buckling and dynamic behavior [26], along with possible related uncertainties and stochastic variations. In this setting, the work authored by J. Chen et al. [24] provides a damage investigation of carbon-fiber-reinforced plastic laminates with fasteners accounting for a complex multiphysics coupling process. A theoretical study on prestressed pipes is also provided in [25], showing that high-strength prestressing wires withstand an internal high water pressure and external load, and a mortar coating protects the wires and cylinder against corrosion. As also highlighted in the work by C. Shen et al. [26], the characteristics of a coarse–fine composite structure and the complexity of dynamics modeling affect the entire system's high precision control performance. In this last work, the authors apply a finite element analysis and theoretical study of the stress and deflection of a two-axis, four-gimbal, coarse–fine composite, UAV electro-optical pod, with useful insights for aerospace applications.

3. Future Developments

Although this Special Issue has been closed, further developments on the theoretical and computational modeling of enhanced structures and composite materials are expected, including their static, dynamic, and buckling responses and fracture mechanics at different scales, which will be useful for many industrial applications.

Author Contributions: Conceptualization, F.T. and R.D.; methodology, F.T. and R.D.; formal analysis, F.T. and R.D.; investigation, F.T. and R.D.; data curation, F.T. and R.D.; writing—original draft preparation, F.T. and R.D.; writing—review and editing, F.T. and R.D. All authors have read and agreed to the published version of the manuscript.

Funding: This research received no external funding.

Conflicts of Interest: The authors declare no conflict of interest.

References

1. Brischetto, S.; Torre, R. 3D Stress Analysis of Multilayered Functionally Graded Plates and Shells under Moisture Conditions. *Appl. Sci.* **2022**, *12*, 512. [CrossRef]
2. Allahyarzadeh, M.; Aliofkhazraei, M.; Rouhaghdam, A.S.; Torabinejad, V. Gradient electrodeposition of Ni-Cu-W (alumina) nanocomposite coating. *Mater. Des.* **2016**, *107*, 74–81. [CrossRef]
3. Jabbari, M.; Hashemitaheri, M.; Mojahedin, A.; Eslami, M.R. Thermal buckling analysis of functionally graded thin circular plate made of saturated porous materials. *J. Therm. Stresses* **2014**, *37*, 202–220. [CrossRef]
4. Chen, D.; Yang, J.; Kitipornchai, S. Buckling and bending analyses of a novel functionally graded porous plate using Chebyshev-Ritz method. *Arch. Civ. Mech. Eng.* **2019**, *19*, 157–170. [CrossRef]
5. Gao, K.; Huang, Q.; Kitipornchai, S.; Yang, J. Nonlinear dynamic buckling of functionally graded porous beams. *Mech. Adv. Mater. Struct.* **2019**, *28*, 418–429. [CrossRef]
6. Allahkarami, F.; Tohidi, H.; Dimitri, R.; Tornabene, R. Dynamic Stability of Bi-Directional Functionally Graded Porous Cylindrical Shells Embedded in an Elastic Foundation. *Appl. Sci.* **2020**, *10*, 1345. [CrossRef]
7. Kiarasi, F.; Babaei, M.; Asemi, K.; Dimitri, R.; Tornabene, F. Three-Dimensional Buckling Analysis of Functionally Graded Saturated Porous Rectangular Plates under Combined Loading Conditions. *Appl. Sci.* **2021**, *11*, 10434. [CrossRef]
8. Dastjerdi, S.; Malikan, M.; Dimitri, R.; Tornabene, F. Nonlocal elasticity analysis of moderately thick porous functionally graded plates in a hygro-thermal environment. *Compos. Struct.* **2021**, *255*, 112925. [CrossRef]
9. Bodaghi, M.; Damanpack, A.R.; Aghdam, M.M.; Shakeri, M. Geometrically non-linear transient thermo-elastic response of FG beams integrated with a pair of FG piezoelectric sensors. *Compos. Struct.* **2014**, *107*, 48–59. [CrossRef]
10. Kulikov, G.M.; Plotnikova, S.V. An analytical approach to three-dimensional coupled thermoelectroelastic analysis of functionally-graded piezoelectric plates. *J. Intell. Mater. Syst. Struct.* **2017**, *28*, 435–450. [CrossRef]
11. Alibeigloo, A. Thermo elasticity solution of functionally-graded, solid, circular, and annular plates integrated with piezoelectric layers using the differential quadrature method. *Mech. Adv. Mater. Struct.* **2018**, *25*, 766–784. [CrossRef]
12. Arefi, M.; Bidgoli, E.M.R.; Dimitri, R.; Bacciocchi, M.; Tornabene, F. Application of sinusoidal shear deformation theory and physical neutral surface to analysis of functionally graded piezoelectric plate. *Compos. Part B Eng.* **2018**, *151*, 35–50. [CrossRef]
13. Jing, H.X.; He, X.T.; Du, D.W.; Peng, D.D.; Sun, J.Y. Vibration Analysis of Piezoelectric Cantilever Beams with Bimodular Functionally-Graded Properties. *Appl. Sci.* **2020**, *10*, 5557. [CrossRef]
14. Zhang, L.; Dong, S.; Du, J.; Lu, Y.L.; Zhao, H.; Feng, L. First-Principles Forecast of Gapless Half-Metallic and Spin-Gapless Semiconducting Materials: Case Study of Inverse Ti2CoSi-Based Compounds. *Appl. Sci.* **2020**, *10*, 782. [CrossRef]
15. Nieto, A.; Bisht, A.; Lahiri, D.; Zhang, C.; Agarwal, A. Graphene reinforced metal and ceramic matrix composites: A review. *Int. Mater. Rev.* **2017**, *62*, 241–302. [CrossRef]
16. Nazarenko, L.; Chirkov, A.Y.; Stolarski, H.; Altenbach, H. On modeling of carbon nanotubes reinforced materials and on influence of carbon nanotubes spatial distribution on mechanical behavior of structural elements. *Int. J. Eng. Sci.* **2019**, *143*, 1–13. [CrossRef]
17. Nematollahi, M.S.; Mohammadi, H.; Dimitri, R.; Tornabene, F. Nonlinear Vibration of Functionally Graded Graphene Nanoplatelets Polymer Nanocomposite Sandwich Beams. *Appl. Sci.* **2020**, *10*, 5669. [CrossRef]
18. Safarpour, M.; Forooghi, A.; Dimitri, R.; Tornabene, F. Theoretical and Numerical Solution for the Bending and Frequency Response of Graphene Reinforced Nanocomposite Rectangular Plates. *Appl. Sci.* **2021**, *11*, 6331. [CrossRef]
19. Griebel, M.; Hamaekers, J. Molecular dynamics simulations of the elastic moduli of polymer–carbon nanotube composites. *Comput. Methods Appl. Mech. Eng.* **2004**, *193*, 1773–1788. [CrossRef]
20. Han, Y.; Elliott, J. Molecular dynamics simulations of the elastic properties of polymer/carbon nanotube composites. *Comput. Mater. Sci.* **2007**, *39*, 315–323. [CrossRef]

21. Daikh, A.A.; Houari, M.S.A.; Karami, B.; Eltaher, M.A.; Dimitri, R.; Tornabene, F. Buckling Analysis of CNTRC Curved Sandwich Nanobeams in Thermal Environment. *Appl. Sci.* **2021**, *11*, 3250. [CrossRef]
22. Gritsenko, D.; Paoli, R. Theoretical Analysis of Fractional Viscoelastic Flow in Circular Pipes: General Solutions. *Appl. Sci.* **2020**, *10*, 9093. [CrossRef]
23. Gritsenko, D.; Paoli, R. Theoretical Analysis of Fractional Viscoelastic Flow in Circular Pipes: Parametric Study. *Appl. Sci.* **2020**, *10*, 9080. [CrossRef]
24. Chen, J.; Bi, X.; Liu, J.; Fu, Z. Damage Investigation of Carbon-Fiber-Reinforced Plastic Laminates with Fasteners Subjected to Lightning Current Components C and D. *Appl. Sci.* **2020**, *10*, 2147. [CrossRef]
25. Zhao, L.; Dou, T.; Cheng, B.; Xia, S.; Yang, J.; Zhang, Q.; Li, M.; Li, X. Theoretical Study and Application of the Reinforcement of Prestressed Concrete Cylinder Pipes with External Prestressed Steel Strands. *Appl. Sci.* **2019**, *9*, 5532. [CrossRef]
26. Shen, C.; Fan, S.; Jiang, X.; Tan, R.; Fan, D. Dynamics Modeling and Theoretical Study of the Two-Axis Four-Gimbal Coarse–Fine Composite UAV Electro-Optical Pod. *Appl. Sci.* **2020**, *10*, 1923. [CrossRef]

Article

Theoretical Study and Application of the Reinforcement of Prestressed Concrete Cylinder Pipes with External Prestressed Steel Strands

Lijun Zhao [1,2], Tiesheng Dou [1,2,*], Bingqing Cheng [1,2], Shifa Xia [1,2], Jinxin Yang [3], Qi Zhang [3], Meng Li [1,2] and Xiulin Li [1]

[1] Division of Materials, China Institute of Water Resources and Hydropower Research (IWHR), Beijing 100038, China; jun15297669998@gmail.com (L.Z.); chengbq1994@gmail.com (B.C.); xiasf@iwhr.com (S.X.); limeng@iwhr.com (M.L.); lixl@iwhr.com (X.L.)
[2] State Key Laboratory of Simulation and Regulation of Water Cycle in River Basin, China Institute of Water Resources and Hydropower Research (IWHR), Beijing 100038, China
[3] Beijing Institute of Water, Beijing 100044, China; yangjinxin@biwmail.com (J.Y.); zhangqi@biwmail.com (Q.Z.)
* Correspondence: doutsh@iwhr.com; Tel.: +86-135-0120-7217

Received: 6 November 2019; Accepted: 13 December 2019; Published: 16 December 2019

Abstract: Prestressed concrete cylinder pipes (PCCPs) can suffer from prestress loss caused by wire-breakage, leading to a reduction in load-carrying capacity or a rupture accident. Reinforcement of PCCPs with external prestressed steel strands is an effective way to enhance a deteriorating pipe's ability to withstand the design load. One of the principal advantages of this reinforcement is that there is no need to drain the pipeline. A theoretical derivation is performed, and this tentative design method could be used to determine the area of prestressed steel strands and the corresponding center spacing in terms of prestress loss. The prestress losses of strands are refined and the normal stress between the strands and the pipe wall are assumed to be distributed as a trigonometric function instead of uniformly. This derivation configures the prestress of steel strands to meet the requirements of ultimate limit states, serviceability limit states, and quasi-permanent limit states, considering the tensile strength of the concrete core and the mortar coating, respectively. This theory was applied to the reinforcement design of a PCCP with broken wires (with a diameter of 2000 mm), and a prototype test is carried out to verify the effect of the reinforcement. The load-carrying capacity of the deteriorating PCCPs after reinforcement reached that of the original design level. The research presented in this paper could provide technical recommendations for the application of the reinforcement of PCCPs with external prestressed steel strands.

Keywords: prestressed concrete cylinder pipe; external prestressed steel strands; theoretical study; wire-breakage

1. Introduction

A prestressed concrete cylinder pipe (PCCP) contains four components, namely, (1) a concrete core, (2) a steel cylinder lined with concrete (LCP) or encased in concrete (ECP), (3) high strength prestressing wires to withstand the internal high water pressure and external load, and (4) a mortar coating to protect the wires and cylinder against corrosion. The promise of the PCCP lies in its high bearing capacity, strong permeability resistance, and cost-effectiveness. Efficiencies in construction and reductions in fabrication costs have led to the extensive use of PCCPs in the USA, Canada, and China, and have also led to the pivotal development of this pipe. However, these pipes may suffer from prestress loss caused by wire-breakage. Wire-breakage or rupturing can result in significant losses to society, making the reinforcement of deteriorating pipes essential.

Reinforcement with external prestressed steel strands is regarded as an efficient way of strengthening bridges and beams that are deteriorating due to increased overloading and progressive structural aging [1–5]. Miyamoto A. [6,7] demonstrated the feasibility of applying this prestressing technique to the strengthening of existing steel bridges. Chen S. [8] proposed a finite element model to investigate the inelastic buckling of continuous composite beams that were prestressed with external tendons. Lou T. [9] also concluded that external prestressing significantly improved the short-term behavior of a composite beam. Tan K. H. [2], Aparicio A. C. [10], Park S. [11], and others have presented a series of prototype tests regarding externally prestressed concrete beams and have verified that external tendons can be used to effectively influence beam behavior.

The reinforcement of a PCCP with external prestressed steel strands involves repairing critical pipes with additional external post-tensioning to increase the longevity of problematic PCCP pipelines. The strands are wrapped outside the pipe with a fixed spacing between each strand, according to the service water pressure [12] (Figure 1). A well-known large-scale application of external prestressed strands is in the Great Man-Made River pipelines in Libya [12]. Most of the pipes in this project have an internal diameter of 4.20 m. Authorities have determined that repair of the critical pipes should proceed, with additional external post-tensioning in areas where pipes had burst. The reinforcement of the external prestressed strands on PCCPs has proven to be effective here. This approach is advantageous due to its ability for construction to proceed with no need to drain the pipeline. However, few theoretical studies have been carried out regarding the prestress losses and the mechanism applying external prestressed strands to strengthen PCCPs.

Figure 1. Structural drawing of a prestressed concrete cylinder pipe (PCCP) strengthened with prestressed steel strands.

This study introduces a theoretical derivation and investigates the prestress loss of steel strands applied to PCCPs. The normal stress between the strands and the pipe wall is assumed to be distributed as a trigonometric function, instead of uniformly, to estimate prestress losses. The area of the steel strands is determined to meet the requirements of ultimate limit states, serviceability limit states, and quasi-permanent limit states, considering the tensile strength of the concrete core and the mortar coating, respectively. An example calculation of this theory and a prototype test is calculated on the same PCCP to verify the feasibility of this theory. The load response of the pipe before and after the reinforcement process is analyzed.

2. Theoretical Derivations

2.1. Calculation of Prestress Loss, $\sigma_{st,l}$

The prestress loss persisted during and after the tensioning operation, and can be divided into two categories, namely, instantaneous loss and long-term loss [13–15]. Instantaneous loss, i.e., short-term loss during the tensioning operation, described the prestress losses caused by friction resistance between the surface of the pipe wall and the steel strands, the anchor deformation, the concrete elastic compression induced by stepwise tensioning operation, and cracks closures. Long-term prestress losses included prestress losses [13,16], while taking into account the materials aging, including the effects of shrinkage and the creep losses of concrete, and the long-term relaxation losses of prestressed steel strands. Types of prestress loss of steel strands applied to PCCPs are illustrated in Figure 2. Since the reinforcement of PCCPs with external prestressed steel strands is a post-tensioning method, the impact of temperature can be removed from consideration when considering the reinforcement of PCCPs with external prestressed steel strands.

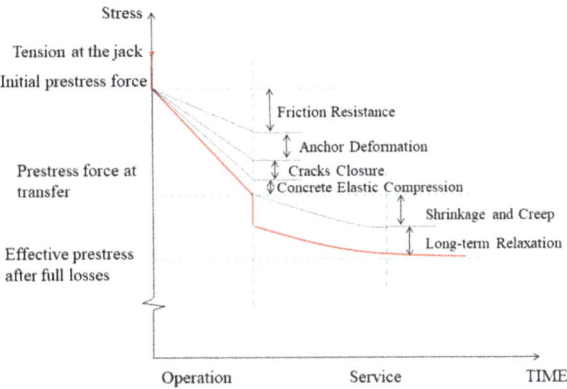

Figure 2. Types of prestress loss of steel strands applied to PCCPs.

2.1.1. Calculation of Retraction Length, l_{re}, and Its Corresponding Retraction Angle, θ_{re}

As far as we know, the stress distribution along the strand is nonlinear. The anchor influenced the prestressed steel strand within a certain length range due to the static friction caused by the retraction of the strand. This length is called the retraction length, l_{re}. Strands within the retraction length showed a displacement opposite to the tension direction, which decreases the prestress. The movement trend is demarcated at point C, and the stress is redistributed from ACB to $A'CB$ (Figure 3).

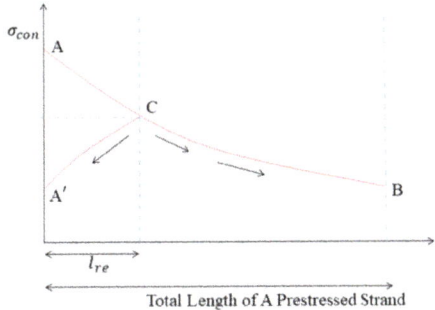

Figure 3. Distribution of strand stress caused by retraction.

The circumferential micro-segment of the prestressed steel strand is regarded as the research object, where the corresponding angle is $d\theta$ (Figure 4). Assuming that the normal stress of the steel strand in the micro-segment is evenly distributed, a differential equation can be established according to the static equilibrium conditions:

$$T \cdot sin\left(\frac{d\theta}{2}\right) + (T + dT) \cdot sin\left(\frac{d\theta}{2}\right) - dP = 0 \qquad (1)$$

where T and P stand for the tension force and the normal pressure of the strand, respectively.

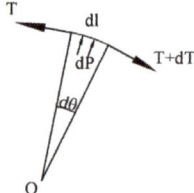

Figure 4. Stress of a micro-segment of a strand.

Higher variables were omitted, taking $\frac{d\theta}{2} = sin(\frac{d\theta}{2})$. Equation (1) can be simplified to $Td\theta = dP$. The equation describing the momentary balance for rotation around the center of curvature, O, can be written as follows:

$$r_{st} \cdot \mu dP + r_{st} dT = 0 \qquad (2)$$

where $T = T_0$ when $\theta = 0$, thus, $T = T_0 e^{-\mu\theta}$. r_{st}, where r_{st} is the calculated radius of the strand wrapped outside the pipe (m) and μ is the friction coefficient between the prestressed strands and the outer surface of the deteriorating pipe. The influencing factors of μ mainly include the type of steel, the type of lubricating grease, the materials wrapped outside, and the quality control of the construction. Here, μ ranges from 0.08 to 0.12, with a mean value of 0.1.

The stress for an arbitrary cross section is calculated as per Equation (3):

$$\sigma = \sigma_{st} e^{-\mu\theta} \qquad (3)$$

where σ_{st} is the tension stress of prestressed steel strands (N/mm^2). $\sigma_{st} = f_{st,t} \cdot \alpha$, in which α is the control coefficient for the tension of the steel strands (N/mm^2). Normally, this value ranges between 0 and 0.75 [17–19]. $f_{st,t}$ is the nominal tensile strength of the prestressed strand (N/mm^2).

The stress at the end section of the retraction length can be written as follows:

$$\sigma_{re} = \sigma_{st} e^{-\mu\theta_{re}}. \qquad (4)$$

The length reduction of the strand caused by the anchor deformation and the clip retraction, Δl_{re}, can be expressed by Equation (5):

$$\Delta l_{re} = \int_0^{\theta_{re}} \frac{\sigma_{12} r_{st}}{E_{st}} d\theta = \frac{2 r_{st} \sigma_{st} \left[1 - e^{-\mu\theta_{re}}(1 + \mu\theta_{re}) \right]}{\mu E_{st}} \qquad (5)$$

where E_{st} is the elastic modulus of the adopted steel strand (N/mm^2) and $e^{-\mu\theta_{re}}$ is expanded into a power series according to the Taylor formula. Only the first three terms of the formula have sufficient precision, since $\mu\theta_{re}$ is adequately small, which is given by Equation (6):

$$e^{-\mu\theta_{re}} = 1 - \mu\theta_{re} + \frac{(\mu\theta_{re})^2}{2}. \qquad (6)$$

Equation (7) can be derived by incorporating Equation (6) into Equation (5) and omitting the high micro $(\mu\theta_{re})^3$:

$$\Delta l_{re} = \frac{\mu r_{st} \sigma_{st} \theta_{re}^2}{E_{st}} \qquad (7)$$

The correspondence between the retraction length, l_{re}, and the retraction angle, θ_{re}, is represented as follows:

$$l_{re} = r_{st} \theta_{re}. \qquad (8)$$

Therefore, the retraction length, l_{re}, and its corresponding angle, θ_{re}, can be given by Equations (8) and (9).

$$l_{re} = \sqrt{\frac{\Delta l_{re} E_{st} r_{st}}{\mu \sigma_{st}}} \tag{9}$$

The various types of anchorage used with steel strands were classified as plug and cone, straight sleeve, contoured sleeve, metal overlay, and split wedge anchorages. The value of Δl_{re} varies with the type of anchor.

2.1.2. Prestress Loss Caused by Friction Resistance, σ_{l1}

The prestress loss caused by the friction resistance, σ_{l1}, can be calculated based on the consideration of two parts, namely, the bending loss and the deviation loss. The radial pressing force, σ_r, is produced between the strand and the pipe wall by prestressed strands, thereby resulting in extrusion friction. The bending loss accounted for a large proportion of the total friction loss.

Based on the assumption of a rigid body, we hypothesized that the pressure between the strand and the pipe wall would be uniformly distributed [20], and that elastic deformation would occur when the two elastic bodies were pressed into contact with each other. The stress between the contact surfaces is ellipsoidal, and its value can be related to the radius of curvature and the elastic modulus of the contact object. It is not accurate enough to consider the contact stress as uniformly distributed under normal contact pressure due to the large tensile force of the prestressed steel strands.

The scope of the bending loss can be related to the retraction length, l_{re}. We can assume that the normal stress between the strands and the pipe wall would be distributed as a trigonometric function [21], as illustrated in Figure 5 and Equation (10).

$$p_{(\alpha)} = p_0 \cos^2\left(\frac{\pi}{\theta}\alpha\right) \tag{10}$$

where $\cos^2\left(\frac{\pi}{\theta}\alpha\right) = \begin{cases} 0 & \alpha = \frac{\theta}{2} \\ 1 & \alpha = 0 \\ 0 & \alpha = -\frac{\theta}{2} \end{cases}$

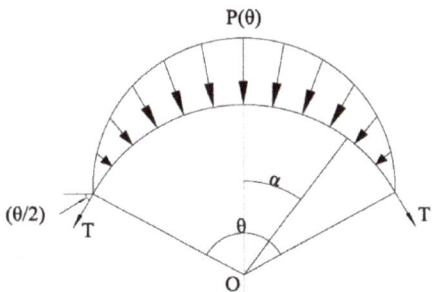

Figure 5. Distribution of the normal stress of the pipe wall excluding the friction.

A balance of forces in the z-direction can be established by Equation (11). From Equation (11), we derived Equation (12). Therefore, the normal stress can be calculated using Equation (13):

$$T \cdot \sin\left(\frac{\theta}{2}\right) + (T + dT) \cdot \sin\left(\frac{\theta}{2}\right) + 2 \int_0^{\frac{\theta}{2}} p_{(\alpha)} \cdot \cos \alpha \, dl = 0 \tag{11}$$

$$p_0 = \frac{\sin\frac{\theta}{2}}{\int_0^{\frac{\theta}{2}} \cos^2\left(\frac{\pi}{\theta}\alpha\right)\cos\alpha\, d\alpha} \times \frac{T}{R} \tag{12}$$

where $\int_0^{\frac{\theta}{2}} \cos^2\left(\frac{\pi}{\theta}\alpha\right)\cos\alpha\, d\alpha = \left(\cos^2\left(\frac{\pi}{\theta}\alpha\right)\sin\alpha\right)\Big|_0^{\frac{\theta}{2}} + \frac{\pi}{\theta}\int_0^{\frac{\theta}{2}}\sin\alpha\sin\left(\frac{2\pi}{\theta}\alpha\right)d\alpha = \frac{2\left(\frac{\pi}{\theta}\right)^2\sin\frac{\theta}{2}}{\left(\frac{2\pi}{\theta}\right)^2-1}$.

$$p_{(\alpha)} = \left(2 - \frac{\theta^2}{2\pi^2}\right)\cdot\cos^2\left(\frac{\pi}{\theta}\alpha\right)\cdot\frac{\sigma_0}{R} \tag{13}$$

The prestress loss related to the bending loss, F, during the tensioning operation is depicted in Equation (14).

$$F = 2\int_0^{\frac{\theta}{2}} \mu p_{(\theta)}\cdot dl = \mu\theta\sigma_0\left(1 - \frac{\theta^2}{4\pi^2}\right) \tag{14}$$

The deviation loss stems from errors in pipe positioning and installation, which causes friction between the force rib and the pipe material, thereby forming contact friction. The deviation loss occupies a small proportion of the total friction loss. The correction coefficient, c_1, is involved here, and the deviation loss is not separately calculated in this paper. As a result, the total prestress loss caused by the friction resistance can be calculated, as displayed in Equation (15).

$$\sigma_{l1} = c_1 F \tag{15}$$

where c_1 is the correction coefficient, accounting for the bending loss and the deviation loss, and is usually in the range of 1 to 1.3 [19].

2.1.3. Prestress Loss Caused by Anchorage Deformation, σ_{l2}

The prestress loss caused by deformation at the end of the anchorage should be taken into consideration. This refers to the prestress loss caused by the deformation of the anchor and the retraction of the clip due to the concentrated stress. A slip at the anchorage depends on the particular prestressing system adopted and is not a function of time. This loss can be written as per Equation (16):

$$\sigma_{l2} = E_{st}\frac{\Delta l_{re}}{2\pi r_{st}}. \tag{16}$$

2.1.4. Prestress Loss Caused by the Elastic Compression of Concrete During Batch Tensioning, σ_{l3}

The prestress loss can be adjusted using certain construction technologies, including ultra-tensioning and repeated tensioning. For example, the tensioning can be started from the middle of the pipeline and gradually pulled symmetrically to both sides when batch tensioning is adopted. The steel strands that would later be tensioned would cause elastic compression deformation of the concrete, which would contribute to the prestress loss of the previously anchored strands. This prestress loss can be simplified by the following formula [21]:

$$\sigma_{l3} = \frac{m-1}{2m}n_y\sigma_{h1} \tag{17}$$

where m is the total number of batches and σ_{h1} is the normal stress of the concrete produced by the combined force of the steel strands at the action point (the center of gravity of all steel strands), which is equal to the sum of the normal stresses of the concrete produced by the batch of steel strands, which is $\Delta\sigma_{h1}$. That is, $\sigma_{h1} = \sum \Delta\sigma_{h1} = m\Delta\sigma_{h1}$. n_y is the combined force of all of the steel strands. $\Delta\sigma_{h1}$ is the normal stress of the concrete generated by the subsequent batch of steel strands at the center of gravity of the first tensioned steel strand, as calculated by the following formula: $\Delta\sigma_{h1} = \frac{n}{m}\left(\frac{1}{A_j} + \frac{e_y \cdot y_i}{I_j}\right)$.

During engineering, ultra-tensioning or repeated tensioning technologies can be utilized in the first several batches of strands, so that the actual effective prestress of the pipe is substantially equal to the design level. After ultra-tensioning or repeated tensioning, the prestress loss caused by the elastic compression of the concrete during batch tensioning, σ_{l3}, is considered to be approximately zero.

2.1.5. Prestress Loss Due to Crack Reduction and Closure, σ_{l4}

In our experiment, due to the restraining effect of the prestressed steel strand, the cracks of the concrete core were reduced to some extent, or even closed. The reduction of the circumference of the pipe led to the prestress loss of the strands. Therefore, the change in the maximum width of the visible cracks in the concrete core can be utilized to estimate the prestress loss caused by crack reduction and closure.

The change in the maximum width of the cracks corresponds to the change in the length of the prestressed steel strands (Figure 6). The prestressed steel strand is in the elastic phase, and the stress is proportional to its strain. Therefore, the prestress loss, σ_{l4}, caused by the crack reduction and closure, can calculated according to the following equation.

$$l_i = \frac{r_{st}}{D/2} w_i \tag{18}$$

$$\frac{\sigma_0}{\sigma_0 - \sigma_{l4}} = \frac{2\pi r_{st} + l_1}{2\pi r_{st} + l_2} \tag{19}$$

where w_i is the maximum width of the visible cracks in the concrete core (m), $i = 1, 2$ for the condition before and after the reinforcement, l_i is the length of prestressed strands corresponding to the maximum width of the cracks in the concrete core before and after the reinforcement (m), and D is the outer diameter of the concrete core (m).

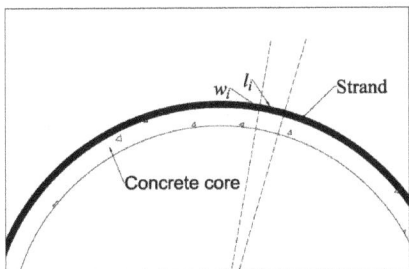

Figure 6. Correspondence between crack width and strand length.

Therefore, the prestress loss due to the crack reduction and closure, σ_{l4}, can be defined as Equation (20):

$$\sigma_{l4} = \frac{\sigma_0 (w_1 - w_2)}{\pi D + w_1} \tag{20}$$

2.1.6. Prestress Loss Caused by Shrinkage and Creep of Concrete, σ_{l5}

The shrinkage and creep of the original concrete pipe were involved in the calculation of the prestressed stress of the prestressing wires. The shrinkage and creep of the reinforced pipe have been basically completed before the reinforcement, and, as such, no further repeated calculations were performed for the prestress loss caused by shrinkage and creep of concrete, i.e., $\sigma_{l5} = 0$.

2.1.7. Prestress Loss Due to Long-Term Relaxation of the Strand, σ_{l6}

The deformation of the strand will change with time, and the stress will decrease accordingly when the strand is subjected to a constant external force, which is the prestress loss due to the long-term relaxation of the strand. The greater the tensile force of the steel strand, the more obvious the stress relaxation effect is. The relaxation generally occurred earlier in the process, without considering the quality of the strands. This effect can be basically completed after one year, and then gradually calmed. The relaxation loss, σ_{l6}, is related to the relaxation coefficient, k, and can calculated according to the following formula:

$$\sigma_{l6} = k\sigma_{st} \tag{21}$$

where k is the relaxation coefficient and is related to the quality of the steel. For the cold-drawn thick steel bar, k is taken as 0.05 for one-time tensioning and 0.035 for ultra-tensioning. As for the steel wires and steel strands, k is considered to be 0.07 for one-time tensioning and 0.045 for ultra-tensioning [17]. For low-relaxation steel wires, the value of k can be taken to be 0.002 when no data are available, which we have learned from our experience.

The total prestress loss of the prestressed wires can calculate by Equation (22):

$$\sigma_{st,l} = \sum_{1}^{6} \sigma_i \tag{22}$$

2.2. Calculation of Area of Prestressed Steel Strands

According to the study by Zarghamee M. [22–25], the cracking of PCCPs under combined loads mainly occurs at (1) the bottom of the inner surface of the concrete core, (2) the top of the inner surface of the concrete core, and (3) the spring-line of the outer surface of the concrete core. Therefore, these three sections are defined as dangerous sections. The area of prestressed steel strands can be determined under the assumption of a complete loss of prestress of prestressing wires.

2.2.1. Stress of PCCPs under Combined Loads

The combined loads acting on the pipe include the vertical earth pressure at the top of the pipe, $F_{sv,k}$, the lateral earth pressure, $F_{ep,k}$, the ground pile load, the weight of the pipe, G_{1k}, the weight of fluid in the pipe, G_{wk}, and the variable load.

The values of $F_{sv,k}$ and $F_{ep,k}$ are calculated according to Marston's theory [17] and Rankine's earth pressure theory [26], respectively. The variable load can be regarded as the ground stacking load, and its standard value is defined as $q_{mk} = 10$ kN/m.

The weight of the pipe can be written as per Equation (23):

$$G_{1k} = \pi r_G (D_i + h_c) h_c \tag{23}$$

where D_i is the inner diameter of the pipe (m), h_c is the thickness of concrete core (m), and r_G is the gravity density of the pipe (kN/m^3).

The weight of fluid in the pipe, G_{wk}, can be calculated by Equation (24):

$$G_{wk} = \frac{r_W \pi D_i^2}{4} \tag{24}$$

where r_w is the gravity density of the fluid in the pipe (kN/m^3).

2.2.2. Calculation of the Area of Prestressed Strands, Considering the Concrete Core Compression of Ultimate Limit States

According to Chinese specifications [27,28], the design requirements for the calculation of ultimate limit states under the external soil load, weight of pipe, weight of fluid, and other variable loads are detailed. The design value of the maximum bending moment of the pipe at the spring-line, M_{max}^l, can be calculated using Equation (25). The value of M_{max}^l is negative, indicating that the outer surface of the concrete core is subjected to tension. The absolute value can be taken when the formula is substituted for reinforcement. The design value of the maximum axial tension of the pipe at the spring-line, N^l, is written as per Equation (26):

$$M^l_{max} = \gamma_0 r \left[k_{vm} \left(\gamma_{G3} F_{sv,k} + \psi_c \gamma_{Q2} q_{vk} D_1 \right) + k_{hm} \gamma_{G4} F_{ep,k} D_1 + k_{wm} \gamma_{G2} G_{wk} + k_{gm} \gamma_{G1} G_{1k} \right] \quad (25)$$

$$N^l = \gamma_0 \left[\psi_c \gamma_{Q1} P_d r \times 10^{-3} - 0.5 \left(F_{sv,k} + \psi_c q_{vk} D_1 \right) \right] \quad (26)$$

where γ_0 is the factor of importance. It varies with the structure and the layout of the pipes. The value of γ_0 is generally 1.1. For side-by-side pipelines, the γ_0 value should be taken as 1.0, in particular. Moreover, the value of γ_0 should also be taken as 1.0 for pipes with storage facilities or those which are used for drainage. r is the calculated radius of the pipe (m), and k_{vm}, k_{hm}, k_{wm}, k_{gm} represent the bending moment coefficient of the bending moment at the spring-line of the pipe under the vertical earth pressure, lateral earth pressure, the weight of fluid inside the pipe, and the weight of the pipe, respectively. These factors were determined according to Appendix E [27]. The k_{gm} of the arc-shaped soil bedding can be adopted according to the data of the bedding angle of 20°. γ_{Gi} and γ_{Qj} are the partial coefficients under the permanent load i and the variable load j. ψ_c is the combination coefficient of the variable loads and usually takes the value of 0.9. D_1 is the outer diameter of the pipe (m). P_d is the designed water pressure (N/mm^2).

The area of the prestressed strands of ultimate limit states should be calculated by Equation (27):

$$A_{st} \geq \frac{\lambda_y}{f_{pyk}} \left(N^l + \frac{M^l_{max}}{d_0} - A_{sc} f'_{yy} \right) \quad (27)$$

where λ_y is the comprehensive adjustment factor of the PCCP, f_{pyk} is the design strength of prestressed strands (N/mm^2), d_0 is the distance from the prestressed strand to the center of gravity of the pipe(m), A_{sc} is the area of the cylinder per unit length (m^2/m), and f'_{yy} is the design strength of the cylinder (N/mm^2).

2.2.3. Checking Calculation of Prestressed Strands Considering the Concrete Core Compression of Serviceability Limit States

The maximum bending moment of the pipe at the top or the bottom, M_{pms}, is calculated as per Equation (28). The value of M_{pms} is negative, indicating that the outer surface of the concrete core is subjected to tension. The absolute value is taken when substituting the following equations. The axial tension of the pipe wall, N_{ps}, is written as per Equation (29):

$$M_{pms} = \gamma_0 \left[k_{vm} \left(F_{sv,k} + \psi_c q_{vk} D_1 \right) + k_{hm} F_{ep,k} D_1 + k_{wm} G_{wk} + k_{gm} G_{1k} \right] \quad (28)$$

$$N_{ps} = \psi_c P_d r \times 10^{-3} \quad (29)$$

where k_{vm}, k_{hm}, k_{wm}, and k_{gm} represent the bending moment coefficient of the bending moment at the top or the bottom of the pipe under the vertical earth pressure, lateral earth pressure, the weight of fluid inside the pipe, and the weight of the pipe, respectively. These factors can be determined according to Appendix E [27]. The k_{gm} of the arc-shaped soil bedding can adopted, according to the data of the bedding angle of 20°.

The maximum tensile stress at the edge of the pipe at the bottom, σ_{ss}, is calculated as per Equation (30).

$$\sigma_{ss} = \frac{N_{ps}}{A_n} + \frac{M_{pms}}{\omega_c W_p} \quad (30)$$

where A_n is the conversion area of the pipe section (including the cylinder, steel strands, and the mortar coating) (m^2/m). ω_c is the conversion coefficient of the elastic resistance moment of the tensioned edge of the pip -wall. W_p is the momentary elastic resistance of the unconverted tension edge of the rectangular section of the pipe wall (m²/m).

The effective prestress of the prestressed steel strands after the prestress loss, σ'_{st}, can be written as $\sigma'_{st} = \sigma_{st} - \sigma_{st,l}$.

Therefore, the area of prestressed strands of serviceability limit states should be calculated by Equation (31):

$$A_{st} \geq (\sigma_{ss} - K\gamma f_{ty})\frac{A_n}{\sigma'_{st}} \tag{31}$$

where K is the influence coefficient of concrete in the tension area, γ is the plastic influence coefficient of concrete in the tension area, and f_{ty} is the standard value of concrete tensile strength.

The area of prestressed strands needs to simultaneously meet the requirements outlined in Equations (27) and (31).

2.2.4. Checking Calculation of the Mortar Coating under Serviceability Limit States

Checking the calculation of mortar at the spring-line of the pipe should be carried out under serviceability limit states.

The maximum bending moment of the pipe at the spring-line, M^l_{pms}, can be calculated by Equation (32). The value of M^l_{pms} is negative, indicating that the mortar coating is subjected to tension. The absolute value is taken when the following equations are substituted. The axial tension of the pipe at the spring-line, N^l_{ps}, can be written as per Equation (33):

$$M^l_{pms} = r\left[k_{vm}(F_{sv,k} + \psi_c q_{vk} D_1) + k_{hm} F_{ep,k} D_1 + k_{vm} G_{wk} + k_{gm} G_{1k}\right] \tag{32}$$

$$N^l_{ps} = \psi_c P_d r \times 10^3 - 0.5(F_{sv,k} + \psi_c q_{vk} D_1) \tag{33}$$

The maximum tensile stress at the edge of the pipe at the spring-line, σ^l_{ss}, can be calculated as per Equation (34):

$$\sigma^l_{ss} = \frac{N^l_{ps}}{A_n} + \frac{M^l_{pms}}{\omega_m W_p} \tag{34}$$

The maximum tensile stress at the edge of the mortar coating at the spring-line, σ^l_{ss}, should be less than its tensile strength (Equation (35)) under serviceability limit states. If not, Sections 2.2.2 and 2.2.3 should be repeated.

$$\sigma^l_{ss} \leq \alpha_m \varepsilon_{mt} E_m \tag{35}$$

where α_m is the design parameter of the mortar coating strain, which is equal to 5. ε_{mt} is the strain of mortar coating when the strength reaches the tensile strength, and can be given as $\varepsilon_{mt} = \frac{f_{mt,k}}{E_m} \geq \frac{0.52\sqrt{f_{mc,k}}}{E_m}$.

2.2.5. Checking Calculation of Mortar Coating under Quasi-Permanent Limit States

Checking the calculation of mortar at the spring-line of the pipe should be carried out under quasi-permanent limit states.

The maximum bending moment of the pipe at the spring-line, M^l_{pml}, can be calculated as per Equation (36). The value of M^l_{pml} is negative, indicating that the mortar coating is subjected to tension. The absolute value is taken when substituting the following equations. The axial tension of the pipe at the spring-line, N^l_{ps}, is written as per Equation (37):

$$M^l_{pml} = r\left[k_{vm}(F_{sv,k} + \psi_{qv} q_{vk} D_1) + k_{hm} F_{ep,k} D_1 + k_{vm} G_{wk} + k_{gm} G_{1k}\right] \tag{36}$$

$$N^l_{pl} = \psi_{qw} P_d r \times 10^3 - 0.5(F_{sv,k} + \psi_{qv} q_{vk} D_1) \tag{37}$$

where ψ_{qv}, ψ_{qw} is the quasi-permanent coefficient of vertical pressure generated by ground vehicle loads and the internal water pressure, respectively.

The maximum tensile stress at the edge of the pipe at the spring-line, σ_{ls}^l, is calculated as per Equation (38):

$$\sigma_{ls}^l = \frac{N_{pl}^l}{A_n} + \frac{M_{pml}^l}{\omega_m W_p}. \tag{38}$$

The maximum tensile stress at the edge of the mortar coating at the spring-line, σ_{ss}^l, should be less than its tensile strength (Equation (39)) under quasi-permanent limit states. If not, we return to Equations (2) and (3).

$$\sigma_{ls}^l \leq \alpha_m' \varepsilon_{mt} E_m \tag{39}$$

where α_m' is the design parameter of strain for mortar coating and is equal to 4.

Above all, the area of prestressed strands per unit length, A_{st}, should be determined.

The prestressed steel strands are spirally wound at equal intervals. Thus, the center spacing of steel strands can be calculated by Equation (40):

$$l_{st} = A \times \frac{1000}{A_{st}} \tag{40}$$

where A is the nominal section area, without polyethylene, of the adopted steel strand.

3. Applications

In order to verify the feasibility of the deduction, an example calculation of the theory and a prototype test were carried out on the same PCCP with broken wires. The specimen was an embedded prestressed concrete cylinder pipe (ECP) and the calculation process used is illustrated in Section 3.2. The center spacing of steel strands, calculated through the deduction, was then applied to the same pipe in a prototype test (Section 3.3).

3.1. Parameters of the Design and Materials

The theory and prototype tests were carried out on the same pipe. The geometric parameters of the adopted pipe are given as Table 1. Key parameters of the materials, involving the concrete, mortar and cylinder, are shown in Table 2.

Table 1. Geometric parameters of the embedded concrete pipe (ECP).

Geometric Parameter	Value	Geometric Parameter	Value
Inner diameter of PCCP, D_i/mm	2000	Net thickness of mortar coating, h_m/mm	25
Thickness of core concrete, h_c/mm	140	Spacing between each wire, l_s/mm	22.1
Outer diameter of cylinder, D_y/mm	2103	Diameter of wires, d_s/mm	6
Thickness of cylinder, t_y/mm	1.5	Number of layers, n	1

Table 2. Key parameters of the materials.

Key Parameter	Value	Key Parameter	Value
Designed 28-day compressive strength of the core concrete, f'_c/(N/mm^2)	44	Modulus of concrete, E_c/(N/mm^2)	3.55×10^4
Standard compressive strength of mortar, $f_{mc,k}$/(N/mm^2)	45	Modulus of mortar, E_m/(N/mm^2)	2.416×10^4
Poisson's ratio of concrete, v_c	0.167	Modulus of cylinder, E_y/(N/mm^2)	2.068×10^5
Poisson's ratio of mortar, v_m	0.2	Modulus of wire, E_s/(N/mm^2)	1.93×10^5
Minimum tensile strength of the prestressed wire, f_{su}/(N/mm^2)	1570	Designed tensile or compressive yield strength of steel cylinder, f_{yy}/(N/mm^2)	227.5
Gross wrapping tensile stress in wire, f_{sg}/(N/mm^2)	$0.75 f_{su}$	Designed tensile strength of steel cylinder at pipe burst, f'_{yy}/(N/mm^2)	215
Design tensile strength of core concrete, f'_t/(N/mm^2)	1.95	Unit weight of the pipe, γ_c/kN/m^3	25
Standard tensile strength of core concrete, f_{ty}/(N/mm^2)	2.75	Unit weight of mortar, γ_m/kN/m^3	23.5
Unit weight of backfill soil, γ_s/kN/m^3	18	Unit weight of water, γ_w/kN/m^3	10

As for the parameters of load, the internal working pressure used was $P_w = 0.6$ N/m^2. The internal transient pressure was $\Delta H_r = \max(0.4 P_w, 276 \text{ kPa}) = 0.276$ N/mm^2. The internal design pressure was $P_d = P_w + \Delta H_r = 0.876$ MPa ≈ 0.9 N/mm^2. The thickness of soil above the top of the pipe was $H = 3$ m. The bedding angle was 90°. The type of installation was trench-type with a positive projecting embankment. The standard value of the ground stacking load was $q_{mk} = 10$ kN/m^2.

The parameters of environment are shown as follows: The average relative humidity of the storage environment was 70% RH, the time in outdoor storage was $t_1 = 270$ d. Burial time after outdoor storage was $t_2 = 1080$ d.

Key parameters of the adopted strand are given in Table 3.

Table 3. Key parameters of the adopted strand.

Key Parameter	Value	Key Parameter	Value
Nominal diameter without polyethylene d_{st}/mm	15.2	Control coefficient for the tensioning of the steel strands, α	0.63
Nominal section area without PE, A/mm^2	140	Tension stress of the prestressed steel strands, σ_{st}/(N/mm^2)	1171.8
Nominal tensile strength, $f_{st,t}$/(N/mm^2)	1860	Standard tensile yield strength, f_{pyk}/(N/mm^2)	1580
Modulus of the strand, E_{st}/(N/mm^2)	1.95×10^5	Designed tensile yield strength, f_{py}/(N/mm^2)	1110 [17]

3.2. Example Calculation

3.2.1. Calculation of Prestress Loss, $\sigma_{st,l}$

For the utilized split wedge without jacking force, $\Delta l_{re} = 6$ mm (measured in the prototype test [29]). Given $\mu = 0.1$, the calculated radius of the strand wrapped outside the pipe, r_{st}, and the retraction length, l_{re}, can be known as follows:

$$r_{st} = \frac{D_i}{2} + h_c + \frac{d_{st}}{2} = 1.1726 \text{ m}, l_{re} = \sqrt{\frac{\Delta l_{re} E_{st} r_{st}}{\mu \sigma_{st}}} = 3.4217 \text{ m}$$

The corresponding angle of the retraction length, θ_{re}, is π, which is consistent with the value of θ, indicating that the assumption is reasonable (Equation (10)).

The prestress loss caused by friction resistance, anchorage deformation, elastic compression of concrete during batch tensioning, crack reduction and closure, shrinkage and creep of concrete, and long-term relaxation of the strand is given in Table 4.

Table 4. The calculation results of the prestress loss.

Item	Value
Prestress loss related to the bending loss, F	276.099
Correction coefficient accounting for the bending loss and the deviation loss, c_1	1.01
Prestress loss caused by friction resistance, $\sigma_{l1}/(\text{N}/\text{mm}^2)$	278.860
Prestress loss caused by anchorage deformation, $\sigma_{l2}/(\text{N}/\text{mm}^2)$	158.802
Prestress loss caused by elastic compression of concrete during batch tensioning, $\sigma_{l3}/(\text{N}/\text{mm}^2)$	0
Maximum width of the visible cracks in the concrete core before the reinforcement, w_1/m	0.0022
Maximum width of the visible cracks in the concrete core after the reinforcement, w_2/m	0.0001
Prestress loss due to the crack reduction and closure, $\sigma_{l4}/(\text{N}/\text{mm}^2)$	0.3434
Prestress loss caused by shrinkage and creep of concrete, $\sigma_{l5}/(\text{N}/\text{mm}^2)$	0
The relaxation coefficient of the strand, k	0.045
Prestress loss due to long-term relaxation of the strand, $\sigma_{l6}/(\text{N}/\text{mm}^2)$	52.731
Total prestress loss of prestressed wires, $\sigma_{st,l}/(\text{N}/\text{mm}^2)$	490.74

3.2.2. Stress of PCCP under Combined Loads

The stress of the adopted PCCP under combined loads, involving the vertical earth pressure at the top of the pipe, the lateral earth pressure, the variable load, weight of the pipe, and weight of water in the pipe, is presented in Table 5.

Table 5. The calculation results of the stress.

Item	Value
Vertical earth pressure at the top of the pipe, $F_{sv,k}/(\text{kN}/\text{m})$	164.245
Lateral earth pressure, $F_{ep,k}/(\text{kN}/\text{m})$	25.026
Variable load, $q_{mk}/(\text{kN}/\text{m})$	10
Weight of the pipe, $G_{1k}/(\text{kN}/\text{m})$	29.157
Weight of water in the pipe/(kN/m)	31.416

3.2.3. Calculation of Area of Prestressed Strands, Considering the Concrete Core Compression of Ultimate Limit States

Assuming that the area of prestressed strands is $A_{st} = 2223 \text{ mm}^2/\text{m}$, the calculation process of the area of prestressed strands, considering the concrete core compression of ultimate limit states, can be depicted in Table 6.

The value of M^l_{max} is negative, indicating that the outer surface of the concrete core is subjected to tension. The absolute value is taken when substituting the formula for reinforcement.

Therefore, $A_{st} \geq \dfrac{\lambda_y}{f_{pyk}} \left(N^l + \dfrac{M^l_{max}}{d_0} - A_{sc} f'_{yy} \right) = 1069.413 \text{ m}^2/\text{m}$

Table 6. The calculation results of the stress.

Item	Value
Thickness of the pipe, T/m	0.171
Calculated radius of the pipe, r/m	1.0855
Outer diameter of the pipe, D_1/m	2.342
Combination coefficient of the variable loads, ψ_c	0.9
The factor of importance for two side-by-side pipelines, γ_0	1
Design value of the maximum bending moment of the pipe at the spring-line, M^l_{max}/(kN·m/m)	−33.998
Design value of the maximum axial tension of the pipe at the spring-line, N^l/(kN/m)	1111.712
Width of calculated section, B/m	1
Ratio of modulus of the strand to the concrete, n_{st}	5.83
Ratio of modulus of the cylinder to the concrete, n_y	5.49
Ratio of modulus of the mortar to the concrete, n_m	0.68
Conversion area of pipe section (including cylinder, steel strands and the mortar coating), A_n/m²	0.1792
Cross sectional area moment of the cross section of the concrete core, mortar, steel cylinder and prestressed steel strand on the inner surface of the pipe wall, S_n/m³	0.01496
Cross sectional area of the cylinder for unit pipe length, A_{sc}/(m²/m)	0.0015
Distance from the mandrel to the inner surface of the pipe wall after the conversion, y_0	0.08347
Distance from the center of the prestressed steel strand to the center of gravity of the pipe wall section, d_0/m	0.06412
Comprehensive adjustment factor for ECP whose diameter is larger than 1600 mm, λ_y	0.9

Notes: $T = h_c + h_m + d_s$, $r = \frac{D_i+T}{2}$, $D_1 = D_i + 2T$, $A_n = Bh_c + (n_y - B)Bt_y + (n_{st} - n_m)A_{st} + n_m B(T - t)$, $S_n = \frac{Bh_c^2}{2} + (n_y-)Bt_y\frac{(D_y-D_i-t_y)}{2} + (n_{st} - n_m) * A_{st} + n_m B(T - h_c)\left(\frac{T-h_c}{2} + h_c\right)$, $A_{sc} = Bt_y$, $y_0 = \frac{A_n}{S_n}$, $d_0 = h_c + \frac{d_{st}}{2} \times 10^{-0} - y_0$.

3.2.4. Checking Calculation of Prestressed Strands Considering the Concrete Core Compression of Serviceability Limit States

The conversion coefficient of the elastic resistance moment of the tensioned edge of the pipe wall can be obtained by interpolation, where $\omega_c = 1.017$ and $\omega_m = 0.9932$. The checking calculation process of the prestressed strands, considering the concrete core compression of serviceability limit states, is depicted in Table 7.

Table 7. The calculation results of the stress.

Item	Value
Maximum bending moment of the pipe at the bottom, M_{pms}/(kN·m/m)	36.316
Maximum bending moment of the pipe at the top, M'_{pms}/(kN·m/m)	23.423
Axial tension of the pipe wall, N_{ps}/(kN/m)	879.255
Elastic resistance moment of the unconverted tension edge of the rectangular section of the pipe wall, W_p/(m²/m)	0.00487
Maximum tensile stress at the edge of the pipe at the bottom, σ_{ss}/(N/mm²)	12.107
Effective prestress of the prestressed steel strands apart from the prestress loss, σ'_{st}/(N/mm²)	681.06
Influence coefficient of concrete in tension area, K	1.2239
Plastic influence coefficient of concrete in tension area, γ	1.75

Note: $K = 0.2449 \frac{M_{pms}}{\omega_c W_p f_{ty}} + 0.5714$.

Therefore, the area of prestressed strands should meet the requirement of $A_{st} \geq (\sigma_{ss} - K\gamma f_{ty})\frac{A_n}{\sigma'_{st}} = 2222.300$ mm²/m.

Above all, the area of prestressed strands is $A_{st} = 2223$ mm²/m.

3.2.5. Checking Calculation of Mortar Coating under Serviceability Limit States

The maximum bending moment of the pipe at the spring-line was $M_{pms}^l = -24.897$ kN·m/m. The value of M_{pms}^l, is negative, indicating that the mortar coating was subjected to tension. The absolute value was taken when substituting the following equations. The axial tension of the pipe at the spring-line was $N_{ps}^l = 769.388$ kN/m The maximum tensile stress at the edge of the pipe at the spring-line was $\sigma_{ss}^l = 9.44$ N/mm². The strain of mortar coating was $\varepsilon_{mt} = \frac{f_{mt,k}}{E_m} \geq \frac{0.52\sqrt{f_{mc,k}}}{E_m} = 0.0001444$. The design parameter of strain for the mortar coating was $\alpha_m = 5$. Thus, $(\alpha_m \varepsilon_{mt} E_m = 17.44$ N/mm²$) > (\sigma_{ss}^l = 9.44$ N/mm²$)$, indicating that the area of prestressed strands is able to meet the tensile requirement of the mortar coating under the serviceability limit states.

3.2.6. Checking Calculation of Mortar Coating under Quasi-Permanent Limit States

ψ_{qv} and ψ_{qw} are the quasi-permanent coefficient of vertical pressure generated by the ground vehicle loads and the internal water pressure, respectively. Here, $\psi_{qv} = 0.5$ and $\psi_{qw} = 0.72$. The maximum bending moment of the pipe at the spring-line was $M_{pml}^l = -23.422$ kN·m/m. The value of M_{pml}^l is negative, indicating that the mortar coating was subjected to tension. The absolute value was taken when substituting the following equations. The axial tension of the pipe at the spring-line was $N_{pl}^l = 602.911$ kN/m. The maximum tensile stress at the edge of the pipe at the spring-line was $\sigma_{ls}^l = 8.21$ N/mm². The design parameter of strain for the mortar coating was $\alpha_m' = 4$. Therefore, $(\alpha_m' \varepsilon_{mt} E_m = 13.95$ N/mm²$) > (\sigma_{ss}^l = 8.21$ N/mm²$)$, indicating that the area of prestressed strands is able to meet the tensile requirement of the mortar coating under quasi-permanent limit states.

Above all, this is reasonable of the calculation result of the area of prestressed strands, which is $A_{st} = 2223$ mm²/m. The center spacing of steel strands was $l_{st} = A \times \frac{1000}{A_{st}} = 62.99$ mm.

3.3. A Prototype Test

A prototype test of ECP reinforced by steel strands with the fixed spacing calculated in Section 3.2 was performed in an assembled apparatus (Figure 7). The apparatus was mainly constituted by two ECPs, whose internal diameters were 2000 mm [29]. The adopted pipes were exactly the same as those given in Section 3.1. The entire test process involved five load stages, namely, (1) increasing the internal water pressure to the working pressure (0–0.6 MPa), (2) cutting the prestressing wires manually until the cracks propagated in the concrete core (0.6 MPa), (3) decreasing the internal water pressure to the artesian pressure (0.6–0.2 MPa), (4) performing the tensioning operation after wrapping the strands externally around the pipe (0.2 MPa), and (5) increasing the internal water pressure to the original level (0.2–0.6 MPa). In most of the actual pipe failures modes, most pipes failed at 4 or 8 o'clock, not at the invert, crown, or spring-lines [29]. The position of 8 o'clock was chosen in this test for convenience (Figure 8).

Post-tensioning was designed with the theory conducted in Section 3.2, indicating that the target tensile strength was equal to 1171.8 MPa and the center spacing of steel strands was taken as 62 mm. To prevent a prestress loss due to the retraction of clips and the stress relaxation of strands, excessive stretching is essential here. The tensioning process is divided into six stages, which were 20%, 25%, 50%, 75%, 100%, and 115%. Tensioning was performed simultaneously from both sides and in a symmetrical manner along the pipeline axis.

The statuses of each component of the pipe and the steel strands were measured by resistance strain gauges along the axial direction at inverted (360°), crown (180°), and spring-line (90°, 270°) orientations (Figure 8). Figure 9 exhibits the hoop strains in the concrete core before and after the reinforcement under the working pressure (0.6 MPa). The strains in the concrete core all showed a drastic drop after the process of tensioning. Moreover, the maximum width of the cracks in the outer concrete core at spring-line reduced from 2.2 mm to 0.1 mm after strengthening, as observed through field observation.

Appl. Sci. **2019**, *9*, 5532

Figure 7. The spot photo of the test apparatus.

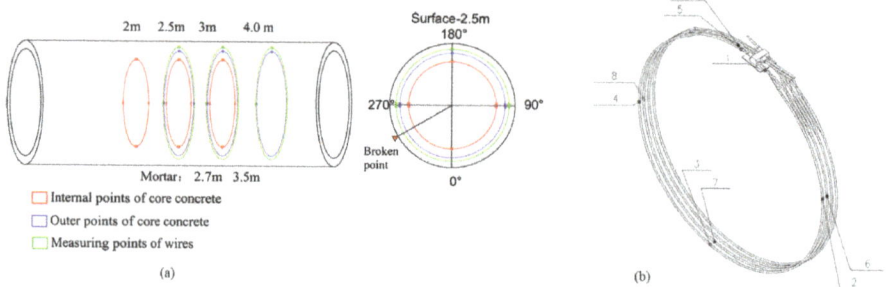

Figure 8. Layout of measuring points of (**a**) the pipe and (**b**) the steel strands.

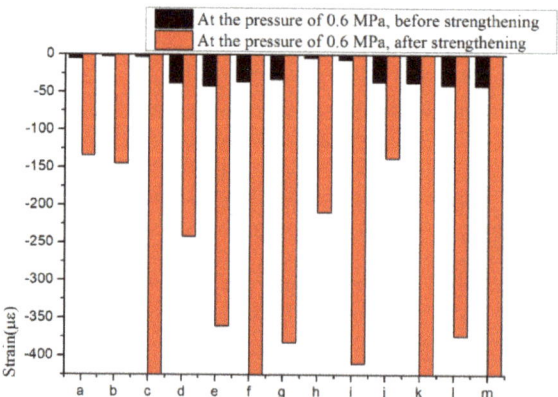

Figure 9. Comparison of strains in core concrete before and after the reinforcement: (**a**) 2.5 m inner concrete core at 0°; (**b**) 2.5 m inner concrete core at 90°; (**c**) 2.5 m inner concrete core at 180°; (**d**) 3 m inner concrete core at 0°; (**e**) 3 m inner concrete core at 90°; (**f**) 3 m inner concrete core at 180°; (**g**) 3 m inner concrete core at 270°; (**h**) 2.5 m outer concrete core at 0°; (**i**) 2.5 m outer concrete core at 90°; (**j**) 3 m outer concrete core at 90°; (**k**) 3 m outer concrete core at 180°; (**l**) 4 m outer concrete core at 90°; (**m**) 4 m outer concrete core at 180°.

The strengthened pipe was capable of sustaining the working pressure and the water tightness property was in a good state. The strains of the steel strands were all below the tensile strain level. The reinforcement of the PCCP with external prestressed steel strands was able to meet the strengthen requirement of the test. The rationality of the derivations in this paper were verified by the effective reinforcement effect with external prestressed steel strands.

4. Conclusions

A theoretical derivation was performed, aiming to determine the appropriate area of prestressed steel strands per unit length, and a prototype test was conducted to verify the rationality of the derivation in this study. The following conclusions can be drawn:

(1) The calculation formula for the prestress loss of different types of steel strands has been derived and the effective prestress of the prestressed steel strands can be determined.
(2) A stress calculation formula of the concrete core under the ultimate limit states and serviceability limit states was determined and used for calculation. The condition of the mortar coating under the serviceability limit and the quasi-permanent limit states was verified, and the reinforcement area of the steel strand was finally determined. This tentative derivation was applied to the reinforced pipe with broken wires (inner diameter of 2000 mm) to calculate the appropriate area of prestressed steel strands.
(3) The crack propagation in the concrete core was constrained by the strands and the test pipe was able to sustain the working pressure after strengthening. In addition, the maximum width of the cracks in the outer concrete core at the spring-line showed some closure because of the contribution of the strands. The bearing capacity of the prototype test was returned to the original design level and the behavior of the pipe was in accordance with the expectation of derivation.

Author Contributions: Conceptualization, T.D., S.X. and L.Z.; Methodology, T.D., L.Z., and B.C.; Software, T.D. and L.Z.; Writing—original draft preparation, L.Z.; Writing—review and editing, B.C., X.L., and M.L.; Funding acquisition, J.Y. and Q.Z.

Funding: This research was supported by the Beijing South-to-North Water Transfer Line Management Office (GXGLC-JSZX -2017-CG01), the National Natural Science Foundation of China (Grant No. 5097911), the Beijing Municipal Science and Technology Commission (Z141100006014058), and the China Institute of Water Resources and Hydropower Research (SM0145B632017).

Acknowledgments: The authors would like to thank the anonymous reviewers for their constructive suggestions to improve the quality of the paper.

Conflicts of Interest: The authors declare no conflict of interest.

References

1. Engelmanna, M.; Wellerb, B. Losses of prestress in post-tensioned glass beams. *Structures* **2019**, *19*, 248–257. [CrossRef]
2. Tan, K.H.; Tjandra, R.A. Strengthening of RC Continuous Beams by External Prestressing. *J. Struct. Eng.* **2007**, *133*, 195–204. [CrossRef]
3. Abdel-Jaber, H.; Glisic, B. Monitoring of long-term prestress losses in prestressed concrete structures using fiber optic sensors. *Struct. Health Monit.* **2019**, *18*, 254–269. [CrossRef]
4. Hou, Y.; Cao, S.; Ni, X.; Li, Y. Research on Concrete Columns Reinforced with New Developed High-Strength Steel under Eccentric Loading. *Materials* **2019**, *12*, 2139. [CrossRef] [PubMed]
5. Song, S.; Zang, H.; Duan, N.; Jiang, J. Experimental Research and Analysis on Fatigue Life of Carbon Fiber Reinforced Polymer (CFRP) Tendons. *Materials* **2019**, *12*, 3383. [CrossRef] [PubMed]
6. Miyamoto, A.; Tei, K.; Nakamura, H.; Bull, J.W. Behavior of prestressed beam strengthened with external tendons. *J. Struct. Eng.* **2000**, *126*, 1033–1044. [CrossRef]
7. Miyamoto, A.; Tei, K.; Gotou, M. *Mechanical Behavior and Design Concept of Prestressed Composite Plate Girders with External Tendons*; Technology Reports-Yamaguchi University: Yamaguchi, Japan, 1995; pp. 233–258.

8. Chen, S.; Jia, Y. Numerical investigation of inelastic buckling of steel–concrete composite. *Thin Wall Struct.* **2010**, *48*, 233–242. [CrossRef]
9. Lou, T.; Lopes, S.M.R.; Lopes, A.V. Numerical modeling of externally prestressed steel concrete composite beams. *J. Constr. Steel Res.* **2016**, *121*, 229–236. [CrossRef]
10. Aparicio, A.C.; Ramos, G.; Casas, J.R. Testing of externally prestressed concrete beams. *Eng. Struct.* **2002**, *24*, 73–84. [CrossRef]
11. Park, S.; Kim, T.; Kim, K.; Hong, S.N. Flexural behavior of steel I-beam prestressed with externally unbonded tendons. *J. Constr. Steel Res.* **2010**, *66*, 125–132. [CrossRef]
12. Elnakhat, H.; Raymond, R. Repair of PCCP by Post Tensioning. In Proceedings of the Pipelines 2006, Chicago, IL, USA, 30 July–2 August 2006; pp. 1–5.
13. Cao, G.; Hu, J.; Kai, Z. Coupling model for calculating prestress loss caused by relaxation loss, shrinkage, and creep of concrete. *J. Cent. South Univ.* **2016**, *23*, 470–478. [CrossRef]
14. Cao, G.; Zhang, W.; Hu, J.; Zhang, K. Experimental study on the long-term behaviour of RBPC T-beams. *Int. J. Civ. Eng.* **2018**, *16*, 887–895. [CrossRef]
15. Baolin, K. The Estimating of the External Prestressing Loss in the Bridge Strengthen System. *J. Hebei Inst. Archit. Sci. Technol.* **2002**, *19*, 27–29.
16. Zhang, J.; Pan, J.; Dong, H. *Analysis of Prestressing Loss on Construction of External Prestressed Wooden Buildings: Advanced Materials Research*; Trans Tech Publications: Bern, Switzerland, 2012.
17. GB 50010-2010. *Code for Design of Concrete Structures*; Industry Press: Beijing, China, 2010.
18. Kang, Y. *Reinforcement Mechanism and Application of External Prestressed Concrete*; Hehai University: Nanjing, China, 2006.
19. Hu, Z. Practical Estimation Method for Prestress Loss of Externally Prestressed Structures. *Bridge Constr.* **2006**, 73–75.
20. Zhang, S.; Zhang, K.Y.; Xie, B.Y.; Fan, Z.L. Research on the Mechanism of Prestressed Loss for Curving Hole of Prestressed Concrete Structure Caused by Frictional Resistance. In *Applied Mechanics and Materials*; Trans Tech Publications: Bern, Switzerland, 2013.
21. Gao, A. *Research on Prestress Loss Based on Friction, Retraction and Natural Frequency*; Southeast University: Nanjing, China, 2016.
22. Erbay, O.O.; Zarghamee, M.S.; Ojdrovic, R.P. Failure Risk Analysis of Lined Cylinder Pipes with Broken Wires and Corroded Cylinder. In *Pipelines 2007: Advances and Experiences with Trenchless Pipeline Projects*; American Society of Civil Engineers: Reston, VA, USA, 2007; pp. 1–10.
23. Zarghamee, M.S.; Dana, W.R. Step-by-step Integration Procedure for Computing State of Stress in Prestressed concrete pipe. *Spec. Publ.* **1991**, *129*, 155–170.
24. Zarghamee, M.S.; Fok, K.; Sikiotis, E.S. Limit states design of prestressed concrete pipe II: Procedure. *J. Struct. Eng.* **1990**, *116*, 2105–2126. [CrossRef]
25. Zarghamee, M.S.; Fok, K. Analysis of prestressed concrete pipe under combined loads. *J. Struct. Eng.* **1990**, *116*, 2022–2039. [CrossRef]
26. American Water Works Association (Ed.) *Concrete Pressure Pipe: M9*; American Water Works Association: Denver, CO, USA, 2008.
27. CECS 140:2011. *Specification for Structural Design of Buried Prestressed Concrete Pipeline and Prestressed Concrete Cylinder Pipeline of Water Supply and Sewerage Engineering*; Beijing General Municipal Engineering Design& Research Institute Co. Ltd.: Beijing, China, 2011.
28. GB 50332. *Structural Design Code for Pipelines of Water Supply and Waste Water Engineering*; Ministry of Housing and Urban-Rural Development of the People's Republic of China: Beijing, China, 2002.
29. Zhao, L.; Dou, T.; Cheng, B.; Xia, S.; Yang, J.; Zhang, Q.; Li, M.; Li, X. Experimental Study on the Reinforcement of Prestressed Concrete Cylinder Pipes with External Prestressed Steel Strands. *Appl. Sci.* **2019**, *9*, 149. [CrossRef]

© 2019 by the authors. Licensee MDPI, Basel, Switzerland. This article is an open access article distributed under the terms and conditions of the Creative Commons Attribution (CC BY) license (http://creativecommons.org/licenses/by/4.0/).

Article

First-Principles Forecast of Gapless Half-Metallic and Spin-Gapless Semiconducting Materials: Case Study of Inverse Ti$_2$CoSi-Based Compounds

Liang Zhang [1], Shengjie Dong [2,*], Jiangtao Du [1], Yi-Lin Lu [3], Hui Zhao [4] and Liefeng Feng [1,*]

1. Tianjin Key Laboratory of Low Dimensional Materials Physics and Preparing Technology, Department of Physics, Faculty of Science, Tianjin University, Tianjin 300350, China; liang_zhang@tju.edu.cn (L.Z.); 2014210001@tju.edu.cn (J.D.)
2. Faculty of Education and Sports, Guangdong Baiyun University, Guangzhou 510450, China
3. College of New Energy, Bohai University, Jinzhou 121007, China; yilinlu@tju.edu.cn
4. Department of Physics, College of Physics and Materials Science, Tianjin Normal University, Tianjin 300387, China; naihuizhao@gmail.com
* Correspondence: shengjiedong@tju.edu.cn (S.D.); fengliefeng@tju.edu.cn (L.F.); Tel.: +86-1862-273-0034 (L.F.)

Received: 23 December 2019; Accepted: 20 January 2020; Published: 22 January 2020

Featured Application: Half of the Si atoms in Ti2.25Co0.75Si are replaced by B, Al, Ga, P, As, and Sb, and the results show that doped Ti2CoSi with an appropriate concentration of impurities could exhibit half-metallic ferromagnetic, gapless half-metallic, and spin-gapless semiconducting states.

Abstract: First-principles calculations were used to investigate several inverse Ti$_2$CoSi-based compounds. Our results indicate that Ti$_2$CoSi could transform from a spin-gapless semiconductor to a half metal if a quarter of the Co atoms are replaced by Ti. Ti$_{2.25}$Co$_{0.75}$Si would keep stable half-metallic properties in a large range of lattice parameter under the effect of hydrostatic strain, and would become a gapless half metal under the effect of tetragonal distortion. Furthermore, we substituted B, Al, Ga, P, As, and Sb for Si in the Ti$_{2.25}$Co$_{0.75}$Si compound. Our results demonstrate that Ti$_{2.25}$Co$_{0.75}$Si$_{0.5}$B$_{0.5}$, Ti$_{2.25}$Co$_{0.75}$Si$_{0.5}$Al$_{0.5}$, and Ti$_{2.25}$Co$_{0.75}$Si$_{0.5}$Ga$_{0.5}$ are half-metallic ferromagnetic materials, and Ti$_{2.25}$Co$_{0.75}$Si$_{0.5}$P$_{0.5}$, Ti$_{2.25}$Co$_{0.75}$Si$_{0.5}$As$_{0.5}$, and Ti$_{2.25}$Co$_{0.75}$Si$_{0.5}$Sb$_{0.5}$ are spin-gapless semiconducting materials. The introduced impurity atoms may adjust the valence electron configuration, change the charge concentration, and shift the location of the Fermi level.

Keywords: first-principles calculation; Heusler compounds; gapless half metals; spin gapless semiconductor

1. Introduction

With the development of nanotechnology and computational materials science, spintronics has developed rapidly in the past 30 years. In order to improve the performance of spin diodes, spin valves, and spin filters, the design of high spin-polarized materials has attracted much attention [1–4]. For half-metallic ferromagnetic compounds with unique electronic structures, one of the two spin channels is semiconducting and the other is metallic. As promising spintronic candidates, they exhibit a complete spin polarization of carriers near the Fermi level [5–12]. A spin-gapless semiconductor is another new kind of spintronic material, which has an almost complete spin polarization and good compatibility with the existing semiconductor industry. By shifting the Fermi energy at the finite gate voltage, the spin-polarized transport properties of spin-gapless semiconductors can be tuned, which has great prospects for future spintronic applications [13,14].

Half metals and spin-gapless semiconductors are special kinds of materials. Because of their unique electronic structures, they have novel spin-dependent electronic properties. The band structure of these materials has only one energy gap for a specific spin direction, which means that the band gap disappears in the opposite spin direction. This phenomenon leads to the application prospects of high carrier spin polarization and spin-controlled electrical and magnetic features. In recent years, in order to develop new technologies involving spintronics, it has been important to search for new materials with these characteristics. Many of these materials are Heusler alloys with a specific crystalline ordering, which is very important for the unique electronic and magnetic properties of these materials. Therefore, it is crucial to study these alloys and compounds with these novel physical properties.

Heusler alloys are named for the German mining engineer and chemist Friedrich Heusler, who investigated Cu–Mn–Al alloys around the year 1900. Heusler compounds are ternary intermetallic compounds, which have been known since 1903 [15]. These alloys were interesting because some of them were ferromagnetic, even though their constituent atoms were non-ferromagnetic as elements. At that time, Cu_2MnAl compounds were proven to be ferromagnetic, although none of the three elements were ferromagnetic. The structure of the alloys was later found to be face-centered cubic (fcc) with a four-atom basis consisting of a single formula unit. Thus, members of this Heusler family with a formula unit A_2BC (known as "full" Heuslers) can be viewed as layers consisting of a square lattice of "A" atoms alternating with layers of "B" and "C" atoms, as shown in Figure 1a. A related Heusler family has members with the formula unit ABC, in which half of the "A" atoms of the full Heusler are replaced by vacancies, as shown in Figure 1b. These are known as "half" Heuslers or semi-Heuslers. A third Heusler family with the formula unit A_2BC (similar to the "full" Heuslers) consists of alloys that can be viewed as "AB" layers alternating with "AC" layers, as shown in Figure 1c. These are known as "inverse" Heuslers.

The properties of Heusler compounds can be altered by element substitution, but no single set of properties can characterize the entire Heusler family. Heusler alloys are expected to play an important role in spintronics, magneto-optical reading–recording devices, magnetic tunnel junctions, or tunneling magnetoresistance devices, due to their various multifunctional magnetic properties [16,17]. In the past years, several inverse binary- and ternary-Heusler alloys with a high Curie temperature, as well as quaternary-Heusler alloys, were theoretically predicated to exhibit spin-gapless semiconducting electronic structures [18,19]. Moreover, the bulk and thin films of Mn_2CoAl and Ti_2MnAl films were experimentally fabricated [20–23]. More recently, with first principle calculations, $Ti_2MnAl_{0.5}Sn_{0.5}$ and $Ti_2MnAl_{0.5}In_{0.5}$ were proven to be half metal and spin-gapless semiconductor, respectively [24].

Magnetic Heusler compounds are widely used in spintronic applications. The Heusler semiconductor can be used as either a spin-conserved tunneling barrier or a spin-transporting spacer in a Heusler ferromagnet–Heusler semiconductor–Heusler ferromagnet heterojunction, which has been extensively investigated in recent years. The magnetic Heusler compound is useful and meaningful for its potential integration into magnetic multilayer devices toward some spin valves, as well as for other opportunities and new phenomena.

In addition to these spintronic applications mentioned above, Heusler compounds, with other excellent properties, were studied extensively in the past several years. For instance, several Heusler compounds can be used as solar cell materials, thermoelectric materials, topological materials, magneto-optical materials, magneto-caloric materials, shape-memory materials, heavy-Fermion materials, superconductors, semiconductors, and so on [25–28].

In this work, the electronic structures and magnetic properties of the Ti_2CoSi inverse Heusler alloy were calculated when a quarter of Co atoms were replaced by Ti. Then, the effects of hydrostatic strain and tetragonal deformation on the electronic structures and magnetic properties of the $Ti_{2.25}Co_{0.75}Si$ compound were studied. Meanwhile, half of the Si atoms in $Ti_{2.25}Co_{0.75}Si$ were replaced by B, Al, Ga, P, As, and Sb, and the results show that doped Ti_2CoSi with an appropriate concentration of impurities could be a half-metallic ferromagnet, a gapless half metal, or a spin-gapless semiconductor.

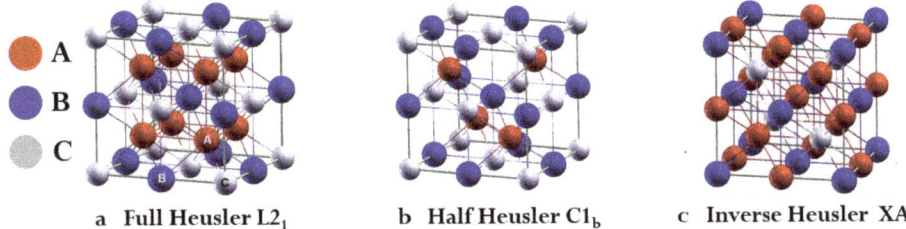

Figure 1. Crystal cell of (**a**) a full Heusler structure, (**b**) half Heusler structure, and (**c**) inverse Heusler structure.

2. Materials and Methods

2.1. Structure of Materials

Structurally, the large Heusler family is described by only two variants, namely: the so-called full-Heusler X_2YZ phase, which is usually crystallized in the Cu_2MnAl-type ($L2_1$) structure, and the half-Heusler XYZ phase with a $C1_b$ structure, where X is a transition metal, Y may be a transition metal or a rare metal, and Z is a main group element. For full-Heusler alloys, if the atomic number of Y is higher than that of X, making the Y element more electronegative than the X element, the inverse Heusler structure of Hg_2CuTi-type can be observed [29]. This structure adopts an F-43m space group, and the atoms obey the following filling rules: X in (0, 0, and 0) and (0.25, 0.25, and 0.25), and Y and Z in (0.5, 0.5, and 0.5) and (0.75, 0.75, and 0.75), respectively [30].

2.2. Computational Methods

For investigating the electronic structures and magnetic properties of these pure and doped inverse Heusler alloys, ab initio calculations were carried out by using the density functional theory (DFT) with the standard generalized gradient approximations (GGA) of Perdew, Burke, and Erzerhof (PBE) to deal with the exchange correlation functional [31–36]. The cutoff energy for the plane wave was set to be 500 eV, and the k point meshes for the Brillouin zone were set to be $12 \times 12 \times 12$. In addition, the convergence for the difference on the total energy was set to be 1×10^{-6} eV/atom [37,38].

3. Results and Discussion

3.1. Crystal Structures and Lattice Parameters

As shown in Figure 2a, the crystal structure of the regular Heusler alloy Ti_2CoSi has a face-centered cubic structure with the following atomic positions: Ti (0, 0, and 0), Ti (0.5, 0.5, and 0.5), Co (0.25, 0.25, and 0.25), and Si (0.75, 0.75, and 0.75). Figure 2b shows that the inverse structure possesses 16 atoms in the unit cell with the following atomic positions: Ti (0, 0, and 0), Co (0.5, 0.5, and 0.5), Ti (0.25, 0.25, and 0.25), and Si (0.75, 0.75, and 0.75). From the previous study, it is known that the inverse Heusler alloy Ti_2CoSi is a spin-gapless semiconductor with an integer magnetic moment of 3 μ_B at the equilibrium lattice constant of 6.03 Å. With the GGA calculations, we found that the inverse Ti_2CoSi alloy evinces gapless semiconducting characteristics at the equilibrium lattice parameter of 6.02 Å, which derives 0.16% less than from the previous investigation, demonstrating that our research is reasonable.

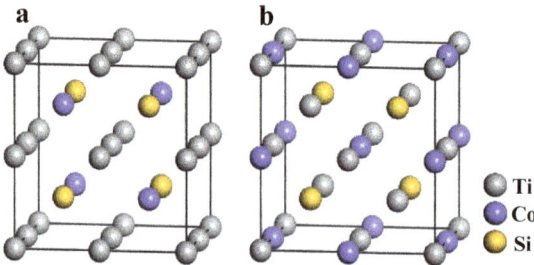

Figure 2. Crystal cell of Ti$_2$CoSi with (**a**) regular cubic Heusler structure and (**b**) an inverse cubic Heusler structure.

For determining the magnetic ground state of these materials, we calculated the energies of the ferromagnetic (FM) and anti-ferromagnetic (AFM) states. The energy differences between the AFM and FM states are listed in Table 1. The positive values illustrate that the energy of the FM state is smaller than that of the AFM state, indicating that the FM state is the magnetic ground state of these materials. At the same time, the minimized energies were obtained by computing the total energy among a large range of lattice parameters. The equilibrium lattice parameters are 6.01 Å and 6.04 Å for Ti$_{2.25}$Co$_{0.75}$Si$_{0.5}$B$_{0.5}$ and Ti$_{2.25}$Co$_{0.75}$Si$_{0.5}$P$_{0.5}$, which is smaller than that of Ti$_{2.25}$Co$_{0.75}$Si, as the radius of Si is larger than that of the B and P atoms. The calculated values are 6.13 Å, 6.13 Å, 6.12 Å, and 6.25 Å for Ti$_{2.25}$Co$_{0.75}$Si$_{0.5}$Al$_{0.5}$, Ti$_{2.25}$Co$_{0.75}$Si$_{0.5}$Ga$_{0.5}$, Ti$_{2.25}$Co$_{0.75}$Si$_{0.5}$As$_{0.5}$, and Ti$_{2.25}$Co$_{0.75}$Si$_{0.5}$Sb$_{0.5}$, respectively.

3.2. Electronic Structures and Magnetic Properties

The electronic band structures of the Ti$_{2.25}$Co$_{0.75}$Si alloy at its equilibrium lattice constant were calculated and are shown in Figure 3a. The spin-up and spin-down electronic bands are indicated by blue and red lines, respectively. When we replaced 25% cobalt with titanium in a cubic cell, a Ti$_{2.25}$Co$_{0.75}$Si alloy was achieved. It can be seen that the spin-up direction shows a metallic behavior and the spin-down direction evinces a semiconducting behavior with an indirect band gap of 0.22 eV around the Fermi level. The valence band maximum (VBM) and conduction band minimum (CBM) are located at R and M points, respectively. Thus, Ti$_{2.25}$Co$_{0.75}$Si is a half-metallic ferromagnetic material with an integer magnetic moment of 7 μ_B (see Table 1). The calculated values of P were 98% for Ti$_{2.25}$Co$_{0.75}$Si$_{0.5}$B$_{0.5}$ and Ti$_{2.25}$Co$_{0.75}$Si$_{0.5}$Ga$_{0.5}$, and 96% for Ti$_{2.25}$Co$_{0.75}$Si$_{0.5}$P$_{0.5}$, and 100% for the other studied compounds in this work, where P = $[N_\uparrow(E_F) - N_\downarrow(E_F)] / [N_\uparrow(E_F) + N_\downarrow(E_F)]$, and $N_{\uparrow,\downarrow}(E_F)$ are the spin-dependent density of states at the Fermi level [39–41].

It is also important to study the effect of the possible lattice distortion during the process of film deposition on the electronic structures. The influences of compression and tetragonalization on the magnetic properties of the Ti$_{2.25}$Co$_{0.75}$Si compound were investigated. Figure 4a presents the position of the CBM and VBM of the spin-down channel as a function of the lattice constant. Upon compressing the lattice strain, the half-metallic properties can be kept between a range from 5.78 Å to 6.35 Å, indicating that the half-metallic characteristics of Ti$_{2.25}$Co$_{0.75}$Si can be preserved when the lattice constants are changed by −4.9%~4.4% in relation to the equilibrium lattice constant. Figure 4b presents the position of the CBM and VBM of the spin-down channel as a function of the c/a ratio. Under the influence of tetragonalization, we found that the alloy experiences the transformation from a magnetic metal to a traditional half metal to a gapless half metal, and finally to a magnetic metal. For a Ti$_{2.25}$Co$_{0.75}$Si compound, a gapless half metal can be obtained when the lattice parameter, c, is expanded by 12%. The corresponding electronic band structures (c/a = 1.12) are plotted in Figure 3b. It can be seen that majority-spin electrons pass through the Fermi level, while the CBM and VBM of the minority spin just touch the Fermi level at R and A points, respectively.

Table 1. The optimized equilibrium lattice parameter, a_o (Å); the energy difference ΔE (eV) between anti-ferromagnetic (AFM) and ferromagnetic (FM) states; and the total magnetic moment μ_{tot} (μ_B) of the studied compounds.

Compound	a_o	ΔE	μ_{tot}
$Ti_{2.25}Co_{0.75}Si$	6.08	1.171	7.00
$Ti_{2.25}Co_{0.75}Si_{0.5}B_{0.5}$	6.01	0.607	5.00
$Ti_{2.25}Co_{0.75}Si_{0.5}Al_{0.5}$	6.13	0.592	5.00
$Ti_{2.25}Co_{0.75}Si_{0.5}Ga_{0.5}$	6.13	0.673	5.00
$Ti_{2.25}Co_{0.75}Si_{0.5}P_{0.5}$	6.04	0.705	8.91
$Ti_{2.25}Co_{0.75}Si_{0.5}As_{0.5}$	6.12	0.830	9.00
$Ti_{2.25}Co_{0.75}Si_{0.5}Sb_{0.5}$	6.25	0.963	9.00

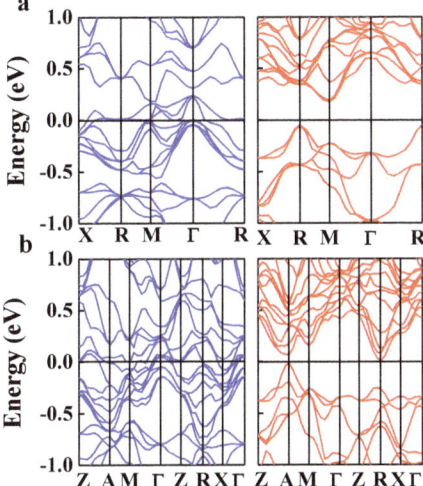

Figure 3. Spin-resolved electronic band structures of $Ti_{2.25}Co_{0.75}Si$ (**a**) at an equilibrium lattice constant and (**b**) with $c/a = 1.12$ lattice distortion. The zero of the energy scale is set at the Fermi energy. The blue and red solid lines represent spin-up and spin-down states, respectively.

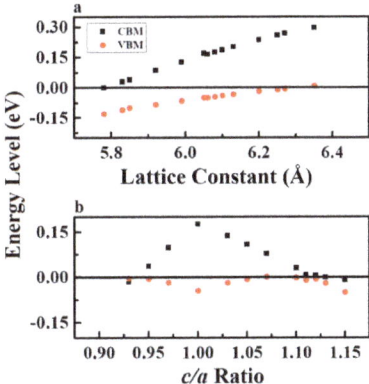

Figure 4. The energy of conduction band minimum (CBM) and valence band maximum (VBM) of the spin-down channel as a function of (**a**) the lattice constant and (**b**) the c/a ratio for $Ti_{2.25}Co_{0.75}Si$.

In the following, we substituted half of the silicon with B, Al, Ga, P, As, and Sb atoms in the cubic cell of $Ti_{2.25}Co_{0.75}Si$ compound. The spin-resolved electronic band structures of the investigated compounds in the equilibrium lattice parameters were computed and are presented in Figures 5 and 6. For $Ti_{2.25}Co_{0.75}Si_{0.5}B_{0.5}$, $Ti_{2.25}Co_{0.75}Si_{0.5}Al_{0.5}$, and $Ti_{2.25}Co_{0.75}Si_{0.5}Ga_{0.5}$, although the Fermi level shifts into a low region of energy with the decrease of the valence electrons, the majority-spin electrons still pass through the Fermi level and the minority-spin bands evince semiconducting characteristics with a band gap, indicating that they are also half metallic. In particular, we found that for $Ti_{2.25}Co_{0.75}Si_{0.5}B_{0.5}$ and $Ti_{2.25}Co_{0.75}Si_{0.5}Ga_{0.5}$, the VBM of the spin-down channel touched the Fermi level at T point, demonstrating two quasi-gapless half metals. We also found that the minority spin band gap of $Ti_{2.25}Co_{0.75}Si_{0.5}B_{0.5}$ was smaller than that of $Ti_{2.25}Co_{0.75}Si_{0.5}Al_{0.5}$ and $Ti_{2.25}Co_{0.75}Si_{0.5}Ga_{0.5}$. The reason for this may be that the atomic radius of the boron atom is relatively small, which leads to a certain chemical pressure, and then reduces the band gap. For the cases of $Ti_{2.25}Co_{0.75}Si_{0.5}P_{0.5}$, $Ti_{2.25}Co_{0.75}Si_{0.5}As_{0.5}$, and $Ti_{2.25}Co_{0.75}Si_{0.5}Sb_{0.5}$, the spin-up bands open a seamless gap, and the Fermi level is located at this zero-gap as a result of the increase of valence electrons. Generally speaking, spin-gapless semiconductors can be divided into four categories, as follows: (i) one spin channel is semiconducting and the other spin channel is gapless; (ii) one spin channel is semiconducting and the VBM touches the Fermi level, while the other spin channel is gapless; (iii) there is a gap for both spin-up and spin-down channels, nonetheless, both the VBM of the spin-up channel and the CBM of the spin-down channel are in contact with the Fermi level; and (iv) one spin channel is gapless, and for the other spin channel, the Fermi level touches the edge of the conduction bands. For $Ti_{2.25}Co_{0.75}Si_{0.5}P_{0.5}$, the spin-up channel is gapless and the CBM touches the Fermi level in the spin-down channel. Thus, it belongs to the fourth type. For $Ti_{2.25}Co_{0.75}Si_{0.5}As_{0.5}$ and $Ti_{2.25}Co_{0.75}Si_{0.5}Sb_{0.5}$, the spin-up channel is gapless, while the spin-down channel is semiconducting. As a consequence, they should belong to the first type. Nevertheless, for $Ti_{2.25}Co_{0.75}Si_{0.5}As_{0.5}$, the CBM of the down spins stand too close to the Fermi level. Thus, the singular electromagnetic transport effect can only be observed at very low temperatures. For $Ti_{2.25}Co_{0.75}Si_{0.5}Sb_{0.5}$, there is also a very small gap in the spin-up channel. Thereby, strictly speaking, it would be a spin-polarized narrow band-gap semiconductor. However, at a finite temperature, the compound would present the typical electromagnetic transport phenomenon of the spin-gapless semiconductor.

Taking $Ti_{2.25}Co_{0.75}Si_{0.5}B_{0.5}$ and $Ti_{2.25}Co_{0.75}Si_{0.5}P_{0.5}$, for example, we discussed the orbital-resolved spin polarization. The total density of the states (DOS) and partial DOS for $Ti_{2.25}Co_{0.75}Si_{0.5}B_{0.5}$ and $Ti_{2.25}Co_{0.75}Si_{0.5}P_{0.5}$ obtained using GGA, are presented in Figure 7. The DOS around the Fermi level is heavily dominated by the 3d states of the Ti and Co atoms. It was found that the 3d states of Co are mainly distributed from -1 to -3 eV, while the $3d$ states of Ti are mainly distributed around and beyond the Fermi level. The electronic configurations of the Ti and Co atoms are $3d^2 4s^2$ and $3d^7 4s^2$, respectively. Thus, the Ti atom with less than half of the filling valence electron shell is mainly distributed at high energy levels, and the Co atom with more than half the filling valence electron shell is mainly distributed at low energy levels. It was also found that the impurity atoms play a key role in adjusting the valence electron concentration of the compound, and shift the Fermi level upwards or downwards. For $Ti_{2.25}Co_{0.75}Si_{0.5}B_{0.5}$, the acceptor doping moves the Fermi level downwards, and thus the Fermi level is submerged in the deeper energy region. For $Ti_{2.25}Co_{0.75}Si_{0.5}P_{0.5}$, the donor doping moves the Fermi level upwards, and thus the Fermi level is able to be pulled out of the filled bands and is located at the nearly zero band gap.

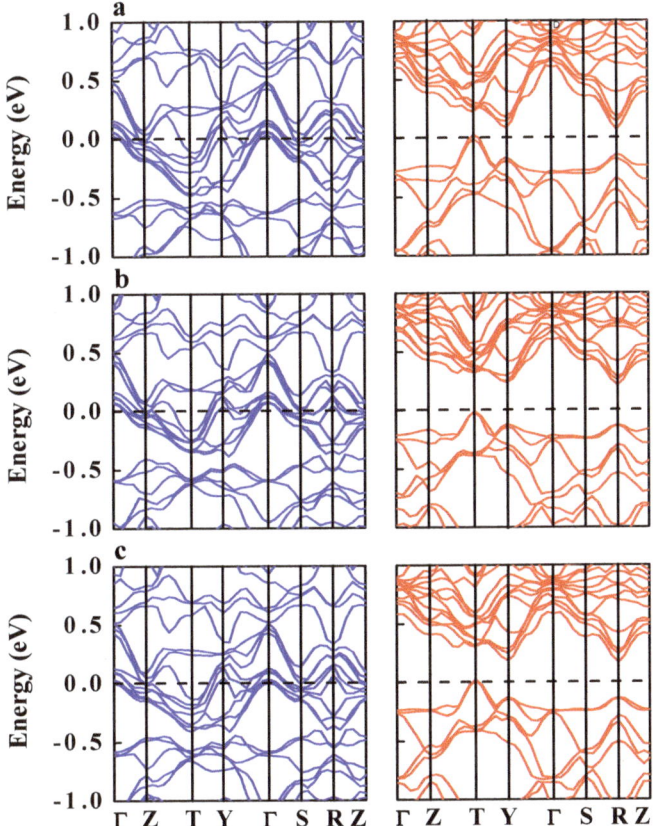

Figure 5. Spin-resolved electronic band structures of (**a**) $Ti_{2.25}Co_{0.75}Si_{0.5}B_{0.5}$, (**b**) $Ti_{2.25}Co_{0.75}Si_{0.5}Al_{0.5}$, and (**c**) $Ti_{2.25}Co_{0.75}Si_{0.5}Ga_{0.5}$ at the equilibrium lattice parameters. The zero of the energy scale is set at the Fermi energy. The blue and red solid lines represent the spin-up and spin-down states, respectively.

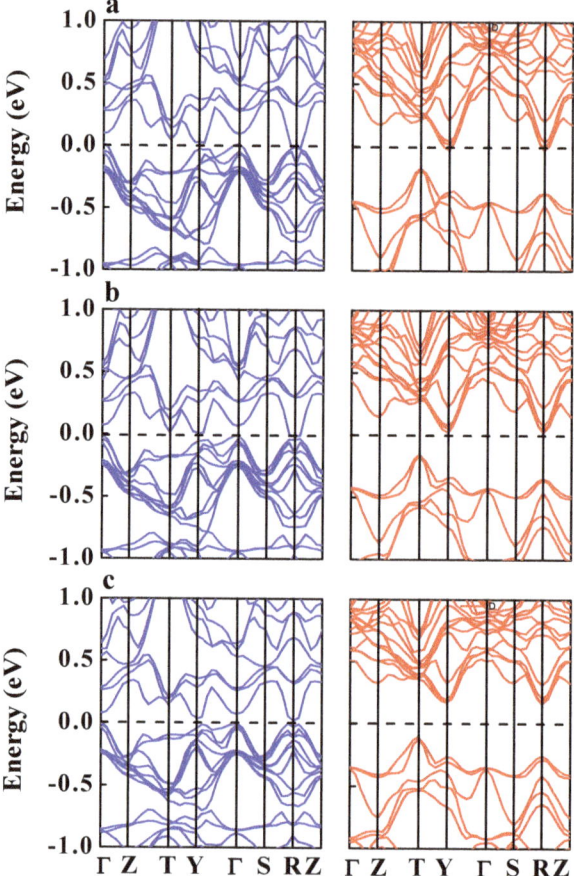

Figure 6. Spin-resolved electronic band structures of (**a**) $Ti_{2.25}Co_{0.75}Si_{0.5}P_{0.5}$, (**b**) $Ti_{2.25}Co_{0.75}Si_{0.5}As_{0.5}$, (**c**) and $Ti_{2.25}Co_{0.75}Si_{0.5}Sb_{0.5}$ at the equilibrium lattice parameters. The zero of the energy scale is set at the Fermi energy. The blue and red solid lines represent the spin-up and spin-down states, respectively.

The results show that the spin-up and spin-down states around the Fermi level are predominantly derived from the Ti-d and Co-d states, and the d–d hybridization is the main origin for spin polarization. In order to understand this matter intuitively, we drew the spin density plots of $Ti_{2.25}Co_{0.75}Si_{0.5}B_{0.5}$ and $Ti_{2.25}Co_{0.75}Si_{0.5}P_{0.5}$ (see Figure 8), which are defined as the difference of spin-up and spin-down states. We can see that the spin density predominantly concentrates on the Ti and Co atoms, evincing that the spin polarization is mainly from the Ti and Co atoms. For these compounds, the magnetic moments are predominantly due to the Ti-d and Co-d electrons. The large exchange splitting of the Ti-d and Co-d states leads to a large magnetic moment.

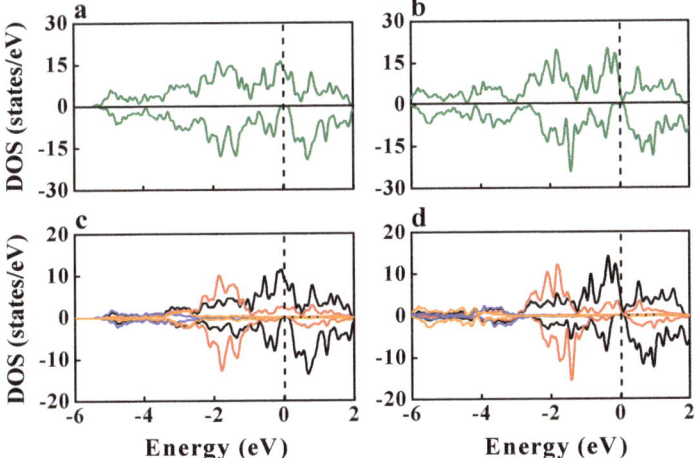

Figure 7. The total density of the states (DOS) of (**a**) $Ti_{2.25}Co_{0.75}Si_{0.5}B_{0.5}$ and (**b**) $Ti_{2.25}Co_{0.75}Si_{0.5}P_{0.5}$ at the equilibrium lattice parameters. The partial DOS of (**c**) $Ti_{2.25}Co_{0.75}Si_{0.5}B_{0.5}$ and (**d**) $Ti_{2.25}Co_{0.75}Si_{0.5}P_{0.5}$ at the equilibrium lattice parameters. The black, red, blue, and orange solid lines indicate the Ti-d, Co-d, Si-p, and B (P)-p states, respectively.

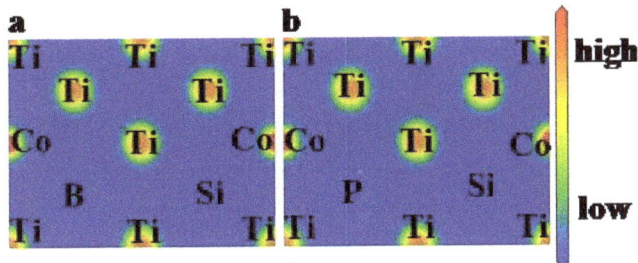

Figure 8. The spin densities of (**a**) $Ti_{2.25}Co_{0.75}Si_{0.5}B_{0.5}$ and (**b**) $Ti_{2.25}Co_{0.75}Si_{0.5}P_{0.5}$ at the equilibrium lattice parameters.

4. Conclusions

In summary, we have determined the structural, electronic, and magnetic properties of inverse Ti_2CoSi-based compounds. By adjusting the concentration of titanium and cobalt, the fabricated $Ti_{2.25}Co_{0.75}Si$ compound is half metallic, that is, single-channel spin polarized. The majority-spin bands are strongly metallic, while the minority-spin bands are semiconductor-like around the Fermi level. This gap has previously been reported by other authors, and is formed as a result of the strong 3d–3d hybridization between the two transition-metal atoms [42–45]. It is thought that this 3d–3d interaction is essential for the formation of the gap at the Fermi level. Under the influence of isotropic hydrostatic pressure, the $Ti_{2.25}Co_{0.75}Si$ alloy can preserve the half-metallic feature between a range of 5.78 Å to 6.35 Å. Nevertheless, it would transform to a gapless half metal upon the tetragonalization of the lattice. Tetragonal deformation could modify the crystal symmetry and vary the position of the edge of the bands, especially at some high-symmetry points in the reciprocal space. Our further calculations on $Ti_{2.25}Co_{0.75}Si_{0.5}B_{0.5}$, $Ti_{2.25}Co_{0.75}Si_{0.5}P_{0.5}$, and other doped compounds demonstrate that the doped foreign impurity is able to adjust the valence electron concentration of the compound and shift the Fermi level upwards or downwards, creating several different electronic structures around the sensitive Fermi level.

Author Contributions: S.D. and L.F. designed the project. L.Z. and J.D. performed the calculations and prepared the manuscript. All of the authors analyzed the data and discussed the results. All authors have read and agreed to the published version of the manuscript.

Funding: This research was funded by the National Natural Science Foundation of China (grant no's. 61804010, 11204209, 60876035, and 61874004), the Natural Science Foundation of Tianjin City (grant no. 17JCYBJC16200), and the Project of Guangdong Baiyun University (grant no. 2017BYKY29).

Conflicts of Interest: The authors declare no conflict of interest.

References

1. Brown, P.J.; Neumann, K.U.; Webster, P.J.; Ziebeck, K.R.A. The magnetization distributions in some Heusler alloys proposed as half-metallic ferromagnets. *J. Phys. Condens. Matter* **2000**, *12*, 1827. [CrossRef]
2. Žutić, I.; Fabian, J.; Sarma, S.D. Spintronics: Fundamentals and applications. *Rev. Mod. Phys.* **2004**, *76*, 323. [CrossRef]
3. Berri, S. First-Principles Calculations to Investigate the Structural, electronic, and half-metallic properties of $Ti_2RhSn_{1-x}Si_x$, $Ti_2RhSn_{1-x}Ge_x$, and $Ti_2RhGe_{1-x}Si_x$ (x = 0, 0.25, 0.5, 0.75, and 1) quaternary Heusler alloys. *J. Supercond. Nov. Magn.* **2019**, *32*, 2219–2228. [CrossRef]
4. Alijani, V.; Winterlik, J.; Fecher, G.H.; Naghavi, S.S.; Felser, C. Quaternary half-metallic Heusler ferromagnets for spintronics applications. *Phys. Rev.* **2011**, *83*, 184428. [CrossRef]
5. Gao, G.Y.; Hu, L.; Yao, K.L.; Luo, B.; Liu, N. Large half-metallic gaps in the quaternary Heusler alloys CoFeCrZ (Z = Al, Si, Ga, Ge): A first-principles study. *J. Alloy Compd.* **2013**, *551*, 539–543. [CrossRef]
6. Singh, M.; Saini, H.S.; Thakur, J.; Reshak, A.H.; Kashyap, M.K. Electronic structure, magnetism and robust half-metallicity of new quaternary Heusler alloy FeCrMnSb. *J. Alloy Compd.* **2013**, *580*, 201–204. [CrossRef]
7. Özdoğan, K.; Şaşıoğlu, E.; Galanakis, I. Slater-Pauling behavior in LiMgPdSn-type multifunctional quaternary Heusler materials: Half-metallicity, spin-gapless and magnetic semiconductors. *J. Appl. Phys.* **2013**, *113*, 193903. [CrossRef]
8. Zhang, Y.J.; Liu, Z.H.; Li, G.T.; Ma, X.Q.; Liu, G.D. Magnetism, band gap and stability of half-metallic property for the quaternary Heusler alloys CoFeTiZ (Z = Si, Ge, Sn). *J. Alloy Compd.* **2014**, *616*, 449–453. [CrossRef]
9. Xiong, L.; Yi, L.; Gao, G.Y. Search for half-metallic magnets with large half-metallic gaps in the quaternary Heusler alloys CoFeTiZ and CoFeVZ (Z = Al, Ga, Si, Ge, As, Sb). *J. Magn. Magn. Mater.* **2014**, *360*, 98–103. [CrossRef]
10. Al-zyadi, J.M.K.; Gao, G.Y.; Yao, K.L. Theoretical investigation of the electronic structures and magnetic properties of the bulk and surface (001) of the quaternary Heusler alloy NiCoMnGa. *J. Magn. Magn. Mater.* **2015**, *378*, 1–6. [CrossRef]
11. Felser, C.; Wollmann, L.; Chadov, S.; Fecher, G.H.; Parkin, S.S. Basics and prospective of magnetic Heusler compounds. *APL Mater.* **2015**, *3*, 041518. [CrossRef]
12. Kundu, A.; Ghosh, S.; Banerjee, R.; Ghosh, S.; Sanyal, B. New quaternary half-metallic ferromagnets with large Curie temperatures. *Sci. Rep.* **2017**, *7*, 1803. [CrossRef] [PubMed]
13. Wang, X.L. Proposal for a new class of materials: Spin gapless semiconductors. *Phys. Rev. Lett.* **2008**, *100*, 156404. [CrossRef] [PubMed]
14. Ouardi, S.; Fecher, G.H.; Felser, C.; Kubler, J. Realization of spin gapless semiconductors: The Heusler compound Mn_2CoAl. *Phys. Rev. Lett.* **2013**, *110*, 100401. [CrossRef]
15. Heusler, F. Über magnetische manganlegierungen. *Verh. Dtsch. Phys. Ges.* **1903**, *5*, 219.
16. Jiang, D.G.; Ye, Y.X.; Yao, W.B.; Zeng, D.W.; Zhou, J.; Liu, L.N.; Wen, Y.F. First-principles calculations of the structural, magnetic, and electronic properties of Fe_2MgB full-Heusler alloy. *J. Electron. Mater.* **2019**, *48*, 7258–7262. [CrossRef]
17. Moodera, J.S.; Nassar, J.; Mathon, G. Spin-tunneling in ferromagnetic junctions. *Annu. Rev. Mater.* **1999**, *29*, 381. [CrossRef]
18. Skaftouros, S.; Ozdogan, K.; Sasioglu, E.; Galanakis, I. Search for spin gapless semiconductors: The case of inverse Heusler compounds. *Appl. Phys. Lett.* **2013**, *102*, 022402. [CrossRef]
19. Xu, G.Z.; Liu, E.K.; Du, Y.; Li, G.J.; Liu, G.D.; Wang, W.H.; Wu, G.H. A new spin gapless semiconductors family: Quaternary Heusler compounds. *Europhys. Lett.* **2013**, *102*, 17007. [CrossRef]

20. Kudryavtsev, Y.V.; Oksenenko, V.A.; Lee, Y.P.; Hyun, Y.H.; Kim, J.B.; Park, J.S.; Park, S.Y.; Dubowik, J. Evolution of the magnetic properties of Co_2MnGa Heusler alloy films: From amorphous to ordered films. *Phys. Rev. B* **2007**, *76*, 024430. [CrossRef]
21. Jamer, M.E.; Assaf, B.A.; Devakul, T.; Heiman, D. Magnetic and transport properties of Mn_2CoAl oriented films. *Appl. Phys. Lett.* **2013**, *103*, 142403. [CrossRef]
22. Xu, G.Z.; Du, Y.; Zhang, X.M.; Zhang, H.G.; Liu, E.K.; Wang, W.H.; Wu, G.H. Magneto-transport properties of oriented Mn_2CoAl films sputtered on thermally oxidized Si substrates. *Appl. Phys. Lett.* **2014**, *104*, 242408. [CrossRef]
23. Feng, W.W.; Fu, X.; Wan, C.H.; Yuan, Z.H.; Han, X.F.; Quang, N.V.; Cho, S. Spin gapless semiconductor like Ti_2MnAl film as a new candidate for spintronics application. *Phys. Status Solidi RRL* **2015**, *9*, 641. [CrossRef]
24. Lukashev, P.; Kharel, P.; Gilbert, S.; Staten, B.; Hurley, N.; Fuglsby, R.; Huh, Y.; Valloppilly, S.; Zhang, W.; Yang, K.; et al. Investigation of spin-gapless semiconductivity and half-metallicity in Ti_2MnAl-based compounds. *Appl. Phys. Lett.* **2016**, *108*, 141901. [CrossRef]
25. Dulal, R.P.; Dahal, B.R.; Forbes, A.; Bhattarai, N.; Pegg, I.L.; Philip, J. Weak localization and small anomalous Hall conductivity in ferromagnetic Weyl semimetal Co_2TiGe. *Sci. Rep.* **2019**, *9*, 3342. [CrossRef] [PubMed]
26. Dahal, B.; Huber, C.; Zhang, W.; Valloppilly, S.; Huh, Y.; Kharel, P.; Sellmyer, D. Effect of partial substitution of In with Mn on the structural, magnetic, and magnetocaloric properties of $Ni_2Mn_{1+x}In_{1-x}$ Heusler alloys. *J. Phys. D Appl. Phys.* **2019**, *52*, 425305. [CrossRef]
27. Casper, F.; Graf, T.; Chadov, S.; Balke, B.; Felser, C. Half-Heusler compounds: Novel materials for energy and spintronic applications. *Semicond. Sci. Technol.* **2012**, *27*, 063001. [CrossRef]
28. Graf, T.; Felser, C.; Parkin, S.S.P. Simple rules for the understanding of Heusler compounds. *Prog. Solid State Chem.* **2011**, *39*, 1–50. [CrossRef]
29. Kandpal, H.C.; Fecher, G.H.; Felser, C. Calculated electronic and magnetic properties of the half-metallic, transition metal based Heusler compounds. *J. Phys. D Appl. Phys.* **2007**, *40*, 1507. [CrossRef]
30. Bradley, A.J.; Rodgers, J.W. The crystal structure of the Heusler alloys. *Proc. R. Soc. Lond. Ser. A* **1934**, *144*, 340. [CrossRef]
31. Payne, M.C.; Teter, M.P.; Allan, D.C.; Arias, T.A.; Joannopoolous, J.D. Iterative minimization techniques for *ab initio* total-energy calculations: Molecular dynamics and conjugate gradients. *Rev. Mod. Phys.* **1992**, *64*, 1065. [CrossRef]
32. Vanderbilt, D. Soft self-consistent pseudopotentials in a generalized eigenvalue formalism. *Phys. Rev. B* **1990**, *41*, 7892. [CrossRef] [PubMed]
33. Milman, V.; Winkler, B.; White, J.A.; Pickard, C.J.; Payne, M.C.; Akhmatskaya, E.V.; Nobes, R.H. Electronic structure, properties, and phase stability of inorganic crystals: A pseudopotential plane-wave study. *Int. J. Quantum Chem.* **2000**, *77*, 895. [CrossRef]
34. Segall, M.D.; Lindan, P.J.D.; Probert, M.J.; Pickard, C.J.; Hasnip, P.J.; Clark, S.J.; Payne, M.C. First-principles simulation: Ideas, illustrations and the CASTEP code. *J. Phys. Condens. Matter* **2002**, *14*, 2717. [CrossRef]
35. Clark, S.J.; Segall, M.D.; Pickard, C.J.; Hasnip, P.J.; Probert, M.I.J.; Refson, K.; Payne, M.C. First principles methods using CASTEP. *Z. Krist. Cryst. Mater.* **2005**, *220*, 567–570. [CrossRef]
36. Hasnip, P.J.; Refson, K.; Probert, M.I.J.; Yates, J.R.; Clark, S.J.; Pickard, C.J. Density functional theory in the solid state. *Philos. Trans. R. Soc. A* **2014**, *372*, 20130270. [CrossRef]
37. Perdew, J.P.; Chevary, J.A.; Vosko, S.H.; Jackson, A.; Pederson, M.R.; Fiolhais, C. Atoms, molecules, solids, and surfaces: Applications of the generalized gradient approximation for exchange and correlation. *Phys. Rev. B* **1992**, *46*, 6671. [CrossRef] [PubMed]
38. Perdew, J.P.; Burke, K.; Ernzerhof, M. Generalized gradient approximation made simple. *Phys. Rev. Lett.* **1996**, *77*, 3865. [CrossRef] [PubMed]
39. Waybright, J.; Halbritter, L.; Dahal, B.; Qian, H.; Huh, Y.; Lukashev, P.; Kharel, P. Structure and magnetism of $NiFeMnGa_xSn_{1-x}$ (x = 0, 0.25, 0.5, 0.75, 1.00) Heusler compounds. *AIP Adv.* **2019**, *9*, 035105. [CrossRef]
40. Wu, B.; Huang, H.; Zhou, G.; Feng, Y.; Chen, Y.; Wang, X. Structure, magnetism, and electronic properties of inverse Heusler alloy $Ti_2CoAl/MgO(100)$ herterojuction: The role of interfaces. *Appl. Sci.* **2018**, *8*, 2336. [CrossRef]
41. Wu, B.; Huang, H.; Li, P.; Zhou, T.; Zhou, G.; Feng, Y.; Chen, Y. First-principles study on the structure, magnetism, and electronic properties in inverse Heusler alloy $Ti_2FeAl/GaAs(100)$ heterojunction. *Superlatt. Microstruct.* **2019**, *133*, 106205. [CrossRef]

42. Fujii, S.; Ishida, S.; Asano, S. A half-metallic band structure and Fe_2MnZ (Z = Al, Si, P). *J. Phys. Soc. Jpn.* **1995**, *64*, 185–191. [CrossRef]
43. Ishida, S.; Fujii, S.; Kashiwagi, S.; Asano, S. Search for half-metallic compounds in Co_2MnZ (Z = III^b, IV^b, V^b element). *J. Phys. Soc. Jpn.* **1995**, *64*, 2152–2157. [CrossRef]
44. Miura, Y.; Shirai, M.; Nagao, K. Ab initio study on stability of half-metallic Co-based full-Heusler alloys. *J. Appl. Phys.* **2006**, *99*, 08J112. [CrossRef]
45. Mebsout, R.; Amari, S.; Méçabih, S.; Abbar, B.; Bouhafs, B. Spin-Polarized Calculations of Magnetic and Thermodynamic Properties of the Full-Heusler MnZ (Z = Al, Ga). *Int. J. Thermophys.* **2013**, *34*, 507–520. [CrossRef]

© 2020 by the authors. Licensee MDPI, Basel, Switzerland. This article is an open access article distributed under the terms and conditions of the Creative Commons Attribution (CC BY) license (http://creativecommons.org/licenses/by/4.0/).

Article

Dynamic Stability of Bi-Directional Functionally Graded Porous Cylindrical Shells Embedded in an Elastic Foundation

Farshid Allahkarami [1], Hasan Tohidi [2], Rossana Dimitri [3] and Francesco Tornabene [3,*]

1. Department of Mechanical Engineering, Bijar Branch, Islamic Azad University, 66515-356 Bijar, Iran; F.allahkarami1366@gmail.com
2. Department of Mechanical Engineering, Saqqez Branch, Islamic Azad University, 66819-73477 Saqqez, Iran; havjin.tohidi@gmail.com
3. Department of Innovation Engineering, Università del Salento, 73100 Lecce, Italy; rossana.dimitri@unisalento.it
* Correspondence: francesco.tornabene@unisalento.it

Received: 18 January 2020; Accepted: 11 February 2020; Published: 16 February 2020

Abstract: This paper investigates the dynamic buckling of bi-directional (BD) functionally graded (FG) porous cylindrical shells for various boundary conditions, where the FG material is modeled by means of power law functions with even and uneven porosity distributions of ceramic and metal phases. The third-order shear deformation theory (TSDT) is adopted to derive the governing equations of the problem via the Hamilton's principle. The generalized differential quadrature (GDQ) method is applied together with the Bolotin scheme as numerical strategy to solve the problem, and to draw the dynamic instability region (DIR) of the structure. A large parametric study examines the effect of different boundary conditions at the extremities of the cylindrical shell, as well as the sensitivity of the dynamic stability to different thickness-to-radius ratios, length-to-radius ratios, transverse and longitudinal power indexes, porosity volume fractions, and elastic foundation constants. Based on results, the dynamic stability of BD-FG cylindrical shells can be controlled efficiently by selecting appropriate power indexes along the desired directions. Furthermore, the DIR is highly sensitive to the porosity distribution and to the extent of transverse and longitudinal power indexes. The numerical results could be of great interest for many practical applications, as civil, mechanical or aerospace engineering, as well as for energy devices or biomedical systems.

Keywords: bi-directional functionally graded; bolotin scheme; dynamic stability; elastic foundation; porosity

1. Introduction

Circular cylindrical shells have an essential role in various fields of engineering applications such as aircraft, pressure vessels, gas turbines, and many other industrial purposes because of their excellent performance. Due to the advancement of the knowledge and technology, in recent years a new category of materials with interesting properties, named as functionally graded materials (FGMs) has been successfully applied. Conventional types of these materials are made of two or more different constituent phases, namely, the ceramic and metal phases, which are distributed gradually according some fixed functions. Since FGMs have some extraordinary properties, namely, a high temperature and a corrosion resistance, as well as an improved residual stress distribution, they are widely studied in many field of the applied sciences and they are adopted as structural components in military, medical, or aerospace industries, as well as in power plants or vessels. Thus, due to their special privileges in comparison with traditional materials, most industries make effort to exert such materials in lieu of ordinary ones [1–4].

A large number of studies in literature has focused on the thermo-mechanical and buckling behavior of FGMs for shell and plate structures. In this context, only some research works associated with FG cylindrical shells will be reviewed here, in line with the perspective developed in the present work. Du et al. [5] investigated the nonlinear forced vibration response of FG cylindrical thin shells, and used the perturbation method and the numerical Poincaré maps to solve the governing equations of the problem. Rahimi et al. [6] studied the vibration of FG cylindrical shells with ring supports. It was found that symmetric and asymmetric boundary conditions affect significantly the vibrations of the structure, with a general increase or decrease, respectively. In a recent work, Ghasemi et al. [7] have studied the agglomeration effect of FG hybrid single-walled carbon nanotubes on the vibration of hybrid laminated cylindrical shell structures. Bich et al. [8] performed the nonlinear static and dynamic buckling of imperfect eccentrically stiffened FG thin circular cylindrical shells subjected to an axial compression. Beni et al. [9] presented a novel formulation based on a modified couple stress theory to study simply supported FG circular cylindrical shells in the framework of thin shell structures, whereby the vibration behavior based on a classical continuum was found to be quite unaffected by the length scale parameters. Da Silva et al. [10] studied the nonlinear vibrations of a simply supported fluid-filled FG cylindrical shell subjected to a lateral time-dependent load and an axial static preloading condition. Bich and Nguyen [11] applied the displacement method to study the nonlinear vibration of FG circular cylindrical shells subjected to an axial and transverse mechanical loading. Ghannad et al. [12] introduced an analytical solution for the deformation and stress response of axisymmetric clamped–clamped thick FG cylindrical shells with variable thickness, while applying the first-order shear deformation theory and the perturbation theory, based on the Donnell's nonlinear large deflection theory. To date, many analytical and numerical approaches have been proposed in literature to handle simple and coupled vibration problems of cylindrical shell structures, including thermo-elastic, piezoelectric, and thermo-piezoelectric multi-field problems (see refs. [13–27], among others). As far as FGMs are concerned, many recent studies about the free vibration and buckling response of conventional and bi-directional FG cylindrical shells have been recently performed in literature [28–33]. A key point of the static and dynamic response of FG shell structures is related to the presence of porosities, which can form during a fabrication process, with possible effects on the global structural response. Indeed, an increasing attention to this aspect has been devoted in the scientific community for a correct interpretation of the mechanical performances of FG materials and structures. For example, in a recent work, Kiran and Kattimani [34] assessed the possible effect of porosity on the vibration behavior and static response of FG magneto-electro-elastic plates, with a clear reduction of the natural frequencies for an increased porosity within the material. In another study, Kiran et al. [35] analyzed the effect of porosity on the structural behavior of skew FG magneto-electro-elastic plates. Barati et al. [36] performed the buckling analysis of higher order graded smart piezoelectric plates with porosities resting on an elastic foundation. It was found that the buckling behavior of piezoelectric plates is significantly influenced by the porosity distribution. In the further works by Wang et al. [37,38], the authors studied the vibration response of longitudinally traveling FG plates with porosities [37] while considering the thermo-mechanical coupled response in [38]. A similar free vibration problem was studied in [39,40] for FG cylindrical shells, by means of the sinusoidal shear deformation theory and the Rayleigh–Ritz method, accounting for the possible presence of defects and porosities. In the context of nanomaterials and nanostructures, some modified couple stress theories have been recently proposed as efficient theoretical tools to study their coupled thermomechanical vibration behavior, also in presence of different levels of porosity, see [41–47], among others.

Up to date, however, there is a general lack of works in literature focusing on the dynamic buckling response of bi-directional (BD)-FG cylindrical shell embedded in a Winkler–Pasternak foundation, including the simultaneous effect of porosity. To this end, we propose the third-order shear deformation theory (TSDT) to model the cylindrical shells with porosities, subjected to an axial compressive excitation. The Hamilton's principle will be employed to determine the governing partial equations of motion, whereby the generalized differential quadrature (GDQ) method is adopted to solve the problem together with the associated boundary conditions into a system of Mathieu–Hill equations. Afterward, the Bolotin method is employed to determine the boundaries of the dynamic instability region (DIR) of BD-FG cylindrical shells. A systematic study focuses on the sensitivity of the dynamic stability behavior to different dimensionless ratios, i.e., the thickness-to-radius ratio, or the length-to-radius ratio, as well as to different boundary conditions, transverse and longitudinal power indexes, porosity volume fractions and foundation constants. The paper is organized as follows. In Section 2 we determine the governing equations of the problem for porous BD-FG cylindrical shells, which are solved numerically according to the GDQ and Bolotin methods, as detailed in Section 3. Section 4 aims at validating the proposed approach and shows the main results from a broad numerical investigation, whereas the final remarks are discussed in Section 5.

2. Governing Equations of the Problem

Let us consider a BD-FG porous cylindrical shell embedded in an elastic foundation with thickness h, radius R, and length L, where two different porosity distributions of the constituent phases are accounted for the analysis, namely an even and an uneven distribution, see Figures 1 and 2. We assume a BD-FG material made of a metal (labeled as m) and a ceramic (labeled as c) in the inner and outer shell surfaces, respectively. While the material properties for a conventional FG model vary continuously along the thickness direction from a ceramic or metal to another one, the basic material properties selected herein, vary also along the shell length from the metal to the ceramic phase. To this end, the volume fractions of the ceramic and metal phases are defined as follows

$$V_c(x,z) = \left(\frac{1}{2} + \frac{z}{h}\right)^{n_z}\left(\frac{x}{L}\right)^{n_x}, \quad V_m(x,z) = 1 - V_c(x,z), \tag{1}$$

where n_z and n_x refer to the non-negative volume fraction exponents defining the profile variation of the material properties along the shell thickness and length directions, respectively. In addition, z and x stand for the radial distance from the mid-plane and longitudinal distance from the origin of the BD-FG cylindrical shell, respectively. The effective material properties (i.e., Yong's modulus, density and Poisson's ratio) of the BD-FG porous cylindrical shell are assumed to change according to a modified power law model with a linear algebraic combination of volume fractions of two basic materials. Two types of BD-FG material models include both even and/or uneven porosity distributions, i.e.,

$$E(x,z) = E_m + (E_c - E_m)\left(\frac{1}{2} + \frac{z}{h}\right)^{n_z}\left(\frac{x}{L}\right)^{n_x} - \frac{\zeta}{2}(E_c + E_m), \tag{2a}$$

$$\rho(x,z) = \rho_m + (\rho_c - \rho_m)\left(\frac{1}{2} + \frac{z}{h}\right)^{n_z}\left(\frac{x}{L}\right)^{n_x} - \frac{\zeta}{2}(\rho_c + \rho_m), \tag{2b}$$

$$\nu(x,z) = \nu_m + (\nu_c - \nu_m)\left(\frac{1}{2} + \frac{z}{h}\right)^{n_z}\left(\frac{x}{L}\right)^{n_x} - \frac{\zeta}{2}(\nu_c + \nu_m), \tag{2c}$$

for even distributions, or

$$E(x,z) = E_m + (E_c - E_m)\left(\frac{1}{2} + \frac{z}{h}\right)^{n_z}\left(\frac{x}{L}\right)^{n_x} - \frac{\xi}{2}(E_c + E_m)\left(1 + \frac{2|z|}{h}\right), \tag{3a}$$

$$\rho(x,z) = \rho_m + (\rho_c - \rho_m)\left(\frac{1}{2} + \frac{z}{h}\right)^{n_z}\left(\frac{x}{L}\right)^{n_x} - \frac{\xi}{2}(\rho_c + \rho_m)\left(1 + \frac{2|z|}{h}\right), \tag{3b}$$

$$\nu(x,z) = \nu_m + (\nu_c - \nu_m)\left(\frac{1}{2} + \frac{z}{h}\right)^{n_z}\left(\frac{x}{L}\right)^{n_x} - \frac{\xi}{2}(\nu_c + \nu_m)\left(1 + \frac{2|z|}{h}\right), \tag{3c}$$

for an uneven distribution. In the all the expression (2) and (3), ζ and ξ denote the volume fraction of an even or uneven porosity inside the phases, respectively. While an even model accounts for porosities evenly distributed across the radial direction, the porosities in an uneven model is mostly concentrated in the shell mid-plane. It is worth noting that the uneven porosity distribution is linearly reduced from a larger value at mid-plane to a smaller value at the top and bottom sides of the structure.

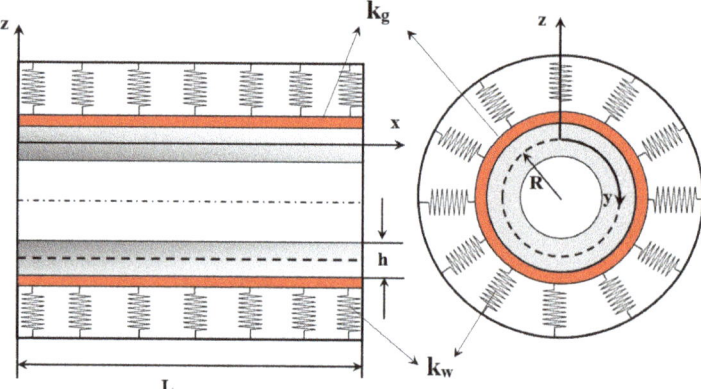

Figure 1. Geometrical scheme of a bi-directional (BD)-functionally graded (FG) porous cylindrical shell embedded in an elastic foundation.

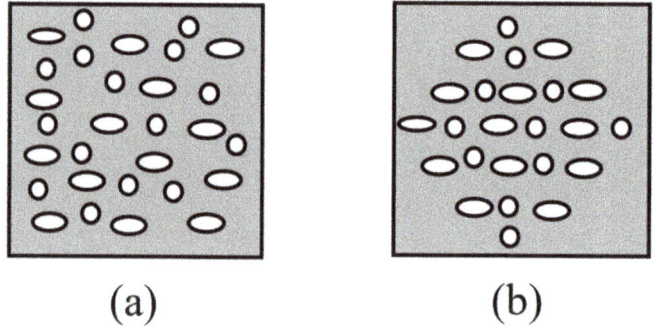

Figure 2. Distribution of porosity: (a) even porosity, (b) uneven porosity.

In what follows, we apply the TSDT, such that the displacement field of an arbitrary point of the shell along the x, y, and z axes, is defined as

$$u(x,y,z,t) = u_0(x,y,t) + z\varphi_x(x,y,t) - \bar{c}z^3\left(\varphi_x(x,y,t) + \frac{\partial w_0(x,y,t)}{\partial x}\right),$$
$$v(x,y,z,t) = v_0(x,y,t) + z\varphi_y(x,y,t) - \bar{c}z^3\left(\varphi_y(x,y,t) + \frac{\partial w_0(x,y,t)}{\partial y}\right), \quad (4)$$
$$w(x,y,z,t) = w_0(x,y,t),$$

where $\bar{c} = \frac{4}{3h^2}$, and u, v, w stand for the longitudinal, circumferential, and transverse (radial) displacement components, respectively. These are determined, in turn, by means of the kinematic quantities u_0, v_0, and w_0 at the middle surface, and the rotations φ_x and φ_y of a transverse normal section about the x and y axis, respectively.

According to the TSDT, the strain components of the cylindrical shell can be written as [48]

$$\varepsilon_{xx} = \frac{\partial u_0(x,y,t)}{\partial x} + z\frac{\partial \varphi_x(x,y,t)}{\partial x} - \bar{c}z^3\left(\frac{\partial \varphi_x(x,y,t)}{\partial x} + \frac{\partial^2 w_0(x,y,t)}{\partial x^2}\right), \quad (5a)$$

$$\varepsilon_{yy} = \frac{\partial v_0(x,y,t)}{\partial y} + z\frac{\partial \varphi_y(x,y,t)}{\partial y} - \bar{c}z^3\left(\frac{\partial \varphi_y(x,y,t)}{\partial y} + \frac{\partial^2 w_0(x,y,t)}{\partial y^2}\right) + \frac{w}{R'} \tag{5b}$$

$$\gamma_{xy} = \frac{\partial u_0(x,y,t)}{\partial y} + \frac{\partial v_0(x,y,t)}{\partial x} + z\left(\frac{\partial \varphi_x(x,y,t)}{\partial y} + \frac{\partial \varphi_y(x,y,t)}{\partial x}\right) \\ -\bar{c}z^3\left(\frac{\partial \varphi_x(x,y,t)}{\partial y} + \frac{\partial \varphi_y(x,y,t)}{\partial x} + 2\frac{\partial^2 w_0(x,y,t)}{\partial y \partial x}\right), \tag{5c}$$

$$\gamma_{xz} = \varphi_x(x,y,t) - 2\bar{c}z^2\left(\varphi_x(x,y,t) + \frac{\partial w_0(x,y,t)}{\partial x}\right) + \frac{\partial w_0(x,y,t)}{\partial x}, \tag{5d}$$

$$\gamma_{yz} = \varphi_y(x,y,t) - 2\bar{c}z^2\left(\varphi_y(x,y,t) + \frac{\partial w_0(x,y,t)}{\partial y}\right) + \frac{\partial w_0(x,y,t)}{\partial y}, \tag{5e}$$

which are related to the stress components as follows

$$\begin{Bmatrix} \sigma_{xx} \\ \sigma_{yy} \\ \sigma_{xz} \\ \sigma_{yz} \\ \sigma_{xy} \end{Bmatrix} = \begin{bmatrix} Q_{11} & Q_{12} & 0 & 0 & 0 \\ Q_{21} & Q_{22} & 0 & 0 & 0 \\ 0 & 0 & Q_{44} & 0 & 0 \\ 0 & 0 & 0 & Q_{55} & 0 \\ 0 & 0 & 0 & 0 & Q_{66} \end{bmatrix} \begin{Bmatrix} \varepsilon_{xx} \\ \varepsilon_{yy} \\ \varepsilon_{xz} \\ \varepsilon_{yz} \\ \varepsilon_{xy} \end{Bmatrix}, \tag{6}$$

where the stiffness Q_{ij} is defined as

$$Q_{11}(x,z) = Q_{22}(x,z) = \frac{E(x,z)}{1-(\nu(x,z))^2}, \quad Q_{12}(x,z) = Q_{21}(x,z) = \nu(x,z)Q_{11}(x,z), \\ Q_{44}(x,z) = Q_{55}(x,z) = Q_{66}(x,z) = \frac{E(x,z)}{2(1+\nu(x,z))}. \tag{7}$$

The strain energy of the BD-FG porous cylindrical shell is expressed as follow

$$\Pi_S = \frac{1}{2}\int_0^L \int_0^{2\pi R} \int_{-\frac{h}{2}}^{\frac{h}{2}} (\sigma_{xx}\varepsilon_{xx} + \sigma_{yy}\varepsilon_{yy} + \tau_{xy}\gamma_{xy} + \tau_{xz}\gamma_{xz} + \tau_{yz}\gamma_{yz})dzdydx, \tag{8}$$

and the kinetic energy of the cylindrical shell reads

$$\Pi_T = \frac{1}{2}\int_0^L \int_0^{2\pi R} \int_{-\frac{h}{2}}^{\frac{h}{2}} \rho(x,z)\left(\left(\frac{\partial u(x,y,z,t)}{\partial t}\right)^2 + \left(\frac{\partial v(x,y,z,t)}{\partial t}\right)^2 + \left(\frac{\partial w(x,y,z,t)}{\partial t}\right)^2\right)dzdydx. \tag{9}$$

The total potential energy corresponding to the axial compressive load $F_a(t)$ together with the Winkler and/or Pasternak elastic foundation, can be written as follow

$$\Pi_E = \frac{1}{2}\int_0^L \int_0^{2\pi R} \int_{-\frac{h}{2}}^{\frac{h}{2}} \left(k_w w_0^2 + k_g\left(\left(\frac{\partial w_0}{\partial x}\right)^2 + \left(\frac{\partial w_0}{\partial y}\right)^2\right) + F_a(t)\left(\frac{\partial w_0}{\partial x}\right)^2\right)dzdydx, \tag{10}$$

where k_w and k_g refer to the Winkler foundation stiffness and shear layer stiffness of the elastic foundation, respectively.

Consistently with the Hamilton's principle, the following governing equations of motion for the BD-FG cylindrical shell are determined

$$\int_{t_1}^{t_2} (\delta\Pi_T - \delta\Pi_S - \delta\Pi_E)dt = 0, \tag{11}$$

where the symbol δ denotes the variation of the energy quantities. By combination of Equations (8)–(10) and Equation (11), after integration by parts, we get the following governing equations of motion, under the assumption of a null value for u_0, v_0, w_0, φ_x, and φ_y.

$$\frac{\partial N_{xx}}{\partial x} + \frac{\partial N_{xy}}{\partial y} = I_0(x)\frac{\partial^2 u_0(x,y,t)}{\partial t^2} + (I_0(x) - \bar{c}I_3(x))\frac{\partial^2 \varphi_x(x,y,t)}{\partial t^2} - \bar{c}I_3(x)\frac{\partial^3 w_0}{\partial x \partial t^2}, \tag{12a}$$

$$\frac{\partial N_{xy}}{\partial y} + \frac{\partial N_y}{\partial y} = I_0(x)\frac{\partial^2 v_0(x,y,t)}{\partial t^2} + (I_0(x) - \bar{c}I_3(x))\frac{\partial^2 \varphi_y(x,y,t)}{\partial t^2} - \bar{c}I_3(x)\frac{\partial^3 w_0(x,y,t)}{\partial y \partial t^2}, \tag{12b}$$

$$\begin{aligned}
&\frac{\partial Q_{xz}}{\partial x} + \frac{\partial Q_{xy}}{\partial y} - 3\bar{c}\left(\frac{\partial R_{xz}}{\partial x} + \frac{\partial R_{xy}}{\partial y}\right) + \bar{c}\left(\frac{\partial^2 P_{xx}}{\partial x^2} + 2\frac{\partial^2 P_{xy}}{\partial x \partial y} + \frac{\partial^2 P_{yy}}{\partial y^2}\right) - \frac{N_{yy}}{R} \\
&- k_w w_0(x,y,t) + k_g\left(\frac{\partial^2 w_0(x,y,t)}{\partial x^2} + \frac{\partial^2 w_0(x,y,t)}{\partial y^2}\right) + F_a \frac{\partial^2 w_0(x,y,t)}{\partial x^2} = I_0(x)\frac{\partial^2 w_0(x,y,t)}{\partial t^2} \\
&- \bar{c}^2 I_6(x)\left(\frac{\partial^4 w_0(x,y,t)}{\partial x^2 \partial t^2} + \frac{\partial^4 w_0(x,y,t)}{\partial y^2 \partial t^2}\right) + \bar{c}I_3(x)\left(\frac{\partial^3 u_0(x,y,t)}{\partial x \partial t^2} + \frac{\partial^3 v_0(x,y,t)}{\partial y \partial t^2}\right) \\
&+ \bar{c}(I_4(x) - \bar{c}I_6(x))\left(\frac{\partial^3 \varphi_x(x,y,t)}{\partial x \partial t^2} + \frac{\partial^3 \varphi_y(x,y,t)}{\partial y \partial t^2}\right),
\end{aligned} \tag{12c}$$

$$\begin{aligned}
&\frac{\partial M_{xx}}{\partial x} + \frac{\partial M_{xy}}{\partial y} - Q_{xz} + 3\bar{c}R_{xz} - \bar{c}\left(\frac{\partial P_{xx}}{\partial x} + \frac{\partial P_{xy}}{\partial y}\right) = (I_0(x) - \bar{c}I_3(x))\frac{\partial^2 u_0(x,y,t)}{\partial t^2} \\
&+ \left(I_2(x) - 2\bar{c}I_4(x) + \bar{c}^2 I_6(x)\right)\frac{\partial^2 \varphi_x(x,y,t)}{\partial t^2} \\
&- \bar{c}(I_4(x) - \bar{c}I_6(x))\frac{\partial^3 w_0(x,y,t)}{\partial x \partial t^2},
\end{aligned} \tag{12d}$$

$$\begin{aligned}
&\frac{\partial M_{yy}}{\partial y} + \frac{\partial M_{xy}}{\partial x} - Q_{yz} + 3\bar{c}R_{yz} - \bar{c}\left(\frac{\partial P_{yy}}{\partial y} + \frac{\partial P_{xy}}{\partial x}\right) = (I_0(x) - \bar{c}I_3(x))\frac{\partial^2 v_0(x,y,t)}{\partial t^2} \\
&+ \left(I_2(x) - 2\bar{c}I_4(x) + \bar{c}^2 I_6(x)\right)\frac{\partial^2 \varphi_y(x,y,t)}{\partial t^2} \\
&- \bar{c}(I_4(x) - \bar{c}I_6(x))\frac{\partial^3 w_0(x,y,t)}{\partial y \partial t^2},
\end{aligned} \tag{12e}$$

Note that the resultants in Equation (12) are computed by integration of the pertaining stress components along the shell structure, i.e.,

$$\left\{\begin{array}{c} N_{xx} \\ N_{yy} \\ N_{xy} \end{array}\right\} = \int_{-\frac{h}{2}}^{\frac{h}{2}} \left\{\begin{array}{c} \sigma_{xx} \\ \sigma_{yy} \\ \tau_{xy} \end{array}\right\} dz, \tag{13}$$

$$\left\{\begin{array}{c} M_{xx} \\ M_{yy} \\ M_{xy} \end{array}\right\} = \int_{-\frac{h}{2}}^{\frac{h}{2}} \left\{\begin{array}{c} \sigma_{xx} \\ \sigma_{yy} \\ \tau_{xy} \end{array}\right\} z\,dz, \tag{14}$$

$$\left\{\begin{array}{c} P_{xx} \\ P_{yy} \\ P_{xy} \end{array}\right\} = \int_{-\frac{h}{2}}^{\frac{h}{2}} \left\{\begin{array}{c} \sigma_{xx} \\ \sigma_{yy} \\ \tau_{xy} \end{array}\right\} z^3\,dz, \tag{15}$$

$$\left\{ \begin{array}{c} Q_{xz} \\ Q_{yz} \end{array} \right\} = \int_{-\frac{h}{2}}^{\frac{h}{2}} \left\{ \begin{array}{c} \tau_{xz} \\ \tau_{yz} \end{array} \right\} dz, \qquad (16)$$

$$\left\{ \begin{array}{c} R_{xz} \\ R_{yz} \end{array} \right\} = \int_{-\frac{h}{2}}^{\frac{h}{2}} \left\{ \begin{array}{c} \tau_{xz} \\ \tau_{yz} \end{array} \right\} z^2 dz. \qquad (17)$$

The generalized inertia moments are defined as

$$(I_0(x), I_1(x), I_2(x), I_4(x), I_5(x), I_6(x)) = \int_{-\frac{h}{2}}^{\frac{h}{2}} (1, z, z^2, z^3, z^4, z^6) \rho(x,z) dz. \qquad (18)$$

Three types of boundary conditions are considered along the shell edges, namely

➢ **Simply-Simply** (S-S) supports

$$x = 0, L \Rightarrow v_0 = w_0 = \varphi_y = M_x = N_x = 0, \qquad (19)$$

➢ **Clamped-Clamped** (C-C) supports

$$x = 0, L \Rightarrow u_0 = v_0 = w_0 = \varphi_x = \varphi_y = 0, \qquad (20)$$

➢ **Clamped-simply** (C-S) supports

$$x = 0 \Rightarrow u_0 = v_0 = w_0 = \varphi_x = \varphi_y = 0, \qquad (21a)$$

$$x = L \Rightarrow v_0 = w_0 = \varphi_y = M_x = N_x = 0, \qquad (21b)$$

3. Solution Procedure

In this section, we want to determine the dynamic stability of BD-FG porous cylindrical shells, where the governing equations of motion are expressed through the following expansion for the kinematic quantities

$$u_0(x,y,t) = U(x) \sin\left(n\frac{y}{R}\right) \overline{U}(t) \qquad (22a)$$

$$v_0(x,y,t) = V(x) \cos\left(n\frac{y}{R}\right) \overline{V}(t) \qquad (22b)$$

$$w_0(x,y,t) = W(x) \sin\left(n\frac{y}{R}\right) \overline{W}(t) \qquad (22c)$$

$$\varphi_x(x,y,t) = \Phi_x(x) \sin\left(n\frac{y}{R}\right) \overline{\Phi}_x(t) \qquad (22d)$$

$$\varphi_y(x,y,t) = \Phi_y(x) \cos\left(n\frac{y}{R}\right) \overline{\Phi}_y(t) \qquad (22e)$$

where n is the circumferential half wave number, $\overline{U}(t), \overline{V}(t), \overline{W}(t), \overline{\Phi}_x(t)$ and $\overline{\Phi}_y(t)$ are the time functions. The admissible displacement functions in Equation (22) satisfy both the equations of motion and their boundary conditions. Afterward, the governing equations of the problem are discretized according to the GDQ method.

Upon substitution of Equation (22a–e) and Equation (5a–e) into Equation (12), after a proper manipulation, we obtain the equations of motion in their final form, as detailed in Appendix A.

The above-mentioned equations of motion are solved numerically in a strong form by means of the GDQ method, as largely discussed in [49] and in a review paper [50] in terms of accuracy, stability and reliability, and successfully applied for many numerical applications, namely, the buckling, free vibration, or dynamic problems of composite structures [51–55], as well as the fracture mechanics problems [56,57] or non-linear transient problems [58,59]. In addition, the Bolotin method [60] is proposed herein to determine the DIRs for the differential equations system, known as Mathieu–Hill system of equations. More details about the basics of the proposed numerical tools are recalled in what follows.

3.1. The GDQ Method

The GDQ method approximates the fundamental system of differential equations, by discretizing the derivatives of a function $J(x)$ respect to a spatial variable at a given discrete grid distribution, by means of the weighting coefficients. For a one-dimensional problem where the whole domain $0 \leq x \leq L$ is discretized in N grids points, the approximation of the nth-order derivatives of J function respect to x variable can be expressed as [49]

$$\frac{d^n J(x_p)}{dx^n} = \sum_{r=1}^{N} \chi_{pr}^{(n)} J(x_r) \quad n = 1, 2, \ldots, N-1, \tag{29}$$

$\chi_{pr}^{(n)}$ being the weighting coefficients, defined as follows [49]

$$\chi_{pq}^{(1)} = \frac{Y(x_p)}{(x_p - x_q) Y(x_q)}, \quad p \neq q; \; p, q = 1, 2, \ldots, N \tag{30}$$

and $Y(x_p)$ is the Lagrangian operator expressed as [49]

$$Y(x_p) = \prod_{p \neq q, \; q=1}^{N} (x_p - x_q). \tag{31}$$

For higher order derivatives of the weighting coefficients it is [49]

$$\chi_{pq}^{(n)} = n \left(\chi_{pp}^{(n-1)} \chi_{pq}^{(1)} - \frac{\chi_{pq}^{(n-1)}}{(x_p - x_q)} \right) \tag{32}$$

It is well known in literature that the type of grid distribution within the domain can affect significantly the accuracy of the proposed method [50]. In what follows we apply a Chebyshev–Gauss–Lobatto non-uniform pattern, due to its great performances, as verified by Shu [49]

$$x_p = \frac{L}{2} \left(1 - \cos\left(\frac{x_p - 1}{N - 1} \pi \right) \right) \quad p = 1, 2, \ldots, N. \tag{33}$$

Thus, the governing differential equations of motion and boundary conditions are discretized according to the GDQ approach, as detailed in Appendix B. Let us denote the periodic axial compressive load as

$$F_a(t) = \alpha F_{cr} + \beta F_{cr} \cos(\omega t) \tag{34}$$

where α and β refer to the static and dynamic load factors, respectively. Furthermore, F_{cr} denotes the critical static load and ω stands for the excitation frequency. By substituting Equation (34) into the third Equation (A3) from the Appendix A, and by combining the discretized equations of motion along with the associated boundary conditions, the problem can be redefined in the following matrix form

$$\begin{pmatrix} M_{bb} & M_{bd} \\ M_{db} & M_{dd} \end{pmatrix} \begin{Bmatrix} \ddot{\Gamma}_b \\ \ddot{\Gamma}_d \end{Bmatrix} + \left(\begin{pmatrix} K_{bb} & K_{bd} \\ K_{db} & K_{dd} \end{pmatrix} + F_{cr}(\alpha + \beta \cos(\omega t)) \begin{pmatrix} K^G_{bb} & K^G_{bd} \\ K^G_{db} & K^G_{dd} \end{pmatrix} \right) \begin{Bmatrix} \Gamma_b \\ \Gamma_d \end{Bmatrix} = \begin{pmatrix} 0 \\ 0 \end{pmatrix} \quad (35)$$

where M, K, and K^G are the mass, stiffness, and geometric stiffness matrixes, respectively, and $\Gamma = \{\overline{U}, \overline{V}, \overline{W}, \overline{\Phi}_x, \overline{\Phi}_y\}^T$ denotes the unknown dynamic displacement vector. In addition, indexes b and d indicate the boundary points and domain points, respectively.

3.2. Bolotin Method

The second order system of differential Equation (35) is known in literature as Mathieu–Hill system of equations due to presence of the periodic coefficient, accordingly. In the present study we propose the Bolotin method [60] to define the boundaries associated to the DIR of the BD-FG porous cylindrical shell. Based on this method, the dynamic displacement vector Γ can be defined in a Fourier series as follows [60]

$$\{\Gamma\} = \sum_{s=1,3,\ldots}^{\infty} \left(\{\vartheta_s\} \sin\left(\frac{s\omega t}{2}\right) + \{v_s\} \cos\left(\frac{s\omega t}{2}\right) \right), \quad (36)$$

where ϑ_s and v_s denote the arbitrary time invariant vectors. It should be mentioned that the first DIR with period $2T$ is generally much meaningful and wider than the secondary one with period T. For this reason, in this work we consider the solutions with period $2T$. By substitution of Equation (36) into Equation (35) and by mathematical manipulation, we get the following first order equation

$$\left| [K] - F_{cr}\left(\alpha \pm \frac{\beta}{2}\right)[K^G] - \frac{\omega^2}{4}[M] \right| = 0, \quad (37)$$

which represents a classical eigenvalue problem. The critical buckling load can be computed from Equation (35) by neglecting the inertia terms and by setting $\alpha + \beta \cos(\omega t) = 1$. Then, solving Equation (37) for some fixed values of α and F_{cr}, the variation of the excitation frequency ω in regards to β can be drawn as DIR for the BD-FG structure.

4. Numerical Investigation

In this section some illustrative example are shown, starting with a preliminary validation of the proposed method with respect to the available literature, and continuing with a parametric investigation of the problem, whose results are evaluated comparatively in order to evaluate the sensitivity of the mechanical response.

4.1. Validation

Due to the general lack of works in the literature on the dynamic buckling behavior of BD-FG porous cylindrical shells, the proposed model is validated, herein, for an axial buckling problem of a simply supported conventional FG cylindrical shell based on two different theories. Thus, for comparative purposes, we select the same material properties and shell geometry as reported in [48,61], and neglect the inertia terms, foundation parameters and porosity effects, while assuming a null value for n_x. In Table 1 we summarize the results in terms of critical axial buckling load F_{cr} for a FG cylindrical shell with $h = 0.001$ m, $L/R = 0.5$, $E_c = 380$ GPa, $E_m = 70$ GPa, $\nu = 0.3$, $\bar{c} = 0$, with a clear excellent agreement between our results and predictions in Ref. [61] based on the first order shear deformation theory. This first numerical example could be considered as limit case, where the TSDT reverts to the FSDT, since it refers to a thin shell structure, just for validation purposes. More accurate results, however, are always expected under a TSDT assumption for increased values of the shell thickness, as done in the next parametric investigation.

Table 1. Comparative evaluation of the critical axial buckling load (MN) for a FG cylindrical shell with $h = 0.001$ m, $L/R = 0.5$, $E_c = 380$ GPa, $E_m = 70$ GPa, $v = 0.3$, $n_x = 0$.

R/h		Khazaeinejad and Najafizadeh [61]	Present
5	Alumina	1.598 (1,1)	1.5975
	$n_z = 1$	0.853 (1,1)	0.8532
	$n_z = 2$	0.662 (1,1)	0.6624
	$n_z = 5$	0.520 (1,1)	0.5197
	$n_z = 10$	0.450 (1,1)	0.4500
	Aluminum	0.294 (1,1)	0.2942
10	Alumina	1.403 (1,1)	1.4029
	$n_z = 1$	0.759 (1,1)	0.7589
	$n_z = 2$	0.589 (1,1)	0.5885
	$n_z = 5$	0.456 (1,1)	0.4557
	$n_z = 10$	0.393 (1,1)	0.3931
	Aluminum	0.258 (1,1)	0.2584
20	Alumina	1.594 (1,1)	1.5936
	$n_z = 1$	0.903 (1,1)	0.9029
	$n_z = 2$	0.698 (1,1)	0.6977
	$n_z = 5$	0.514 (1,1)	0.5140
	$n_z = 10$	0.430 (1,1)	0.4295
	Aluminum	0.293 (1,1)	0.2935
30	Alumina	1.566 (2,1)	1.5664
	$n_z = 1$	0.826 (2,1)	0.8262
	$n_z = 2$	0.642 (2,1)	0.6419
	$n_z = 5$	0.511 (2,1)	0.5108
	$n_z = 10$	0.449 (2,1)	0.4486
	Aluminum	0.289 (2,1)	0.2885
100	Alumina	1.443 (3,1)	1.4428
	$n_z = 1$	0.782 (3,1)	0.7822
	$n_z = 2$	0.606 (3,1)	0.6064
	$n_z = 5$	0.469 (3,1)	0.4681
	$n_z = 10$	0.404 (3,1)	0.4008
	Aluminum	0.266 (3,1)	0.2657
300	Alumina	1.443 (5,1)	1.4431
	$n_z = 1$	0.787 (5,1)	0.7841
	$n_z = 2$	0.610 (5,1)	0.6079
	$n_z = 5$	0.468 (5,1)	0.4683
	$n_z = 10$	0.402 (5,1)	0.4017
	Aluminum	0.266 (5,1)	0.2658

As further comparative study, Table 2 compares the dimensionless critical buckling load ($P_{cr} = F_{cr}L^2/\pi^2 D_m$; $D_m = E_m h^3/12(1-v_m^2)$) for a FG cylindrical shell with $h = 0.001$ m, $E_c = 380$ GPa, $E_m = 70$ GPa, $v = 0.3$, $\bar{c} = 4/3h^2$. Based on Table 2, it is worth noticing the high precision between our results and predictions by Bagherizadeh et al. [48], which confirms the accuracy of the GDQ method. This method is thus proposed in the following parametric study, as efficient numerical tool to solve the problem.

Table 2. Comparative evaluation of the dimensionless critical axial buckling load for a FG cylindrical shell with $h = 0.001$ m, $E_m = 70$ GPa, $v = 0.3$, $n_z = 2$.

Z	h/R	Bagherizadeh et al. [48]	Present
50	0.01	79.9296 (4,5)	79.9295
	0.025	79.48684 (4,3)	79.4868
	0.05	78.79842 (4,3)	78.7984
300	0.01	479.5066 (10,5)	479.5065
	0.025	476.3834 (10,3)	476.3834
	0.05	470.8775 (11,1)	470.8775
900	0.01	1438.157 (18,3)	1438.1576
	0.025	1428.611 (18,2)	1428.6108
	0.05	1412.380 (19,1)	1412.3802

4.2. Parametric Study

We refer to a BD-FG cylindrical structure with constituent phases of properties listed in Table 3, where the following dimensionless parameters are considered to compute the dimensionless structural excitation frequencies

$$\Omega = \omega R \sqrt{\frac{\rho_m}{E_m}}, K_g = \frac{k_g R^2}{E_m h^3}, K_w = \frac{k_w R^4}{E_m h^3}. \tag{38}$$

Table 3. Material properties of the BD-FG cylindrical shell.

Constituent Phases	Materi	Properties		
		E (GPa)	ρ (Kg/m^3)	v
c	SiC	427	3100	0.17
m	Al	70	2702	0.3

We determine the DIR, and highlight the effects of different parameters such as the thickness-to-radius ratio (h/R), the length-to-radius ratio (L/R), the static load factor, the boundary conditions, the power law indexes (n_x, n_z), the type and volume fraction of porosity, and the foundation parameters, on the dynamic buckling behavior of the BD-FG cylindrical shell.

In Figure 3 we plot the variation of the DIR for different thickness-to-radius ratios (h/R), where a clear shift of the DIR is observed for increasing h/R ratios. This means that the DIR becomes wider for a certain value of the dynamic load factor, and occurs with a sort of delay. An increased h/R ratio from 0.01 to 0.1 yields a global shift of the DIR origin point towards high excitation frequencies.

Figure 4 shows the sensitivity of the DIR for varying L/R ratios, while keeping fixed $h/R = 0.01$. In detail, for $L/R = 1$ the structure has a wider DIR in comparison with the other values, whereby an increasing L/R ratio yields the DIR to take place at lower excitation frequencies. Based on the plots in Figure 4, we can observe a reduction of about 41.96% in the excitation frequencies corresponding to the origin of the instability region, for a L/R ranging between 1 and 10. When the L/R ratio features higher magnitudes, the bending resistance gradually reduces and yields an increased bending deformation.

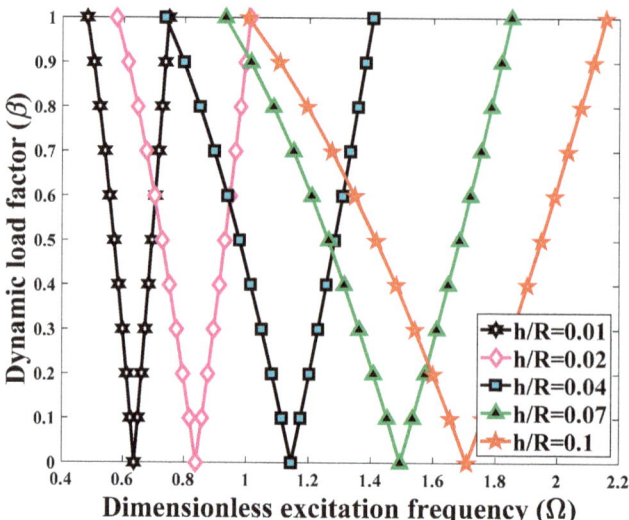

Figure 3. Effect of the thickness-to-radius ratio on the dynamic instability region (DIR) for a BD-FG cylindrical shell with $R = 0.5$ m, $L/R = 1$, $n_x = n_z = 1$, $\alpha = 0.3$.

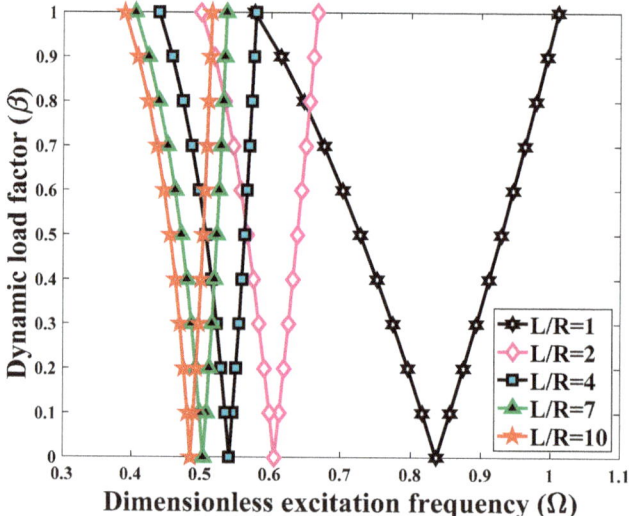

Figure 4. Effect of the length-to-radius ratio on the DIR for a BD-FG cylindrical shell with $R = 0.5$ m, $h/R = 0.02$, $n_x = n_z = 1$, $\alpha = 0.3$.

In Figure 5, we evaluate the effect of the static load factor on the instability region of the BD-FG cylindrical shell. As expected, in absence of a static load on the structure, the width of DIR gets smaller, whereby for an increased static load factor, it becomes gradually greater for a fixed value of dynamic load factor (i.e., $\beta = 1$), and its origin tends to move on the left side. This proves the sensitivity of the structural instability to the static load factor.

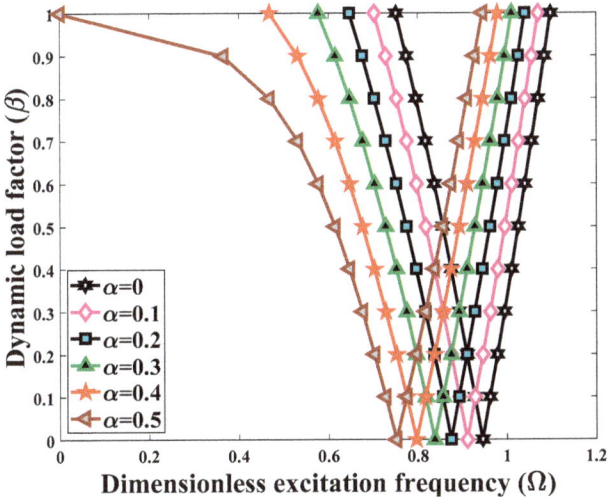

Figure 5. Effect of the static load factor on the DIR for a BD-FG cylindrical shell with $R = 0.5$ m, $h/R = 0.02$, $L/R = 1$, $n_x = n_z = 1$, $\alpha = 0.3$.

As also visible in Figure 6, we evaluate the impact of different boundary conditions on the DIR of the cylindrical shell. Here, we consider three different boundary conditions, namely, S-S, C-S, and C-C boundary conditions. Based on the plots of Figure 7, it is worth noting that a C-C boundary conditions yields higher values of the dimensionless excitation frequencies than those ones provided by a S-S or C-S supports, due to an increased stiffness of the structure. Furthermore, the origin of the instability region tends to get away from the origin. Once the dynamic load factor β reaches the unit value, the width of the DIR for S-S boundary condition becomes smaller, compared to the other boundary conditions. This means that, for lower values of dimensionless excitation frequency, a BD-FG cylindrical shell with S-S supports tends to become more unstable compared to the other boundary conditions.

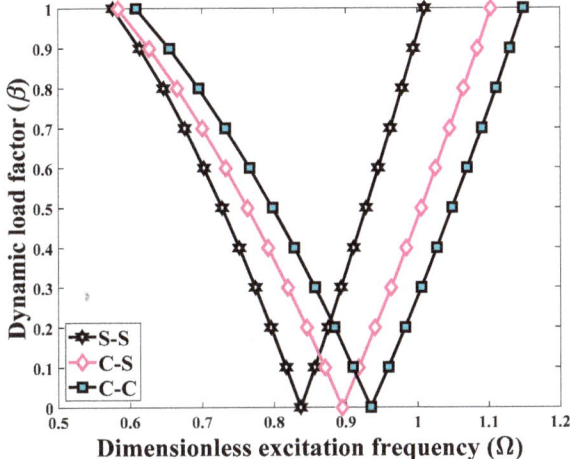

Figure 6. Effect of boundary conditions on the DIR for a BD-FG cylindrical shell with $R = 0.5$ m, $h/R = 0.02$, $L/R = 1$, $n_x = n_z = 1$, $\alpha = 0.3$.

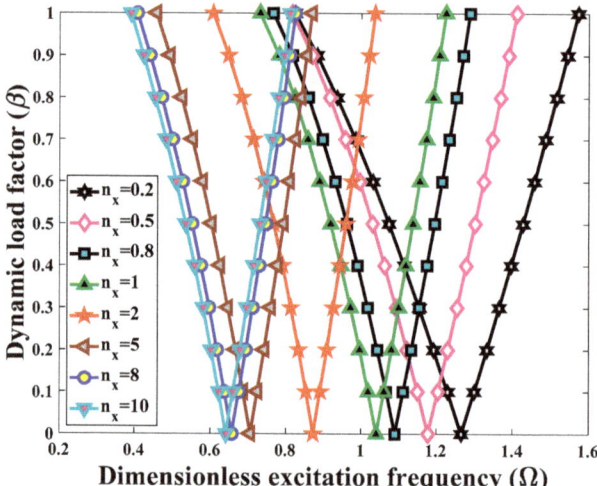

Figure 7. Effect of the longitudinal power index on the DIR for a FG cylindrical shell with $R = 0.5$ m, $h/R = 0.02$, $L/R = 1$, $n_z = 0$, $\alpha = 0.3$.

A further investigation is devoted to study the influence of the power law index along the length n_x on the dynamic buckling behavior of one-directional FG cylindrical shell, as depicted Figure 7. In such a case, the DIR takes place at lower frequencies owning to an increased magnitude of the power law index. The effect of an increased dimensionless excitation frequency related to the origin of the DIR is meaningful within the range $0.2 \leq n_x \leq 5$. For greater values of n_x, the variation in frequency corresponding to the origin of DIR becomes less remarkable. A one-directional FG cylindrical shell with $n_x = 10$ or $n_x = 8$ is more sensitive to the dynamic instability for lower excitation frequencies compared with those ones with $n_x \leq 5$.

For a conventional FG cylindrical shell, we also investigate the effect of the transverse power index n_z on the DIR, as plotted in Figure 8. It can be mentioned that for a constant value of the dynamic load factor, the enhancement of n_z yields a reduction width of the dynamic instability region, especially for lower excitation frequencies. In addition, the origin DIR moves to a lower dimensionless excitation frequency. Comparing the results from Figures 7 and 8, it can be concluded that, a double increase of both n_x, and n_z, leads to a reduction in the excitation frequency. Nevertheless, n_x plays an important role in the reduction of the excitation frequency and in the increase of the structural instability. For $\beta = 0$, for example, we note a reduction of the excitation frequency equal to 49.3% and 44.49%, for and increasing value of n_x and n_z from 0.2 up to 10, respectively.

Figure 9 shows the effect of the transverse and longitudinal volume fraction indexes on the DIR. In detail, for an increase of these two parameters, the origin of DIR moves to higher dimensionless excitation frequency, and the DIR width declines. In conclusion, the double increase of n_x and n_z leads a metal phase reinforcement, with a overall decrease of the structural stiffness and an increase in the structural instability. BD-FG cylindrical shells with lower values of n_z and n_x, are less sensitive to the dynamic instability, due to their higher stiffness. The contrary occurs for higher values of n_x and n_z due to an increased deformability of the structure. The importance of applying BD-FG materials is highlighted when the variation of material properties is considered in a single direction or more directions simultaneously. This issue can be beneficial for the fabrication and design purposes of modern FG structures. Subsequently, the dynamic stability of BD-FG cylindrical shells can be controlled selecting appropriate power indexes corresponding to the desired direction.

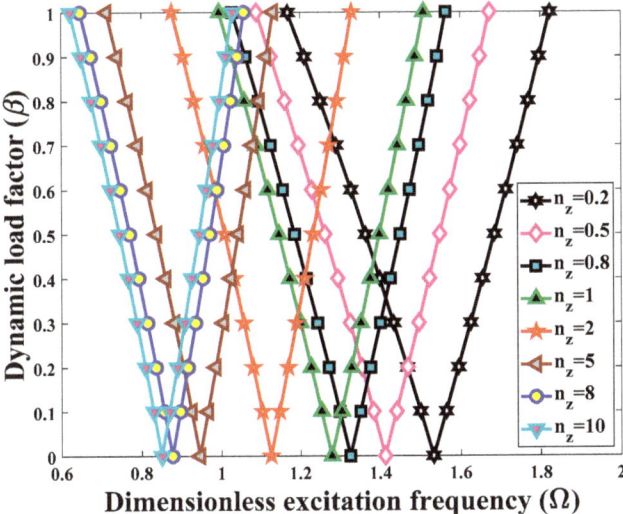

Figure 8. Effect of the transverse power index on the DIR for a BD-FG cylindrical shell with $R = 0.5$ m, $h/R = 0.02$, $L/R = 1$, $n_x = 0$, $\alpha = 0.3$.

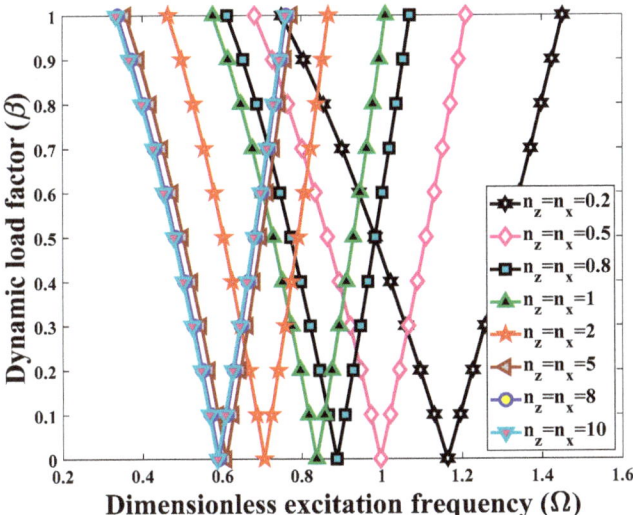

Figure 9. Effect of the longitudinal and transverse power indexes on the DIR for a FG cylindrical shell with $R = 0.5$ m, $h/R = 0.02$, $L/R = 1$, $\alpha = 0.3$.

Moreover, Figure 10 shows the effect of an even porosity volume fraction on the dimensionless excitation frequency of a BD-FG porous cylindrical shell. Note that the dimensionless excitation frequency decreases for increasing values of n_x and n_z. In addition, for a fixed value of n_x, n_z, the excitation frequency tends to converge to a common point. As expected, at the intersection point, the effect of an even porosity volume fraction, ζ, on the dimensionless excitation frequency is almost negligible. However, for different values of n_x, n_z, the effect of even porosity between the ceramic and metal phases can change significantly, such that before the intersection point, an increased even porosity volume fraction increases the dimensionless excitation frequency, and the contrary occurs after the intersection point, with a gradual decrease in the excitation frequency for an enhanced ζ. The additional Figures 11 and 12 show the different response for different values of ζ, while assuming the same value for n_x and n_z. In detail, under the assumption $n_x = n_z = 0.15$ (Figure 11), it is visible that an increased porosity ζ moves the DIR towards higher excitation frequencies. A reversed behavior occurs in Figure 12 under the assumption $n_x = n_z = 1.5$, since an increased value of ζ causes a general shift of the DIR to lower excitation frequencies. This confirms the effect of either the even porosity volume fraction ζ and the power indexes n_x and n_z on the stability response of the structure.

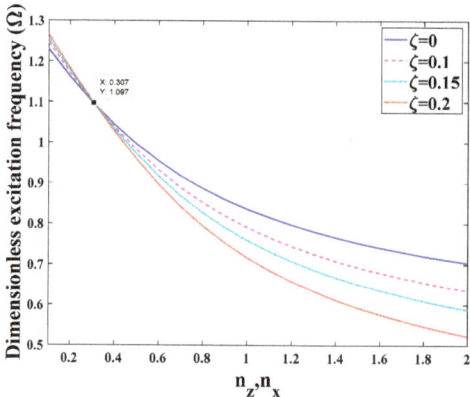

Figure 10. Effect of the even porosity volume fraction ζ on the dimensionless excitation frequency versus the longitudinal and transverse power indexes for a BD-FG porous cylindrical shell with $R = 0.5$ m, $h/R = 0.02$, $L/R = 1$, $\alpha = 0.3$, $\beta = 0$.

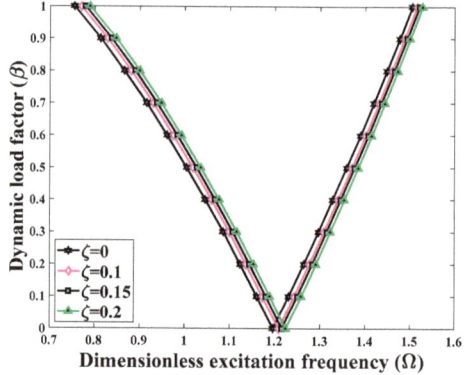

Figure 11. Effect of the even porosity volume fraction ζ on the DIR for a BD-FG porous cylindrical shell with $R = 0.5$ m, $h/R = 0.02$, $L/R = 1$, $\alpha = 0.3$, $n_x = n_z = 0.15$.

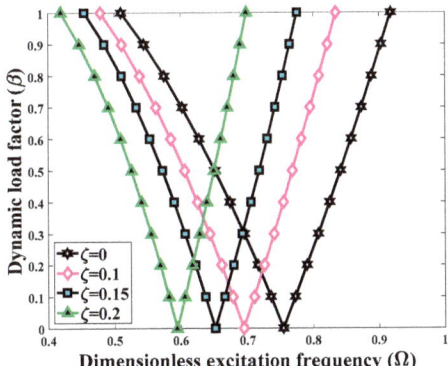

Figure 12. Effect of the even porosity volume fraction ζ on the DIR for a BD-FG porous cylindrical shell with $R = 0.5$ m, $h/R = 0.02$, $L/R = 1$, $\alpha = 0.3$, $n_x = n_z = 0.15$.

In Figures 13–15 we repeat the parametric analysis to evaluate the effect of an uneven porosity between two phases of the second ceramic and second metal. In detail, Figure 13 shows the variation of the dimensionless excitation frequency versus n_x, n_z, for different values of ξ. Figure 14 is devoted to check for the influence of ξ on the DIR of the structure for $n_x = n_z = 0.15$. Additionally, in this case, we can observe as the DIR moves to the right side by increasing ξ and it takes place at higher excitation frequencies. Nevertheless, by assuming $n_x = n_z = 1.5$, a different trend is noticed for the DIR in Figure 15, since an increased value of ξ yields the DIR to occur at lower excitation frequencies and its width gets smaller.

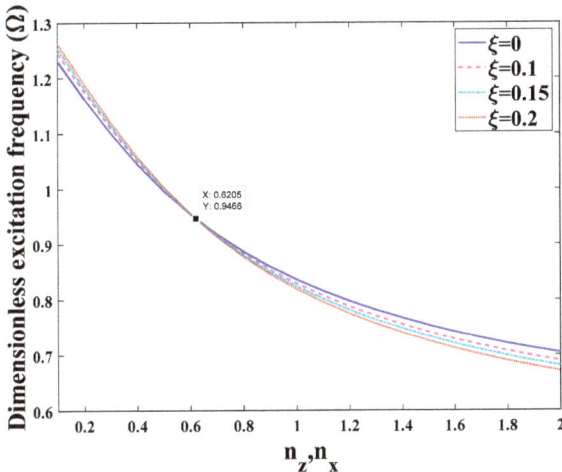

Figure 13. Effect of the uneven porosity volume fraction ξ on the dimensionless excitation frequency versus the longitudinal and transverse power indexes for a BD-FG porous cylindrical shell with $R = 0.5$ m, $h/R = 0.02$, $L/R = 1$, $\alpha = 0.3$, $\beta = 0$.

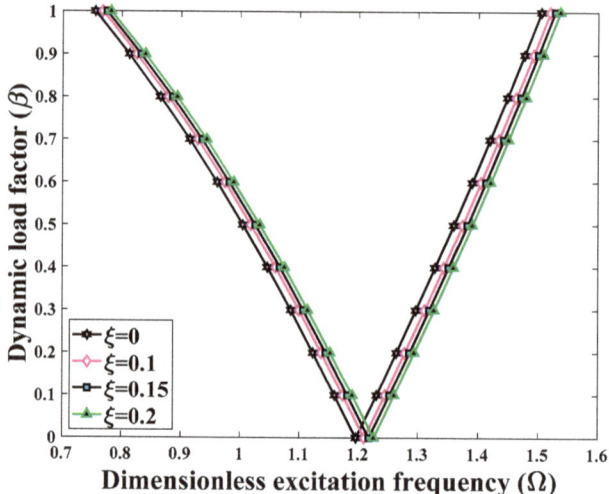

Figure 14. Effect of the uneven porosity volume fraction ξ on the DIR for a BD-FG porous cylindrical shell with $R = 0.5$ m, $h/R = 0.02$, $L/R = 1$, $\alpha = 0.3$, $n_x = n_z = 0.15$.

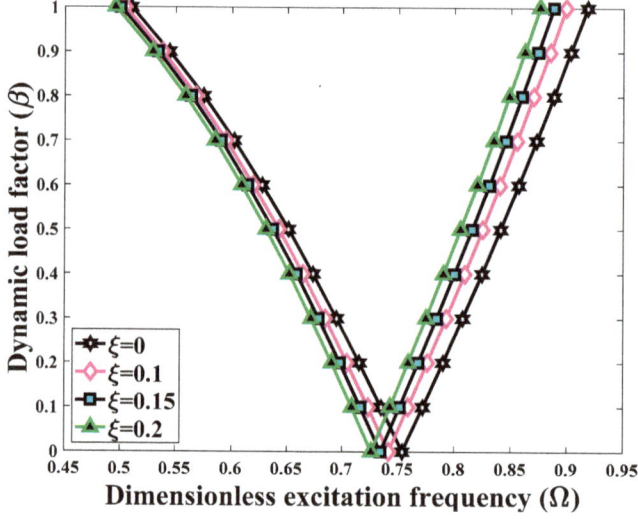

Figure 15. Effect of the uneven porosity volume fraction ξ on the DIR for a BD-FG porous cylindrical shell with $R = 0.5$ m, $h/R = 0.02$, $L/R = 1$, $\alpha = 0.3$, $n_x = n_z = 0.15$.

The last parametric analysis considers the possible sensitivity of the response to the elastic foundation. For this reason, Figures 16 and 17 plot the variation of the DIR with the Winkler or the Pasternak foundation coefficients, respectively. A noteworthy increase in stiffness emerges from both figures, where the origin of the DIR moves towards higher values of frequency. According to a comparative evaluation of the results, it seems that the best dynamic behavior of the cylindrical shell is reached for a structure surrounded by a Pasternak elastic foundation. Hence, the effect of the Pasternak elastic coefficient is more remarkable than the Winkler-based one, where a BD-FG cylindrical shell becomes more stable if embedded in a Pasternak foundation.

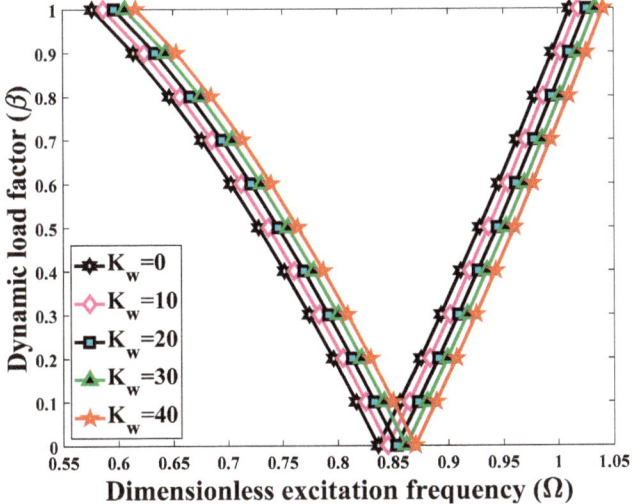

Figure 16. Effect of the Winkler coefficient on the DIR for a BD-FG cylindrical shell with $R = 0.5$ m, $h/R = 0.02$, $L/R = 1$, $\alpha = 0.3$, $K_g = 0$, $n_x = n_z = 1.5$.

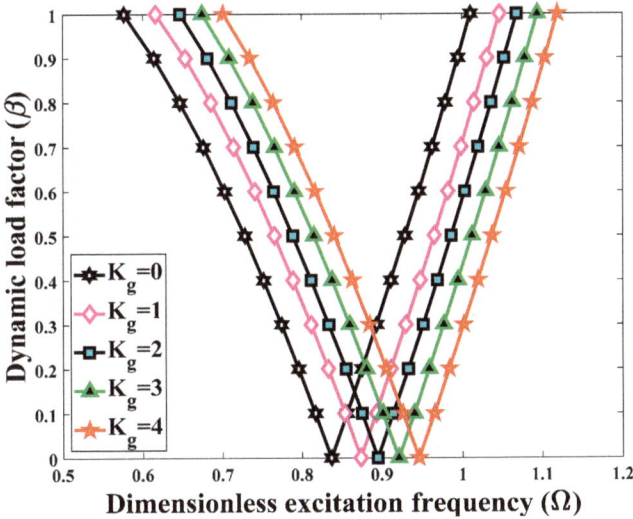

Figure 17. Effect of the Pasternak coefficient on the DIR for a BD-FG cylindrical shell with $R = 0.5$ m, $h/R = 0.02$, $L/R = 1$, $\alpha = 0.3$, $K_w = 0$, $n_x = n_z = 1$.

5. Conclusions

This work investigates the dynamic stability of BD-FG cylindrical shells embedded in an elastic foundation, including possible effects related to porosity. The material properties of BD-FG porous cylindrical shells are computed according to a modified BD power law model. Using the Hamilton's principle, we determine the governing equations of the problem, under the classical TSDT assumptions. The aforementioned equations are rewritten into a system of Mathieu–Hill equations, according to a GDQ approach. The work is also devoted to determine the DIR of the structure while applying the Bolotin method. After a preliminary validation of the proposed formulation, with respect to the available literature, we perform a large numerical investigation to check for the sensitivity of the response both in terms of excitation frequencies and DIRs, for different thickness-to-radius ratios, length-to-radius ratios, boundary conditions, transverse and longitudinal power law indexes, even and uneven porosity volume fractions, and foundation parameters. Based on the systematic numerical investigation, the main conclusions can be summarized as follows

- An increased thickness-to-radius ratio causes a general shift of the DIR origin towards higher excitation frequencies. Moreover, the DIR gets wider at a certain value of the dynamic load factor.
- An increased length-to-radius dimensionless ratio moves the DIR origin towards lower excitation frequencies, whereas the DIR gets smaller.
- A simultaneous increase of longitudinal and transverse power indexes yields an overall decrease in the excitation frequencies associated with the DIR origin.
- The control of the dynamic instability for a BD-FG cylindrical shell, is convenient for an appropriate selection of the power indexes.
- BD-FG cylindrical shells with a simply support at both ends are more unstable than the clamped-clamped or clamped-simply supported structures, because of their higher deformability.
- The effect of coefficients and type of porosity on the structural DIR depend on the extent of the longitudinal and transverse power law indexes. There exists a certain value for these indexes, for which the excitation frequencies corresponding to the DIR can invert their behavior.
- A general increase in the elastic foundation coefficients yields higher excitation structural frequencies especially when a Pasternak foundation is assumed instead of a Winkler foundation. Anyway, the presence of an elastic foundation makes the structure stiffer and more stable.

Author Contributions: Conceptualization, F.A., H.T., R.D. and F.T.; Formal analysis, F.A., H.T., R.D. and F.T.; Investigation, F.A., H.T. and F.T.; Validation, H.T., R.D. and F.T.; Writing—Original Draft, F.A., H.T., R.D. and F.T.; Writing—Review & Editing, R.D. and F.T. All authors have read and agreed to the published version of the manuscript.

Funding: This research received no external funding.

Conflicts of Interest: The authors declare no conflict of interest.

Appendix A

Here below are the equations of motion in their final form

$$\begin{aligned}
&\left(\frac{dA_{11}(x)}{dx}\right)\left(\frac{dU(x)}{dx}\right) + A_{11}(x)\left(\frac{d^2U(x)}{dx^2}\right) - \frac{n^2 A_{66}(x)U(x)}{R^2} \\
&- \frac{nV(x)}{R}\left(\frac{dA_{12}(x)}{dx}\right) - \frac{nA_{12}(x)}{R}\left(\frac{dV(x)}{dx}\right) - \frac{nA_{66}(x)}{R}\left(\frac{dV(x)}{dx}\right) \\
&+ \frac{W(x)}{R}\left(\frac{dA_{12}(x)}{dx}\right) + \frac{A_{12}(x)}{R}\left(\frac{dW(x)}{dx}\right) - \bar{c}\left(\frac{dD_{11}(x)}{dx}\right)\left(\frac{d^2W(x)}{dx^2}\right) \\
&- \bar{c}D_{11}(x)\left(\frac{d^3W(x)}{dx^3}\right) + \frac{\bar{c}n^2 W(x)}{R^2}\left(\frac{dD_{12}(x)}{dx}\right) + \frac{2\bar{c}n^2 D_{66}(x)}{R^2}\left(\frac{dW(x)}{dx}\right) \\
&+ \frac{\bar{c}n^2 D_{12}(x)}{R^2}\left(\frac{dW(x)}{dx}\right) + \left(\frac{dB_{11}(x)}{dx}\right)\left(\frac{d\Phi_x(x)}{dx}\right) + B_{11}(x)\left(\frac{d^2\Phi_x(x)}{dx^2}\right) \\
&- \bar{c}D_{11}(x)\left(\frac{d^2\Phi_x(x)}{dx^2}\right) - \bar{c}\left(\frac{dD_{11}(x)}{dx}\right)\left(\frac{d\Phi_x(x)}{dx}\right) - \frac{n^2 B_{66}(x)\Phi_x(x)}{R^2} + \frac{\bar{c}n^2 D_{66}(x)\Phi_x(x)}{R^2} \\
&- \frac{n\Phi_y(x)}{R}\left(\frac{dB_{12}(x)}{dx}\right) - \frac{nB_{12}(x)}{R}\left(\frac{d\Phi_y(x)}{dx}\right) - \frac{nB_{66}(x)}{R}\left(\frac{d\Phi_y(x)}{dx}\right) \\
&+ \frac{\bar{c}nD_{12}(x)}{R}\left(\frac{d\Phi_y(x)}{dx}\right) + \frac{\bar{c}n\Phi_y(x)}{a}\left(\frac{dD_{12}(x)}{dx}\right) + \frac{\bar{c}nD_{66}(x)}{a}\left(\frac{d\Phi_y(x)}{dx}\right) \\
&= \frac{1}{\overline{U}(t)}\left(I_0(x)U(x)\frac{d^2\overline{U}(t)}{dt^2} + (I_0(x) - \bar{c}I_3(x))\Phi_x(x)\frac{d^2\overline{\Phi}_x(t)}{dt^2}\right. \\
&\left. - \bar{c}I_3(x)\left(\frac{dW(x)}{dx}\right)\left(\frac{d^2\overline{W}(t)}{dt^2}\right)\right)
\end{aligned} \quad (A1)$$

$$\begin{aligned}
&\frac{nU(x)}{R}\left(\frac{\mathrm{d}A_{66}(x)}{\mathrm{d}x}\right)+\frac{nA_{66}(x)}{R}\left(\frac{\mathrm{d}U(x)}{\mathrm{d}x}\right)+\frac{nA_{12}(x)}{R}\left(\frac{\mathrm{d}U(x)}{\mathrm{d}x}\right)\\
&+\left(\frac{\mathrm{d}A_{66}(x)}{\mathrm{d}x}\right)\left(\frac{\mathrm{d}V(x)}{\mathrm{d}x}\right)+A_{66}(x)\left(\frac{\mathrm{d}^2V(x)}{\mathrm{d}x^2}\right)-\frac{n^2A_{11}(x)V(x)}{R^2}\\
&-\frac{2\bar{c}n}{R}\left(\frac{\mathrm{d}D_{66}(x)}{\mathrm{d}x}\right)\left(\frac{\mathrm{d}W(x)}{\mathrm{d}x}\right)-\frac{2\bar{c}nD_{66}(x)}{R}\left(\frac{\mathrm{d}^2W(x)}{\mathrm{d}x^2}\right)+\frac{nA_{11}(x)W(x)}{R^2}\\
&+\frac{\bar{c}n^3D_{11}(x)W(x)}{R^3}-\frac{\bar{c}nD_{12}(x)}{R}\left(\frac{\mathrm{d}^2W(x)}{\mathrm{d}x^2}\right)\\
&+\frac{n\Phi_x(x)}{R}\left(\frac{\mathrm{d}B_{66}(x)}{\mathrm{d}x}\right)+\frac{nB_{66}(x)}{R}\left(\frac{\mathrm{d}\Phi_x(x)}{\mathrm{d}x}\right)-\frac{\bar{c}n\Phi_x(x)}{R}\left(\frac{\mathrm{d}D_{66}(x)}{\mathrm{d}x}\right)\\
&-\frac{\bar{c}nD_{66}(x)}{R}\left(\frac{\mathrm{d}\Phi_x(x)}{\mathrm{d}x}\right)+\frac{nB_{12}(x)}{R}\left(\frac{\mathrm{d}\Phi_x(x)}{\mathrm{d}x}\right)-\frac{\bar{c}nD_{12}(x)}{R}\left(\frac{\mathrm{d}\Phi_x(x)}{\mathrm{d}x}\right)\\
&+\left(\frac{\mathrm{d}B_{66}(x)}{\mathrm{d}x}\right)\left(\frac{\mathrm{d}\Phi_y(x)}{\mathrm{d}x}\right)+B_{66}(x)\left(\frac{\mathrm{d}^2\Phi_y(x)}{\mathrm{d}x^2}\right)-\bar{c}\left(\frac{\mathrm{d}D_{66}(x)}{\mathrm{d}x}\right)\left(\frac{\mathrm{d}\Phi_y(x)}{\mathrm{d}x}\right)\\
&-\bar{c}D_{66}(x)\left(\frac{\mathrm{d}^2\Phi_y(x)}{\mathrm{d}x^2}\right)-\frac{n^2B_{11}(x)\Phi_y(x)}{R^2}+\frac{\bar{c}n^2D_{11}(x)\Phi_y(x)}{R^2}\\
&=\frac{1}{\bar{V}(t)}\Big(I_0(x)V(x)\frac{\mathrm{d}^2\bar{V}(t)}{\mathrm{d}t^2}+(I_0(x)-\bar{c}I_3(x))\Phi_y(x)\frac{\mathrm{d}^2\bar{\Phi}_y(t)}{\mathrm{d}t^2}\\
&\qquad -\frac{\bar{c}n}{R}I_3(x)W(x)\frac{\mathrm{d}^2\bar{W}(t)}{\mathrm{d}t^2}\Big)
\end{aligned} \tag{A2}$$

$$\begin{aligned}
&-\frac{2\bar{c}n^2U(x)}{R^2}\left(\frac{\mathrm{d}D_{66}(x)}{\mathrm{d}x}\right)-\frac{2\bar{c}n^2D_{66}(x)}{R^2}\left(\frac{\mathrm{d}U(x)}{\mathrm{d}x}\right)\\
&-\frac{\bar{c}n^2D_{12}(x)}{R^2}\left(\frac{\mathrm{d}U(x)}{\mathrm{d}x}\right)-\frac{A_{12}(x)}{R}\left(\frac{\mathrm{d}U(x)}{\mathrm{d}x}\right)\\
&+\bar{c}\left(\frac{\mathrm{d}^2D_{11}(x)}{\mathrm{d}x^2}\right)\left(\frac{\mathrm{d}U(x)}{\mathrm{d}x}\right)+2\bar{c}\left(\frac{\mathrm{d}D_{11}(x)}{\mathrm{d}x}\right)\left(\frac{\mathrm{d}^2U(x)}{\mathrm{d}x^2}\right)+\bar{c}D_{11}(x)\left(\frac{\mathrm{d}^3U(x)}{\mathrm{d}x^3}\right)\\
&-\frac{\bar{c}nV(x)}{R}\left(\frac{\mathrm{d}^2D_{12}(x)}{\mathrm{d}x^2}\right)-\frac{2\bar{c}n}{R}\left(\frac{\mathrm{d}D_{12}(x)}{\mathrm{d}x}\right)\left(\frac{\mathrm{d}V(x)}{\mathrm{d}x}\right)-\frac{\bar{c}nD_{12}(x)}{R}\left(\frac{\mathrm{d}^2V(x)}{\mathrm{d}x^2}\right)\\
&-\frac{2\bar{c}n}{R}\left(\frac{\mathrm{d}D_{66}(x)}{\mathrm{d}x}\right)\left(\frac{\mathrm{d}V(x)}{\mathrm{d}x}\right)-\frac{2\bar{c}nD_{66}(x)}{R}\left(\frac{\mathrm{d}^2V(x)}{\mathrm{d}x^2}\right)+\frac{\bar{c}n^3D_{11}(x)V(x)}{R^3}+\frac{nA_{11}(x)V(x)}{R^2}\\
&+\frac{2\bar{c}^2n^2}{R^2}\left(\frac{\mathrm{d}G_{12}(x)}{\mathrm{d}x}\right)\left(\frac{\mathrm{d}W(x)}{\mathrm{d}x}\right)+\frac{4\bar{c}^2n^2}{R^2}\left(\frac{\mathrm{d}G_{66}(x)}{\mathrm{d}x}\right)\left(\frac{\mathrm{d}W(x)}{\mathrm{d}x}\right)-\frac{\bar{c}^2n^4G_{11}(x)W(x)}{R^4}\\
&+\frac{\bar{c}^2n^2W(x)}{R^2}\left(\frac{\mathrm{d}^2G_{12}(x)}{\mathrm{d}x^2}\right)+\frac{2\bar{c}^2n^2G_{12}(x)}{R^2}\left(\frac{\mathrm{d}^2W(x)}{\mathrm{d}x^2}\right)-\frac{9\bar{c}^2n^2F_{66}(x)W(x)}{R^2}\\
&+\frac{4\bar{c}^2n^2G_{66}(x)}{R^2}\left(\frac{\mathrm{d}^2W(x)}{\mathrm{d}x^2}\right)+\frac{6\bar{c}n^2C_{66}(x)W(x)}{R^2}-\frac{2\bar{c}n^2D_{11}(x)W(x)}{R^3}\\
&-\frac{n^2A_{66}(x)W(x)}{R^2}+\frac{\bar{c}W(x)}{R}\left(\frac{\mathrm{d}^2D_{12}(x)}{\mathrm{d}x^2}\right)+\frac{2\bar{c}}{R}\left(\frac{\mathrm{d}D_{12}(x)}{\mathrm{d}x}\right)\left(\frac{\mathrm{d}W(x)}{\mathrm{d}x}\right)\\
&+\frac{2\bar{c}D_{12}(x)}{R}\left(\frac{\mathrm{d}^2W(x)}{\mathrm{d}x^2}\right)+A_{66}(x)\left(\frac{\mathrm{d}^2W(x)}{\mathrm{d}x^2}\right)+\left(\frac{\mathrm{d}A_{66}(x)}{\mathrm{d}x}\right)\left(\frac{\mathrm{d}W(x)}{\mathrm{d}x}\right)\\
&-2\bar{c}^2\left(\frac{\mathrm{d}G_{11}(x)}{\mathrm{d}x}\right)\left(\frac{\mathrm{d}^3W(x)}{\mathrm{d}x^3}\right)-\bar{c}^2G_{11}(x)\left(\frac{\mathrm{d}^4W(x)}{\mathrm{d}x^4}\right)-\frac{A_{11}(x)W(x)}{R^2}\\
&-6\bar{c}\left(\frac{\mathrm{d}C_{66}(x)}{\mathrm{d}x}\right)\left(\frac{\mathrm{d}W(x)}{\mathrm{d}x}\right)-6\bar{c}C_{66}(x)\left(\frac{\mathrm{d}^2W(x)}{\mathrm{d}x^2}\right)\\
&+\frac{2\bar{c}^2n^2G_{66}(x)}{R^2}\left(\frac{\mathrm{d}\Phi_x(x)}{\mathrm{d}x}\right)+\frac{2\bar{c}^2n^2\Phi_x(x)}{R^2}\left(\frac{\mathrm{d}G_{66}(x)}{\mathrm{d}x}\right)-\frac{\bar{c}n^2F_{12}(x)}{R^2}\left(\frac{\mathrm{d}\Phi_x(x)}{\mathrm{d}x}\right)\\
&-\frac{2\bar{c}n^2\Phi_x(x)}{R^2}\left(\frac{\mathrm{d}F_{66}(x)}{\mathrm{d}x}\right)-\frac{2\bar{c}n^2F_{66}(x)}{R^2}\left(\frac{\mathrm{d}\Phi_x(x)}{\mathrm{d}x}\right)+\frac{\bar{c}^2n^2G_{12}(x)}{R^2}\left(\frac{\mathrm{d}\Phi_x(x)}{\mathrm{d}x}\right)\\
&+\frac{\bar{c}D_{12}(x)}{R}\left(\frac{\mathrm{d}\Phi_x(x)}{\mathrm{d}x}\right)+\left(\frac{\mathrm{d}A_{66}(x)}{\mathrm{d}x}\right)\Phi_x(x)+A_{66}(x)\left(\frac{\mathrm{d}\Phi_x(x)}{\mathrm{d}x}\right)\\
&-\bar{c}^2G_{11}(x)\left(\frac{\mathrm{d}^3\Phi_x(x)}{\mathrm{d}x^3}\right)-\frac{B_{12}(x)}{R}\left(\frac{\mathrm{d}\Phi_x(x)}{\mathrm{d}x}\right)-6\bar{c}\left(\frac{\mathrm{d}C_{66}(x)}{\mathrm{d}x}\right)\Phi_x(x)\\
&-6\bar{c}C_{66}(x)\left(\frac{\mathrm{d}\Phi_x(x)}{\mathrm{d}x}\right)+9\bar{c}^2\left(\frac{\mathrm{d}F_{66}(x)}{\mathrm{d}x}\right)\Phi_x(x)+9\bar{c}^2F_{66}(x)\left(\frac{\mathrm{d}\Phi_x(x)}{\mathrm{d}x}\right)\\
&+\bar{c}F_{11}(x)\left(\frac{\mathrm{d}^3\Phi_x(x)}{\mathrm{d}x^3}\right)+\bar{c}\left(\frac{\mathrm{d}^2F_{11}(x)}{\mathrm{d}x^2}\right)\left(\frac{\mathrm{d}\Phi_x(x)}{\mathrm{d}x}\right)+2\bar{c}\left(\frac{\mathrm{d}F_{11}(x)}{\mathrm{d}x}\right)\left(\frac{\mathrm{d}^2\Phi_x(x)}{\mathrm{d}x^2}\right)\\
&-\bar{c}^2\left(\frac{\mathrm{d}^2G_{11}(x)}{\mathrm{d}x^2}\right)\left(\frac{\mathrm{d}\Phi_x(x)}{\mathrm{d}x}\right)-2\bar{c}^2\left(\frac{\mathrm{d}G_{11}(x)}{\mathrm{d}x}\right)\left(\frac{\mathrm{d}^2\Phi_x(x)}{\mathrm{d}x^2}\right)\\
&+\frac{2\bar{c}^2n}{R}\left(\frac{\mathrm{d}G_{12}(x)}{\mathrm{d}x}\right)\left(\frac{\mathrm{d}\Phi_y(x)}{\mathrm{d}x}\right)-\frac{\bar{c}^2n^3G_{11}(x)\Phi_y(x)}{R^3}-\frac{9\bar{c}^2nF_{66}(x)\Phi_y(x)}{R}\\
&+\frac{2\bar{c}^2n}{R}\left(\frac{\mathrm{d}G_{66}(x)}{\mathrm{d}x}\right)\left(\frac{\mathrm{d}\Phi_y(x)}{\mathrm{d}x}\right)+\frac{\bar{c}^2nG_{12}(x)}{R}\left(\frac{\mathrm{d}^2\Phi_y(x)}{\mathrm{d}x^2}\right)+\frac{\bar{c}n^3F_{11}(x)\Phi_y(x)}{R^3}
\end{aligned} \tag{A3}$$

$$\begin{aligned}
&-\frac{\bar{c}nD_{11}(x)\Phi_y(x)}{R^2} + \frac{2\bar{c}^2 nG_{66}(x)}{R}\left(\frac{d^2\Phi_y(x)}{dx^2}\right) + \frac{6\bar{c}nC_{66}(x)\Phi_y(x)}{R} \\
&-\frac{2\bar{c}n}{R}\left(\frac{dF_{66}(x)}{dx}\right)\left(\frac{d\Phi_y(x)}{dx}\right) - \frac{2\bar{c}nF_{66}(x)}{R}\left(\frac{d^2\Phi_y(x)}{dx^2}\right) - \frac{\bar{c}n\Phi_y(x)}{R}\left(\frac{d^2F_{12}(x)}{dx^2}\right) \\
&-\frac{2\bar{c}n}{R}\left(\frac{dF_{12}(x)}{dx}\right)\left(\frac{d\Phi_y(x)}{dx}\right) - \frac{\bar{c}nF_{12}(x)}{R}\left(\frac{d^2\Phi_y(x)}{dx^2}\right) \\
&+\frac{\bar{c}^2 n\Phi_y(x)}{R}\left(\frac{d^2G_{12}(x)}{dx^2}\right) + \frac{nB_{11}(x)\Phi_y(x)}{R^2} - \frac{nA_{66}(x)\Phi_y(x)}{R} \\
&-k_w W(x) + k_g\left(\frac{d^2W(x)}{dx^2} - \frac{n^2}{R^2}W(x)\right) - F_a \frac{d^2W(x)}{dx^2} = \frac{1}{\overline{W}(t)}\left(I_0(x)W(x)\frac{d^2\overline{W}(t)}{dt^2}\right.\\
&-\bar{c}^2 I_6(x)\left(\frac{d^2W(x)}{dx^2}\frac{d^2\overline{W}(t)}{dt^2} - \frac{n^2}{R^2}W(x)\frac{d^2\overline{W}(t)}{dt^2}\right) + \bar{c}I_3(x)\left(\frac{dU(x)}{dx}\frac{d^2\overline{U}(t)}{dt^2}\right.\\
&\left.\left.-\frac{n}{R}V(x)\frac{d^2\overline{V}(t)}{dt^2}\right) + \bar{c}(I_4(x) - \bar{c}I_6(x))\left(\frac{d\Phi_x(x)}{dx}\frac{d^2\overline{\Phi}_x(t)}{dt^2} - \frac{n}{R}\Phi_y(x)\frac{d^2\overline{\Phi}_y(t)}{dt^2}\right)\right)
\end{aligned}$$

$$\begin{aligned}
&\frac{\bar{c}n^2 D_{66}(x)U(x)}{R^2} - \frac{n^2 B_{66}(x)U(x)}{R^2} + \left(\frac{dB_{11}(x)}{dx}\right)\left(\frac{dU(x)}{dx}\right) + B_{11}(x)\left(\frac{d^2U(x)}{dx^2}\right) \\
&-\bar{c}\left(\frac{dD_{11}(x)}{dx}\right)\left(\frac{dU(x)}{dx}\right) - \bar{c}D_{11}(x)\left(\frac{d^2U(x)}{dx^2}\right) \\
&-\frac{nB_{66}(x)}{R}\left(\frac{dV(x)}{dx}\right) - \frac{nV(x)}{R}\left(\frac{dB_{12}(x)}{dx}\right) - \frac{nB_{12}(x)}{R}\left(\frac{dV(x)}{dx}\right) \\
&+\frac{\bar{c}nD_{12}(x)}{R}\left(\frac{dV(x)}{dx}\right) + \frac{\bar{c}nD_{66}(x)}{R}\left(\frac{dV(x)}{dx}\right) + \frac{\bar{c}nV(x)}{R}\left(\frac{dD_{12}(x)}{dx}\right) \\
&-\frac{\bar{c}W(x)}{R}\left(\frac{dD_{12}(x)}{dx}\right) - \frac{\bar{c}D_{12}(x)}{R}\left(\frac{dW(x)}{dx}\right) - \frac{c^2 n^2 G_{12}(x)}{R^2}\left(\frac{dW(x)}{dx}\right) \\
&+\frac{cn^2 F_{12}(x)}{R^2}\left(\frac{dW(x)}{dx}\right) + \frac{\bar{c}n^2 W(x)}{R^2}\left(\frac{dF_{12}(x)}{dx}\right) \\
&-\frac{2\bar{c}^2 n^2 G_{66}(x)}{R^2}\left(\frac{dW(x)}{dx}\right) + \frac{2\bar{c}n^2 F_{66}(x)}{R^2}\left(\frac{dW(x)}{dx}\right) \\
&-\frac{c^2 n^2 W(x)}{R^2}\left(\frac{dG_{12}(x)}{dx}\right) - A_{66}(x)\left(\frac{dW(x)}{dx}\right) - 9\bar{c}^2 F_{66}(x)\left(\frac{dW(x)}{dx}\right) \\
&+\bar{c}^2\left(\frac{dG_{11}(x)}{dx}\right)\left(\frac{d^2W(x)}{dx^2}\right) + \bar{c}^2 G_{11}(x)\left(\frac{d^3W(x)}{dx^3}\right) \\
&+\frac{W(x)}{R}\left(\frac{dB_{12}(x)}{dx}\right) + \frac{B_{12}(x)}{R}\left(\frac{dW(x)}{dx}\right) - \bar{c}\left(\frac{dF_{11}(x)}{dx}\right)\left(\frac{d^2W(x)}{dx^2}\right) \\
&-\bar{c}F_{11}(x)\left(\frac{d^3W(x)}{dx^3}\right) + 6\bar{c}C_{66}(x)\left(\frac{dW(x)}{dx}\right) \\
&-\frac{n^2 C_{66}(x)\Phi_x(x)}{R^2} + \frac{2\bar{c}n^2 F_{66}(x)\Phi_x(x)}{R^2} - \frac{\bar{c}^2 n^2 G_{66}(x)\Phi_x(x)}{R^2} + \left(\frac{dC_{11}(x)}{dx}\right)\left(\frac{d\Phi_x(x)}{dx}\right) \\
&+C_{11}(x)\left(\frac{d^2\Phi_x(x)}{dx^2}\right) - A_{66}(x)\Phi_x(x) - 9\bar{c}^2 F_{66}(x)\Phi_x(x) + \bar{c}^2\left(\frac{dG_{11}(x)}{dx}\right)\left(\frac{d\Phi_x(x)}{dx}\right) \\
&+\bar{c}^2 G_{11}(x)\left(\frac{d^2\Phi_x(x)}{dx^2}\right) - 2\bar{c}\left(\frac{dF_{11}(x)}{dx}\right)\left(\frac{d\Phi_x(x)}{dx}\right) \\
&-2\bar{c}F_{11}(x)\left(\frac{d^2\Phi_x(x)}{dx^2}\right) + 6\bar{c}C_{66}(x)\Phi_x(x) - \frac{nC_{66}(x)}{R}\left(\frac{d\Phi_y(x)}{dx}\right) \\
&-\frac{n\Phi_y(x)}{R}\left(\frac{dC_{12}(x)}{dx}\right) - \frac{nC_{12}(x)}{R}\left(\frac{d\Phi_y(x)}{dx}\right) - \frac{\bar{c}^2 nG_{66}(x)}{a}\left(\frac{d\Phi_y(x)}{dx}\right) \\
&+\frac{2\bar{c}nF_{66}(x)}{R}\left(\frac{d\Phi_y(x)}{dx}\right) + \frac{2\bar{c}nF_{12}(x)}{R}\left(\frac{d\Phi_y(x)}{dx}\right) - \frac{\bar{c}^2 n\Phi_y(x)}{R}\left(\frac{dG_{12}(x)}{dx}\right) \\
&+\frac{2\bar{c}n\Phi_y(x)}{R}\left(\frac{dF_{12}(x)}{dx}\right) - \frac{\bar{c}^2 nG_{12}(x)}{R}\left(\frac{d\Phi_y(x)}{dx}\right) \\
&=\frac{1}{\overline{\Phi}_x(t)}\left((I_0(x) - \bar{c}I_3(x))U(x)\frac{d^2\overline{U}(t)}{dt^2} + (I_2(x) - 2\bar{c}I_4(x) + \bar{c}^2 I_6(x))\Phi_x(x)\frac{d^2\overline{\Phi}_x(t)}{dt^2}\right.\\
&\left.-\bar{c}(I_4(x) - \bar{c}I_6(x))\frac{dW(x)}{dx}\frac{d^2\overline{W}(t)}{dt^2}\right),
\end{aligned} \quad (A4)$$

$$
\begin{aligned}
&\frac{nB_{12}(x)}{R}\left(\frac{dU(x)}{dx}\right)+\frac{nU(x)}{R}\left(\frac{dB_{66}(x)}{dx}\right)+\frac{nB_{66}(x)}{R}\left(\frac{dU(x)}{dx}\right)\\
&-\frac{\bar{c}nD_{12}(x)}{R}\left(\frac{dU(x)}{dx}\right)-\frac{\bar{c}nU(x)}{R}\left(\frac{dD_{66}(x)}{dx}\right)-\frac{\bar{c}nD_{66}(x)}{R}\left(\frac{dU(x)}{dx}\right)\\
&+\left(\frac{dB_{66}(x)}{dx}\right)\left(\frac{dV(x)}{dx}\right)+B_{66}(x)\left(\frac{d^2V(x)}{dx^2}\right)-\frac{n^2B_{11}(x)V(x)}{R^2}\\
&+\frac{\bar{c}n^2D_{11}(x)V(x)}{R^2}-\bar{c}\left(\frac{dD_{66}(x)}{dx}\right)\left(\frac{dV(x)}{dx}\right)-\bar{c}D_{66}(x)\left(\frac{d^2V(x)}{dx^2}\right)\\
&+\frac{nB_{11}(x)W(x)}{R^2}-\frac{nA_{66}(x)W(x)}{R}-\frac{\bar{c}nF_{12}(x)}{R}\left(\frac{d^2W(x)}{dx^2}\right)-\frac{9\bar{c}^2nF_{66}(x)W(x)}{R}\\
&-\frac{2\bar{c}nF_{66}(x)}{R}\left(\frac{d^2W(x)}{dx^2}\right)+\frac{\bar{c}n^3F_{11}(x)W(x)}{R^3}-\frac{\bar{c}^2n^3G_{11}(x)W(x)}{R^3}\\
&-\frac{2\bar{c}n}{R}\left(\frac{dF_{66}(x)}{dx}\right)\left(\frac{dW(x)}{dx}\right)+\frac{2\bar{c}^2nG_{66}(x)}{R}\left(\frac{d^2W(x)}{dx^2}\right)\\
&+\frac{6\bar{c}nC_{66}(x)W(x)}{R}+\frac{2\bar{c}^2n}{a}\left(\frac{dG_{66}(x)}{dx}\right)\left(\frac{dW(x)}{dx}\right)+\frac{\bar{c}^2nG_{12}(x)}{R}\left(\frac{d^2W(x)}{dx^2}\right)\\
&-\frac{\bar{c}nD_{11}(x)W(x)}{R^2}+\frac{nC_{12}(x)}{R}\left(\frac{d\Phi_x(x)}{dx}\right)\\
&+\frac{n\Phi_x(x)}{R}\left(\frac{dC_{66}(x)}{dx}\right)+\frac{nC_{66}(x)}{R}\left(\frac{d\Phi_x(x)}{dx}\right)+\frac{\bar{c}^2n\Phi_x(x)}{R}\left(\frac{dG_{66}(x)}{dx}\right)\\
&-\frac{2\bar{c}nF_{12}(x)}{R}\left(\frac{d\Phi_x(x)}{dx}\right)-\frac{2\bar{c}nF_{66}(x)}{R}\left(\frac{d\Phi_x(x)}{dx}\right)+\frac{\bar{c}^2nG_{66}(x)}{R}\left(\frac{d\Phi_x(x)}{dx}\right)\\
&+\frac{\bar{c}^2nG_{12}(x)}{R}\left(\frac{d\Phi_x(x)}{dx}\right)-\frac{2\bar{c}n\Phi_x(x)}{R}\left(\frac{dF_{66}(x)}{dx}\right)\\
&+\left(\frac{dC_{66}(x)}{dx}\right)\left(\frac{d\Phi_y(x)}{dx}\right)+C_{66}(x)\left(\frac{d^2\Phi_y(x)}{dx^2}\right)-A_{66}(x)\Phi_y(x)-\frac{n^2C_{11}(x)\Phi_y(x)}{R^2}\\
&+\frac{2\bar{c}n^2F_{11}(x)\Phi_y(x)}{R^2}-\frac{\bar{c}^2n^2G_{11}(x)\Phi_y(x)}{R^2}-2\bar{c}\left(\frac{dF_{66}(x)}{dx}\right)\left(\frac{d\Phi_y(x)}{dx}\right)\\
&-2\bar{c}F_{66}(x)\left(\frac{d^2\Phi_y(x)}{dx^2}\right)+6\bar{c}C_{66}(x)\Phi_y(x)-9\bar{c}^2F_{66}(x)\Phi_y(x)\\
&+\bar{c}^2\left(\frac{dG_{66}(x)}{dx}\right)\left(\frac{d\Phi_y(x)}{dx}\right)+\bar{c}^2G_{66}(x)\left(\frac{d^2\Phi_y(x)}{dx^2}\right)\\
&=\frac{1}{\overline{\Phi}_y(t)}\Big(\big(I_0(x)-\bar{c}I_3(x)\big)V(x)\frac{d^2\overline{V}(t)}{dt^2}+\big(I_2(x)-2\bar{c}I_4(x)+\bar{c}^2I_6(x)\big)\Phi_y(x)\frac{d^2\overline{\Phi}_y(t)}{dt^2}\\
&\qquad-\bar{c}(I_4(x)-\bar{c}I_6(x))\tfrac{n}{R}\frac{d^2\overline{W}(t)}{dt^2}\Big),
\end{aligned}
\qquad\text{(A5)}
$$

where

$$
\big(A_{ij}(x),B_{ij}(x),C_{ij}(x),D_{ij}(x),F_{ij}(x),G_{ij}(x)\big)=\int_{-\frac{h}{2}}^{\frac{h}{2}}Q_{ij}(x,z)\big(1,z,z^2,z^3,z^4,z^6\big)dz \qquad\text{(A6)}
$$

Appendix B

In what follows we rewrite the equations of motion (A1)–(A5) in a discretized form, according to the GDQ method.

$$
\begin{aligned}
&\left(\sum_{r=1}^{N}\chi_{pr}^{(1)}A_{11}(x_r)\right)\left(\sum_{r=1}^{N}\chi_{pr}^{(1)}U(x_r)\right)+A_{11}(x_p)\left(\sum_{r=1}^{N}\chi_{pr}^{(2)}U(x_r)\right)-\frac{n^2A_{66}(x_p)U(x_p)}{R^2}\\
&-\frac{nV(x_p)}{R}\left(\sum_{r=1}^{N}\chi_{pr}^{(1)}A_{12}(x_r)\right)-\frac{nA_{12}(x_p)}{R}\left(\sum_{r=1}^{N}\chi_{pr}^{(1)}V(x_r)\right)-\frac{nA_{66}(x_p)}{R}\left(\sum_{r=1}^{N}\chi_{pr}^{(1)}V(x_r)\right)\\
&+\frac{W(x_p)}{R}\left(\sum_{r=1}^{N}\chi_{pr}^{(1)}A_{12}(x_r)\right)+\frac{A_{12}(x_p)}{R}\left(\sum_{r=1}^{N}\chi_{pr}^{(1)}W(x_r)\right)-\bar{c}\left(\sum_{r=1}^{N}\chi_{pr}^{(1)}D_{11}(x_r)\right)\left(\sum_{r=1}^{N}\chi_{pr}^{(2)}W(x_r)\right)\\
&-\bar{c}D_{11}(x_p)\left(\sum_{r=1}^{N}\chi_{pr}^{(3)}W(x_r)\right)+\frac{\bar{c}n^2W(x_p)}{R^2}\left(\sum_{r=1}^{N}\chi_{pr}^{(1)}D_{12}(x_r)\right)+\frac{2\bar{c}n^2D_{66}(x_p)}{R^2}\left(\sum_{r=1}^{N}\chi_{pr}^{(1)}W(x_r)\right)\\
&+\frac{\bar{c}n^2D_{12}(x_p)}{R^2}\left(\sum_{r=1}^{N}\chi_{pr}^{(1)}W(x_r)\right)+\left(\sum_{r=1}^{N}\chi_{pr}^{(1)}B_{11}(x_r)\right)\left(\sum_{r=1}^{N}\chi_{pr}^{(1)}\Phi_x(x_r)\right)
\end{aligned}
\qquad\text{(A7)}
$$

$$
\begin{aligned}
&+ B_{11}(x_p)\left(\sum_{r=1}^{N}\chi_{pr}^{(2)}\Phi_x(x_r)\right) - \bar{c}D_{11}(x_p)\left(\sum_{r=1}^{N}\chi_{pr}^{(2)}\Phi_x(x_r)\right) \\
&- \bar{c}\left(\sum_{r=1}^{N}\chi_{pr}^{(1)}D_{11}(x_r)\right)\left(\sum_{r=1}^{N}\chi_{pr}^{(1)}\Phi_x(x_r)\right) - \frac{n^2 B_{66}(x_p)\Phi_x(x_p)}{R^2} + \frac{\bar{c}n^2 D_{66}(x_p)\Phi_x(x_p)}{R^2} \\
&- \frac{n\Phi_y(x_p)}{R}\left(\sum_{r=1}^{N}\chi_{pr}^{(1)}B_{12}(x_r)\right) - \frac{nB_{12}(x_p)}{R}\left(\sum_{r=1}^{N}\chi_{pr}^{(1)}\Phi_y(x_r)\right) - \frac{nB_{66}(x_p)}{R}\left(\sum_{r=1}^{N}\chi_{pr}^{(1)}\Phi_y(x_r)\right) \\
&+ \frac{\bar{c}nD_{12}(x_p)}{R}\left(\sum_{r=1}^{N}\chi_{pr}^{(1)}\Phi_y(x_r)\right) + \frac{\bar{c}n\Phi_y(x_p)}{a}\left(\sum_{r=1}^{N}\chi_{pr}^{(1)}D_{12}(x_r)\right) + \frac{\bar{c}nD_{66}(x_p)}{a}\left(\sum_{r=1}^{N}\chi_{pr}^{(1)}\Phi_y(x_r)\right) \\
&= \frac{1}{\overline{U}(t)}\left(I_0(x_p)U(x_p)\frac{d^2\overline{U}(t)}{dt^2} + (I_0(x_p)-\bar{c}I_3(x_p))\Phi_x(x_p)\frac{d^2\overline{\Phi}_x(t)}{dt^2}\right. \\
&\left. - \bar{c}I_3(x_p)\left(\sum_{r=1}^{N}\chi_{pr}^{(1)}W(x_r)\right)\left(\frac{d^2\overline{W}(t)}{dt^2}\right)\right)
\end{aligned}
$$

$$
\begin{aligned}
&\frac{nU(x_p)}{R}\left(\sum_{r=1}^{N}\chi_{pr}^{(1)}A_{66}(x_r)\right) + \frac{nA_{66}(x_p)}{R}\left(\sum_{r=1}^{N}\chi_{pr}^{(1)}U(x_r)\right) + \frac{nA_{12}(x_p)}{R}\left(\sum_{r=1}^{N}\chi_{pr}^{(1)}U(x_r)\right) \\
&+ \left(\sum_{r=1}^{N}\chi_{pr}^{(1)}A_{66}(x_r)\right)\left(\sum_{r=1}^{N}\chi_{pr}^{(1)}V(x_r)\right) + A_{66}(x_p)\left(\sum_{r=1}^{N}\chi_{pr}^{(2)}V(x_r)\right) - \frac{n^2 A_{11}(x_p)V(x_p)}{R^2} \\
&- \frac{2\bar{c}n}{R}\left(\sum_{r=1}^{N}\chi_{pr}^{(1)}D_{66}(x_r)\right)\left(\sum_{r=1}^{N}\chi_{pr}^{(1)}W(x_r)\right) - \frac{2\bar{c}nD_{66}(x_p)}{R}\left(\sum_{r=1}^{N}\chi_{pr}^{(2)}W(x_r)\right) + \frac{nA_{11}(x_p)W(x_p)}{R^2} \\
&+ \frac{\bar{c}n^3 D_{11}(x_p)W(x_p)}{R^3} - \frac{\bar{c}nD_{12}(x_p)}{R}\left(\sum_{r=1}^{N}\chi_{pr}^{(2)}W(x_r)\right) \\
&+ \frac{n\Phi_x(x_p)}{R}\left(\sum_{r=1}^{N}\chi_{pr}^{(1)}B_{66}(x_r)\right) + \frac{nB_{66}(x_p)}{R}\left(\sum_{r=1}^{N}\chi_{pr}^{(1)}\Phi_x(x_r)\right) - \frac{\bar{c}n\Phi_x(x_p)}{R}\left(\sum_{r=1}^{N}\chi_{pr}^{(1)}D_{66}(x_r)\right) \\
&- \frac{\bar{c}nD_{66}(x_p)}{R}\left(\sum_{r=1}^{N}\chi_{pr}^{(1)}\Phi_x(x_r)\right) + \frac{nB_{12}(x_p)}{R}\left(\sum_{r=1}^{N}\chi_{pr}^{(1)}\Phi_x(x_r)\right) \\
&- \frac{\bar{c}nD_{12}(x_p)}{R}\left(\sum_{r=1}^{N}\chi_{pr}^{(1)}\Phi_x(x_r)\right) + \left(\sum_{r=1}^{N}\chi_{pr}^{(1)}B_{66}(x_r)\right)\left(\sum_{r=1}^{N}\chi_{pr}^{(1)}\Phi_y(x_r)\right) \\
&+ B_{66}(x_p)\left(\sum_{r=1}^{N}\chi_{pr}^{(2)}\Phi_y(x_r)\right) - \bar{c}\left(\sum_{r=1}^{N}\chi_{pr}^{(1)}D_{66}(x_r)\right)\left(\sum_{r=1}^{N}\chi_{pr}^{(1)}\Phi_y(x_r)\right) \\
&- \bar{c}D_{66}(x_p)\left(\sum_{r=1}^{N}\chi_{pr}^{(2)}\Phi_y(x_r)\right) - \frac{n^2 B_{11}(x_p)\Phi_y(x_p)}{R^2} + \frac{\bar{c}n^2 D_{11}(x_p)\Phi_y(x_p)}{R^2} \\
&= \frac{1}{\overline{V}(t)}\left(I_0(x)V(x_p)\frac{d^2\overline{V}(t)}{dt^2} + (I_0(x_p)-\bar{c}I_3(x_p))\Phi_y(x_p)\frac{d^2\overline{\Phi}_y(t)}{dt^2}\right. \\
&\left. - \frac{\bar{c}n}{R}I_3(x_p)W(x_p)\frac{d^2\overline{W}(t)}{dt^2}\right)
\end{aligned}
\tag{A8}
$$

$$
\begin{aligned}
&- \frac{2\bar{c}n^2 U(x_p)}{R^2}\left(\sum_{r=1}^{N}\chi_{pr}^{(1)}D_{66}(x_r)\right) - \frac{2\bar{c}n^2 D_{66}(x_p)}{R^2}\left(\sum_{r=1}^{N}\chi_{pr}^{(1)}U(x_r)\right) \\
&- \frac{\bar{c}n^2 D_{12}(x_p)}{R^2}\left(\sum_{r=1}^{N}\chi_{pr}^{(1)}U(x_r)\right) - \frac{A_{12}(x_p)}{R}\left(\sum_{r=1}^{N}\chi_{pr}^{(1)}U(x_r)\right) \\
&+ \bar{c}\left(\sum_{r=1}^{N}\chi_{pr}^{(2)}D_{11}(x_r)\right)\left(\sum_{r=1}^{N}\chi_{pr}^{(1)}U(x_r)\right) + 2\bar{c}\left(\sum_{r=1}^{N}\chi_{pr}^{(1)}D_{11}(x_r)\right)\left(\sum_{r=1}^{N}\chi_{pr}^{(2)}U(x_r)\right) \\
&+ \bar{c}D_{11}(x_p)\left(\sum_{r=1}^{N}\chi_{pr}^{(3)}U(x_r)\right) - \frac{\bar{c}nV(x_p)}{R}\left(\sum_{r=1}^{N}\chi_{pr}^{(2)}D_{12}(x_r)\right) \\
&- \frac{2\bar{c}n}{R}\left(\sum_{r=1}^{N}\chi_{pr}^{(1)}D_{12}(x_r)\right)\left(\sum_{r=1}^{N}\chi_{pr}^{(1)}V(x_r)\right) - \frac{\bar{c}nD_{12}(x_p)}{R}\left(\sum_{r=1}^{N}\chi_{pr}^{(2)}V(x_r)\right) \\
&- \frac{2\bar{c}n}{R}\left(\sum_{r=1}^{N}\chi_{pr}^{(1)}D_{66}(x_r)\right)\left(\sum_{r=1}^{N}\chi_{pr}^{(1)}V(x_r)\right) - \frac{2\bar{c}nD_{66}(x_p)}{R}\left(\sum_{r=1}^{N}\chi_{pr}^{(2)}V(x_r)\right) \\
&+ \frac{\bar{c}n^3 D_{11}(x_p)V(x_p)}{R^3} + \frac{nA_{11}(x_p)V(x_p)}{R^2} + \frac{2\bar{c}^2 n^2}{R^2}\left(\sum_{r=1}^{N}\chi_{pr}^{(1)}G_{12}(x_r)\right)\left(\sum_{r=1}^{N}\chi_{pr}^{(1)}W(x_r)\right)
\end{aligned}
$$

$$
\begin{aligned}
&+ \frac{4\bar{c}^2 n^2}{R^2}\left(\sum_{r=1}^{N}\chi_{pr}^{(1)}G_{66}(x_r)\right)\left(\sum_{r=1}^{N}\chi_{pr}^{(1)}W(x_r)\right) - \frac{\bar{c}^2 n^4 G_{11}(x_p)W(x_p)}{R^4}\\
&+ \frac{\bar{c}^2 n^2 W(x_p)}{R^2}\left(\sum_{r=1}^{N}\chi_{pr}^{(2)}G_{12}(x_r)\right) + \frac{2\bar{c}^2 n^2 G_{12}(x_p)}{R^2}\left(\sum_{r=1}^{N}\chi_{pr}^{(2)}W(x_r)\right) - \frac{9\bar{c}^2 n^2 F_{66}(x_p)W(x_p)}{R^2}\\
&+ \frac{4\bar{c}^2 n^2 G_{66}(x_p)}{R^2}\left(\sum_{r=1}^{N}\chi_{pr}^{(2)}W(x_r)\right) + \frac{6\bar{c}n^2 C_{66}(x_p)W(x_p)}{R^2} - \frac{2\bar{c}n^2 D_{11}(x_p)W(x_p)}{R^3}\\
&- \frac{n^2 A_{66}(x_p)W(x_p)}{R^2} + \frac{\bar{c}W(x_p)}{R}\left(\sum_{r=1}^{N}\chi_{pr}^{(2)}D_{12}(x_r)\right) + \frac{2\bar{c}}{R}\left(\sum_{r=1}^{N}\chi_{pr}^{(1)}D_{12}(x_r)\right)\left(\sum_{r=1}^{N}\chi_{pr}^{(1)}W(x_r)\right)\\
&+ \frac{2\bar{c}D_{12}(x_p)}{R}\left(\sum_{r=1}^{N}\chi_{pr}^{(2)}W(x_r)\right) + A_{66}(x_p)\left(\sum_{r=1}^{N}\chi_{pr}^{(2)}W(x_r)\right) + \left(\sum_{r=1}^{N}\chi_{pr}^{(1)}A_{66}(x_r)\right)\left(\sum_{r=1}^{N}\chi_{pr}^{(1)}W(x_r)\right)\\
&- 2\bar{c}^2\left(\sum_{r=1}^{N}\chi_{pr}^{(1)}G_{11}(x_r)\right)\left(\sum_{r=1}^{N}\chi_{pr}^{(3)}W(x_r)\right) - c^2 G_{11}(x_p)\left(\sum_{r=1}^{N}\chi_{pr}^{(4)}W(x_r)\right) - \frac{A_{11}(x_p)W(x_p)}{R^2}\\
&- 6\bar{c}\left(\sum_{r=1}^{N}\chi_{pr}^{(1)}C_{66}(x_r)\right)\left(\sum_{r=1}^{N}\chi_{pr}^{(1)}W(x_r)\right) - 6\bar{c}C_{66}(x_p)\left(\sum_{r=1}^{N}\chi_{pr}^{(2)}W(x_r)\right)\\
&+ \frac{2\bar{c}^2 n^2 G_{66}(x_p)}{R^2}\left(\sum_{r=1}^{N}\chi_{pr}^{(1)}\Phi_x(x_r)\right) + \frac{2\bar{c}^2 n^2 \Phi_x(x_p)}{R^2}\left(\sum_{r=1}^{N}\chi_{pr}^{(1)}G_{66}(x_r)\right)\\
&- \frac{\bar{c}n^2 F_{12}(x_p)}{R^2}\left(\sum_{r=1}^{N}\chi_{pr}^{(1)}\Phi_x(x_r)\right) - \frac{2\bar{c}n^2 \Phi_x(x_p)}{R^2}\left(\sum_{r=1}^{N}\chi_{pr}^{(1)}F_{66}(x_r)\right)\\
&- \frac{2\bar{c}n^2 F_{66}(x_p)}{R^2}\left(\sum_{r=1}^{N}\chi_{pr}^{(1)}\Phi_x(x_r)\right) + \frac{\bar{c}^2 n^2 G_{12}(x_p)}{R^2}\left(\sum_{r=1}^{N}\chi_{pr}^{(1)}\Phi_x(x_r)\right)\\
&+ \frac{\bar{c}D_{12}(x_p)}{R}\left(\sum_{r=1}^{N}\chi_{pr}^{(1)}\Phi_x(x_p)\right) + \left(\sum_{r=1}^{N}\chi_{pr}^{(1)}A_{66}(x_r)\right)\Phi_x(x_p) + A_{66}(x_p)\left(\sum_{r=1}^{N}\chi_{pr}^{(1)}\Phi_x(x_r)\right)\\
&- \bar{c}^2 G_{11}(x_p)\left(\sum_{r=1}^{N}\chi_{pr}^{(3)}\Phi_x(x_r)\right) - \frac{B_{12}(x_p)}{R}\left(\sum_{r=1}^{N}\chi_{pr}^{(1)}\Phi_x(x_r)\right) - 6\bar{c}\left(\sum_{r=1}^{N}\chi_{pr}^{(1)}C_{66}(x_r)\right)\Phi_x(x_p)\\
&- 6\bar{c}C_{66}(x_p)\left(\sum_{r=1}^{N}\chi_{pr}^{(1)}\Phi_x(x_r)\right) + 9\bar{c}^2\left(\sum_{r=1}^{N}\chi_{pr}^{(1)}F_{66}(x_r)\right)\Phi_x(x_p) \quad\quad\quad (A9)\\
&+ 9\bar{c}^2 F_{66}(x_p)\left(\sum_{r=1}^{N}\chi_{pr}^{(1)}\Phi_x(x_r)\right) + \bar{c}F_{11}(x)\left(\sum_{r=1}^{N}\chi_{pr}^{(3)}\Phi_x(x_r)\right)\\
&+ \bar{c}\left(\sum_{r=1}^{N}\chi_{pr}^{(2)}F_{11}(x_r)\right)\left(\sum_{r=1}^{N}\chi_{pr}^{(1)}\Phi_x(x_r)\right) + 2\bar{c}\left(\sum_{r=1}^{N}\chi_{pr}^{(1)}F_{11}(x_r)\right)\left(\sum_{r=1}^{N}\chi_{pr}^{(2)}\Phi_x(x_r)\right)\\
&- \bar{c}^2\left(\sum_{r=1}^{N}\chi_{pr}^{(2)}G_{11}(x_r)\right)\left(\sum_{r=1}^{N}\chi_{pr}^{(1)}\Phi_x(x_r)\right) - 2\bar{c}^2\left(\sum_{r=1}^{N}\chi_{pr}^{(1)}G_{11}(x_r)\right)\left(\sum_{r=1}^{N}\chi_{pr}^{(2)}\Phi_x(x_r)\right)\\
&+ \frac{2\bar{c}^2 n}{R}\left(\sum_{r=1}^{N}\chi_{pr}^{(1)}G_{12}(x_r)\right)\left(\sum_{r=1}^{N}\chi_{pr}^{(1)}\Phi_y(x_r)\right) - \frac{\bar{c}^2 n^3 G_{11}(x_p)\Phi_y(x_p)}{R^3}\\
&- \frac{9\bar{c}^2 n F_{66}(x_p)\Phi_y(x_p)}{R} + \frac{2\bar{c}^2 n}{R}\left(\sum_{r=1}^{N}\chi_{pr}^{(1)}G_{66}(x_r)\right)\left(\sum_{r=1}^{N}\chi_{pr}^{(1)}\Phi_y(x_r)\right)\\
&+ \frac{\bar{c}^2 n G_{12}(x_p)}{R}\left(\sum_{r=1}^{N}\chi_{pr}^{(2)}\Phi_y(x_r)\right) + \frac{\bar{c}n^3 F_{11}(x_p)\Phi_y(x_p)}{R^3}\\
&- \frac{\bar{c}n D_{11}(x_p)\Phi_y(x_p)}{R^2} + \frac{2\bar{c}^2 n G_{66}(x_p)}{R}\left(\sum_{r=1}^{N}\chi_{pr}^{(2)}\Phi_y(x_r)\right)\\
&+ \frac{6\bar{c}n C_{66}(x_p)\Phi_y(x_p)}{R} - \frac{2\bar{c}n}{R}\left(\sum_{r=1}^{N}\chi_{pr}^{(1)}F_{66}(x_r)\right)\left(\sum_{r=1}^{N}\chi_{pr}^{(1)}\Phi_y(x_r)\right)\\
&- \frac{2\bar{c}n F_{66}(x_p)}{R}\left(\sum_{r=1}^{N}\chi_{pr}^{(2)}\Phi_y(x_r)\right) - \frac{\bar{c}n \Phi_y(x_p)}{R}\left(\sum_{r=1}^{N}\chi_{pr}^{(2)}F_{12}(x_r)\right)\\
&- \frac{2\bar{c}n}{R}\left(\sum_{r=1}^{N}\chi_{pr}^{(1)}F_{12}(x_r)\right)\left(\sum_{r=1}^{N}\chi_{pr}^{(1)}\Phi_y(x_r)\right) - \frac{\bar{c}n F_{12}(x_p)}{R}\left(\sum_{r=1}^{N}\chi_{pr}^{(2)}\Phi_y(x_r)\right)\\
&+ \frac{\bar{c}^2 n \Phi_y(x_p)}{R}\left(\sum_{r=1}^{N}\chi_{pr}^{(2)}G_{12}(x_r)\right) + \frac{n B_{11}(x_p)\Phi_y(x_p)}{R^2} - \frac{n A_{66}(x_p)\Phi_y(x_p)}{R} - k_w W(x_p)\\
&+ k_g\left(\sum_{r=1}^{N}\chi_{pr}^{(2)}W(x_r) - \frac{n^2}{R^2}W(x_p)\right) - F_a \sum_{r=1}^{N}\chi_{pr}^{(2)}W(x_r) = \frac{1}{\overline{W}(t)}\left(I_0(x_p)W(x_p)\frac{d^2\overline{W}(t)}{dt^2}\right)
\end{aligned}
$$

$$
\begin{aligned}
&-\bar{c}^2 I_6(x_p)\left(\sum_{r=1}^{N} \chi_{pr}^{(2)} W(x_r)\frac{d^2\overline{W}(t)}{dt^2} - \frac{n^2}{R^2}W(x_p)\frac{d^2\overline{W}(t)}{dt^2}\right) + \bar{c}I_3(x_p)\left(\sum_{r=1}^{N}\chi_{pr}^{(1)}U(x_r)\frac{d^2\overline{U}(t)}{dt^2}\right.\\
&\left. - \frac{n}{R}V(x_p)\frac{d^2\overline{V}(t)}{dt^2}\right) + \bar{c}\big(I_4(x_p) - \bar{c}I_6(x_p)\big)\left(\sum_{r=1}^{N}\chi_{pr}^{(1)}\Phi_x(x_r)\frac{d^2\overline{\Phi}_x(t)}{dt^2}\right.\\
&\left. - \frac{n}{R}\Phi_y(x_p)\frac{d^2\overline{\Phi}_y(t)}{dt^2}\right)
\end{aligned}
$$

$$
\begin{aligned}
&-\frac{n^2 B_{66}(x_p)U(x_p)}{R^2} + \frac{\bar{c}n^2 D_{66}(x_p)U(x_p)}{R^2} + \left(\sum_{r=1}^{N}\chi_{pr}^{(1)}B_{11}(x_r)\right)\left(\sum_{r=1}^{N}\chi_{pr}^{(1)}U(x_r)\right)\\
&+B_{11}(x_p)\left(\sum_{r=1}^{N}\chi_{pr}^{(2)}U(x_r)\right) - \bar{c}\left(\sum_{r=1}^{N}\chi_{pr}^{(1)}D_{11}(x_r)\right)\left(\sum_{r=1}^{N}\chi_{pr}^{(1)}U(x_r)\right) - \bar{c}D_{11}(x_p)\left(\sum_{r=1}^{N}\chi_{pr}^{(2)}U(x_r)\right)\\
&-\frac{nB_{66}(x_p)}{R}\left(\sum_{r=1}^{N}\chi_{pr}^{(1)}V(x_r)\right) - \frac{nV(x_p)}{R}\left(\sum_{r=1}^{N}\chi_{pr}^{(1)}B_{12}(x_r)\right) - \frac{nB_{12}(x_p)}{R}\left(\sum_{r=1}^{N}\chi_{pr}^{(1)}V(x_r)\right)\\
&+\frac{\bar{c}nD_{12}(x_p)}{R}\left(\sum_{r=1}^{N}\chi_{pr}^{(1)}V(x_r)\right) + \frac{\bar{c}nD_{66}(x_p)}{R}\left(\sum_{r=1}^{N}\chi_{pr}^{(1)}V(x_r)\right) + \frac{\bar{c}nV(x_p)}{R}\left(\sum_{r=1}^{N}\chi_{pr}^{(1)}D_{12}(x_r)\right)\\
&-\frac{\bar{c}W(x_p)}{R}\left(\sum_{r=1}^{N}\chi_{pr}^{(1)}D_{12}(x_r)\right) - \frac{\bar{c}D_{12}(x_p)}{R}\left(\sum_{r=1}^{N}\chi_{pr}^{(1)}W(x_r)\right) - \frac{c^2 n^2 G_{12}(x_p)}{R^2}\left(\sum_{r=1}^{N}\chi_{pr}^{(1)}W(x_r)\right)\\
&+\frac{cn^2 F_{12}(x_p)}{R^2}\left(\sum_{r=1}^{N}\chi_{pr}^{(1)}W(x_r)\right) + \frac{\bar{c}n^2 W(x_p)}{R^2}\left(\sum_{r=1}^{N}\chi_{pr}^{(1)}F_{12}(x_r)\right)\\
&-\frac{2\bar{c}^2 n^2 G_{66}(x_p)}{R^2}\left(\sum_{r=1}^{N}\chi_{pr}^{(1)}W(x_r)\right) + \frac{2\bar{c}n^2 F_{66}(x_p)}{R^2}\left(\sum_{r=1}^{N}\chi_{pr}^{(1)}W(x_r)\right)\\
&-\frac{c^2 n^2 W(x_p)}{R^2}\left(\sum_{r=1}^{N}\chi_{pr}^{(1)}G_{12}(x_r)\right) - A_{66}(x_p)\left(\sum_{r=1}^{N}\chi_{pr}^{(1)}W(x_r)\right) - 9\bar{c}^2 F_{66}(x_p)\left(\sum_{r=1}^{N}\chi_{pr}^{(1)}W(x_r)\right)\\
&+\bar{c}^2\left(\sum_{r=1}^{N}\chi_{pr}^{(1)}G_{11}(x_r)\right)\left(\sum_{r=1}^{N}\chi_{pr}^{(2)}W(x_r)\right) + \bar{c}^2 G_{11}(x_p)\left(\sum_{r=1}^{N}\chi_{pr}^{(3)}W(x_r)\right)\\
&+\frac{W(x_p)}{R}\left(\sum_{r=1}^{N}\chi_{pr}^{(1)}B_{12}(x_r)\right) + \frac{B_{12}(x_p)}{R}\left(\sum_{r=1}^{N}\chi_{pr}^{(1)}W(x_r)\right) - \bar{c}\left(\sum_{r=1}^{N}\chi_{pr}^{(1)}F_{11}(x_r)\right)\left(\sum_{r=1}^{N}\chi_{pr}^{(2)}W(x_r)\right)\\
&-\bar{c}F_{11}(x_p)\left(\sum_{r=1}^{N}\chi_{pr}^{(3)}W(x_r)\right) + 6\bar{c}C_{66}(x_p)\left(\sum_{r=1}^{N}\chi_{pr}^{(1)}W(x_r)\right)\\
&-\frac{n^2 C_{66}(x_p)\Phi_x(x_p)}{R^2} + \frac{2\bar{c}n^2 F_{66}(x_p)\Phi_x(x_p)}{R^2} - \frac{\bar{c}^2 n^2 G_{66}(x_p)\Phi_x(x_p)}{R^2}\\
&+\left(\sum_{r=1}^{N}\chi_{pr}^{(1)}C_{11}(x_r)\right)\left(\sum_{r=1}^{N}\chi_{pr}^{(1)}\Phi_x(x_r)\right) + C_{11}(x_p)\left(\sum_{r=1}^{N}\chi_{pr}^{(2)}\Phi_x(x_r)\right) - A_{66}(x_p)\Phi_x(x_p)\\
&-9\bar{c}^2 F_{66}(x_p)\Phi_x(x_p) + \bar{c}^2\left(\sum_{r=1}^{N}\chi_{pr}^{(1)}G_{11}(x_r)\right)\left(\sum_{r=1}^{N}\chi_{pr}^{(1)}\Phi_x(x_r)\right)\\
&+\bar{c}^2 G_{11}(x_p)\left(\sum_{r=1}^{N}\chi_{pr}^{(2)}\Phi_x(x_r)\right) - 2\bar{c}\left(\sum_{r=1}^{N}\chi_{pr}^{(1)}F_{11}(x_r)\right)\left(\sum_{r=1}^{N}\chi_{pr}^{(1)}\Phi_x(x_r)\right)\\
&-2\bar{c}F_{11}(x_p)\left(\sum_{r=1}^{N}\chi_{pr}^{(2)}\Phi_x(x_r)\right) + 6\bar{c}C_{66}(x_p)\Phi_x(x_p) - \frac{nC_{66}(x_p)}{R}\left(\sum_{r=1}^{N}\chi_{pr}^{(1)}\Phi_y(x_r)\right)\\
&-\frac{n\Phi_y(x_p)}{R}\left(\sum_{r=1}^{N}\chi_{pr}^{(1)}C_{12}(x_r)\right) - \frac{nC_{12}(x_p)}{R}\left(\sum_{r=1}^{N}\chi_{pr}^{(1)}\Phi_y(x_r)\right)\\
&-\frac{\bar{c}^2 n G_{66}(x_p)}{a}\left(\sum_{r=1}^{N}\chi_{pr}^{(1)}\Phi_y(x_r)\right) + \frac{2\bar{c}nF_{66}(x_p)}{R}\left(\sum_{r=1}^{N}\chi_{pr}^{(1)}\Phi_y(x_r)\right)\\
&+\frac{2\bar{c}nF_{12}(x_p)}{R}\left(\sum_{r=1}^{N}\chi_{pr}^{(1)}\Phi_y(x_r)\right) - \frac{\bar{c}^2 n\Phi_y(x_p)}{R}\left(\sum_{r=1}^{N}\chi_{pr}^{(1)}G_{12}(x_r)\right)\\
&+\frac{2\bar{c}n\Phi_y(x_p)}{R}\left(\sum_{r=1}^{N}\chi_{pr}^{(1)}F_{12}(x_r)\right) - \frac{\bar{c}^2 nG_{12}(x_p)}{R}\left(\sum_{r=1}^{N}\chi_{pr}^{(1)}\Phi_y(x_r)\right)\\
&= \frac{1}{\overline{\Phi}_x(t)}\left(\big(I_0(x_p) - \bar{c}I_3(x_p)\big)U(x_p)\frac{d^2\overline{U}(t)}{dt^2} + \big(I_2(x_p) - 2\bar{c}I_4(x_p)\right.\\
&\left.+\bar{c}^2 I_6(x_p)\big)\Phi_x(x_p)\frac{d^2\overline{\Phi}_x(t)}{dt^2} - \bar{c}\big(I_4(x_p) - \bar{c}I_6(x_p)\big)\sum_{r=1}^{N}\chi_{pr}^{(1)}W(x_r)\frac{d^2\overline{W}(t)}{dt^2}\right),
\end{aligned}
\qquad\text{(A10)}
$$

$$\begin{aligned}
&\frac{nB_{12}(x_p)}{R}\left(\sum_{r=1}^{N}\chi_{pr}^{(1)}U(x_r)\right)+\frac{nU(x_p)}{R}\left(\sum_{r=1}^{N}\chi_{pr}^{(1)}B_{66}(x_r)\right)+\frac{nB_{66}(x_p)}{R}\left(\sum_{r=1}^{N}\chi_{pr}^{(1)}U(x_r)\right)\\
&-\frac{\bar{c}nD_{12}(x_p)}{R}\left(\sum_{r=1}^{N}\chi_{pr}^{(1)}U(x_r)\right)-\frac{\bar{c}nU(x_p)}{R}\left(\sum_{r=1}^{N}\chi_{pr}^{(1)}D_{66}(x_r)\right)-\frac{\bar{c}nD_{66}(x_p)}{R}\left(\sum_{r=1}^{N}\chi_{pr}^{(1)}U(x_r)\right)\\
&+\left(\sum_{r=1}^{N}\chi_{pr}^{(1)}B_{66}(x_r)\right)\left(\sum_{r=1}^{N}\chi_{pr}^{(1)}V(x_r)\right)+B_{66}(x_p)\left(\sum_{r=1}^{N}\chi_{pr}^{(2)}V(x_r)\right)-\frac{n^2B_{11}(x_p)V(x_p)}{R^2}\\
&+\frac{\bar{c}n^2D_{11}(x_p)V(x_p)}{R^2}-\bar{c}\left(\sum_{r=1}^{N}\chi_{pr}^{(1)}D_{66}(x_r)\right)\left(\sum_{r=1}^{N}\chi_{pr}^{(1)}V(x_r)\right)-\bar{c}D_{66}(x_p)\left(\sum_{r=1}^{N}\chi_{pr}^{(2)}V(x_r)\right)\\
&+\frac{nB_{11}(x_p)W(x_p)}{R^2}-\frac{nA_{66}(x_p)W(x_p)}{R}-\frac{\bar{c}nF_{12}(x_p)}{R}\left(\sum_{r=1}^{N}\chi_{pr}^{(2)}W(x_r)\right)-\frac{9\bar{c}^2nF_{66}(x_p)W(x_p)}{R}\\
&-\frac{2\bar{c}nF_{66}(x_p)}{R}\left(\sum_{r=1}^{N}\chi_{pr}^{(2)}W(x_r)\right)+\frac{\bar{c}n^3F_{11}(x_p)W(x_p)}{R^3}-\frac{\bar{c}^2n^3G_{11}(x_p)W(x_p)}{R^3}\\
&-\frac{2\bar{c}n}{R}\left(\sum_{r=1}^{N}\chi_{pr}^{(1)}F_{66}(x_r)\right)\left(\sum_{r=1}^{N}\chi_{pr}^{(1)}W(x_r)\right)+\frac{2\bar{c}^2nG_{66}(x_p)}{R}\left(\sum_{r=1}^{N}\chi_{pr}^{(2)}W(x_r)\right)\\
&+\frac{6\bar{c}nC_{66}(x_p)W(x_p)}{R}+\frac{2\bar{c}^2n}{a}\left(\sum_{r=1}^{N}\chi_{pr}^{(1)}G_{66}(x_r)\right)\left(\sum_{r=1}^{N}\chi_{pr}^{(1)}W(x_r)\right)\\
&+\frac{\bar{c}^2nG_{12}(x_p)}{R}\left(\sum_{r=1}^{N}\chi_{pr}^{(2)}W(x_r)\right)-\frac{\bar{c}nD_{11}(x_p)W(x_p)}{R^2}+\frac{nC_{12}(x_p)}{R}\left(\sum_{r=1}^{N}\chi_{pr}^{(1)}\Phi_x(x_r)\right)\\
&+\frac{n\Phi_x(x_p)}{R}\left(\sum_{r=1}^{N}\chi_{pr}^{(1)}C_{66}(x_r)\right)+\frac{nC_{66}(x_p)}{R}\left(\sum_{r=1}^{N}\chi_{pr}^{(1)}\Phi_x(x_r)\right)\\
&+\frac{\bar{c}^2n\Phi_x(x_p)}{R}\left(\sum_{r=1}^{N}\chi_{pr}^{(1)}G_{66}(x_r)\right)-\frac{2\bar{c}nF_{12}(x_p)}{R}\left(\sum_{r=1}^{N}\chi_{pr}^{(1)}\Phi_x(x_r)\right)\\
&-\frac{2\bar{c}nF_{66}(x_p)}{R}\left(\sum_{r=1}^{N}\chi_{pr}^{(1)}\Phi_x(x_r)\right)+\frac{\bar{c}^2nG_{66}(x_p)}{R}\left(\sum_{r=1}^{N}\chi_{pr}^{(1)}\Phi_x(x_r)\right)\\
&+\frac{\bar{c}^2nG_{12}(x_p)}{R}\left(\sum_{r=1}^{N}\chi_{pr}^{(1)}\Phi_x(x_r)\right)-\frac{2\bar{c}n\Phi_x(x_p)}{R}\left(\sum_{r=1}^{N}\chi_{pr}^{(1)}F_{66}(x_r)\right)\\
&+\left(\sum_{r=1}^{N}\chi_{pr}^{(1)}C_{66}(x_r)\right)\left(\sum_{r=1}^{N}\chi_{pr}^{(1)}\Phi_y(x_r)\right)+C_{66}(x_p)\left(\sum_{r=1}^{N}\chi_{pr}^{(2)}\Phi_y(x_r)\right)\\
&-A_{66}(x_p)\Phi_y(x_p)-\frac{n^2C_{11}(x_p)\Phi_y(x_p)}{R^2}+\frac{2\bar{c}n^2F_{11}(x_p)\Phi_y(x_p)}{R^2}\\
&-\frac{\bar{c}^2n^2G_{11}(x_p)\Phi_y(x_p)}{R^2}-2\bar{c}\left(\sum_{r=1}^{N}\chi_{pr}^{(1)}F_{66}(x_r)\right)\left(\sum_{r=1}^{N}\chi_{pr}^{(1)}\Phi_y(x_r)\right)\\
&-2\bar{c}F_{66}(x_p)\left(\sum_{r=1}^{N}\chi_{pr}^{(2)}\Phi_y(x_r)\right)+6\bar{c}C_{66}(x_p)\Phi_y(x_p)-9\bar{c}^2F_{66}(x_p)\Phi_y(x_p)\\
&+\bar{c}^2\left(\sum_{r=1}^{N}\chi_{pr}^{(1)}G_{66}(x_r)\right)\left(\sum_{r=1}^{N}\chi_{pr}^{(1)}\Phi_y(x_r)\right)+\bar{c}^2G_{66}(x_p)\left(\sum_{r=1}^{N}\chi_{pr}^{(2)}\Phi_y(x_r)\right)\\
&=\frac{1}{\Phi_y(t)}\left(\left(I_0(x_p)-\bar{c}I_3(x_p)\right)V(x_p)\frac{d^2\overline{V}(t)}{dt^2}+\left(I_2(x_p)-2\bar{c}I_4(x_p)+\bar{c}^2I_6(x_p)\right)\Phi_y(x_p)\frac{d^2\overline{\Phi}_y(t)}{dt^2}\right.\\
&\left.-\bar{c}\left(I_4(x_p)-\bar{c}I_6(x_p)\right)\frac{n}{R}\frac{d^2\overline{W}(t)}{dt^2}\right),
\end{aligned}$$ (A11)

References

1. Birman, V.; Byrd, L.W. Modeling and analysis of functionally graded materials and structures. *Appl. Mech. Rev.* **2007**, *60*, 195–216. [CrossRef]
2. Miyamoto, Y.; Kaysser, W.A.; Rabin, B.H.; Kawasaki, A.; Ford, R.G. *Functionally Graded Materials: Design, Processing and Applications*; Springer Science & Business Medi: New York, NY, USA, 2013.
3. Noda, N. Thermal stresses in functionally graded materials. *J. Therm. Stress.* **1999**, *22*, 477–512. [CrossRef]
4. Shen, H.S. *Functionally Graded Materials: Nonlinear Analysis of Plates and Shells*; CRC Press: Boca Raton, FL, USA, 2016.
5. Du, C.; Li, Y.; Jin, X. Nonlinear forced vibration of functionally graded cylindrical thin shells. *Thin-Walled Struct.* **2014**, *78*, 26–36. [CrossRef]

6. Rahimi, G.H.; Ansari, R.; Hemmatnezhad, M. Vibration of functionally graded cylindrical shells with ring support. *Sci. Iran.* **2011**, *18*, 1313–1320. [CrossRef]
7. Ghasemi, A.R.; Mohandes, M.; Dimitri, R.; Tornabene, F. Agglomeration effects on the vibrations of CNTs/fiber/polymer/metal hybrid laminates cylindrical shell. *Compos. Part B Eng.* **2019**, *167*, 700–716. [CrossRef]
8. Bich, D.H.; van Dung, D.; Nam, V.H.; Phuong, N.T. Nonlinear static and dynamic buckling analysis of imperfect eccentrically stiffened functionally graded circular cylindrical thin shells under axial compression. *Int. J. Mech. Sci.* **2013**, *74*, 190–200. [CrossRef]
9. Beni, Y.T.; Mehralian, F.; Zeighampour, H. The modified couple stress functionally graded cylindrical thin shell formulation. *Mech. Adv. Mater. Struct.* **2016**, *23*, 791–801. [CrossRef]
10. da Silva, F.M.A.; Montes, R.O.P.; Goncalves, P.B.; del Prado, Z.J.G.N. Nonlinear vibrations of fluid-filled functionally graded cylindrical shell considering a time-dependent lateral load and static preload. *Proc. Inst. Mech. Eng. Part C J. Mech. Eng. Sci.* **2016**, *230*, 102–119. [CrossRef]
11. Bich, D.H.; Nguyen, N.X. Nonlinear vibration of functionally graded circular cylindrical shells based on improved Donnell equations. *J. Sound Vib.* **2012**, *331*, 5488–5501. [CrossRef]
12. Ghannad, M.; Rahimi, G.H.; Nejad, M.Z. Elastic analysis of pressurized thick cylindrical shells with variable thickness made of functionally graded materials. *Compos. Part B Eng.* **2013**, *45*, 388–396. [CrossRef]
13. Jafari, A.A.; Khalili, S.M.R.; Tavakolian, M. Nonlinear vibration of functionally graded cylindrical shells embedded with a piezoelectric layer. *Thin-Walled Struct.* **2014**, *79*, 8–15. [CrossRef]
14. Jin, G.; Xie, X.; Liu, Z. The Haar wavelet method for free vibration analysis of functionally graded cylindrical shells based on the shear deformation theory. *Compos. Struct.* **2014**, *108*, 435–448. [CrossRef]
15. Liu, Y.Z.; Hao, Y.X.; Zhang, W.; Chen, J.; Li, S.B. Nonlinear dynamics of initially imperfect functionally graded circular cylindrical shell under complex loads. *J. Sound Vib.* **2015**, *348*, 294–328. [CrossRef]
16. Mehralian, F.; Beni, Y.T. Size-dependent torsional buckling analysis of functionally graded cylindrical shell. *Compos. Part B Eng.* **2016**, *94*, 11–25. [CrossRef]
17. Sheng, G.G.; Wang, X. Nonlinear response of fluid-conveying functionally graded cylindrical shells subjected to mechanical and thermal loading conditions. *Compos. Struct.* **2017**, *168*, 675–684. [CrossRef]
18. Sheng, G.G.; Wang, X. The dynamic stability and nonlinear vibration analysis of stiffened functionally graded cylindrical shells. *Appl. Math. Model.* **2018**, *56*, 389–403. [CrossRef]
19. Zhang, Y.; Huang, H.; Han, Q. Buckling of elastoplastic functionally graded cylindrical shells under combined compression and pressure. *Compos. Part B Eng.* **2015**, *69*, 120–126. [CrossRef]
20. Sofiyev, A.H.; Hui, D. On the vibration and stability of FGM cylindrical shells under external pressures with mixed boundary conditions by using FOSDT. *Thin-Walled Struct.* **2019**, *134*, 419–427. [CrossRef]
21. Sun, J.; Xu, X.; Lim, C.W.; Qiao, W. Accurate buckling analysis for shear deformable FGM cylindrical shells under axial compression and thermal loads. *Compos. Struct.* **2015**, *123*, 246–256. [CrossRef]
22. Huang, H.; Han, Q.; Wei, D. Buckling of FGM cylindrical shells subjected to pure bending load. *Compos. Struct.* **2011**, *93*, 2945–2952. [CrossRef]
23. Wali, M.; Hentati, T.; Jarraya, A.; Dammak, F. Free vibration analysis of FGM shell structures with a discrete double directors shell element. *Compos. Struct.* **2015**, *125*, 295–303. [CrossRef]
24. Mohammadi, M.; Arefi, M.; Dimitri, R.; Tornabene, F. Higher-Order Thermo-Elastic Analysis of FG-CNTRC Cylindrical Vessels Surrounded by a Pasternak Foundation. *Nanomaterials* **2019**, *9*, 79. [CrossRef] [PubMed]
25. Tornabene, F.; Brischetto, S.; Fantuzzi, F.; Viola, E. Numerical and exact models for free vibration analysis of cylindrical and spherical shell panels. *Compos. Part B Eng.* **2015**, *3675*, 231–250. [CrossRef]
26. Arefi, M.; Mohammadi, M.; Tabatabaeian, A.; Dimitri, R.; Tornabene, F. Two-dimensional thermo-elastic analysis of FG-CNTRC cylindrical pressure vessels. *Steel Compos. Struct.* **2018**, *27*, 525–536.
27. Nejati, M.; Dimitri, R.; Tornabene, F.; Yas, M.H. Thermal buckling of nanocomposite stiffened cylindrical shells reinforced by functionally Graded wavy Carbon NanoTubes with temperature-dependent properties. *Appl. Sci.* **2017**, *7*, 1223. [CrossRef]
28. Aragh, B.S.; Hedayati, H. Static response and free vibration of two-dimensional functionally graded metal/ceramic open cylindrical shells under various boundary conditions. *Acta Mech.* **2012**, *223*, 309–330. [CrossRef]
29. Zafarmand, H.; Hassani, B. Analysis of two-dimensional functionally graded rotating thick disks with variable thickness. *Acta Mech.* **2014**, *225*, 453–464. [CrossRef]

30. Ebrahimi, M.J.; Najafizadeh, M.M. Free vibration analysis of two-dimensional functionally graded cylindrical shells. *Appl. Math. Model.* **2014**, *38*, 308–324. [CrossRef]
31. Allahkarami, F.; Satouri, S.; Najafizadeh, M.M. Mechanical buckling of two-dimensional functionally graded cylindrical shells surrounded by Winkler–Pasternak elastic foundation. *Mech. Adv. Mater. Struct.* **2016**, *23*, 873–887. [CrossRef]
32. Satouri, S.; Kargarnovin, M.H.; Allahkarami, F.; Asanjarani, A. Application of third order shear deformation theory in buckling analysis of 2D-functionally graded cylindrical shell reinforced by axial stiffeners. *Compos. Part B Eng.* **2015**, *79*, 236–253. [CrossRef]
33. Li, L.; Hu, Y. Torsional vibration of bi-directional functionally graded nanotubes based on nonlocal elasticity theory. *Compos. Struct.* **2017**, *172*, 242–250. [CrossRef]
34. Kiran, M.C.; Kattimani, S.C. Assessment of porosity influence on vibration and static behaviour of functionally graded magneto-electro-elastic plate: A finite element study. *Eur. J. Mech.-A/Solids* **2018**, *71*, 258–277. [CrossRef]
35. Kiran, M.C.; Kattimani, S.C.; Vinyas, M. Porosity influence on structural behaviour of skew functionally graded magneto-electro-elastic plate. *Compos. Struct.* **2018**, *191*, 36–77. [CrossRef]
36. Barati, M.R.; Sadr, M.H.; Zenkour, A.M. Buckling analysis of higher order graded smart piezoelectric plates with porosities resting on elastic foundation. *Int. J. Mech. Sci.* **2016**, *117*, 309–320. [CrossRef]
37. Wang, Y.Q.; Wan, Y.H.; Zhang, Y.F. Vibrations of longitudinally traveling functionally graded material plates with porosities. *Eur. J. Mech.-A/Solids* **2017**, *66*, 55–68. [CrossRef]
38. Wang, Y.Q.; Zu, J.W. Vibration behaviors of functionally graded rectangular plates with porosities and moving in thermal environment. *Aerosp. Sci. Technol.* **2017**, *69*, 550–562. [CrossRef]
39. Wang, Y.; Wu, D. Free vibration of functionally graded porous cylindrical shell using a sinusoidal shear deformation theory. *Aerosp. Sci. Technol.* **2017**, *66*, 83–91. [CrossRef]
40. Kiani, Y.; Dimitri, R.; Tornabene, F. Free vibration of FG-CNT reinforced composite skew cylindrical shells using the Chebyshev-Ritz formulation. *Compos. Part B Eng.* **2018**, *147*, 169–177. [CrossRef]
41. Ghadiri, M.; SafarPour, H. Free vibration analysis of size-dependent functionally graded porous cylindrical microshells in thermal environment. *J. Therm. Stress.* **2017**, *40*, 55–71. [CrossRef]
42. Barati, M.R.; Zenkour, A.M. Vibration analysis of functionally graded graphene platelet reinforced cylindrical shells with different porosity distributions. *Mech. Adv. Mater. Struct.* **2019**, *26*, 1580–1588. [CrossRef]
43. Malikan, M.; Tornabene, F.; Dimitri, R. Nonlocal three-dimensional theory of elasticity for buckling behavior of functionally graded porous nanoplates using volume integrals. *Mater. Res. Express* **2018**, *5*, 095006. [CrossRef]
44. Malikan, M.; Tornabene, F.; Dimitri, R. Effect of sinusoidal corrugated geometries on the vibrational response of viscoelastic nanoplates. *Appl. Sci.* **2018**, *8*, 1432. [CrossRef]
45. Arefi, M.; Bidgoli, E.M.R.; Dimitri, R.; Tornabene, F. Free vibrations of functionally graded polymer composite nanoplates reinforced with graphene nanoplatelets. *Aerosp. Sci. Technol.* **2018**, *81*, 108–117. [CrossRef]
46. Jouneghani, F.Z.; Dimitri, R.; Tornabene, F. Structural response of porous FG nanobeams under hygro-thermo-mechanical loadings. *Compos. Part B Eng.* **2018**, *152*, 71–78. [CrossRef]
47. Arefi, M.; Bidgoli, E.M.R.; Dimitri, R.; Bacciocchi, M.; Tornabene, F. Nonlocal bending analysis of curved nanobeams reinforced by graphene nanoplatelets. *Compos. Part B Eng.* **2019**, *166*, 1–12. [CrossRef]
48. Bagherizadeh, E.; Kiani, Y.; Eslami, M.R. Mechanical buckling of functionally graded material cylindrical shells surrounded by Pasternak elastic foundation. *Compos. Struct.* **2011**, *93*, 3063–3071. [CrossRef]
49. Shu, C. *Differential Quadrature and Its Application in Engineering*; Springer Science & Business Media: New York, NY, USA, 2012.
50. Tornabene, F.; Fantuzzi, N.; Ubertini, F.; Viola, E. Strong formulation finite element method based on differential quadrature: A survey. *Appl. Mech. Rev.* **2015**, *67*, 1–55. [CrossRef]
51. Tornabene, F.; Fantuzzi, N.; Bacciocchi, M.; Dimitri, R. Free vibrations of composite oval and elliptic cylinders by the generalized differential quadrature method. *Thin-Walled Struct.* **2015**, *97*, 114–129. [CrossRef]
52. Yas, M.; Nejati, M.; Asanjarani, A. Free vibration analysis of continuously graded fiber reinforced truncated conical shell via third-order shear deformation theory. *J. Solid Mech.* **2016**, *8*, 212–231.
53. Kamarian, S.; Salim, M.; Dimitri, R.; Tornabene, F. Free vibration analysis conical shells reinforced with agglomerated Carbon Nanotubes. *Int. J. Mech. Sci.* **2016**, *108–109*, 157–165. [CrossRef]

54. Liu, G.R.; Wu, T.Y. In-plane vibration analyses of circular arches by the generalized differential quadrature rule. *Int. J. Mech. Sci.* **2001**, *43*, 2597–2611. [CrossRef]
55. Tornabene, F.; Dimitri, R. A numerical study of the seismic response of arched and vaulted structures made of isotropic or composite materials. *Eng. Struct.* **2018**, *159*, 332–366. [CrossRef]
56. Dimitri, R.; Tornabene, F.; Zavarise, G. Analytical and numerical modeling of the mixed-mode delamination process for composite moment-loaded double cantilever beams. *Compos. Struct.* **2018**, *187*, 535–553. [CrossRef]
57. Dimitri, R.; Tornabene, F. Numerical Study of the Mixed-Mode Delamination of Composite Specimens. *J. Compos. Sci.* **2018**, *2*, 30. [CrossRef]
58. Tomasiello, S. Differential quadrature method: Application to initial-boundary-value problems. *J. Sound Vib.* **1998**, *218*, 573–585. [CrossRef]
59. Tornabene, F.; Dimitri, R.; Viola, E. Transient dynamic response of generally-shaped arches based on a GDQ-time-stepping method. *Int. J. Mech. Sci.* **2016**, *114*, 277–314. [CrossRef]
60. Bolotin, V.V. The dynamic stability of elastic systems. *Am. J. Phys.* **1965**, *33*, 752–753. [CrossRef]
61. Khazaeinejad, P.; Najafizadeh, M.M. Mechanical buckling of cylindrical shells with varying material properties. *Proc. Inst. Mech. Eng. Part C J. Mech. Eng. Sci.* **2010**, *224*, 1551–1557. [CrossRef]

© 2020 by the authors. Licensee MDPI, Basel, Switzerland. This article is an open access article distributed under the terms and conditions of the Creative Commons Attribution (CC BY) license (http://creativecommons.org/licenses/by/4.0/).

Article

Dynamics Modeling and Theoretical Study of the Two-Axis Four-Gimbal Coarse–Fine Composite UAV Electro-Optical Pod

Cheng Shen, Shixun Fan *, Xianliang Jiang, Ruoyu Tan and Dapeng Fan

College of Intelligence Science and Technology, National University of Defense Technology, Changsha 410073, China; shensicheng1996@sina.cn (C.S.); jxl123gfkd@163.com (X.J.); tyirorie@hotmail.com (R.T.); fdp@nudt.edu.cn (D.F.)
* Correspondence: shixunfan@nudt.edu.cn; Tel.: +86-158-7428-5588

Received: 12 February 2020; Accepted: 7 March 2020; Published: 11 March 2020

Abstract: In the UAV electro-optical pod of the two-axis four-gimbal, the characteristics of a coarse–fine composite structure and the complexity of dynamics modeling affect the entire system's high precision control performance. The core goal of this paper is to solve the high precision control of a two-axis four-gimbal electro-optical pod through dynamic modeling and theoretical study. In response to this problem, we used finite element analysis (FEA) and stress study of the key component to design the structure. The gimbals adopt the aerospace material 7075-t3510 aluminum alloy in order to meet the requirements of an ultralight weight of less than 1 kg. According to the Euler rigid body dynamics model, the transmission path and kinematics coupling compensation matrix between the two-axis four-gimbal structures are obtained. The coarse–fine composite self-correction drive equation in the Cartesian system is derived to solve the pre-selection and check problem of the mechatronic under high-precision control. Finally, the modeling method is substituted into the disturbance observer (DOB) disturbance suppression experiment, which can monitor and compensate for the motion coupling between gimbal structures in real time. Results show that the disturbance suppression impact of the DOB method with dynamics model is increased by up to 90% compared to PID (Proportion Integration Differentiation method) and is 25% better than the traditional DOB method.

Keywords: two-axis four-gimbal; electro-optical pod; dynamics modeling; coarse–fine composite

1. Introduction

The UAV electro-optical pod system is widely used in ship-borne, vehicular, and airborne equipment and also plays a necessary role in recent information technology equipment [1]. It can accept the region target image information, accurately identify the target motion state, and guide decision making. Previous studies in the literature [2–6] used the PIOGRAM diagram method to explain the kinematics principle of the stable mechanism and pointed out the geometric coupling problem of a two-axis two-gimbal structure. Through special bearing and motor design, previous studies in the literature [7,8] constructed a two-axis two-gimbal stable tracking platform with large field-of-view visual axis. One study [9] applied the Euler dynamics theorem to establish the equation of the visual axis stabilization structure. However, the two-axis two-gimbal structure is suitable for a stable platform with low speed and low demand for stability precision, which may cause too much error or even self-locking when working under normal conditions. In the UAV two-axis four-gimbal electro-optical pod, the characteristics of a coarse–fine composite structure and the complexity of dynamics modeling affect the high precision control performance of the system. Therefore, it is necessary to adopt new dynamics modeling and theory to study the two-axis four-gimbal coarse–fine composite electro-optical pod for use in a UAV.

Appl. Sci. **2020**, *10*, 1923; doi:10.3390/app10061923 www.mdpi.com/journal/applsci

A previous study [10] adopted external prestressed steel, which is applied to concrete cylinder pipes. This derivation configures the prestress of steel strands to meet the requirements of ultimate limit states, serviceability limit states, and quasi-permanent limit states, considering the tensile strength of the concrete core and the mortar coating, respectively. Another study [11] presents a simplified mathematical model for the analysis of varying compliance vibrations of a rolling bearing. The results of the parametric analysis demonstrate that, with the proper choice of the size of the internal radial clearance and external radial load, the level of the varying compliance vibrations in a rolling bearing can be theoretically reduced to zero. In the literature [12,13], an aluminum conductor steel-reinforced cable and a racing tire are modeled to study their vibrations and finite element analysis. The above modeling methods are worth being referred to. However, these methods do not study the two-axis four-gimbal electro-optical pod for use in a UAV, and there is lack of experiments on dynamics modeling of a coarse–fine composite structure platform.

Another previous study [14] analyzed the equal-acceleration model of a two-axis four-gimbal maneuvering target. Taking the equivalent sinusoidal movement and the uniform linear movement as examples, the system was simulated. The results show that the precision of the coarse–fine composite control is higher than that of single-detector control, and the two-axis four-gimbal structure is simple and suitable for engineering implementation. Reference [15] presents the magnetic field analysis for the double layer Halbach array voice coil motor. The analytical model is built by adopting Fourier analysis and proves the feasibility of the analytic method with the equivalent structure. Reference [16] is an analysis and modeling the fast steering mirror. A detailed analysis was provided to show the proposed approach and improve disturbance suppression performance with only a slight weakening of the target tracking ability. The proposed feed-forward control was effectively verified through a series of comparative simulations and experiments. Besides, the method was applied in a real ship-based project. However, this dynamic modeling and the theoretical study of these methods are applicable to medium or large platforms and devices. It is of little significance to the design of an ultralight two-axis four-gimbal coarse–fine UAV electro-optical pod.

In this paper, the dynamics modeling and theoretical study of the two-axis four-gimbal coarse–fine composite UAV electro-optical pod is deeply analyzed. In response to this problem, finite element analysis (FEA) and theoretical analysis of the stress and deflection of the key structural component are used to design the structure. According to the Euler rigid body dynamics model, the transmission path and motion coupling compensation matrix between two-axis four-gimbal are obtained, and suitable aerospace materials were used for analysis. Finally, the simulation verifies the correctness of the model.

2. Structure Design

As shown in Figure 1, the two-axis four-gimbal coarse–fine composite structure can be simplified to a cantilever beam. Because the integrated shafting structure requires high precision, and there are deflection errors in actual processing and manufacturing, it is necessary to check the mechanical parameters of the uniaxial structure to observe whether it meets the performance requirements of the cantilever beam.

2.1. Bending Internal Force and Deflection

To better clarify the simplified model, the structure of the two-axis four-gimbal electro-optical pod structure is divided into five key components for analysis. As shown in Figures 1 and 2, S1 is the spherical cover, S2 is the outer pitch gimbal that is the core component of the simplified model of the cantilever beam, S3 is the fine-stage components (think of it as the load q in the middle of S2), S4 is the voice coil motor that outputs constant torque F, S5 is the end cover that is on the left side and the fixed end of the cantilever beam. What is more, because the rotation angle between the gimbals is relatively small, the torque change is ignored, and its maximum value is taken for analysis.

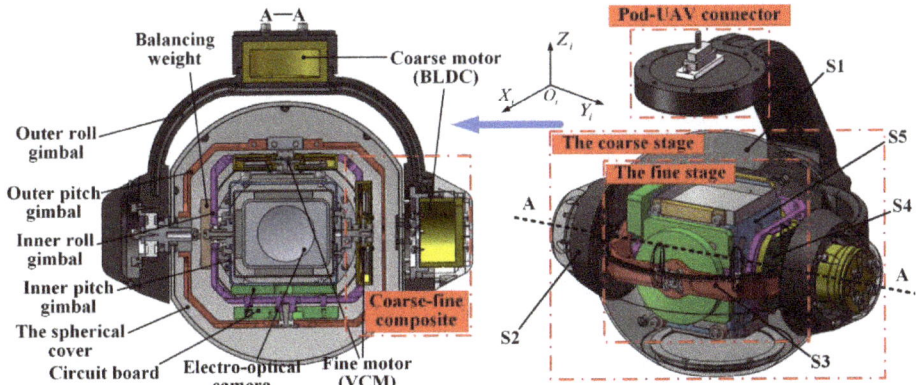

Figure 1. Ultralight two-axis four-gimbal electro-optical pod and coarse–fine composite system. BLDC: brushless direct current motor; VCM: voice coil motor.

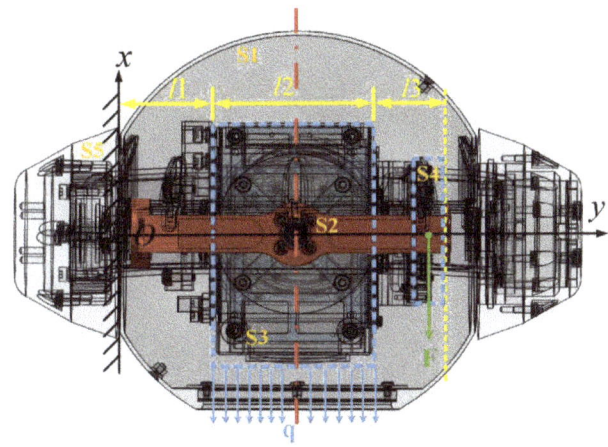

Figure 2. The clarifications of the model as simplified to a cantilever beam.

First, the bending internal force of plane bending under external force is analyzed. Moreover, the internal force diagram of bending moment and shear force is drawn by force analysis. In additions, the core problem is to check the deflection error of the simplified model of the cantilever beam.

In Figures 2 and 3, suppose that the connection between S2 and S5 is the origin O. Then, establish the Cartesian coordinate system Oxy. The distance of $l1$, $l2$, and $l3$ are shown in Figure 2. $l1$ is the distance between the fixed end of the left end cover and the fine-stage components. $l2$ is the length of fine-stage components. $l3$ is the distance between the fine-stage components and the voice coil motor (VCM). Span H is the sum of $l1$, $l2$, and $l3$. F is the VCM output constant torque. q is the load that is enforced by the fine-stage components S2 in the middle of the cantilever beam S2 (outer pitch gimbal).

In Figure 3, the x-axis is the length of the outer pitch gimbal, which is the simplified model of the cantilever beam. F_x-x and M-x represent the changing states of shear force F_x and bending moment M at different positions, X, of in the cantilever beam. What is more, the internal force diagram of bending moment and shear force is drawn. Figure 3 shows the change of bending moment M and shear force F_x with x.

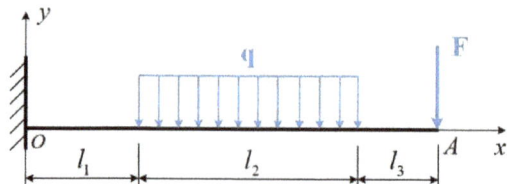

Figure 3. Force analysis of coarse–fine composite structure.

Because the deformation of the cantilever beam is very small, the change of the beam's span length after deformation is ignored. Since the example reports a cantilever beam, at the O-point (the fixed end), the bending moment is max. Figure 4a,b presents the changing states of the shear force F_x and bending moment M. Moreover, the material of the beam works within the elastic range of the beam, so the deflection and angle of the beam are linear with the load acting on the beam. Using static equilibrium analysis of material mechanics, the deflection of cantilever beam is calculated by the superposition principle. Because of its complicated force, it is divided into three force forms to solve the equations, respectively, which are finally superimposed together to obtain the deflection curve equation of the coarse–fine composite structure.

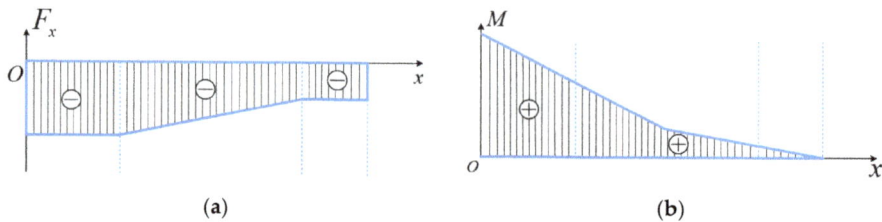

Figure 4. The internal force diagram. (a) $F_x - x$; (b) $M - x$.

In Figure 5, due its complexity, the force of the system is divided into three force forms in order to use the superposition principle to solve the equation. What is more, the rigid displacement of the free end A of the cantilever beam is selective analysis. The key is to decompose the load q in the middle into two loads q starting at the origin A. The first is down, the second is up.

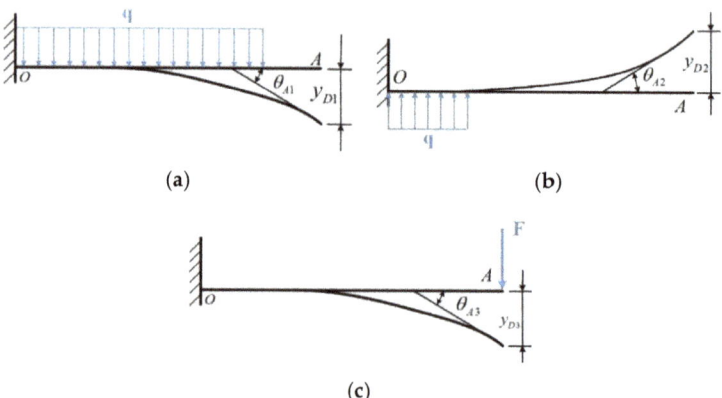

Figure 5. Superposition principle method static equilibrium analysis. (a) Form 1; (b) Form 2; (c) Form 3.

As shown in Table 1, the structure distance data obtained after Solidworks software simulation is analyzed. The deflection curve equation of each superposition diagram is as follows:

$$y_{D1} = y_1 + \theta_{A1} \cdot l_3 = \frac{q(l_1+l_2)^4}{8EI} + \frac{q(l_1+l_2)^3}{6EI} \cdot l_3 = \frac{3q(l_1+l_2)^4 + 4q(l_1+l_2)^3 l_3}{24EI}, \quad (1)$$

$$y_{D2} = y_2 + \theta_{A2} \cdot (l_2+l_3) = -\frac{q(l_1)^4}{8EI} - \frac{q(l_1)^3}{6EI} \cdot (l_2+l_3) = -\frac{3q(l_1)^4 + 4q(l_1)^3(l_2+l_3)}{24EI}, \quad (2)$$

$$y_{D3} = y_3 = \frac{F(l_1+l_2+l_3)^3}{3EI}, \quad (3)$$

Table 1. Structure distance data obtained by Solidworks software.

Length	Value/mm
l_1	40
l_2	43
l_3	33
Span H	116

By superimposing Equations (1)–(3), we can obtain

$$y_D = y_{D1} + y_{D2} + y_{D3} = \frac{3q(l_1+l_2)^4 + 4q(l_1+l_2)^3 l_3}{24EI} - \frac{3q(l_1)^4 + 4q(l_1)^3(l_2+l_3)}{24EI} + \frac{F(l_1+l_2+l_3)^3}{3EI}, \quad (4)$$

were E = the elastic modulus of the material, N/mm^2; I = the cross-sectional area of the material, mm^2; and q = standard values for distributed loads, kN/m. As shown in Table 2, after Solidworks software simulation, the free end force q and the average distributed load F were obtained as follows:

Table 2. Structure free and force and average distributed load data by Solidworks software.

Parameter	Value
F	0.00022 kN
q	0.0085 kN/m

It is made of aluminum alloy nonferrous metal with excellent comprehensive performance and its brand name is 7075-t3510. According to the data, the elastic modulus of 7075 aluminum alloy is $E = 71.7 Gpa$. The coarse–fine composite structure adopted the calculation method of moment of inertia of circular section. All known parameters are substituted into the equation of the deflection curve derived from the superposition.

$$y_D = y_{D1} + y_{D2} + y_{D3} \approx 1.67 \times 10^{-6} mm, \quad (5)$$

The coarse–fine stage composite structure is the overall plane bending of the main structure. Therefore, the allowable deflection is less than H/1500. The calculated result is

$$\frac{H}{1500} = \frac{0.116}{1500} \approx 0.77 \times 10^{-4} mm > 1.67 \times 10^{-6} mm. \quad (6)$$

To sum up, the deflection of the coarse–fine composite structure is checked to meet the specified deviation requirement. According to the internal force diagram of bending moment and shear force, the bending internal force of the cantilever beam under the action of external force is within the normal range.

2.2. Finite Element Analysis

In the Figure 6, finite element analysis (FEA) was carried out for the force of the key structural component, and the objects were meshed and solved by FEA. We then analyzed whether the stress, strain, and displacement parameters met the requirements.

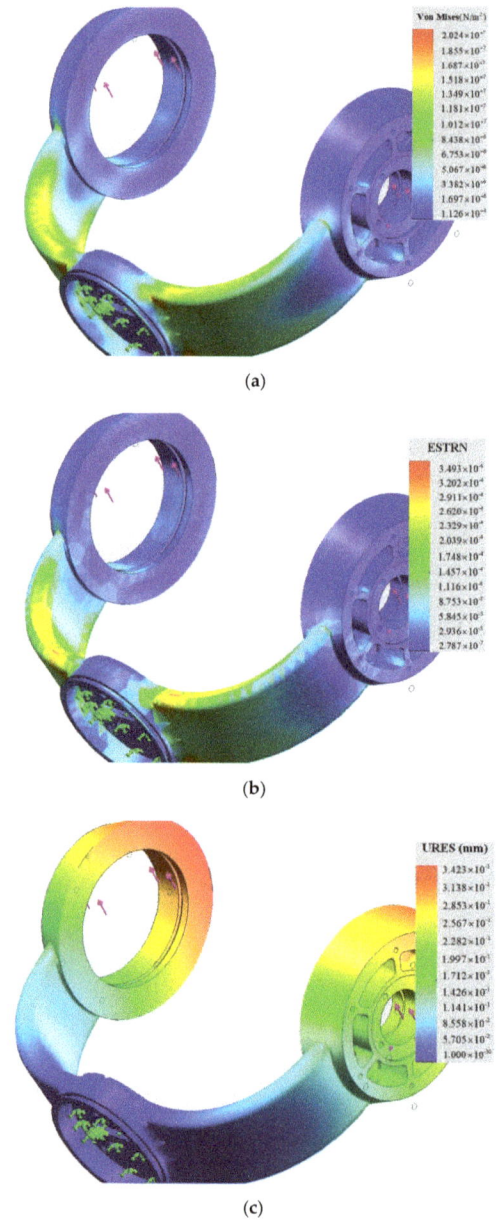

Figure 6. Finite Element Analysis (FEA) of key structure component (outer roll gimbal). (**a**) Stress analysis; (**b**) strain analysis; (**c**) displacement analysis.

It can be seen from the grading on the right of Figure 6a–c and Table 3 that the more red the structure is, the more dangerous it is. The increase of stress, strain, and displacement in the drilling position is large and relatively concentrated, but it still meets the needs for the normal working of the structure within the safety range. The results of finite element analysis still prove that it can meet the requirements of operation, and the outer roll gimbal is safe and reliable as a whole.

Table 3. Finite element analysis data obtained by Solidworks software.

Parameter	Value	
	Max	Min
Stress	$2.02 \times 10^7 N/m^2$	$1.13 \times 10^4 N/m^2$
Strain	3.49×10^{-4}	2.79×10^{-7}
Displacement	$3.42 \times 10^{-1} mm$	$1.00 \times 10^{-30} mm$
The deformation ratio	49.3719	

2.3. Design of Limit Structure of Rotation Angle

According to the Euler transformation of the fixed-point rotation of a rigid body, the Euler angle has no limit. However, in the coarse–fine composite structure of the two-axis four-gimbal electro-optical pod, the rotation angle of each gimbal is limited due to the external dimension, load weight, and the center of mass imbalance of the gimbal.

As shown in Figure 7a, in order to ensure the normal operation of the UAV's electro-optical pod in a safe range, a limit stopper is used to limit the rotation angle of each gimbal structure. A balancing weight is used to allocate the overall mass to prevent the occurrence of center of mass imbalance. As shown in Figure 7a,b, the rotation angle of inner pitch gimbal limiting stopper is +7°~−7° (a total of 14°), and the rotation angle of inner roll gimbal limiting stopper is +12°~−12° (a total of 28°). In this angle range, the operation of the UAV's electro-optical pod is normal and safe.

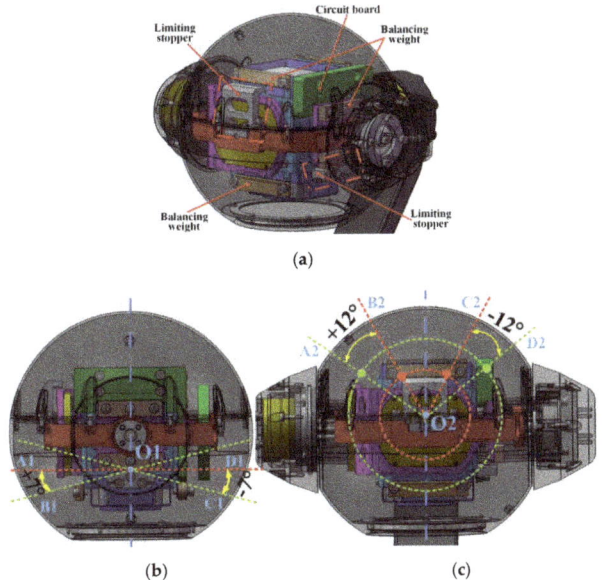

Figure 7. The diagram of rotation angle of each gimbal of the electro-optical pod. (**a**) The electro-optical stopper; (**b**) the rotation angle of the inner pitch gimbal limiting stopper; (**c**) the rotation angle of the inner roll gimbal limiting stopper.

3. Dynamics Modeling

3.1. Coarse–Fine Composite Analysis

As shown in Figure 1, the structure of the two-axis four-gimbal electro-optical pod is more complicated. It is therefore an effective solution to study the coarse–fine composite structure first. The working principle of the coarse–fine composite structure involves the definition of multiple coordinate systems, which are respectively explained as follows.

A   Inertial coordinate system ($\{i\}$, $O_i X_i Y_i Z_i$)
B   UAV coordinate system ($\{d\}$, $O_d X_d Y_d Z_d$)
C   Coarse motor stator coordinate system ($\{u\}$, $O_u X_u Y_u Z_u$)
D   Fine motor rotor coordinate system ($\{g\}$, $O_g X_g Y_g Z_g$)

The coarse motor is fixedly connected with the guide rail through the threaded connection, without considering the damping effect between the structures. There is geometric eccentricity e_r in the shafting structure of the coarse–fine stage mechatronic system, which will cause coaxiality error and affect the high precision control performance of the electro-optical pod. As shown in Figure 8a, the geometric eccentricity of the shafting structure is caused by force deformation, uneven cutting force, and chip formation of the cutting edge.

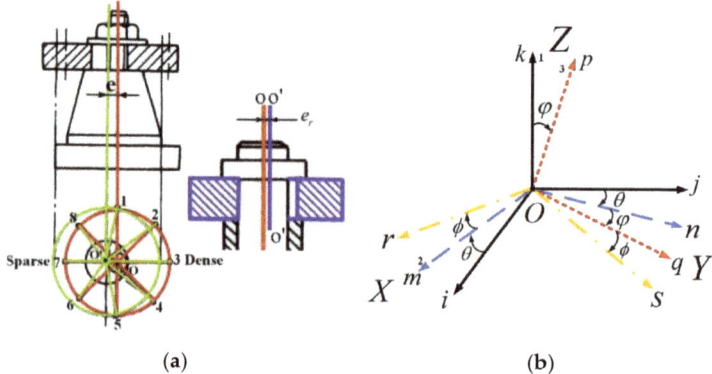

Figure 8. (a) Geometric eccentricity error of coarse–fine composite structure; (b) Euler transformation diagram for fixed-point rotation.

In Figure 8b, according to the transformation matrix of fixed-point rotation in the Cartesian coordinate system, the Euler transformation [17] is analyzed as shown in Figure 8b, and the Euler angle is θ, ϕ, φ. The first step is to rotate the θ angle about the k axis, so that the i axis rotates to the m position and the j axis rotates to the n position; Cartesian coordinate system $Oijk \rightarrow Omnk$. The second step is to rotate the ϕ angle about the m axis, so that the n axis rotates to the q position and the k axis rotates to the p position; Cartesian coordinate system $Omnk \rightarrow Omqp$. The third step is to rotate the φ angle about the p axis, so that the m axis rotates to the r position and the q axis rotates to the s position; Cartesian coordinate system $Omqp \rightarrow Orsp$. Finally, the Euler transformation of fixed-point rotation is completed.

In the Cartesian coordinate system, after the system $\{u\}$ rotates the θ_1 angle around the X_i axis, the system $\{i\}$ is used as the reference system to observe the position of the system $\{u\}$. Then, the system $\{u\}$

rotates the θ_2 angle around Z_i axis. Euler transformation of coarse–fine composite structure can be calculated as the rotation transformation matrix, denoted as

$$E^{X\theta} = A^\theta = \begin{pmatrix} 1 & 0 & 0 \\ 0 & \cos\theta_1 & -\sin\theta_1 \\ 0 & \sin\theta_1 & \cos\theta_1 \end{pmatrix}, \quad (7)$$

$$E^{Z\theta} = A^\theta = \begin{pmatrix} \cos\theta_2 & -\sin\theta_2 & 0 \\ \sin\theta_2 & \cos\theta_2 & 0 \\ 0 & 0 & 1 \end{pmatrix}. \quad (8)$$

According to Euler transformation law of rigid body fixed point rotation, it can be obtained from Equations (7) and (8) that

$$E = E^{k\theta} \cdot E^{m\phi} \cdot E^{p\varphi} = \begin{pmatrix} \cos\theta & -\sin\theta & 0 \\ \sin\theta & \cos\theta & 0 \\ 0 & 0 & 1 \end{pmatrix} \cdot \begin{pmatrix} 1 & 0 & 0 \\ 0 & \cos\phi & -\sin\phi \\ 0 & \sin\phi & \cos\phi \end{pmatrix} \cdot \begin{pmatrix} \cos\varphi & -\sin\varphi & 0 \\ \sin\varphi & \cos\varphi & 0 \\ 0 & 0 & 1 \end{pmatrix}. \quad (9)$$

Since the stator of the coarse motor is connected with the guide rail by thread, there is no fixed-point rotation for the system $\{u\}$ against the system $\{i\}$. Only the installation error of rotation along the Y-axis exists. The kinematic coupling equation shows that

$$\omega_u = \begin{pmatrix} \omega_{ux} \\ \omega_{uy} \\ \omega_{uz} \end{pmatrix} = \begin{pmatrix} \cos\theta_u & 0 & \sin\theta_u \\ 0 & 1 & 0 \\ -\sin\theta_u & 0 & \cos\theta_u \end{pmatrix} \begin{pmatrix} \omega_{ix} \\ \omega_{iy} \\ \omega_{iz} \end{pmatrix} + \begin{pmatrix} 0 \\ \dot{\theta}_u \\ 0 \end{pmatrix}. \quad (10)$$

Due to the Euler transformation law of rigid body fixed point rotation, the kinematics coupling equations of the system $\{u\}$ against the system $\{v\}$ and the system $\{v\}$ against the system $\{g\}$ are

$$\omega_v = \begin{pmatrix} \omega_{vx} \\ \omega_{vy} \\ \omega_{vz} \end{pmatrix} = E^{k\theta v} E^{m\phi v} E^{p\varphi v} \begin{pmatrix} \omega_{ux} \\ \omega_{uy} \\ \omega_{uz} \end{pmatrix} + \begin{pmatrix} \dot{\theta}_{vx} \\ \dot{\theta}_{vy} \\ \dot{\theta}_{vz} \end{pmatrix}, \quad (11)$$

$$\omega_g = \begin{pmatrix} \omega_{gx} \\ \omega_{gy} \\ \omega_{gz} \end{pmatrix} = E^{k\theta g} E^{m\phi g} E^{p\varphi g} \begin{pmatrix} \omega_{vx} \\ \omega_{vy} \\ \omega_{vz} \end{pmatrix} + \begin{pmatrix} \dot{\theta}_{gx} \\ \dot{\theta}_{gy} \\ \dot{\theta}_{gz} \end{pmatrix}, \quad (12)$$

The symbols used in the equation and Figure 9 are defined as follows: $\dot{\theta}_u$ = the angular velocity vector of the coarse motor stator relative to the inertial coordinate system; $\dot{\theta}_{vx}, \dot{\theta}_{vy}, \dot{\theta}_{vz}$ = the angular velocity vector of the coarse motor rotor relative to the coarse motor stator coordinate system; $\dot{\theta}_{gx}, \dot{\theta}_{gy}, \dot{\theta}_{gz}$ = the angular velocity vector of the fine motor rotor relative to the coarse motor rotor coordinate system; $\omega_u, \omega_v, \omega_g$ = the angular velocity and its components on the coordinate axis; and LOS = the line of sight.

In Figure 9, the external environment disturbance is included. In order to simplify the analysis process of the coarse–fine stage visual axis stabilization, this paper mainly discusses the conduction path and characteristics of UAV motion to the mechatronic system. Therefore, the disturbance input of the external environment is analyzed as an inertial coordinate system, and the whole process of motion coupling of the coarse–fine mechatronic system is obtained through the transformation of Cartesian coordinate along the system $\{u\}$, system $\{v\}$, and system $\{g\}$. The Euler angle θ, ϕ, φ is determined separately in order to determine the relationship between the angular velocities at each stage and the inertial space.

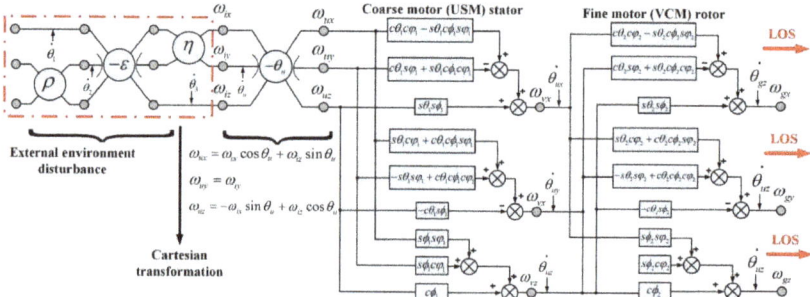

Figure 9. The coarse–fine composite structure of electro-optical pod kinematic coupling Piogram.

3.2. Two-Axis Four-Gimbal Structure

Based on the analysis of the transmission path and kinematics coupling compensation matrix of the coarse–fine composite structure, the dynamics modeling and theoretical study of the ultralight two-axis four-gimbal electro-optical pod are studied. The working principle of the two-axis four-gimbal electro-optical pod involves the definition of multiple coordinate systems, which are respectively explained as shown in Figure 10.

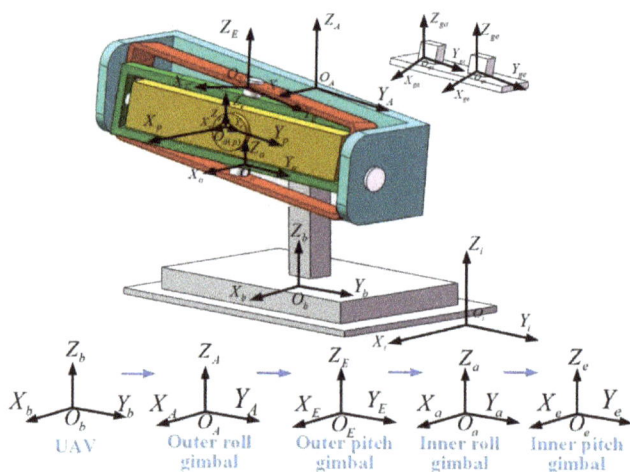

Figure 10. Simplified Cartesian coordinate system of the two-axis four-gimbal electro-optical pod.

The inner gimbal rotates in a small range, and the outer gimbal follows the macro-field control of the inner gimbal in a large range. At the same time, the feedback error of the outer gimbal is compensated by the inner gimbal so that the inner gimbal can offset the disturbance of rolling and pitching. Finally, the two-axis four-gimbal electro-optical pod maintains the stability of the visual axis to achieve high-precision coarse–fine composite control.

A. Direct Connection Stabilization

The gyroscope is sensitive to the angular velocity of the inner pitch system {e} and the inner roll system {a} relative to the inertial system {i}. Therefore, make $\omega_{Ye} = \omega_{Ze} = 0$, and then the structure can keep the visual axis of the detector stabilization.

$$\begin{pmatrix} \dot{\theta}_e \\ \dot{\theta}_a \end{pmatrix} = \begin{Bmatrix} \omega_{gyro\_Y} \\ \omega_{gyro\_X} \end{Bmatrix} = \begin{pmatrix} \sin\theta_a \cos\theta_E \\ \cot\theta_e \cos\theta_a \cos\theta_E + \sin\theta_E \end{pmatrix} \omega_{ZA} + \begin{pmatrix} -\cos\theta_a & -\sin\theta_E \sin\theta_a \\ \cot\theta_e \sin\theta_a & -\cot\theta_e \cos\theta_a \sin\theta_E - \cos\theta_E \end{pmatrix} \begin{pmatrix} \omega_{YE} \\ \omega_{XA} \end{pmatrix}, \quad (13)$$

According to Equation (13), the structure can keep the visual axis of the stabilization.

B. Indirect Connection Stabilization

The angular velocity of gyroscope sensitive the outer pitch axis system {E} and the outer roll axis system {a} relative to the inertial system {i} is $\omega_{gyro\_X} = \omega_{XA}, \omega_{gyro\_Y} = \omega_{YE}$.

$$\begin{pmatrix} \dot{\theta}_e \\ \dot{\theta}_a \end{pmatrix} = \begin{pmatrix} 1 & 0 \\ 0 & -\sec\theta_e \end{pmatrix} \begin{pmatrix} \omega_{Ye} \\ \omega_{Ze} \end{pmatrix} + \begin{pmatrix} \sin\theta_a \cos\theta_E \\ \cot\theta_e \cos\theta_a \cos\theta_E + \sin\theta_E \end{pmatrix} \omega_{ZA} + \begin{pmatrix} -\cos\theta_a & -\sin\theta_E \sin\theta_a \\ \cot\theta_e \sin\theta_a & -\cot\theta_e \cos\theta_a \sin\theta_E - \cos\theta_E \end{pmatrix} \begin{pmatrix} \omega_{gyro\_Y} \\ \omega_{gyro\_X} \end{pmatrix}, \quad (14)$$

According to Equation (14), the structure can keep the visual axis of the stabilization.

Assuming that, in the case of sensitive motion of pitch and roll gyroscopes, their sensitivity values are $\omega_{gyro\_Y}, \omega_{gyro\_Z}$, respectively, $\dot{\theta}_a$ represents the angular velocity vector of the inner roll gimbal relative to the outer pitch gimbal, and $\dot{\theta}_e$ represents the angular velocity vector of the inner roll gimbal relative to the inner roll gimbal; $\omega_A, \omega_E, \omega_a, \omega_e$, respectively, represent the angular velocity of the two-axis four-gimbal structure and the components of its three coordinate axes.

According to the Euler dynamical theorem and Coriolis rotation law,

$$\frac{dH}{dt} = \frac{\partial H}{\partial t} + \omega \times H, \quad (15)$$

where $H = \begin{pmatrix} H_X & H_Y & H_Z \end{pmatrix}^T$ = moment of momentum; $\frac{dH}{dt}$ = absolute derivatives (rate of change) of vector H; $\frac{\partial H}{\partial t} = \frac{dH_X}{dt}i + \frac{dH_Y}{dt}j + \frac{dH_Z}{dt}k$ = relative derivatives of vector H; and i, j, k = unit vectors of coordinate axis of the body reference system, respectively.

According to the moment of momentum theorem,

$$\frac{dH}{dt} = M, \quad (16)$$

where $M = \begin{pmatrix} M_X & M_Y & M_Z \end{pmatrix}^T$ = the external addition torque vector of the rigid body. Under the assumption that the all three axes are principal axes of inertia, the following equation can be established:

$$\begin{cases} I_X \dot{\omega}_X + (I_Z - I_Y)\omega_Y \omega_Z = M_X \\ I_Y \dot{\omega}_Y + (I_X - I_Z)\omega_X \omega_Z = M_Y \\ I_Z \dot{\omega}_Z + (I_Y - I_X)\omega_X \omega_Y = M_Z \end{cases}, \quad (17)$$

where I_X, I_Y, I_Z = the moment of inertia of the rigid body around the coordinate axis of the follower reference system.

The moment of momentum theorem is applicable to the calculation of larger angular velocity. However, the electro-optical pod of two-axis four-gimbal structure is compact and ultralight, with a mass less than 1 kg and a stabilization precision is 20 μrad. When the design is carried out in combination with the actual situation, the values of some parameters are ignored. Based on the space dynamics [18], the coarse–fine composite self-correction drive equation are derived.

A. Inner Pitch Gimbal

$$M_{Ye} \approx (x_e \times \dot{v}_b)m_e + [x_e \times (\dot{\omega}_a \times y_e)]m_e + [x_e \times (\dot{\omega}_E \times y_a)]m_e + [x_e \times (\dot{\omega}_A \times y_E)]m_e + I_{Ye}\dot{\omega}_e + (\omega_e \times I_{Ye}\omega_e), \quad (18)$$

B. Inner Roll Gimbal

$$\begin{aligned}
M_{Xa} &\approx (x_a \times \dot{v}_b)m_a + [x_a \times (\dot{\omega}_E \times y_a)]m_a + [x_a \times (\dot{\omega}_A \times y_E)]m_a + (y_e \times \dot{v}_b)m_e \\
&+ [y_e \times (\dot{\omega}_a \times y_e)]m_e + [y_e \times (\dot{\omega}_e \times y_e)]m_e + I_{Xa}\dot{\omega}_a + [\omega_a \times (I_{Xa}\omega_a)] + M_{Ye}
\end{aligned} \quad (19)$$

C. Outer Pitch Gimbal

$$\begin{aligned}
M_{YE} &\approx (x_E \times \dot{v}_b)m_E + [x_E \times (\dot{\omega}_A \times y_E)]m_E + (y_a \times \dot{v}_b)m_a + [y_a \times (\dot{\omega}_E \times y_a)]m_a \\
&+ [y_a \times (\dot{\omega}_a \times y_a)]m_a + (y_a \times \dot{v}_b)m_e + [y_a \times (\dot{\omega}_E \times y_e)]m_e + [y_a \times (\dot{\omega}_a \times y_e)]m_e \\
&+ [y_a \times (\dot{\omega}_e \times y_e)]m_e + I_{YE}\dot{\omega}_E + [\omega_E \times (I_{YE}\omega_E)] + M_{Xa} + M_{Ye}
\end{aligned} \quad (20)$$

D. Outer Roll Gimbal

$$\begin{aligned}
M_{XA} &\approx (x_A \times \dot{v}_b)m_A + (y_E \times \dot{v}_b)m_E + [y_E \times (\dot{\omega}_A \times y_E)]m_E + [y_E \times (\dot{\omega}_E \times y_E)]m_E \\
&+ (y_E \times \dot{v}_b)m_a + [y_E \times (\dot{\omega}_A \times y_a)]m_a + [y_E \times (\dot{\omega}_E \times y_a)]m_a + [y_E \times (\dot{\omega}_a \times y_a)]m_a \\
&+ (y_E \times \dot{v}_b)m_e + [y_E \times (\dot{\omega}_A \times y_e)]m_e + [y_E \times (\dot{\omega}_E \times y_e)]m_e + [y_E \times (\dot{\omega}_a \times y_e)]m_e \\
&+ [y_E \times (\dot{\omega}_e \times y_e)]m_e + I_{XA}\dot{\omega}_A + [\omega_A \times (I_{XA}\omega_A)] + M_{YE} + M_{Xa} + M_{Ye}
\end{aligned} \quad (21)$$

where ω_i = angular velocity of inner pitch{e}, inner roll{a}, outer pitch{e}, outer roll{a} relative to inertial gimbal system{i}; v_b = the speed of the UAV gimbal {b} relative to the inertial gimbal; m_i = the quality of inner pitch, inner roll, outer pitch and outer roll gimbal; $I_{..} = I_{Ye}, I_{Xa}, I_{YE}, I_{XA}$ is the rotational inertia of the inner pitch, the inner roll, the outer pitch and the outer roll gimbal along their respective rotation axis; x_i = the vector distance from the origin of four gimbal coordinate systems and UAV coordinate systems to their respective centroids is designated as the inner pitch x_e, inner roll x_E, outer pitch x_a, outer roll x_A, and UAV x_b; y_i = the vector displacement between the rotation axis of the inner pitch gimbal and the inner roll gimbal is y_e, the vector displacement between the rotation axis of the inner roll and the outer pitch gimbal is y_a, the vector displacement between the rotation axis of the outer pitch and the outer roll gimbal is y_E; and $M_{..}$ = the torque of the four gimbals relative to the rotation axis in the inertial coordinate system is the output torque of the four motors.

In order to further study the Euler rigid body dynamics model mechanism of the ultralight two-axis four-gimbal electro-optical pod, Figure 11 is drawn. In Figure 11, the torques due to gimbal kinematics and those due to geometrical coupling have been combined. The key problem is to ensure the high precision control of the structure visual axis.

When both $\theta_e = 0$ and $\theta_a = 0$, the inner gimbal angle is zero. The disturbing moment of the two axes is minimized. Through the following performance of the outer gimbal, the mutual perpendicularity between the inner gimbals can be guaranteed, so as to eliminate the geometric constraint coupling brought by the outer gimbal to the visual axis and realize the interference isolation, proving once again that the system can decouple two stabilization channels from the perspective of kinematics. The coupling interference of geometric constraints can be eliminated, and the control precision can be improved.

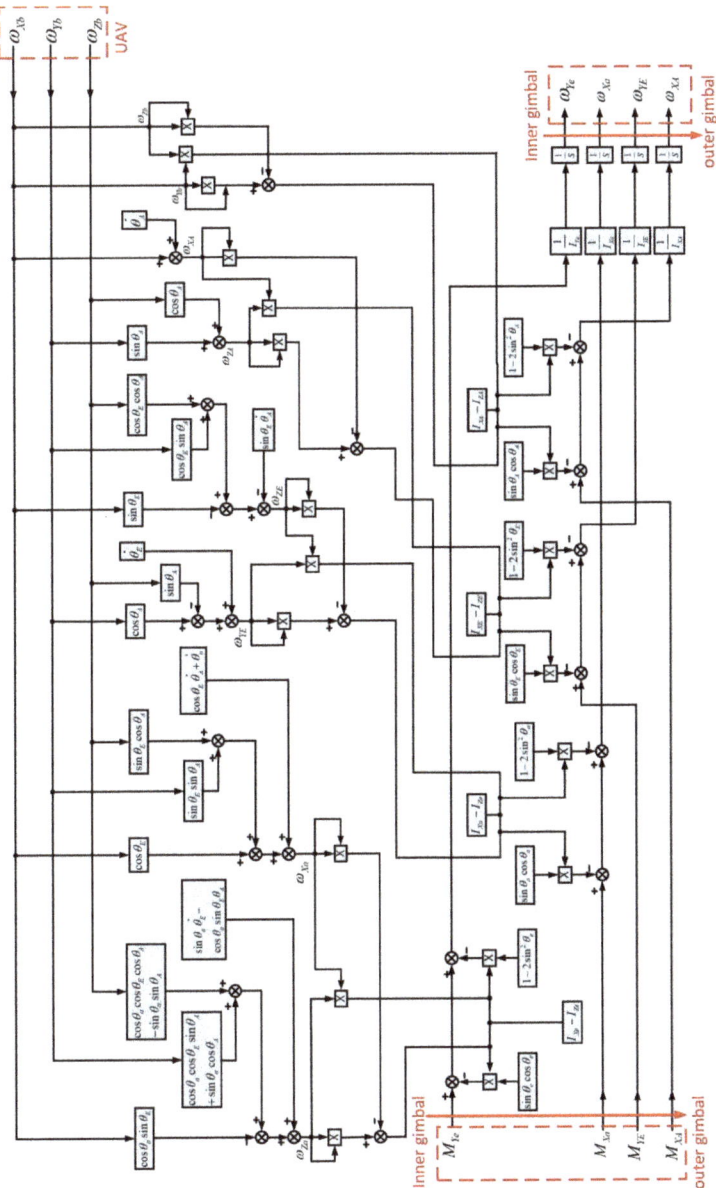

Figure 11. The ultralight two-axis four-gimbal electro-optical pod torque relationships.

3.3. Comparison Validation

Because the motor adopts a direct drive way, it does not consider slip failure. The impact caused by friction is small, so it is assumed that the transmission efficiency of the final stage of the pod is $\eta = 99\%$. The parameters of gimbal rotation angular velocity, UAV maneuvering acceleration, and gimbal angular velocity are

$$\dot{\omega}_{Ye} = \dot{\omega}_{Xa} = \dot{\omega}_{YE} = \dot{\omega}_{XA} = \varepsilon = 120°/s^2 \approx 2 rad/s^2 \approx 0.318 r/s^2, \tag{22}$$

$$\dot{v}_b = a \leq 5g \approx 50 m/s^2, \tag{23}$$

$$\omega_{Ye} = \omega_{Xa} = \omega_{YE} = \omega_{XA} = \omega_{max} = 60°/s \approx 1 rad/s \approx 0.159 r/s, \tag{24}$$

As shown in the coarse–fine composite drive self-correction equation, the cross-product term value is small, and the included angle is small as it approaches zero infinitely and is greater than zero. According to the trig function, if $\theta \to 0, \cos\theta \to 1$. What is more, $a \times b = |a| \cdot |b| \cdot \cos\langle a, b \rangle$. Therefore, the term $\cos\langle a, b \rangle$ can be ignored as the constant 1.

As shown in Table 4, the moment of inertia and mass data of gimbals at all stages when the electro-optical pod rotates at $0°$ are presented. The moment of inertia and the distance between the center of mass and the origin are analyzed when the electro-optical pod rotates at different angles. We then calculate the coarse–fine composite forecast torque (Equations (18)–(21)), full payload, and equivalent dynamics load calculates torque (Equations (22)–(24)) of the electro-optical pod structure. Our results are shown in Figure 12.

Table 4. Rotational inertia when the initial rotation angle is $0°$ and quality simulation data.

Gimbal	Rotational Inertia (Including Load/kg·m^2)			Mass (Including Load/kg)
	X	Y	Z	
Inner pitch e	0.57×10^{-3}	0.58×10^{-3}	0.48×10^{-3}	0.471
Inner roll a	0.89×10^{-3}	0.84×10^{-3}	0.93×10^{-3}	0.681
Outer pitch E	0.99×10^{-3}	0.94×10^{-3}	0.11×10^{-2}	0.740
Outer roll A	0.26×10^{-1}	0.88×10^{-2}	0.27×10^{-1}	1.925

It can be observed from Figure 12 that the difference of the coarse–fine composite forecast torque, full payload, and equivalent dynamics load calculates the torque. This is due to the influence of friction, wind load, and conductor's interference torque. The load of the inner pitch gimbal is less at the center and the influence of disturbance is minimal, so the difference with the real value is not large. The gimbal is extended one level outward, the bearing load increase, the shape is more irregular, the circuit board leads are complex, and other factors cause the error to increase within a certain range of the true value.

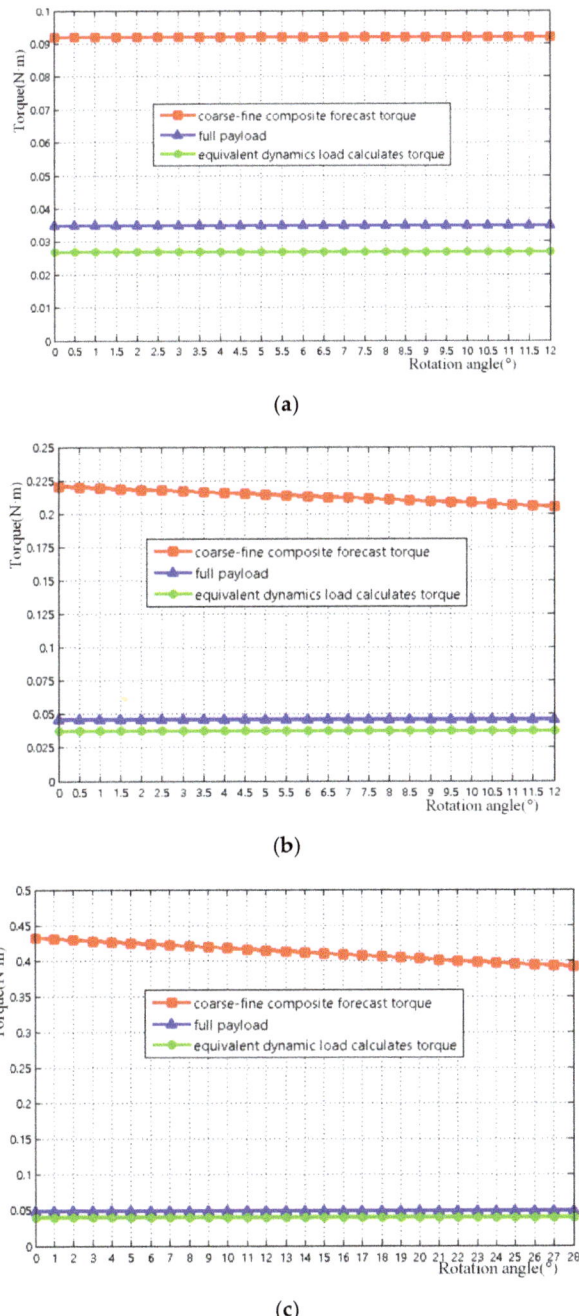

Figure 12. *Cont.*

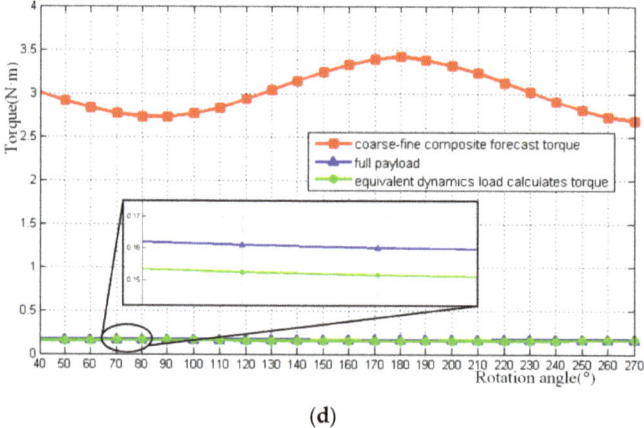

(d)

Figure 12. The ultralight two-axis four-gimbal electro-optical pod torque comparison validation. (**a**) Inner roll gimbal; (**b**) inner pitch gimbal; (**c**) outer roll gimbal; (**d**) outer pitch gimbal.

4. Experiment

In view of the problem that the coupling effect between two-axes four-gimbal seriously affects the stability precision, it is necessary to combine the coupling relationship of rotational inertia of each gimbal axis for disturbance suppression analysis. Because of the effectiveness of the interference observer (DOB) in suppressing external interference [19,20], in this paper, an interference observer suitable for the ultralight two-axis four-gimbal electro-optical pod is studied.

As shown in Figure 13, the control object is set as the ultralight two-axis four-gimbal system, and the minimum phase system under ideal state is adopted. The nominal inverse model of the controlled object is $Js + B$. Based on the kinematic coupling analysis and modeling, a DOB disturbance observer is used to study self-correcting disturbance suppression. The traditional DOB controller is improved to a time-varying DOB controller with rotational inertia. At the same time, the results of the moment of inertia analysis after the modeling mentioned in this paper are substituted into the nominal inverse model J_n of each gimbal control loop. Realize the real-time change of J_n following the change of gimbal angle θ. The control loop of the outer roll gimbal A was given sinusoidal interference as an example, and a Matlab Simulink simulation comparison experiment was carried out.

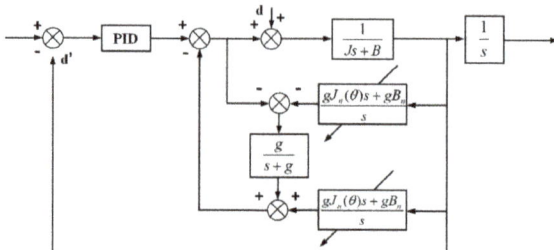

Figure 13. Disturbance suppression output diagram of outer roll gimbal A.

The rotational inertia of the rotating axis changes in real time. First, the parameters of the traditional PID controller are set as follows: $K_p = 20, K_i = 6$. As can be seen from Table 4, the initial parameters of rotational inertia are set as $J = J_{XA} = 0.176 \times 10^{-1} \text{kg} \cdot \text{m}^2$. The parameters of the disturbance observer and its low-pass filter are set as follows: $B = 0.002, g = 200, B_n = 0.002$. From the motion coupling and modeling analysis, it can be known that the coupling rotational inertia on the outer roll gimbal A is

$$J = (\cos\theta_E \cos^2\theta_e - \sin\theta_E \cos\theta_a \sin\theta_e \cos\theta_e)J_{Xe} + \\ (\cos\theta_E \sin^2\theta_e - \sin\theta_E \sin\theta_a \cos\theta_a \cos^2\theta_e)J_{Ze} + \cos\theta_E J_{XE} + J_{XA} \quad , \tag{25}$$

Figures 14 and 15 verify the optimality of the velocity loop's traditional PID control, the traditional DOB disturbance suppression control, the improved DOB self-correcting disturbance suppression control, and the low-pass filter parameter selection. Set the system input amplitude to 0. The input amplitude of sine wave disturbance is 10 rad/s, and the frequency is 8 Hz. As shown in Figure 14, in order to facilitate the observation of the experimental results, the output value of the improved DOB was taken as negative gain output Scope which was distinguished from the other three waveforms. Further observation of the experimental results shows that

$$\Delta X < \Delta Y, \tag{26}$$

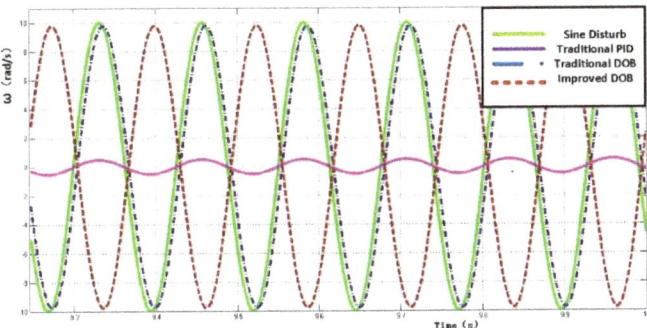

Figure 14. Disturbance suppression output diagram of outer roll gimbal A.

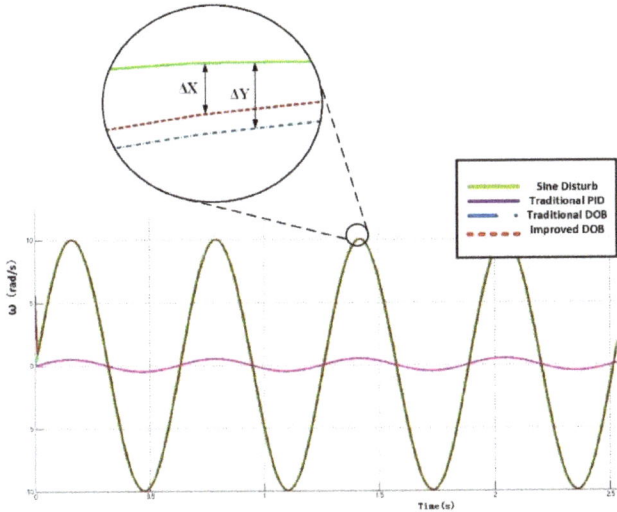

Figure 15. Comparison diagram of disturbance suppression output of outer roll gimbal A.

The results show that the disturbance suppression impact of DOB method with dynamics model is increased by up to 90% better than PID. As shown in Figure 13, this is defined as

$$e = d - d', \tag{27}$$

As shown in Figure 16, by comparing the two figures, it can be known that the estimated deviation of traditional DOB disturbance suppression is $e = 0.51$ and the estimated deviation of the improved DOB disturbance suppression is $e = 0.43$. The results show that the disturbance suppression impact of DOB method with dynamics model is increased by up to 25% compared to the traditional DOB.

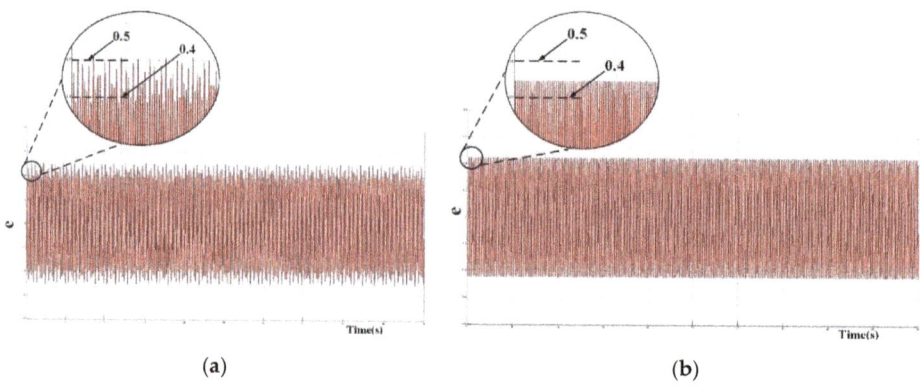

Figure 16. Comparison diagram of disturbance suppression difference output of outer roll gimbal A. (**a**) The output deviation of traditional DOB disturbance suppression; (**b**) the output deviation of improved DOB disturbance suppression.

5. Conclusions

This paper represents an in-depth study on the dynamics modeling and theoretical study of the two-axis four-gimbal coarse–fine composite electro-optical pod. Our conclusions are as follows.

A In the UAV electro-optical pod of the two-axis four-gimbal, the characteristics of the coarse–fine composite structure and the complexity of dynamics modeling affect the entire system's high-precision control performance. The core goal of this paper is solve the high precision control of two-axis four-gimbal electro-optical pod through dynamic modeling and theoretical study. FEA and theoretical analysis of the stress and deflection of the key structure component was used to design the structure. The gimbal structure adopts 7075-t3510 aluminum alloy, which is an aerospace material that meets the requirements of an ultralight electro-optical pod weighing less than 1 kg.

B According to the Euler rigid body dynamics model, the transmission path and kinematics coupling compensation matrix for the two-axis four-gimbal are obtained. The coarse–fine composite drive correction equation of the inner-outer gimbals is derived to solve the pre-selection and check problem of the coarse–fine motors under high-precision control.

C The modeling method is substituted into the DOB disturbance suppression experiment, which can monitor and compensate for the motion coupling between gimbal structures in real time. Our results show that the disturbance suppression impact of the DOB method with dynamics model is up to 90% better than PID and 25% better than traditional DOB.

D This manuscript is based on the dynamics modeling and theoretical study of the two-axis four-gimbal coarse–fine composite UAV electro-optical pod. This manuscript is valuable for all researchers interested in the coarse–fine composite, two-axis four-gimbal structures, and ultralight electro-optical pods.

Author Contributions: C.S. designed the Euler rigid dynamics model and high precision macro–micro composite DOB control algorithm and carried out experimental research on the effect of the modeling and algorithm. D.F. and S.F. guided the research and proposed the ideas and revisions of the paper. X.J. provide help with control and simulation. R.T. provide structure model. C.S. and S.F. revised the paper. All authors have read and agreed to the published version of the manuscript.

Funding: This research received no external funding.

Acknowledgments: The author would like to thank all the teachers and colleagues who provided inspirations and equipment in the experiment. The author would like to thank all the anonymous reviewers for their meticulous comments and helpful suggestions.

Conflicts of Interest: The authors declare no conflict of interest.

References

1. Fan, D.P.; Zhang, Z.Y.; Fan, S.X.; Li, Y. Research of the basic principles of E-O stabilization and stabilization and tracking devices. *Opt. Precis. Eng.* **2006**, *14*, 673–680.
2. Rue, A.K. Precision Stabilization Systems. *IEEE Trans. Aerosp. Electron. Syst.* **1974**, AES-10. [CrossRef]
3. Rue, A.K. Stabilization of precision electro-optical pointing and tracking systems. *IEEE Trans. Aerosp. Electron. Syst.* **1969**, AES-5, 805–819. [CrossRef]
4. Rue, A.K. Calibration of Precision Gimbaled Pointing Systems. *IEEE Trans. Aerosp. Electron. Syst.* **1970**, AES-6, 697–706. [CrossRef]
5. Rue, A.K. Correction to Stabilization of precision electro-optical pointing and tracking systems. *IEEE Trans. Aerosp. Electron. Syst.* **1970**, AES-6, 855–857.
6. Rue, A.K. Confidence Limits for the Pointing Error of Gimbaled Sensors. *IEEE Trans. Aerosp. Electron. Syst.* **1966**, AES-2, 648–654.
7. Preliasco, R.J. Wide-look-angle gimbal for an Airborne Electro-Optical Systems. *Proc. SPIE* **1998**, 104–111. [CrossRef]
8. Royalty, J. Development of Kinematics for Gimbaled Mirror and Prism Systems. *SPIE* **1990**, *1304*, 262–274.
9. Kennedy, P.J.; Kennedy, R.L. Direct versus Indirect Line of Sight (LOS) Stabilization. *IEEE Trans. Control Syst. Technol.* **2003**, *11*, 3–15. [CrossRef]
10. Zhao, L.J.; Dou, T.S.; Cheng, B.Q.; Xia, S.F.; Yang, J.X.; Zhang, Q.; Li, M.; Li, X.L. Theoretical Study and Application of the Reinforcement of Prestressed Concrete Cylinder Pipes with External Prestressed Steel Strands. *Appl. Sci.* **2019**, *9*, 5532. [CrossRef]
11. Tomović, R. A Simplified Mathematical Model for the Analysis of Varying Compliance Vibrations of a Rolling Bearing. *Appl. Sci.* **2020**, *10*, 670. [CrossRef]
12. Ma, P.F.; Li, Y.J.; Han, J.C.; He, C.; Xiao, W.K. Finite Element Modeling and Stress Analysis of a Six-Splitting Mid-Phase Jumper. *Appl. Sci.* **2020**, *10*, 644. [CrossRef]
13. Harsh, D.; Shyrokau, B. Tire Model with Temperature Effects for Formula SAE Vehicle. *Appl. Sci.* **2019**, *9*, 5328. [CrossRef]
14. Guo, L.D. Compound-Axis Macro-Micro Control and Modeling of Laser Weapon Tracking System. *J. Jilin Univ. (Eng. Technol. Ed.)* **2011**, *41*, 48.
15. Choi, Y.M.; Gweon, D.G. A high-Precision Dual-Servo Stage Using Halbach Linear Active Magnetic Bearings. *IEEE/ASME Trans. Mechatron.* **2011**, *16*, 925–931. [CrossRef]
16. Xia, Y.X.; Bao, Q.L.; Liu, Z.D. A New Disturbance Feedforward Control Method for Electro-Optical Tracking System Line-Of-Sight Stabilization on Moving Platform. *Sensors* **2018**, *18*, 4350. [CrossRef] [PubMed]
17. Pio, R.L. Euler angle transformations. *IEEE Trans. Autom. Control* **1966**, AC-11, 707–715. [CrossRef]
18. Thomson, W.T. *Introduction to Space Dynamics*; Wiley: New York, NY, USA, 1963; pp. 101–108.
19. Cong, J.W.; Tian, D.P.; Shen, H.H. Research on coupling self-correcting interference suppression control of airborne photoelectric platform. *Electr. Mech. Eng.* **2019**, *36*, 749–754.
20. Chen, X.G.; Cai, M.; Dai, N. Disturbance suppression method of airborne photoelectric stabilized platform based on DOB observer. *Electro-Optic Control* **2019**, *11*, 1–6.

 © 2020 by the authors. Licensee MDPI, Basel, Switzerland. This article is an open access article distributed under the terms and conditions of the Creative Commons Attribution (CC BY) license (http://creativecommons.org/licenses/by/4.0/).

Article

Damage Investigation of Carbon-Fiber-Reinforced Plastic Laminates with Fasteners Subjected to Lightning Current Components C and D

Jian Chen [1], Xiaolei Bi [2], Juan Liu [2] and Zhengcai Fu [1,*]

1. Key Laboratory of Control of Power Transmission and Conversion, Ministry of Education, Shanghai Jiao Tong University, Shanghai 200240, China; chen_jian@sjtu.edu.cn
2. State Key Laboratory of Safety and Control for Chemicals, Qingdao Safety Engineering Institute, SINOPEC, Qingdao 266000, China; bxlluck@163.com (X.B.); liuj.qday@sinopec.com (J.L.)
* Correspondence: zcfu@sjtu.edu.cn

Received: 24 February 2020; Accepted: 19 March 2020; Published: 21 March 2020

Abstract: The damage induced by lightning strikes in carbon-fiber-reinforced plastic (CFRP) laminates with fasteners is a complex multiphysics coupling process. To clarify the effects of different lightning current components on the induced damage, components C and D were used in simulated lightning strike tests. Ultrasonic C-scans and stereomicroscopy were used to evaluate the damage in the tested specimens. In addition, the electrothermal coupling theory was adopted to model the different effects of the arc and the current flowing through the laminate (hereinafter referred to as the conduction current) on CFRP laminates with fasteners under different lightning current components. Component C, which has a low current amplitude and a long duration, ablated and gasified the fastener and caused less damage to the CFRP laminate. Under component C, the heat produced by the arc played a leading role in damage generation. Component D, which has a high current amplitude and a short duration, caused serious surface and internal damage in the CFRP laminate and little damage to the fastener. Under component D, the damage was mainly caused by the Joule heat generated by the conduction current.

Keywords: Carbon-fiber-reinforced plastics (CFRPs); fastener; arc; Joule heat; finite element analysis (FEA)

1. Introduction

Carbon-fiber-reinforced plastics (CFRPs) have excellent mechanical properties and are widely used in various industries [1–3]. With the massive expansion of wind power and the rapid growth in the number of aircraft, the chances of wind turbine blades and aircraft being struck by lightning have inevitably increased substantially. Because the destructive effects of lightning strikes often lead to serious consequences, research on lightning damage in CFRPs has received unprecedented attention [4–8].

In structural design, depending on the excellent formability of composite materials, the number of fasteners can be reduced by optimizing the design; however, completely eliminating the need for fasteners is difficult. These fasteners have greater electrical conductivity than the other materials in CFRP laminates. Therefore, when a lightning strike occurs, the current is discharged through the fasteners first and then into each layer of the CFRP laminate, which leads to fiber breakage, resin degradation, and internal delamination [9]. In addition, the temperature and air pressure inside the fastener hole will change dramatically during the lightning strike [10], which leads to damage around the hole, loss of fastener support, and weakening of the mechanical and electrical properties of the CFRP [11]. Therefore, research on the damage in CFRPs with fasteners subjected to lightning strikes is the key point of composite lightning protection [10,12].

The duration of a lightning strike is extremely short, during which a large amount of energy is released in an instant, resulting in extremely high temperatures. Observing the damage characteristics near the attachment point of the lightning strike in real time with instruments is difficult. Therefore, finite element analysis (FEA) has become an effective method for studying the damage evolution process of CFRP laminates with fasteners under lightning strikes and for verifying the correctness of the test results. Chemartin et al. [13] used the finite volume method in the time domain and unstructured mesh to establish the mechanism model of CFRPs with fasteners, and simulated the spark phenomenon in fasteners. The research showed that sparking may occur when the current density is greater than 10 kA/mm^2. Kirchdoerfer [10] simulated the gas conditions and related local geometric changes in the interior space around the fastener during lightning strikes, and discussed the importance of chemical change modeling in future work. Meanwhile, the electrothermal coupling model is used to investigate the damage of CFRP under lightning strike. Muñoz et al. [14] developed a finite element model to consider the damage sources observed in a lightning strike, such as thermal damage caused by Joule heat. Yin et al. [11] established a three-dimensional electrothermal coupling model of ablation damage of CFRP with fasteners based on the relationship of the energy balance in s lightning strike. The results indicated that fasteners distributed the lightning current to each layer, and a larger conduction current dispersion area led to less damage to the laminate. Abdelal et al. [15] proposed a physical model to predict lightning strike damage for composite materials. The finite element method of non-linear material model was used to analyze composite materials with copper mesh protection. Ogasawara et al. [16] proposed a electrothermal coupling model of angle ply composite laminates, which considered the anisotropic thermoelectric behavior of layer and unidirectional composite laminates. However, their work neglected the arc heat effect in numerical simulation. Dong et al. [17] considered the influence of the arc heat effect and replaced it with heat flux in the models, while the damage to CFRP with fasteners was not mentioned. On the basis of previous work, this paper used the electrothermal coupling module in COMSOL to design simulation models to explore the arc heat and conduction current effects on the damage of CFRP with fasteners, as well as the damage difference under different lightning current conditions.

Experimental investigations are often used to study the damage in CFRP laminates with fasteners subjected to lightning strikes. Previous studies have found a relationship between the damaged area and the mounting depth of the fasteners during a lightning strike. The shallower the mounting depth, the larger the surface damage area [18]. CFRP laminates with fasteners show penetrating damage under lightning strikes, with damage occurring on both sides of the specimen [19,20]. The lightning current component D is influential in developing out-gassing, whereas no out-gassing is observed when component C is used [21]. To make the simulated lightning strikes in the laboratory more closely approximate natural lightning strikes, it is important to ensure that the lightning current waveforms and current amplitudes used in the tests meet the standard requirements [19]. However, in the study of lightning damage in CFRP laminates with fasteners, many of the waveforms and amplitudes used in previous studies did not meet the standard requirements [22,23], and the roles of the arc and conduction current in the process of damage were not clearly distinguished.

In this work, simulated lightning strike tests were performed on CFRP laminates with fasteners using lightning current components C and D, which comply with the standards. Ultrasonic C-scans and stereomicroscopy were used to evaluate the differences in specimen damage under the two components, and an electrothermal coupling model was adopted to verify the test results to study the different effects of the conduction current and arc after the action of lightning current components C and D. The results were compared with the lightning damage of pure CFRP under components C and D [17].

2. Materials and Methods

2.1. Specimen Preparation

The material used in this work was a unidirectional carbon fiber prepreg (TC35/FRD-Y360). The specimen dimensions were 250 mm (length) × 250 mm (width) × 2 mm (thickness), and the stacking sequence was $[0°/90°/0°/90°/0°/90°/0°/90°/0°/90°/0°/90°/0°/90°/0°]$, creating a total of 15 plies. The diameter of the mounting hole was 8 mm, and the fastener material was stainless steel. Compared with titanium alloy, stainless steel has good electrical conductivity and a lower melting point, which will result in more severe damage during lightning strikes, which is beneficial for observation and analysis. The specimens used in this work were unpainted and unprotected. This structure allowed us to focus on the details of CFRP response to high energy discharge alone [19]. The assembly of the fastener and CFRP laminate is shown in Figure 1a. The tight fitting of the fastener and the CFRP laminate mounting hole can reduce the contact resistance between the two elements. The discharge electrode with a diameter of 8 mm was made of tungsten–copper alloy (W80Cu20) and was positioned directly above the centre point of the specimen, separated from the specimen by a distance of approximately 3 mm. The four sides of the specimen were fixed on a metal plate with detachable copper strips. For reliable grounding, four copper braids were connected at the four corners of the metal plate. The simulated lightning current was injected into the specimen as an arc discharge and then flowed through the metal plate and out through the copper braid. The clamp and connection are shown in Figure 1b.

Figure 1. (a) Assembly diagram of the CFRP with a fastener; (b) the clamp and connection.

2.2. Test setup and waveform

The lightning current waveforms described in [22] and [23] include four components (Figure 2). Components A, B, C, and D represent the first return stroke, intermediate current, continuing current, and the subsequent return stroke, respectively. These four components can be divided into two categories: (1) short-duration, small-transferred-charge, high-action-integral, high-current-amplitude components A and D; and (2) long-duration, large-transferred-charge, low-action-integral, low-current-amplitude components B and C. To make this study universal, lightning current components C and D were used in the simulated lightning strike test of the CFRP laminates with fasteners. The current amplitude of component C was 200 A, the duration of which could reach 1 s, and the transferred charge was 200 C; the actual test waveform is shown in Figure 3a. The current amplitude of component D was 100 kA, the duration could reach 500 μs, and the action integral was 2.5×10^5 A^{2}s; the actual test waveform is shown in Figure 3b.

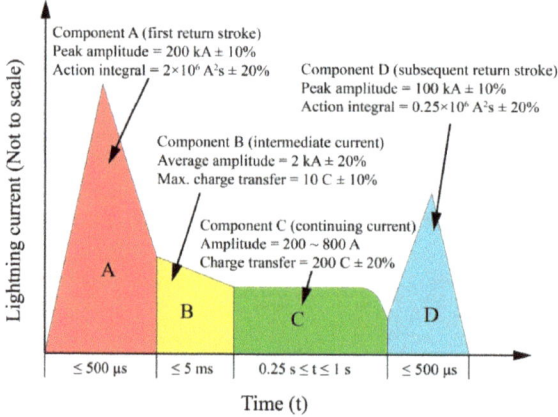

Figure 2. Simulated lightning current waveforms in [23].

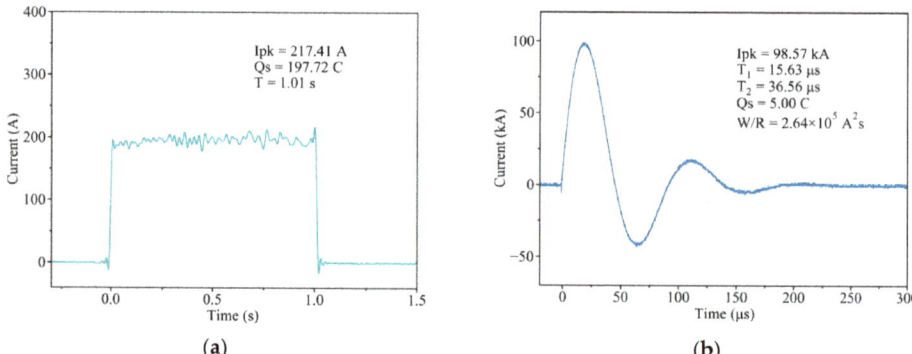

Figure 3. Lightning current waveforms used in the present study: (**a**) component C; (**b**) component D.

3. FEA

3.1. Electrothermal Coupling Theory

Because of the short duration of the lightning strike, measuring and observing the damage evolution process is difficult. Therefore, FEA is an effective method for analyzing the process. COMSOL was successfully applied to the finite element simulation of CFRP damage during lightning strikes [10]. The electrothermal coupling module in COMSOL provides a method for analyzing such problems. The module considers both the effect of the electrical conductivity with respect to temperature and the effect of the electric field with respect to the current density.

Herein, a steady-state electrical simulation analysis and a transient thermal simulation analysis are performed in sequence [24]. The lightning current flowing through the CFRP laminate will generate Joule heat, causing the resin to pyrolyze and gasify, which is a typical electrothermal coupling process [25–27]. During this process, the following charge conservation equations need to be followed:

$$\frac{\partial \rho_e}{\partial t} + \nabla \cdot \mathbf{j} = 0 \tag{1}$$

where ρ_e is the charge density and $\mathbf{j}$ is the current density inside the material.

The relationship between the current and Joule heat can be expressed as follows:

$$P_e = j \cdot E = \frac{j^2}{\sigma} \quad (2)$$

where P_e is Joule heat per unit volume and σ and E are the electric conductivity and electric field intensity per unit volume, respectively.

The heat transfer in the material conforms to Fourier's law. The governing equation of heat balance can be expressed as follows [14,27,28]:

$$\rho C_p \frac{\partial T}{\partial t} = \nabla \cdot (k \nabla T) + \frac{j^2}{\sigma} \quad (3)$$

where C_p is the specific heat capacity at constant pressure, ρ is the density, T is the absolute temperature, and k is the thermal conductivity.

The heat transferred to the material by the arc is equivalent to the heat flux [17], and the heat flux is expressed as follows:

$$Q(r,t) \approx 10J(r,t) \quad (4)$$

where t is the time, Q is the heat flux of the lightning arc, and J is the current density on the top of the fastener.

3.2. Finite Element Modeling

During the lightning strike, the temperature changes drastically, and the thermal conductivity, electrical conductivity, density, and specific heat of the material change significantly with respect to the temperature [15,29,30]. Table 1 shows the material parameters measured by thermogravimetric analysis (NETZSCH STA449F3, Germany) and laser flash thermal conductivity analysis (LFA467, Germany).

Table 1. Physical properties of the specimen.

Tempe-rature (°C)	Density (kg/m^3)	Specific Heat (J/(kg·K))	Thermal Conductivity		Electrical Conductivity		
			Longi-tudinal (W/(m·K))	Transverse/through-thickness (W/(m·K))	Longi-tudinal (S/m)	Transverse(S/m)	Through-thickness (S/m)
25	1472	1176	6.578	0.723	17,800	10.4	2.8
300	1472	2048	9.617	0.633	17,800	10.4	2.8
500	1110	1454	7.166	0.423	17,800	2000	2000
510	1110	1454	7.166	0.423	17,800	2000	2000
3316 [1]	1110	2146	7.166	0.423	17,800	20,000	20,000
>3316 [2]	1110	5875	1.0×10^8	1.0×10^8	1.0×10^8	1.0×10^8	1.0×10^8

[1,2]- These parameters are not directly measured but obtained through extrapolation [15,27,31,32].

The parameters of stainless steel fasteners are shown in Table 2.

Table 2. Physical properties of stainless steel.

Density (kg/m^3)	Specific Heat (J/(kg·K))	Thermal Conductivity (W/(m·K))	Electrical Conductivity (S/m)
7850	4.75×10^2	44.5	4.032×10^6

Some assumptions were proposed: no clearance between fastener and composite, and the contact electrical resistance and thermal resistance on the interface were ignored; the delamination caused by Joule heat and the thermal stress inside the specimen was not considered. The stacking sequences and dimensions of the CFRP laminate used in the finite element modeling were the same as those of the specimen. The model is shown in Figure 4. The whole model had 28,157 elements, 28,055 hexahedra,

2,200 edge elements, and 136 vertex elements. The electrical potential boundary of the model was set as follows: the electrical potential of the two parallel sides was 0 V. Heat exchange between the specimen and the environment occurred when the specimen was directly exposed to the air. Therefore, the thermal boundary of the model was set as follows: the thermal radiation emissivity of the top and bottom surfaces was 0.9; the four sides of the model were thermally insulated and did not exchange heat with the external environment; and the environmental temperature was 25 °C. The conduction current and heat flux of components D and C were imported into COMSOL and applied on the entire top surface of the fastener. For modeling of different fiber orientations, rotated coordinate systems were used and the material properties of CFRP were assigned to each layer. For each layer, the size of elements increased from the center to the four sides to improve the efficiency and ensure the accuracy at the same time. The maximum and minimum side lengths were 10 mm and 3 mm, respectively. The average element quality was about 0.9. The total time of the transient solver was set according to the duration of the waveform (300 μs for component D and 1 s for component C). In this work, the relative tolerance of the simulation model was 0.01, the tolerance factor was 0.1, the maximum number of iterations was 10, and the termination technique was the tolerance. The simulation was done on a Dell Precision Tower 780 workstation equipped with two E5-2603 CPUs and 48 GB memory, and the longest time consumed about 11 h for a calculation case. In the process of calculation, the temperature distribution and the physical parameters of the materials were monitored by the domain point probes arranged in the model to ensure the convergence of the model.

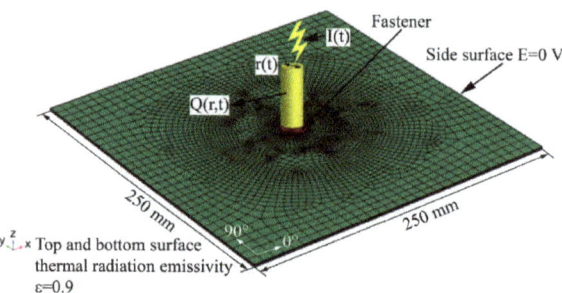

Figure 4. Finite element model and boundary conditions.

In the model, r(t) is the arc channel radius; Q(r, t) is the heat flux obtained by Equation (4), which is used to represent the role of the arc; and I(t) is the conduction current injected into the top of the fastener.

The heat flux radius is assumed to be the same as the fastener radius. When a lightning strike occurs, the lightning current is divided into two parts: one part is attached to the top of the fastener in the form of an arc, whereas the other part is conducted in the fastener and the laminate in the form of conduction current. Therefore, the damage in CFRP laminates with fasteners is caused by the combination of the arc and conduction current. To understand the effect of the arc and conduction current on CFRP damage and to understand the difference in CFRP damage under different lightning current components, we designed three simulation models for each lightning current component, as shown in Table 3.

Table 3. Simulation model.

Simulation Model	Function
Conduction current + heat flux	Simulate the combined effect of the arc and conduction current
Conduction current	Simulate the effect of the conduction current
Heat flux	Simulate the effect of the arc

4. Results

4.1. Surface Damage

Figure 5a,b show the surface damage in the specimens under the action of components C and D, respectively.

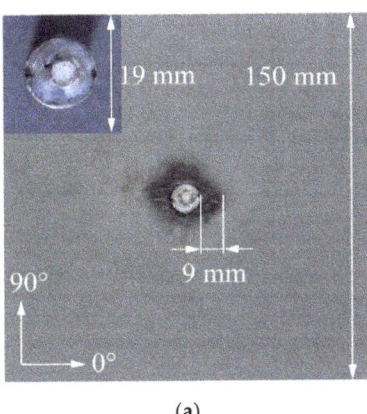
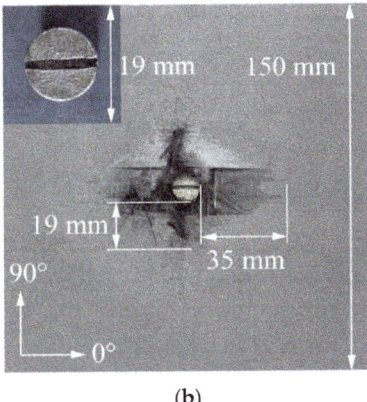

(a) (b)

Figure 5. Surface damage in the specimens after lightning strike: (**a**) component C; (**b**) component D.

For component C, serious ablation damage occurred on the top of the fastener, which was still tightly connected to the laminate. The damage in the laminate was located within a distance of 9 mm around the fastener. Resin discoloration was observed but very few fibers were exposed or warped.

For component D, no obvious damage was observed on the top of the fastener, but the fastener was loose and slightly sunken. Fiber tufts, fiber breakage, resin sublimation, and resin discoloration appeared on the surface of the laminate. These forms of damage extended 35 mm along the fiber direction and 19 mm orthogonal to the fiber direction. In addition, flaky fiber shedding was observed along the fiber direction.

4.2. Damage in the Fastener Hole

After the lightning strike tests, each specimen was cut, as shown in Figure 6c, and the cross-section of the hole was observed with a stereomicroscope (Leica/M125, Germany). Figure 6a,b are cross-sectional photos of the specimen after the action of components C and D, respectively. In these figures, the regularly spaced vertical lines are the shadows left by the synthesis of multiple cross-sectional photos, which do not represent the damage.

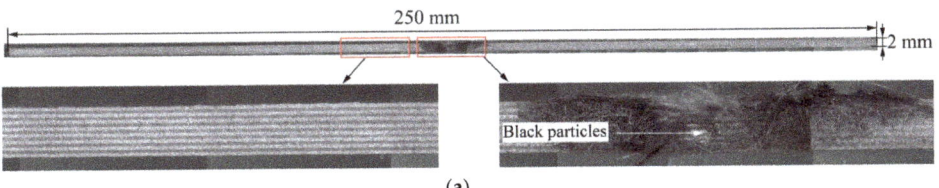

(a)

Figure 6. *Cont.*

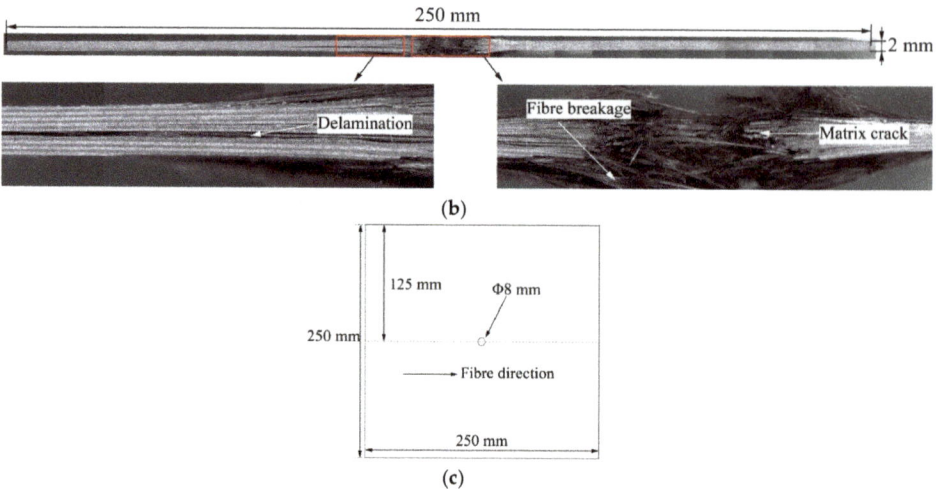

Figure 6. Damage in the fastener hole of the specimens after lightning strikes: (**a**) component C; (**b**) component D. (**c**) Specimen cutting diagram for microscopic evaluation.

For component C, black particles cover the surface of the inner wall of the fastener hole and few fiber breakages or cracks are observed between the layers, which indicates that the temperature of the inner wall of the hole was not high when the specimen was subjected to a lightning strike and that the amount of resin vaporized by the lightning strike was not substantial.

For component D, Figure 6b shows extended delamination damage away from the hole. Resin matrix cracks and fiber breakages are observed around the hole, and black particles also appear in the hole. As the fastener penetrated the laminate, the delamination damage near the hole became extremely serious.

4.3. Internal Damage

The internal damage in the CFRP laminate can be non-destructively evaluated with an ultrasonic scanning device (KSI V400E, Germany). The frequency of the ultrasonic wave was 40MHz, and the pulse–echo mode was used for scanning. Ultrasound C-scan, which is sensitive to delamination, is based on the reflection of ultrasonic energy from the intermediate interface. When the ultrasonic wave encounters the damaged interface, the reflected energy in the form of pulse–echo amplitude is different from the undamaged condition [33]. The reflected ultrasonic signals are converted into image signals with different gray values.

Figure 7a,b show the internal damage morphology under the action of components C and D, respectively. These figures show that the damage expands from the fastener to the delamination boundary after the lightning strikes. The damage within the delamination boundary in Figure 7 consists of two zones. One is the thermal decomposition damage and ablation damage, where the material is pyrolyzed and vaporized (marked with dark red, hereinafter referred to as the decomposition damage). The other is speculated to be delamination (marked with blue), where the damage of the interlayer structure is caused by the internal pressure generated by the rapid expansion of pyrolysis gas [34].

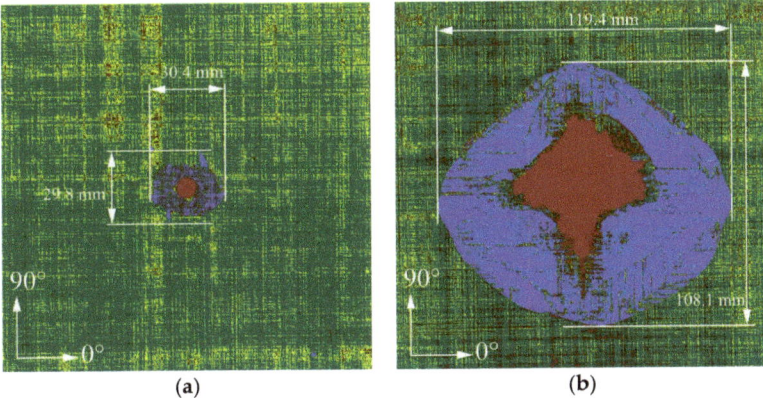

Figure 7. Ultrasonic C-scan results of specimens after lightning strikes: (**a**) component C; (**b**) component D.

For component C, the damage area inside the specimen is small, the damage lengths in the 0° direction and 90° directions are approximately 30 mm, and the difference between the two directions is small (in Figure 7a, the damage length in the 0° direction is 30.4 mm, whereas the damage length in the 90° direction is 29.8 mm). This finding indicates that a small amount of resin in the specimen has undergone pyrolysis and gasification during the lightning strike.

For component D, a large area of damage is observed in the specimen, and the damage lengths in the 0° and 90° directions are greater than 100 mm, which indicates that during the lightning strike, component D induces more resin pyrolysis and gasification than component C. Moreover, an internal explosion occurs under component D. The rhomboid-shaped damage area might be caused by the orthogonal stacking structure of the specimen.

4.4. Finite Element Simulation Results

During a lightning strike, the Joule heat generated by the conduction current flowing through the laminate will cause a continuous rise in the CFRP temperature [35]. A temperature contour exceeding a specific threshold value is used to characterize the damage area in each layer [16]. The threshold value usually adopts the decomposition temperature of the resin. When the temperature exceeds this temperature, the area is considered to be the decomposition damage area. According to Table 1, the decomposition temperature of the resin is set to 300 °C [29]. Therefore, the area surrounded by the 300 °C temperature contour in the FEA represents the same effect as the dark-red area (decomposition damage) under the C-scan.

Figure 8(a1–a3) are the temperature profiles of the specimen under component C in the conduction current + heat flux, conduction current, and heat flux simulation models, respectively. Figure 8(a1,a2) indicate that the decomposition damage shape and size of the laminates are similar to each other and that the damage shape is circular. Such decomposition damage shapes, which are similar to the dark-red area in Figure 7a, do not reflect the anisotropy of the CFRP laminates because component C has a long duration, providing Joule heat in sufficient time to spread in all directions. The first layer (0° direction) and the second layer (90° direction) are two orthogonal layers, and their temperature distributions are shown in the upper-right corner and lower-right corner of Figure 8(a1), respectively. The decomposition damage shapes of the two layers are almost identical.

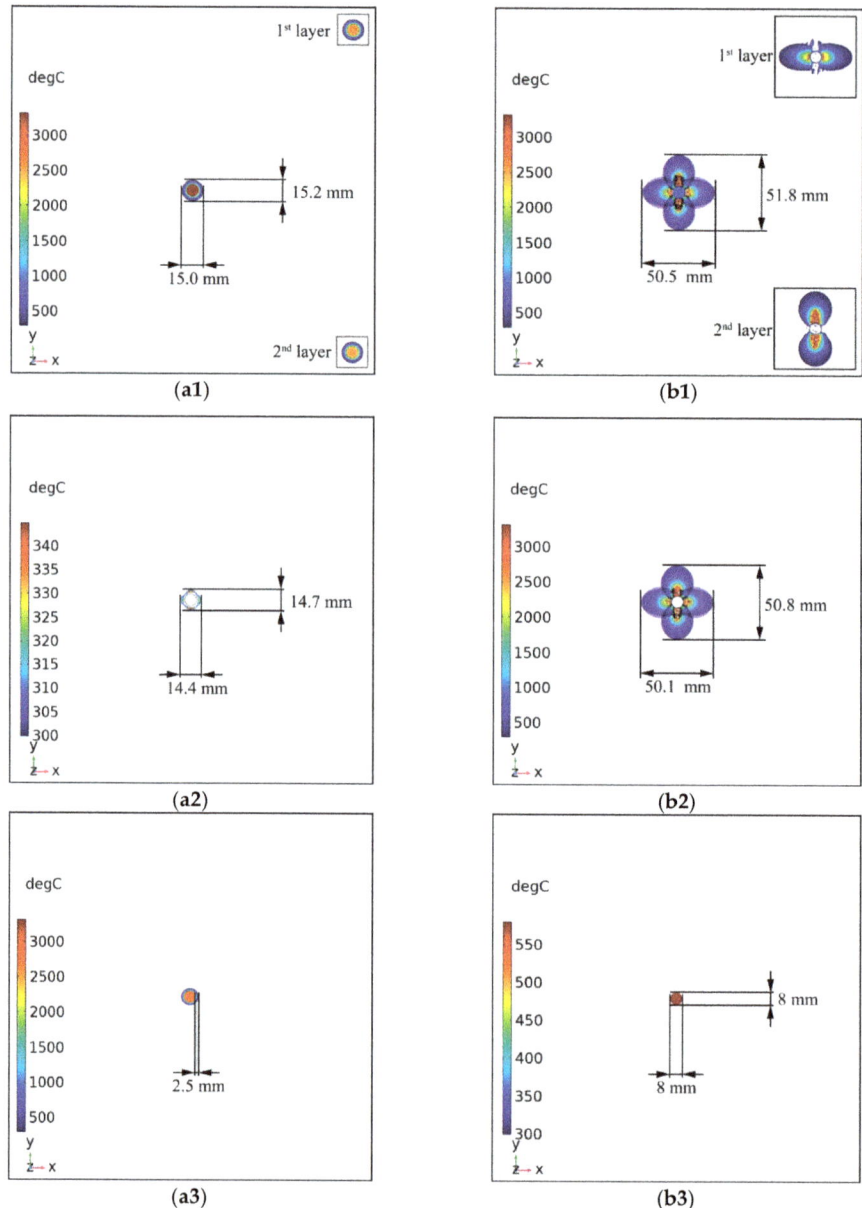

Figure 8. Temperature profiles of specimens under different simulation models and different lightning current components: component C in the (**a1**) conduction current + heat flux model, (**a2**) conduction current model, and (**a3**) heat flux model; component D in the (**b1**) conduction current + heat flux model, (**b2**) conduction current model, and (**b3**) heat flux model.

In addition, the temperature of the top of the fastener (in the center of Figure 8(a1)) exceeds 3000 °C after the combination of the conduction current and the heat flux. This temperature is much higher than the melting temperature of the fastener and causes severe ablation damage to the fastener. Figure 8(a3) shows the temperature distribution when using the heat flux model. In this model, the damage to the laminate caused by heat flux is approximately 2.5mm, and the temperature of the fastener is similar to that in the conduction current + heat flux model, which means that the ablation damage of the fastener under the action of component C has an important relationship with the heat flux.

Figure 9(a1–a3) are the cross-sectional temperature profiles of the specimen under component C in three simulation models. The heat diffusion can be inferred from the temperature gradient of the fasteners in Figure 9(a1,a3); heat is transferred to the CFRP laminate along with the fastener, causing the material temperature around the fastener to increase. In Figure 9(a2), the temperature of the fastener is far less than 300 °C and the decomposition damage occurs in a small area around the fastener. Because the damage occurs in the conduction current model, it is completely caused by the Joule heat generated by the conduction current. According to the results of Figure 5a, Figure 8(a1–a3), and Figure 9(a1–a3), a small area of CFRP laminate damage and serious fastener damage will clearly occur under component C. Thus, arc heating (heat flux) is the main cause of fastener damage under component C.

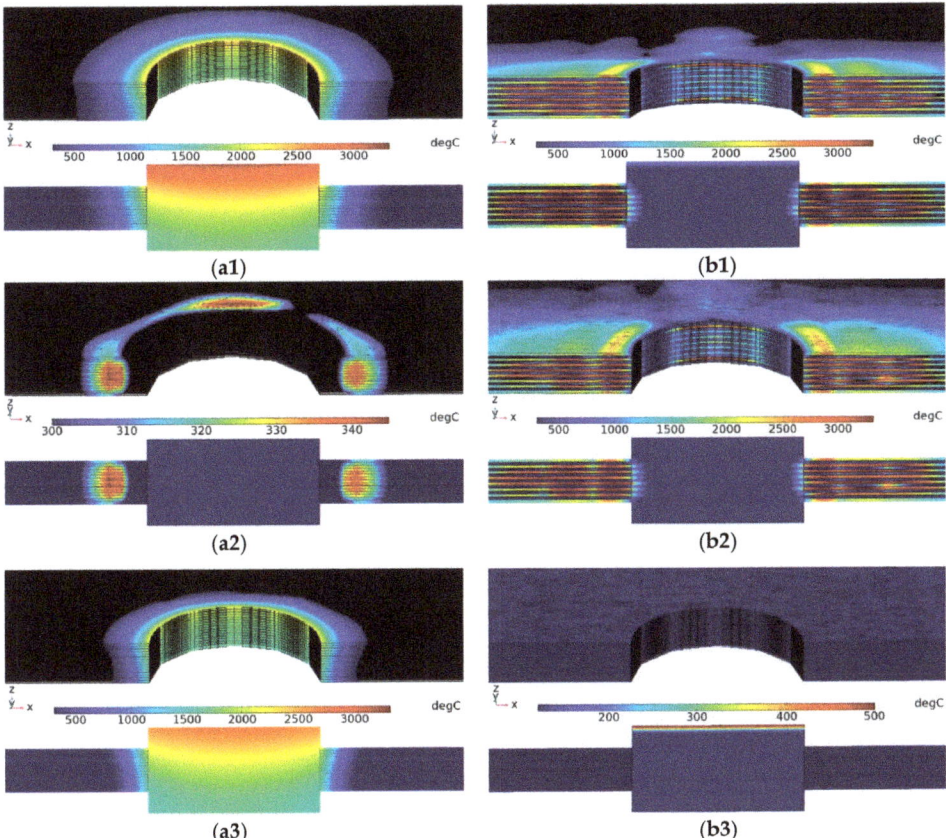

Figure 9. Cross-sectional temperature profiles of the fastener hole under different simulation models and different lightning current components: component C in the (**a1**) conduction current + heat flux model, (**a2**) conduction current model, and (**a3**) heat flux model; component D in the (**b1**) conduction current + heat flux model, (**b2**) conduction current model, and (**b3**) heat flux model.

Figure 8(b1–b3) are the temperature profiles of the specimen under component D in the conduction current + heat flux, conduction current, and heat flux simulation models, respectively. Figure 8(b1,b2) show that the decomposition damage shape and size of the specimen were similar, and that the decomposition damage was centered on the fastener and fanned out along the 0° and 90° directions.

The decomposition damage in both directions exceeded 50 mm. This decomposition damage shape, which resembles the dark-red area in Figure 7b, reflects the anisotropy of the CFRP laminates because component D has a short duration, which does not provide sufficient time for the Joule heat to spread evenly in all directions. The first layer (0° direction) and the second layer (90° direction) are the two orthogonal layers, and their temperature distributions are shown in the upper-right corner and lower-right corner of Figure 8(a1), respectively. Both layers showed that the decomposition damage in the fiber direction was substantially greater than that perpendicular to the fiber direction. In addition, Figure 8(b1,b3) show that no noticeable temperature rise occurred at the top of the fasteners from the heat flux. Figure 8(b3) is the temperature distribution when the heat flux model is used. In this model, the highest temperature at the top of the fastener is approximately 550 °C, which does not ablate the fastener. There are no visible signs of damage around the fastener, which means that the heat flux has little contribution to the damage of the CFRP with fasteners when component D is applied.

Figure 9(b1–b3) are the cross-sectional temperature profiles of the specimen under component D in three simulation models. The cross-sections of the specimens in Figure 9(b1,b2) show that the Joule heat generated by the conduction current caused the massive decomposition damage area in the laminate, whereas the fasteners remained at a lower temperature. According to the results of Figure 5b, Figure 8(b1–b3), and Figure 9(b1–b3), large CFRP laminate decomposition damage and minor fastener damage will occur under component D. Thus, the Joule heat generated by the conduction current is the leading factor for CFRP laminate decomposition damage under component D.

Figure 10 compares the damage profiles in the conduction current + heat flux model (marked with the solid red line) with the thermal decomposition damage and ablation damage profiles in the ultrasonic C-scans (marked with the black dotted line). The damage in the specimen caused by component D is much greater than that caused by component C. Figure 10a,b show that the FEA results are in good agreement with the experimental results. Therefore, it is reliable to study CFRP laminates with fasteners by FEA.

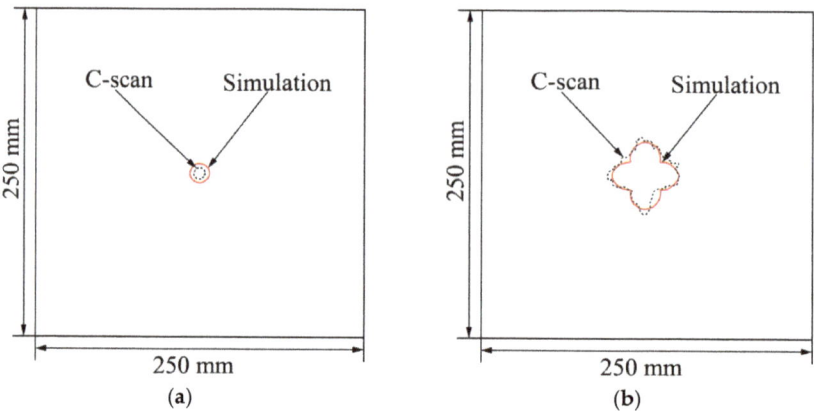

Figure 10. Damage comparison between simulated lightning strikes and finite element simulations: (**a**) component C; (**b**) component D.

5. Discussion

Dong et al. [17] considered the Joule heat effect and arc heat effect when investigating the damage of CFRP laminates without fasteners under components D and C. Comparing their results with ours, it can be found that there is a significant difference in damage between CFRP laminates without fasteners and CFRP laminates with fasteners with the same lightning current components.

For CFRP laminates without fasteners, the results showed that the Joule heat effect and the arc heat effect could cause damage. Component D controls the area of in-plane damage, while the sequential injection of component C after D aggravates the in-plane damage and tends to increase the damage depth. The Joule heat effect plays a leading role in component D, while the arc heat effect plays a dominant role in component C.

For CFRP laminates with fasteners, when the lightning strike acts on the fastener, because the conductivity of the fastener is approximately 1000 times greater than that of the carbon fibers [36], the lightning current will be discharged through the fastener first and then redistributed in each layer of the laminate in the form of a conduction current. The amplitude of the conduction current in each layer is related to the stacking sequence, the grounding mode, and the electrical conductivity. During the test, the specimen is grounded on four sides. In this grounding method, the conduction current can be scattered along each layer to the ground boundary. According to the surface damage (Figure 5), the fastener hole damage (Figure 6), and the internal damage (Figure 7), the damages induced by components C and D differ markedly. Laminates subjected to component C sustained less damage than those subjected to component D. However, the fasteners were severely ablated under component C and not under component D.

When the gap between the discharge electrode and the laminate is broken down, an arc will be generated, forming a high-temperature ionization region at 30,000 K [15]. For component C, the high-temperature arc, which has a long duration, acts on the fastener. The accumulation of heat makes the fastener gasify. Figure 5a shows that the fastener has sustained serious ablation damage. For component D, because of the short duration, the heat of the high-temperature arc seldom accumulates in the fastener; thus, Figure 5b shows that the fastener has minor ablation damage under component D.

Figure 5b shows that resin damage occurs on the surface of the laminate far from the fastener after component D is applied. This damage should be caused by dielectric surface discharge [12]. The dielectric breakdown model considers that the discharge direction is determined by the local electrical potential gradient. During the lightning strike, a large amount of induced charge will accumulate on the surface of the specimen. When the electric field intensity generated by the accumulated charge exceeds the critical value, discharge will occur on the surface of the material, causing damage to the resin. For component C (Figure 5a), due to the low amplitude of the discharge voltage, dielectric breakdown is less likely to occur; thus, the resin damage on the specimen surface should be caused by the metal droplets splashing and attaching to the surface after the fastener has melted.

During a lightning strike, the resin on the inner wall of the fastener hole is pyrolyzed and gasified, which generates high-temperature, high-pressure gas [37]. The heated gas expands rapidly, producing a high-speed gas flow doped with black particles, which may be the product of carbon fiber sublimation and pyrolysis carbonization at high temperatures [38]. Some of the black particles entered the fastener nut through the gap, whereas the rest remained on the inner wall of the fastener hole (Figure 6).

Figure 7 shows that there is a great difference in the internal damage of the laminates under different lightning components.

For component C, the current amplitude is 200 A. Due to the high current density around the fastener, a large amount of Joule heat can be generated. Moreover, the arc heat will spread to the laminate through the fastener (see Figure 9(a1,a3)). The combined effect of the Joule heat and arc heat results in resin pyrolysis. In the zone far from the fastener, the current density and the arc heat decay rapidly, and no internal damage in this zone can be observed under the C-scan. The internal damage caused by component C is concentrated in a small area around the fastener, as shown in Figures 7a and 8(a1).

For component D, the fastener spreads the conduction current to all the layers, and this current is approximately 6.7 kA per layer. These conduction currents can generate large amounts of Joule heat, allowing the resin to pyrolyze over large areas and release gas. According to Figure 6b, Figure 7b, and Figure 9b, it can be speculated that these high-temperature, high-pressure gases trapped inside the CFRP laminate may cause an internal explosion when the gas pressure reaches a critical level, which will result in serious delamination damage.

6. Conclusions

In this work, components C and D were used in simulated lightning strike tests. Ultrasonic C-scans and stereomicroscopy were used to evaluate the damage sustained by the specimens during the lightning strike tests. Moreover, the electrothermal coupling theory was adopted to model the different effects of the arc heat and Joule heat. The following conclusions are drawn.

(1) The damage to the laminate was concentrated around the fasteners. The conduction current flowed through the fastener to all layers and caused damage in each layer. With increasing distance from the fastener, the current density and the arc heat decayed.

(2) Component D, which had a high current amplitude and a short duration, led to serious surface and internal damage in the CFRP laminate and little damage to the fastener. The damage was mainly caused by the Joule heat generated by the conduction current. Component C, which had a low current amplitude and a long duration, ablated and gasified the fastener and caused less damage to the CFRP laminate. In this process, the arc heating produced by the arc played a leading role.

(3) The temperature profiles in the conduction current + heat flux model were analogous to the thermal decomposition damage and ablation damage profiles from the C-scan. Therefore, the conduction current + heat flux model is reasonable in FEA of CFRPs with fasteners subjected to lightning strikes. The simulation results show an obvious anisotropy in Component D but not in Component C, because component C has sufficient time for the Joule heat to spread in all directions, whereas component D lacks sufficient time.

This work evaluated the damage in CFRP laminates with fasteners subjected to lightning current components C and D and found that the damage under different lightning current components presented unique characteristics. Due to the variety of lightning strikes in nature, the damage induced by other lightning current components and multi-components needs further study on the basis of the research results of the single lightning current component.

Author Contributions: Conceptualization, J.C. and Z.F.; data curation, J.C.; formal analysis, J.C., X.B., and J.L.; funding acquisition, Z.F.; investigation, J.C., X.B., and J.L.; methodology, J.C.; project administration, Z.F.; resources, Z.F.; supervision, Z.F.; validation, J.C.; visualization, J.C., X.B., and J.L.; writing—original draft, J.C.; writing—review and editing, Z.F. All authors have read and agreed to the published version of the manuscript.

Funding: This research was funded by National Key R&D Program of China, grant number 2017YFC1501506.

Conflicts of Interest: The authors declare no conflict of interest.

References

1. Zhengchun, D.; Mengrui, Z.; Zhiguo, W.; Jianguo, Y. Design and application of composite platform with extreme low thermal deformation for satellite. *Compos. Struct.* **2016**, *152*, 693–703. [CrossRef]
2. Mishnaevsky, L.; Branner, K.; Petersen, H.N.; Beauson, J.; McGugan, M.; Sørensen, B.F. Materials for Wind Turbine Blades: An Overview. *Materials* **2017**, *10*, 1285. [CrossRef] [PubMed]
3. Chen, C.-C.; Chen, S.-L. Strengthening of Reinforced Concrete Slab-Column Connections with Carbon Fiber Reinforced Polymer Laminates. *Appl. Sci.* **2020**, *10*, 265. [CrossRef]
4. Chen, J.; Fu, Z.; Zhao, Y. Resistance behaviour of carbon fibre-reinforced polymers subjected to lightning strikes: Experimental investigation and application. *Adv. Compos. Lett.* **2019**, *28*, 1–10. [CrossRef]
5. WANG, F.; MA, X.; ZHANG, Y.; Jia, S. Lightning Damage Testing of Aircraft Composite-Reinforced Panels and Its Metal Protection Structures. *Appl. Sci.* **2018**, *8*, 1791. [CrossRef]
6. Guo, Y.; Jia, Y. Thermal damage characteristics of CFRP laminates subjected to lightning strike. In Proceedings of the 21st International Conference on Composite, Xi'an, China, 20–25 August 2017.
7. Dong, Q.; Guo, Y.; Chen, J.; Yao, X.; Yi, X.; Ping, L.; Jia, Y. Influencing factor analysis based on electrical–Thermal-pyrolytic simulation of carbon fiber composites lightning damage. *Compos. Struct.* **2016**, *140*, 1–10. [CrossRef]
8. Foster, P.; Abdelal, G.; Murphy, A. Understanding how arc attachment behaviour influences the prediction of composite specimen thermal loading during an artificial lightning strike test. *Compos. Struct.* **2018**, *192*, 671–683. [CrossRef]

9. Hirano, Y.; Katsumata, S.; Iwahori, Y.; Todoroki, A. Artificial lightning testing on graphite/epoxy composite laminate. *Compos. Part A Appl. Sci. Manuf.* **2010**, *41*, 1461–1470. [CrossRef]
10. Kirchdoerfer, T.; Liebscher, A.; Ortiz, M. CTH shock physics simulation of non-linear material effects within an aerospace CFRP fastener assembly due to direct lightning attachment. *Compos. Struct.* **2018**, *189*, 357–365. [CrossRef]
11. Yin, J.J.; Li, S.L.; Yao, X.L.; Chang, F.; Li, L.K.; Zhang, X.H. Lightning Strike Ablation Damage Characteristic Analysis for Carbon Fiber/Epoxy Composite Laminate with Fastener. *Appl. Compos. Mater.* **2016**, *23*, 821–837. [CrossRef]
12. Kawakami, H. Lightning Strike Induced Damage Mechanisms of Carbon Fiber Composites. Ph.D. Thesis, University of Washington, Seattle, WA, USA, 2011.
13. Chemartin, L.; Lalande, P.; Tristant, F. Modeling and simulation of sparking in fastening assemblies. In Proceedings of the International Conference on Lightning and Static Electricity, Seattle, WA, USA, 18–20 September 2013.
14. Muñoz, R.; Delgado, S.; González, C.; López-Romano, B.; Wang, D.-Y.; LLorca, J. Modeling Lightning Impact Thermo-Mechanical Damage on Composite Materials. *Appl. Compos. Mater.* **2014**, *21*, 149–164. [CrossRef]
15. Abdelal, G.; Murphy, A. Nonlinear numerical modelling of lightning strike effect on composite panels with temperature dependent material properties. *Compos. Struct.* **2014**, *109*, 268–278. [CrossRef]
16. Ogasawara, T.; Hirano, Y.; Yoshimura, A. Coupled thermal–Electrical analysis for carbon fiber/epoxy composites exposed to simulated lightning current. *Compos. Part A Appl. Sci. Manuf.* **2010**, *41*, 973–981. [CrossRef]
17. Dong, Q.; Wan, G.; Guo, Y.; Zhang, L.; Wei, X.; Yi, X.; Jia, Y. Damage analysis of carbon fiber composites exposed to combined lightning current components D and C. *Compos. Sci. Technol.* **2019**, *179*, 1–9. [CrossRef]
18. Hirano, Y.; Reurings, C.; Iwahori, Y. Damage resistance of graphite/epoxy laminates with a fastener subjected to artificial lightning. In Proceedings of the 18th International Conference on Composites Materials, Jeju, Korea, 21–26 August 2011.
19. Feraboli, P.; Miller, M. Damage resistance and tolerance of carbon/epoxy composite coupons subjected to simulated lightning strike. *Compos. Part A Appl. Sci. Manuf.* **2009**, *40*, 954–967. [CrossRef]
20. Feraboli, P.; Kawakami, H. Damage of Carbon/Epoxy Composite Plates Subjected to Mechanical Impact and Simulated Lightning. *J. Aircr.* **2010**, *47*, 999–1012. [CrossRef]
21. Evans, S.; Jenkins, M.; Cole, M.; Haddad, M.; Carr, D.; Clark, D.; Stone, C.; Fay, A.; Mills, R.; Blair, D. An introduction to a new aerospace lightning direct effects research programme and the significance of Zone 2A waveform components on sparking joints. In Proceedings of the 33rd International Conference on Lightning Protection, Estoril, Lisboa, Portugal, 25–30 September 2016.
22. *Department of Defense Interface Standard-Electromagnetic Environmental Effects Requirements for System*; MIL-STD-464A; Defense Threat Reduction Agency: Alexandria, VA, USA, 2002.
23. *Aircraft Lightning Environment and Related Test Waveforms*; SAE ARP5412B; Society of Automotive Engineers: Warrendale, PA, USA, 2013.
24. Wang, Y.; Zhupanska, O.I. Modeling of thermal response and ablation in laminated glass fiber reinforced polymer matrix composites due to lightning strike. *Appl. Math. Model.* **2018**, *53*, 118–131. [CrossRef]
25. Chemartin, L.; Lalande, P.; Peyrou, B.; Chazottes, A.; Lago, F. Direct Effects of Lightning on Aircraft Structure: Analysis of the Thermal, Electrical and Mechanical Constraints. *J. Aerosp. Lab.* **2012**, *AL05-09*, 1–15.
26. Wang, Y. Multiphysics analysis of lightning strike damage in laminated carbon/glass fiber reinforced polymer matrix composite materials: A review of problem formulation and computational modeling. *Compos. Part A Appl. Sci. Manuf.* **2017**, *101*, 543–553. [CrossRef]
27. Guo, Y.; Dong, Q.; Chen, J.; Yao, X.; Yi, X.; Jia, Y. Comparison between temperature and pyrolysis dependent models to evaluate the lightning strike damage of carbon fiber composite laminates. *Compos. Part A Appl. Sci. Manuf.* **2017**, *97*, 10–18. [CrossRef]
28. Lago, F.; Gonzalez, J.J.; Freton, P.; Uhlig, F.; Lucius, N.; Piau, G.P. A numerical modelling of an electric arc and its interaction with the anode: Part III. Application to the interaction of a lightning strike and an aircraft in flight. *J. Phys. D Appl. Phys.* **2006**, *39*, 2294–2310. [CrossRef]
29. Lee, J.; Lacy, T.E.; Pittman, C.U.; Mazzola, M.S. Thermal response of carbon fiber epoxy laminates with metallic and nonmetallic protection layers to simulated lightning currents. *Polym. Compos.* **2018**, *39*, E2149–E2166. [CrossRef]

30. Gao, S.-P.; Lee, H.M.; Gao, R.X.-K.; Lim, Q.F.; Thitsartarn, W.; Liu, E.-X.; Png, C.E. Effective modeling of multidirectional CFRP panels based on characterizing unidirectional samples for studying the lightning direct effect. In Proceedings of the 32nd URSI General Assembly and Scientific Symposium, Montreal, QC, Canada, 19–26 August 2017.
31. Chen, H.; Wang, F.S.; Ma, X.T.; Yue, Z.F. The coupling mechanism and damage prediction of carbon fiber/epoxy composites exposed to lightning current. *Compos. Struct.* **2018**, *203*, 436–445. [CrossRef]
32. Dong, Q.; Wan, G.; Ping, L.; Guo, Y.; Yi, X.; Jia, Y. Coupled thermal-mechanical damage model of laminated carbon fiber/resin composite subjected to lightning strike. *Compos. Struct.* **2018**, *206*, 185–193. [CrossRef]
33. Shen, Q.; Omar, M.; Dongri, S. Ultrasonic NDE Techniques for Impact Damage Inspection on CFRP Laminates. *JMSR* **2011**, *1*. [CrossRef]
34. Kamiyama, S.; Hirano, Y.; Ogasawara, T. Delamination analysis of CFRP laminates exposed to lightning strike considering cooling process. *Compos. Struct.* **2018**, *196*, 55–62. [CrossRef]
35. Fu, S.; Guo, Y.; Shi, L.; Zhou, Y. Investigation on temperature behavior of CFRP during lightning strike using experiment and simulation. *Polym. Compos.* **2019**, *592*, 16. [CrossRef]
36. Che, H.; Gagné, M.; Rajesh, P.S.M.; Klemberg-Sapieha, J.E.; Sirois, F.; Therriault, D.; Yue, S. Metallization of Carbon Fiber Reinforced Polymers for Lightning Strike Protection. *J. Mater. Eng. Perform.* **2018**, *27*, 5205–5211. [CrossRef]
37. Teulet, P.; Billoux, T.; Cressault, Y.; Masquère, M.; Gleizes, A.; Revel, I.; Lepetit, B.; Peres, G. Energy balance and assessment of the pressure build-up around a bolt fastener due to sparking during a lightning impact. *Eur. Phys. J. Appl. Phys.* **2017**, *77*, 2080101–2080113. [CrossRef]
38. Sun, J.; Yao, X.; Tian, X.; Chen, J.; Wu, Y. Damage Characteristics of CFRP Laminates Subjected to Multiple Lightning Current Strike. *Appl. Compos. Mater.* **2018**, *30*, 156. [CrossRef]

© 2020 by the authors. Licensee MDPI, Basel, Switzerland. This article is an open access article distributed under the terms and conditions of the Creative Commons Attribution (CC BY) license (http://creativecommons.org/licenses/by/4.0/).

Article

Vibration Analysis of Piezoelectric Cantilever Beams with Bimodular Functionally-Graded Properties

Hong-Xia Jing [1], Xiao-Ting He [1,2,*], Da-Wei Du [1], Dan-Dan Peng [1] and Jun-Yi Sun [1,2]

1. School of Civil Engineering, Chongqing University, Chongqing 400045, China; 201716131083@cqu.edu.cn (H.-X.J.); 20145190@cqu.edu.cn (D.-W.D.); 201816021014@cqu.edu.cn (D.-D.P.); sunjunyi@cqu.edu.cn (J.-Y.S.)
2. Key Laboratory of New Technology for Construction of Cities in Mountain Area (Chongqing University), Ministry of Education, Chongqing 400045, China
* Correspondence: hexiaoting@cqu.edu.cn; Tel.: +86-(0)23-65120720

Received: 23 July 2020; Accepted: 10 August 2020; Published: 11 August 2020

Abstract: Piezoelectric materials have been found to have many electromechanical applications in intelligent devices, generally in the form of the flexible cantilever element; thus, the analysis to the corresponding cantilever is of importance, especially when advanced mechanical properties of piezoelectric materials should be taken into account. In this study, the vibration problem of a piezoelectric cantilever beam with bimodular functionally-graded properties is solved via analytical and numerical methods. First, based on the equivalent modulus of elasticity, the analytical solution for vibration of the cantilever beam is easily derived. By the simplified mechanical model based on subarea in tension and compression, as well as on the layer-wise theory, the bimodular functionally-graded materials are numerically simulated; thus, the numerical solution of the problem studied is obtained. The comparison between the theoretical solution and numerical study is carried out, showing that the result is reliable. This study shows that the bimodular functionally-graded properties may change, to some extent, the dynamic response of the piezoelectric cantilever beam; however, the influence could be relatively small and unobvious.

Keywords: piezoelectric effect; bimodular model; functionally-graded materials; cantilever; vibration

1. Introduction

The piezoelectric effect is among the most exploited transduction mechanisms for multiscale electromechanical applications, such as sensors, actuators and energy conversion devices, in which piezoelectric vibrational energy harvesting is attractive. For piezoelectric materials, there is a wide spectrum, from piezoelectric ceramics, perovskite structured lead zirconate titanate (PZT), to piezoelectric polymer films, polyvinylidene fluoride (PVDF). Among the piezoelectric materials, piezoelectric ceramics (PZT) is currently regarded as the most promising material system of piezoelectric vibrational energy harvesting devices since it can produce large output power, effective electromechanical coupling and high mechanical strain under an applied electric field [1,2]. Usually, piezoelectric sensors are a laminated original made of ceramic slice; thus, it is easy to result in stress concentration and also to develop interfacial microcracks. For overcoming this difficulty, functionally-graded piezoelectric materials (FGPM), whose properties of materials continuously change along certain direction, are developed. In FGPM, the obvious interface is disappeared, thus effectively avoiding the damage caused by the stress concentration at the interface. Studies concerning FGPM and corresponding structures made of FGPM have attracted the interests of scholars from all over the world [3–8].

In the existing works, the vibration problems of piezoelectric structures with functionally-graded properties have been extensively studied, and some valuable results were obtained. Based on the first-order shear theory, Mahinzare et al. [9] studied the free vibration of a rotating circular nanoplate composed of two directional functionally-graded piezo materials (two directional FGPM). The steady-state forced vibration of functionally-graded piezoelectric beams was investigated by Yao and Shi [10]. Shakeri and Mirzaeifar [11] made static and dynamic analysis of thick functionally-graded plates with piezoelectric layers based on the layerwise finite element model. Ebrahimi [12] investigated, analytically, the vibrations and dynamic response of functionally-graded plate, which is integrated with piezoelectric layers in thermal environment. Using a numerical method, Chen et al. [13] investigated the transient response and natural vibration of FGPM curved beam. Li et al. [14] studied the free vibration of FGM beams with surface-bonded piezoelectric layers, which is statically thermal post-buckled and subjected to both voltage and temperature rise. Huang and Shen [15] investigated the dynamic response and vibration of FGM plates with piezoelectric actuators in thermal environments. Fu et al. [16] made a nonlinear analysis of free vibration, dynamic stability and buckling for the FGPM beams in thermal environment. Li and Shi [17] studied the free vibration of a FGPM beam by using state-space based differential quadrature. Considering that there have been many studies in this field, it is not necessary to review them in detail here.

Compared with the FGPM, the bimodular effect of materials is relatively less known. However, many investigations have indicated that some materials [18,19], such as graphite, plastics, ceramics, concrete, steel, polymeric materials, powder metallurgy materials and some composites, will perform different elastic properties when they are in tension and compression; that is, they have different moduli when tensioned and compressed and thus are named bimodular materials [20] (see Figure 1), in which σ is the stress, ε is the strain and E^+ and E^- represents the tensile modulus of elasticity and compressive one, respectively. In 1982, the Elasticity Theory of Different Moduli, by Ambartsumyan [21], was published, in which the constitutive model for bimodular materials and the corresponding application in structural analysis were introduced systematically. The publication of this book marks that the idea of bimodular materials entered the field of vision of scholars from all over the world. Thereafter, bimodular problems for materials and structures have been investigated extensively [22–25]. These works indicated that the bimodular effect of materials will modify, to some extent, the mechanical behaviors of structures. Unfortunately, due to the complexity in analysis, the bimodular effect of materials is often neglected. Although some works have been carried out to combine the functionally-graded properties with bimodular effect of materials, for example, [26], the existing works appear insufficient. In fact, not only pure functionally-graded materials but also functionally-graded piezoelectric materials may have a certain degree of bimodular effect. Simple neglect will inevitably lead to analytical errors, which could further influence the design application of the electromechanical devices based on piezoelectric effect.

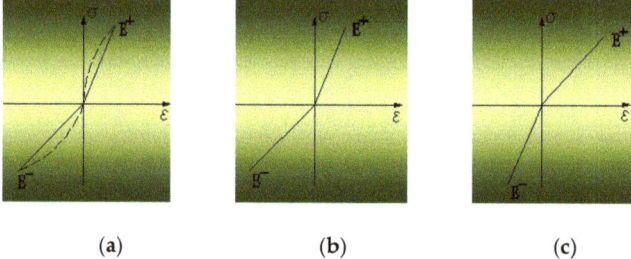

Figure 1. Constitutive model for bimodular materials: (**a**) nonlinear model under actual state; (**b**) bilinear model when $E^+ > E^-$; (**c**) bilinear model when $E^+ < E^-$.

He et al. [27] considered the bimodular effect during the analysis of functionally-graded piezoelectric materials and structures for the first time, and presented a two-dimensional analytical solution for a FGPM bimodular cantilever beam. Thereafter, aiming at the static problem of a bimodular FGPM cantilever, He et al. [28] neglected some unimportant factors to derive the one-dimensional theoretical solution and, based on the model of tension-compression subarea, conducted a two-dimensional numerical simulation. However, the existing works appear insufficient since there is no dynamic solution to the corresponding problem. At present, for a cantilever-type structure which is extensively adopted in piezoelectric sensors devices, the bimodular functionally-graded effect on its vibration has not been investigated. In addition, for piezoelectric polymer elements, the vibration problem, due to flexible and lightweight characters of the elements, is particularly acute, which also deserves further research.

In this paper, the free damping vibration problem of a piezoelectric cantilever beam with bimodular functionally-graded properties is analyzed by using analytical and numerical methods. The paper is organized as follows: In Section 2, the problem studied is briefly described and the constitutive relation for bimodular functionally-graded piezoelectric materials is presented. In Section 3, we derive the equivalent modulus of elasticity and obtain the analytical solution of the problem described. The numerical simulation for the problem is performed step by step in Section 4 and the corresponding comparisons and discussions are given in Section 5. Based on the results obtained in this study, some main conclusions are presented in Section 6.

2. The Problem Description

An FGPM orthotropic cantilever beam with different tensile and compressive properties is considered here, and the corresponding Cartesian coordinates system (x, y, z) is established, as depicted in Figure 2, where the right end of the beam is free and the left end is fully fixed; h is the rectangular section height of the beam, b is the section width and l is the beam length ($h \ll l$). By exerting a concentrated load at the right end of the beam, an initial displacement v_0 is generated at this end of the beam, as depicted in Figure 2. Then, by removing the load suddenly, the beam will freely vibrate until it ceases due to the existence of damping.

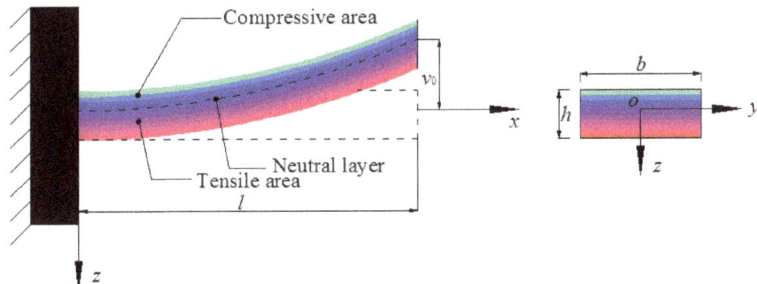

Figure 2. Scheme of the functionally-graded piezoelectric materials (FGPM) bimodular cantilever beam.

During the bending vibration, the beam will continuously present downward bending and upwards bending up to the last cease. In the downward (or upward) bending, the tensile and compressive areas, bound by the neutral layer, will take turns to generate; this physical phenomenon is different from the corresponding static problem, which has an unchanged tensile area and compressive area [27,28]. Therefore, unlike a bending beam in static analysis, in the vibration problem here, it seems that there is no definite tensile or compressive area in the beam. For the convenience of the following analysis, however, we assume in this study that the tensile or compressive area is defined in the present of initial displacement, i.e., the upper part of the beam is in compression and the lower part is in tension, as shown in Figure 2.

Note that the physical parameters of materials of the beam are also functions of coordinates due to the functionally-graded property. In present study, we assume physical parameters vary only along the thickness direction. Thus, all the material parameters are assumed to change with z, according to the following relations.

$$\begin{aligned} s_{ij}^+ &= s_{ij}^0 F^+(z), d_{ij}^+ = d_{ij}^0 F^+(z), \lambda_{ij}^+ = \lambda_{ij}^0 F^+(z), \\ s_{ij}^- &= s_{ij}^0 F^-(z), d_{ij}^- = d_{ij}^0 F^-(z), \lambda_{ij}^- = \lambda_{ij}^0 F^-(z) \end{aligned} \tag{1}$$

in which, $F^+(z) = e^{\alpha_1 z/h}$, $F^-(z) = e^{\alpha_2 z/h}$ are the tensile and compressive gradient functions, respectively; superscript "+" represents tension and "−" compression; $d_{ij}^{+/-}, \lambda_{ij}^{+/-}, s_{ij}^{+/-}$ are piezoelectric coefficient, dielectric coefficient and elastic coefficient, respectively; $d_{ij}^0, \lambda_{ij}^0, s_{ij}^0$ are values at the neutral layer of the corresponding material parameters. It should be noted here that the neutral layer is defined at $z = 0$, whose determination has been reported in our previous study on static problem [27,28]. The proposal of the neutral layer stems from the static problem but is still adopted in dynamic counterpart; otherwise, the so-called subarea in tension and compression cannot be realized. Furthermore, note that a set of very small electrodes are adhered discontinuously to the lower and upper surfaces of the beam and the beam is then poled along the direction of z.

Suppose that, in a two-dimensional problem, the stress and strain components is denoted by $\sigma_x^{+/-}, \sigma_z^{+/-}, \tau_{zx}^{+/-}$ and $\varepsilon_x^{+/-}, \varepsilon_z^{+/-}, \gamma_{zx}^{+/-}$, respectively; the electrical displacement and the electrical field components by $D_x^{+/-}, D_z^{+/-}$ and $E_x^{+/-}, E_z^{+/-}$, respectively. Therefore, the physical equations are

$$\begin{Bmatrix} \varepsilon_x^{+/-} \\ \varepsilon_z^{+/-} \\ \gamma_{zx}^{+/-} \end{Bmatrix} = \begin{bmatrix} s_{11}^{+/-} & s_{13}^{+/-} & 0 \\ s_{13}^{+/-} & s_{33}^{+/-} & 0 \\ 0 & 0 & s_{44}^{+/-} \end{bmatrix} \begin{Bmatrix} \sigma_x^{+/-} \\ \sigma_z^{+/-} \\ \tau_{zx}^{+/-} \end{Bmatrix} + \begin{bmatrix} 0 & d_{31}^{+/-} \\ 0 & d_{33}^{+/-} \\ d_{15}^{+/-} & 0 \end{bmatrix} \begin{Bmatrix} E_x^{+/-} \\ E_z^{+/-} \end{Bmatrix} \tag{2}$$

and

$$\begin{Bmatrix} D_x^{+/-} \\ D_z^{+/-} \end{Bmatrix} = \begin{bmatrix} 0 & 0 & d_{15}^{+/-} \\ d_{31}^{+/-} & d_{33}^{+/-} & 0 \end{bmatrix} \begin{Bmatrix} \sigma_x^{+/-} \\ \sigma_z^{+/-} \\ \tau_{zx}^{+/-} \end{Bmatrix} + \begin{bmatrix} \lambda_{11}^{+/-} & 0 \\ 0 & \lambda_{33}^{+/-} \end{bmatrix} \begin{Bmatrix} E_x^{+/-} \\ E_z^{+/-} \end{Bmatrix}, \tag{3}$$

in which superscript "+/−" denotes "tension/compression", similar to Equation (1). Until now, Equations (1)–(3) construct the materials relation considering piezoelectric effect as well as bimodular functionally-graded properties.

3. Equivalent Modulus of Elasticity and Analytical Solution

For a relatively shallow beam, the stress and strain along x direction, $\sigma_x^{+/-}$ and $\varepsilon_x^{+/-}$, are dominant, while other stresses and strains along z direction, $\sigma_z^{+/-}$ and $\tau_{zx}^{+/-}$ as well as $\varepsilon_z^{+/-}$ and $\gamma_{zx}^{+/-}$, are less important. Thus, the constitutive relation of FGPM with bimodular effect, that is, Equations (2) and (3), may be further simplified as,

$$\begin{cases} \varepsilon_x^{+/-} = s_{11}^{+/-} \sigma_x^{+/-} + d_{31}^{+/-} E_z^{+/-} \\ \varepsilon_z^{+/-} = 0 \\ \gamma_{zx}^{+/-} = 0 \end{cases} \tag{4}$$

and

$$\begin{cases} D_x^{+/-} = \lambda_{11}^{+/-} E_x^{+/-} \\ D_z^{+/-} = d_{31}^{+/-} \sigma_x^{+/-} + \lambda_{33}^{+/-} E_z^{+/-} \end{cases}. \tag{5}$$

In existing studies for the two-dimensional problem [27,29], $D_x \gg D_z$ may be found; thus, it may be assumed that $D_z \approx 0$ in a one-dimensional problem, especially if a long and shallow beam is considered here. From the second one of Equation (5), we have

$$E_z^{+/-} = -\frac{d_{31}^{+/-}}{\lambda_{33}^{+/-}} \sigma_x^{+/-} \tag{6}$$

Plugging Equation (6) into the first one of Equation (4) yields

$$\varepsilon_x^{+/-} = \left[\frac{s_{11}^{+/-}\lambda_{33}^{+/-} - (d_{31}^{+/-})^2}{\lambda_{33}^{+/-}}\right]\sigma_x^{+/-} = \frac{\sigma_x^{+/-}}{E^*}, \tag{7}$$

in which E^* represents the equivalent modulus of elasticity, that is

$$E^* = \frac{\lambda_{33}^{+/-}}{s_{11}^{+/-}\lambda_{33}^{+/-} - (d_{31}^{+/-})^2}. \tag{8}$$

We note that Equation (8) may be rewritten as the form $E^* = [s_{11}^{+/-} - (d_{31}^{+/-})^2/\lambda_{33}^{+/-}]^{-1}$, clearly showing the piezoelectric effect on the elastic modulus. Then, $E^* = 1/s_{11}^{+/-}$ is obtained when $s_{11}^{+/-} \gg (d_{31}^{+/-})^2/\lambda_{33}^{+/-}$; this, exactly, stands for the reciprocal relation of stiffness coefficient and flexibility coefficient. Meanwhile, the existence of the term $(d_{31}^{+/-})^2/\lambda_{33}^{+/-}$ also reveals the well-known piezoelectric stiffening effect. It is this term that the resulting equivalent modulus becomes larger than $1/s_{11}^{+/-}$, thus stiffening the mechanical performance of piezoelectric materials and structures under external loading.

Now, the vibration equation of free damping of the bimodular FGPM cantilever beam may be easily obtained, by only replacing E in a classical equation [30] by E^*,

$$E^*I_y\frac{\partial^4 v(x,t)}{\partial x^4} + \overline{m}\frac{\partial^2 v(x,t)}{\partial t^2} + c\frac{\partial v(x,t)}{\partial t} = 0. \tag{9}$$

in which $\overline{m}$ is the uniformly-distributed mass, $v(x,t)$ is the displacement along z direction, t is the time variable, E^*I_y is the equivalent bending stiffness of the beam, I_y is the moment of inertia with respect to y axis, c is the damping parameter, and $c = 2\xi\overline{m}\omega$, in which, ξ is the damping ratio and ω is the undamped frequency. Equation (9) may be solved under the following boundary conditions:

$$v(x,t) = 0 \text{ and } \frac{\partial v(x,t)}{\partial x} = 0, \text{ at } x = 0 \tag{10}$$

$$EI_y\frac{\partial^2 v(x,t)}{\partial x^2} = 0 \text{ and } EI_y\frac{\partial^3 v(x,t)}{\partial x^3} = 0, \text{ at } x = l. \tag{11}$$

and the conditions of initial values

$$v(x,t) = v_0 \text{ and } \frac{\partial v(x,t)}{\partial t} = 0, \text{ at } x = l, t = 0 \tag{12}$$

The variable separation is first needed to solve the Equation (9); suppose that

$$v(x,t) = \phi(x)Y(t) \tag{13}$$

Plugging Equation (13) into Equation (9), we may obtain the following two equations:

$$\frac{d^4\phi(x)}{dt^4} - a^4\phi(x) = 0 \tag{14}$$

and

$$\frac{d^2Y(t)}{dt^2} + \frac{c}{\overline{m}}\frac{dY(t)}{dt} + \omega^2 Y(t) = 0 \tag{15}$$

in which a is an unknown constant: $\omega^2 = a^4 E^* I_y/\overline{m}$. Thus, the partial differential Equation (9) is transformed two ordinary differential Equations (14) and (15), and their solutions under defined boundary conditions or initial conditions, $\phi(x)$ and $Y(t)$, may be easily derived. Since the solving process is readily found in any a textbook on dynamic problems of bending beams, here we do not repeat the details, only directly presenting the final solution; they are

$$\phi(x) = \cos ax - \cosh ax - \frac{(\cos aL + \cosh aL)}{(\sin aL + \sinh aL)}(\sin ax - \sinh ax) \tag{16}$$

and

$$Y(t) = \left[Y(0) \cos \omega_D t + \left(\frac{dY(0)/dt + Y(0)\xi\omega}{\omega_D} \right) \sin \omega_D t \right] e^{(-\xi\omega t)} \tag{17}$$

in which $Y(0)$ is the corresponding mode amplitude for initial displacement v_0, and ω_D is the natural damped frequency,

$$\omega_D = \omega\sqrt{1-\xi^2} \tag{18}$$

Obviously, the proposal of the equivalent modulus of elasticity for bimodular FGPM beams plays an important role; this equivalent modulus may be used in the analysis of similar bimodular FGPM structures.

4. Numerical Simulation

In order to simulate the free vibration of the beam, a transient load is applied on the right end of the beam at the beginning, thus generating an initial displacement v_0 at the end, and then release suddenly, the beam will vibrate up to the cease due to the damping, as shown in Figure 2. In this section, the software ABAQUS is used to simulate the free damping vibration of the bimodular FGPM cantilever beam.

4.1. Constitutive Equation of Piezoelectrical Materials

First, we need to describe the input of the piezoelectrical materials parameters in the software. The piezoelectrical materials model in ABAQUS follows the e-form constitutive equation, such that

$$\begin{cases} \sigma_{ij} = c^E_{ijkl}\varepsilon_{kl} - e_{kij}E_k \\ D_i = e_{ijk}\varepsilon_{kl} + \lambda^\varepsilon_{ik}E_k \end{cases}, \tag{19}$$

where the stress component is denoted by σ_{ij}; the strain component is denoted by ε_{ij}; the electrical displacement component is denoted by D_i; the electrical field strength are denoted by E_k; the stiffness coefficient matrix is denoted by c^E_{ijkl}; the dielectric constant matrix is denoted by λ^ε_{ik}.

In piezoelectrical materials, there is a polarization direction which corresponds to z direction in x-y-z coordinate system (i.e., 3-direction in matrix). In ABAQUS, we need to transform the flexibility coefficient matrix into the stiffness coefficient matrix, such that

$$[s_{ij}] = \begin{bmatrix} s_{11} & s_{12} & s_{13} & 0 & 0 & 0 \\ s_{21} & s_{22} & s_{23} & 0 & 0 & 0 \\ s_{31} & s_{32} & s_{33} & 0 & 0 & 0 \\ 0 & 0 & 0 & s_{66} & 0 & 0 \\ 0 & 0 & 0 & 0 & s_{44} & 0 \\ 0 & 0 & 0 & 0 & 0 & s_{44} \end{bmatrix} \Rightarrow [c_{ij}] = \begin{bmatrix} c_{11} & c_{12} & c_{13} & 0 & 0 & 0 \\ c_{21} & c_{22} & c_{23} & 0 & 0 & 0 \\ c_{31} & c_{32} & c_{33} & 0 & 0 & 0 \\ 0 & 0 & 0 & c_{66} & 0 & 0 \\ 0 & 0 & 0 & 0 & c_{44} & 0 \\ 0 & 0 & 0 & 0 & 0 & c_{44} \end{bmatrix}. \tag{20}$$

Note that the above definition for the variation form of functionally-graded materials and the flexibility coefficient $s_{ij} = s^0_{ij}e^{\alpha_i z/h}$ (where $i = 1,2$ due to the bimodular effect), the stiffness coefficient will thus vary with $c_{ij} = c^0_{ij}e^{-\alpha_i z/h}$, otherwise s_{ij} and c_{ij} cannot satisfy $[s_{ij}][c_{ij}] = [s^0_{ij}]e^{\alpha_i z/h}[c^0_{ij}]e^{-\alpha_i z/h} = [E]$ (here $[E]$ is an unit matrix). The piezoelectrical strain constants should, at the same time, be $e_{ij} = d_{ij}c_{ij}^E = d_{ij}^0 e^{\alpha_i z/h} c_{ij}^0 e^{-\alpha_i z/h} = e_{ij}^0$, where c^E is the constant matrix of elastic stiffness in the state of short circuit; thus, the piezoelectrical stress constants matrix is

$$[e_{ij}] = [e_{ij}^0] = \begin{bmatrix} 0 & 0 & 0 & 0 & e_{15} & 0 \\ 0 & 0 & 0 & 0 & 0 & e_{15} \\ e_{31} & e_{31} & e_{33} & 0 & 0 & 0 \end{bmatrix}. \quad (21)$$

The constitutive equation for piezoelectric materials is, in the form of matrix, expressed as

$$\begin{bmatrix} \sigma_{11} \\ \sigma_{22} \\ \sigma_{33} \\ \sigma_{12} \\ \sigma_{13} \\ \sigma_{23} \end{bmatrix} = \begin{bmatrix} c_{11} & c_{12} & c_{13} & 0 & 0 & 0 \\ c_{12} & c_{11} & c_{13} & 0 & 0 & 0 \\ c_{13} & c_{13} & c_{33} & 0 & 0 & 0 \\ 0 & 0 & 0 & c_{66} & 0 & 0 \\ 0 & 0 & 0 & 0 & c_{44} & 0 \\ 0 & 0 & 0 & 0 & 0 & c_{44} \end{bmatrix} \begin{bmatrix} \varepsilon_{11} \\ \varepsilon_{22} \\ \varepsilon_{33} \\ \gamma_{12} \\ \gamma_{13} \\ \gamma_{23} \end{bmatrix} - \begin{bmatrix} 0 & 0 & e_{31} \\ 0 & 0 & e_{31} \\ 0 & 0 & e_{33} \\ 0 & 0 & 0 \\ e_{15} & 0 & 0 \\ 0 & e_{15} & 0 \end{bmatrix} \begin{bmatrix} E_1 \\ E_2 \\ E_3 \end{bmatrix} \quad (22)$$

and

$$\begin{bmatrix} D_1 \\ D_2 \\ D_3 \end{bmatrix} = \begin{bmatrix} 0 & 0 & 0 & 0 & e_{15} & 0 \\ 0 & 0 & 0 & 0 & 0 & e_{15} \\ e_{31} & e_{31} & e_{33} & 0 & 0 & 0 \end{bmatrix} \begin{bmatrix} \varepsilon_{11} \\ \varepsilon_{22} \\ \varepsilon_{33} \\ \gamma_{12} \\ \gamma_{13} \\ \gamma_{23} \end{bmatrix} + \begin{bmatrix} \lambda_{11} & 0 & 0 \\ 0 & \lambda_{11} & 0 \\ 0 & 0 & \lambda_{33} \end{bmatrix} \begin{bmatrix} E_1 \\ E_2 \\ E_3 \end{bmatrix} \quad (23)$$

According to Voigt notation, the vector components (1, 2, 3, 4, 5 and 6) correspond to the second-order tensor mark of double-subscript (11, 22, 33, 13, 23 and 12), respectively. Thus, the above two equations can be written as,

$$\begin{bmatrix} \sigma_{11} \\ \sigma_{22} \\ \sigma_{33} \\ \sigma_{12} \\ \sigma_{13} \\ \sigma_{23} \end{bmatrix} = \begin{bmatrix} D_{1111} & D_{1122} & D_{1133} & 0 & 0 & 0 \\ D_{2211} & D_{2222} & D_{2233} & 0 & 0 & 0 \\ D_{3311} & D_{3322} & D_{3333} & 0 & 0 & 0 \\ 0 & 0 & 0 & D_{1212} & 0 & 0 \\ 0 & 0 & 0 & 0 & D_{1313} & 0 \\ 0 & 0 & 0 & 0 & 0 & D_{1313} \end{bmatrix} \begin{bmatrix} \varepsilon_{11} \\ \varepsilon_{22} \\ \varepsilon_{33} \\ \gamma_{12} \\ \gamma_{13} \\ \gamma_{23} \end{bmatrix} - \begin{bmatrix} 0 & 0 & e_{311} \\ 0 & 0 & e_{322} \\ 0 & 0 & e_{333} \\ 0 & 0 & 0 \\ e_{113} & 0 & 0 \\ 0 & e_{223} & 0 \end{bmatrix} \begin{bmatrix} E_1 \\ E_2 \\ E_3 \end{bmatrix} \quad (24)$$

and

$$\begin{bmatrix} q_1 \\ q_2 \\ q_3 \end{bmatrix} = \begin{bmatrix} 0 & 0 & 0 & 0 & e_{113} & 0 \\ 0 & 0 & 0 & 0 & 0 & e_{113} \\ e_{311} & e_{322} & e_{333} & 0 & 0 & 0 \end{bmatrix} \begin{bmatrix} \varepsilon_{11} \\ \varepsilon_{22} \\ \varepsilon_{33} \\ \gamma_{12} \\ \gamma_{13} \\ \gamma_{23} \end{bmatrix} + \begin{bmatrix} D_{11} & 0 & 0 \\ 0 & D_{11} & 0 \\ 0 & 0 & D_{33} \end{bmatrix} \begin{bmatrix} E_1 \\ E_2 \\ E_3 \end{bmatrix} \quad (25)$$

where, D_{ijkl} represents the modulus of elasticity, D_{ij} represents the dielectric coefficient and q_i represents the electrical displacement component. By the comparison of the above two sets of equations, the corresponding relationship among constants is readily found, which can be used to input values of the constants. For example, c_{11} should be input at the location of D_{1111}; e_{31} should be input at the location of e_{311} and λ_{11} should be input at the location of D_{11}.

4.2. Initial Modeling

(i) Establishment of structures

Suppose that the length l of a FGPM cantilever beam is set to be 50 mm, the section width b to be 10 mm and the section height h to be 2 mm.

(ii) Determination of tension-compression subarea under initial displacement

For the determination of the tensile and compressive height of the beam, a set of functionally-graded indexes should be chosen first in the light of our previous study on static analysis [28]. In this study, we consider the following two groups of gradient indexes: (a) $\alpha_1 = -2$, $\alpha_2 = -3$ and (b) $\alpha_1 = 2$, $\alpha_2 = 3$ to correspond to the two different cases: $E^+(z) > E^-(z)$ as well as $E^-(z) > E^+(z)$, respectively, as shown in Figure 3, in which E_0 is the electric modulus value on the neutral layer, in case (a), $h_1 = 0.6$ mm, $h_2 = 1.4$ mm and in case (b), $h_1 = 1.4$ mm, $h_2 = 0.6$ mm.

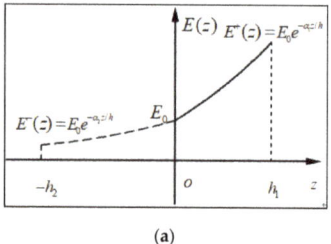 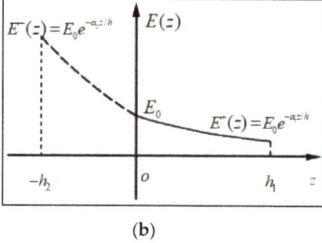

(a) (b)

Figure 3. $E^-(z)$: (a) $\alpha_1 = -2$, $\alpha_2 = -3$; (b) $\alpha_1 = 2$, $\alpha_2 = 3$.

As is the case in most commercial software, there will be the limitations for implementing FGMs in ABAQUS; for example, it seems to be inconvenient to realize the change of material properties along a certain direction as a continuous function. To overcome the shortcomings, an alternative implementation was proposed in the context of the commercial finite element package ABAQUS [31]. In this study, however, the layer-wise model was still used to simulate the functionally-graded properties, since this practice is conventional and well-known. Without losing the computational accuracy, we divided the beam into a moderate number of layers along the thickness direction; the physical parameters of the material on each layer are considered to be the same, thus indirectly realizing the continuous variation of properties of materials along the thickness direction if the numbers for layering are sufficient. To this end, bound by the neutral layer, the upper and lower areas of the beam are equally divided into 40 layers along the thickness direction, each layer being 0.05 mm thick, as depicted in Figure 4. It is easy to see that in case (a), there are 28 layers in the compressive area and 12 layers in the tensile area, while in case (b), the layering is the opposite, that is, 12 layers in the compressive area and 28 layers in the tensile area. The coordinate origin is still on the neutral layer.

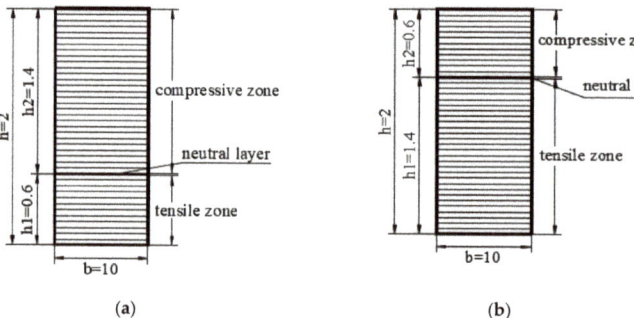

(a) (b)

Figure 4. Sketch of layering on cross section of the beam (unit: mm): (a) $\alpha_1 = -2$, $\alpha_2 = -3$; (b) $\alpha_1 = 2$, $\alpha_2 = 3$.

(iii) Input of properties of materials

The material PbZrTiO$_3$-4 (generally abbreviated as PZT-4) is selected as our materials simulated. Table 1 shows the material constant on the neutral layer $z = 0$ which are directly input into ABAQUS.

Material constants up or down the neutral layer may be computed and input into the program, by using the layering pattern of tension-compression established in Step (ii).

Table 1. Elastic, piezoelectric and dielectric constants of PZT-4 materials [32].

Elastic Constant (10^{-12} m$^2 \cdot$N^{-1})					Piezoelectric Constant (10^{-12} C$\cdot$N^{-1})			Dielectric Constant (10^{-8} F$\cdot$m^{-1})	
s_{11}^0	s_{12}^0	s_{13}^0	s_{33}^0	s_{44}^0	d_{31}^0	d_{33}^0	d_{15}^0	λ_{11}^0	λ_{33}^0
12.4	−3.98	−5.52	16.1	39.1	−135	300	525	1.301	1.151

(iv) Boundary conditions

The left end of the beam is fully fixed and the right end is free; this agrees with the mechanical model presented in Figure 2. Besides, the upper and lower surfaces of the beam are open circuited.

(v) Mesh division

An 8-node linear piezoelectric brick C3D8E is used, and the mesh size is set to be 1 mm × 1 mm, in which the size ratio is 0.001 and also the global seed is set.

4.3. Analysis of Frequency Extraction

While solving linear dynamic problems, the mode superposition method is used in ABAQUS. Before the dynamic analysis, we need to extract the frequency of the computational example to obtain the vibration mode and natural frequency of the structure.

First of all, according to the initial model established in Section 4.2, the density of materials is defined as 7.5×10^3 kg/m^3 in a property function module and also the frequency extraction analysis is defined in the analysis step. In ABAQUS, there are two kinds of method in the frequency extraction: one is the Lanczos method, which is suitable for the larger model and needs to extract the multi-order mode; the other is the subspace iterative method. In this computation, the latter is selected due to the characteristic of structural unit.

In the analysis of linear dynamic problems using the mode superposition method, a sufficient number of modes are required in the frequency extraction. The criterion follows that the total effective mass in the main direction of motion exceeds 90% of the movable mass in the model. To this end, we calculate and extract the frequency whose eigenvalue numbers are 10, 15 and 20, respectively, to select the appropriate eigenvalue. By viewing DAT file, we extract the data of frequency, participation factors and effective mass for 10, 15 and 20 eigenvalues, as shown in Tables 2–4. Here, below, are listed the data from the case $\alpha_1 = -2$, $\alpha_2 = -3$, which is very significant.

The main motion direction of the beam is along the z-axis direction, the participation factors in Tables 2–4 reflect that the first-order mode acts mainly in the z-direction. The total mass of this model is 7.500×10^{-3} kg. Since the constrained nodes account for a small proportion of all nodes, it may be approximated that the movable mass in the model is equal to the total of the model. Table 5 shows that, when eigenvalue numbers are 10, 15 and 20, respectively, the ratio of effective mass in z-direction to total motion mass. It is easy to see that when the eigenvalue numbers are 15 and 20, the ratios are uniformly greater than 90%, which both satisfy the basic requirement for sufficient number of modes. However, considering the hardware factors and amount of calculation, the 15th order mode is adopted here. Besides, for another case $\alpha_1 = 2$, $\alpha_2 = 3$, we still select the 15th order mode, the ratio reads $6.993 \times 10^{-3} / (7.500 \times 10^{-3}) = 93.24\%$ in this case.

Table 2. Frequency, participation factor and effective mass for 10 eigenvalues.

Mode Number	Frequency/Hz (Cycles Time)	Participation Factor (z-component)	Effective Mass/Kg (z-component)
1	326.490	1.565	4.562×10^{-3}
2	1767.500	6.332×10^{-11}	7.519×10^{-24}
3	2029.300	−0.868	1.418×10^{-3}
4	2688.300	2.535×10^{-10}	8.065×10^{-23}
5	5634.600	0.510	4.941×10^{-4}
6	7994.700	8.990×10^{-9}	1.007×10^{-19}
7	9720.000	-2.243×10^{-8}	1.114×10^{-18}
8	10,916.000	−0.366	2.549×10^{-4}
9	13,994.000	-4.504×10^{-6}	1.983×10^{-14}
10	14,237.000	1.494×10^{-2}	8.313×10^{-7}
Total			6.729×10^{-3}

Table 3. Frequency, participation factor and effective mass for 15 eigenvalues.

Mode Number	Frequency/Hz (Cycles Time)	Participation Factor (z-component)	Effective Mass/Kg (z-component)
1	326.490	1.565	4.562×10^{-3}
2	1767.500	6.332×10^{-11}	7.519×10^{-24}
3	2029.300	−0.868	1.418×10^{-3}
4	2688.300	2.527×10^{-10}	8.016×10^{-23}
5	5634.600	0.510	4.941×10^{-4}
... ...	... ...	... ...	... ...
12	20,587.000	-7.667×10^{-7}	5.125×10^{-16}
13	22,745.000	-1.099×10^{-6}	3.228×10^{-15}
14	26,094.000	−0.244	1.042×10^{-4}
15	28,006.000	-2.242×10^{-5}	3.993×10^{-13}
Total			6.989×10^{-3}

Table 4. Frequency, participation factor and effective mass for 20 eigenvalues.

Mode Number	Frequency/Hz (Cycles Time)	Participation Factor (z-component)	Effective Mass/Kg (z-component)
1	326.490	1.565	4.562×10^{-3}
2	1767.500	6.331×10^{-11}	7.517×10^{-24}
3	2029.300	−0.868	1.418×10^{-3}
4	2688.300	2.527×10^{-10}	8.016×10^{-23}
5	5634.600	0.510	4.941×10^{-4}
... ...	... ...	... ...	... ...
17	36,310.000	4.508×10^{-7}	1.586×10^{-16}
18	37,469.000	4.075×10^{-7}	5.328×10^{-16}
19	42,285.000	1.940×10^{-2}	1.246×10^{-6}
20	45,307.000	0.236	3.677×10^{-5}
Total			7.101×10^{-3}

Table 5. Ratios of z-direction mass to total mass under different eigenvalue numbers.

Eigenvalue Numbers	Effective z-Direction Movable Mass to Total Motion Mass
10	$6.729 \times 10^{-3} / (7.500 \times 10^{-3}) = 89.72\%$
15	$6.989 \times 10^{-3} / (7.500 \times 10^{-3}) = 93.19\%$
20	$7.101 \times 10^{-3} / (7.500 \times 10^{-3}) = 94.68\%$

4.4. Simulation for Free Damping Vibration

As shown in Table 3, we have finished the frequency extraction of 15 eigenvalues for the case $\alpha_1 = -2$, $\alpha_2 = -3$, in which the maximum frequency reads 28,006 Hz; thus, the corresponding period is $1/28006$ s $= 0.00003571$ s. Since the time increment in the analysis step of transient modal should be less than this period value, the time increment is determined as 3×10^{-5} s. The details are shown below.

(i) Establishment of dynamic analysis step of transient modal

Although the material properties considered in this study are bimodular, functionally-graded and piezoelectric—that is, it is not linear—the nonlinear behavior of the material itself has little impact on the dynamic response of the structure. Besides, the motion attributes to the small deflection bending vibration thus there is no geometrical nonlinearity here. Furthermore, the damping of the structural system is relatively small, and the frequency involved in the analysis is low. Therefore, the system may be regarded as linear, which is suitable for linear transient dynamic analysis.

Frequency extraction analysis step is followed by this step. To observe the attenuation process of the vibration, the analytical step time is set to be 0.5 s and the time increment is determined as 3×10^{-5} s, as indicated above.

(ii) Setting of the damping

Direct modal damping is used here and the value of the damping ratio is 0.03. The starting mode order is 1 and the terminating mode order is 15.

(iii) Setting of the output of historical variable

In the output of historical variable, the displacement component is selected.

(iv) Definition of load

In order to make the model produce the initial displacement of 2 mm, a short-term load is applied on the model, in which the time of duration of the load is determined by the variation of load amplitudes with time and the amplitude is given in tabulate. In the time period 0–0.005 s, the amplitude is 1; and in the later time period, the amplitude is set as 0. The smoothness parameter is set as 0.25. Lastly, along the negative direction of z-axis, the load is applied and the magnitude of the load is 25 N.

(v) Submission of analysis and post-processing

After a job is established, we may submit the job and calculate and output the results. Figures 5 and 6 show cloud diagrams of stress and displacement of 15th order mode at the end of the transient modal analysis, in which Figure 5 is for the case $\alpha_1 = -2$, $\alpha_2 = -3$ and Figure 6 is for the case $\alpha_1 = 2$, $\alpha_2 = 3$.

Note that although the numerical implementation in ABAQUS is three-dimensional in the form, we still put out the final results in the form of two-dimensional case, including three stresses, σ_x, σ_z and τ_{xz}, as well as two displacements, u and w, which are typical in two-dimensional plane problem. In fact, the studied problem is a two-dimensional plane problem, even in some cases (for example, a slim beam), the problem may be further simplified as a one-dimensional problem, like our analytical solution based on Euler–Bernoulli beam theory.

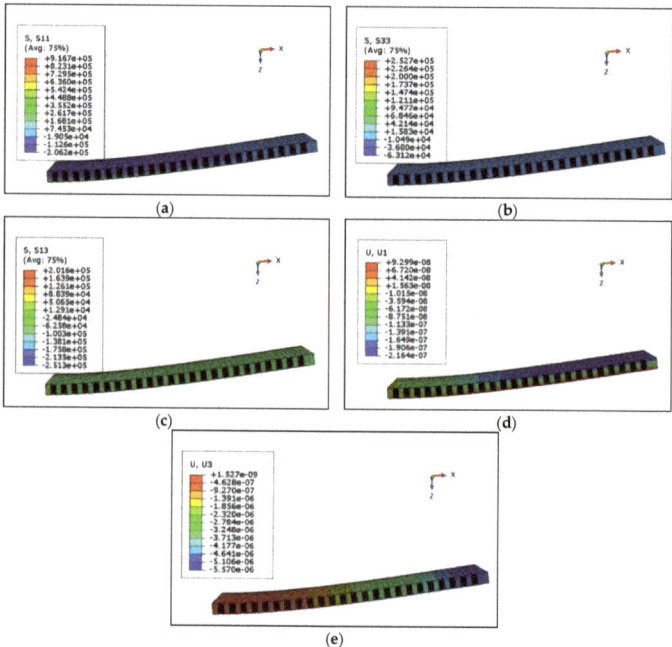

Figure 5. Cloud diagram of stress and displacement for the case $\alpha_1 = -2, \alpha_2 = -3$: (**a**) σ_x; (**b**) σ_z; (**c**) τ_{xz}; (**d**) u; (**e**) w.

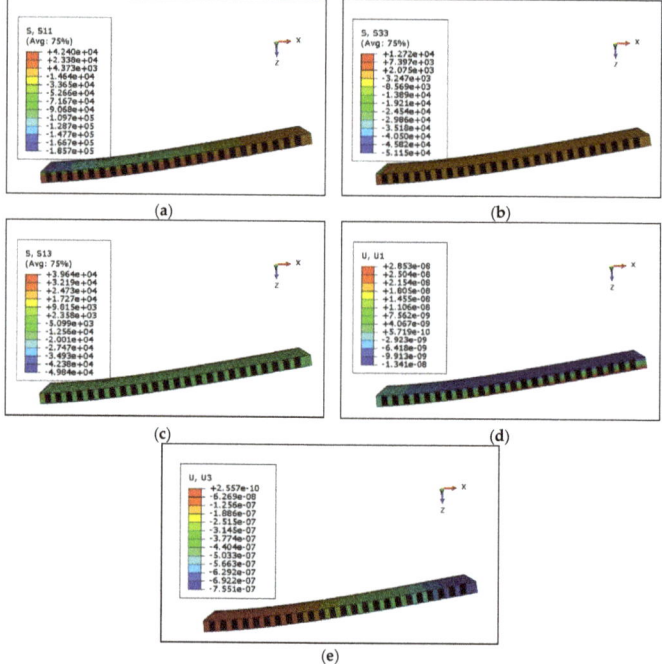

Figure 6. Cloud diagram of stress and displacement for the case $\alpha_1 = 2, \alpha_2 = 3$: (**a**) σ_x; (**b**) σ_z; (**c**) τ_{xz}; (**d**) u; (**e**) w.

5. Comparisons and Discussion

5.1. Comparison of Theoretical and Numerical Results

For the purpose of the full comparison between the theoretical results and numerical ones, both under the two cases of different modulus—that is, the tensile elastic modulus is greater than the compressive one $E^+(z) > E^-(z)$, or reversely, $E^+(z) < E^-(z)$—we should select an object of study from the beam, for example, the upper surface of the free end, as our study object. In order to investigate the displacement variation with time under two cases of different modulus, we take the following two cases of functionally-grade index: (a) $\alpha_1 = -2$, $\alpha_2 = -3$, which corresponds to $E^+(z) > E^-(z)$, and the case (b) $\alpha_1 = 2$, $\alpha_2 = 3$, which corresponds to $E^+(z) < E^-(z)$. The numerical results are taken from the 15th order mode. For the theoretical results, we should compute the equivalent modulus of elasticity first. For this purpose, substituting Equation (1) into Equation (8), we have

$$E^* = \frac{\lambda_{33}^{+/-}}{s_{11}^{+/-}\lambda_{33}^{+/-} - (d_{31}^{+/-})^2} = \frac{\lambda_{33}^0}{s_{11}^0 \lambda_{33}^0 - (d_{31}^0)^2} e^{-\alpha_i z/h} \quad (26)$$

Substituting the values of s_{11}^0, d_{31}^0 and λ_{33}^0 from Table 1 into the above equation and also noting that the upper surface is in compression; thus, $\alpha_i = \alpha_2$, the values of the equivalent modulus of elasticity, may be determined and is listed in Table 6. Note that for the first case, $E^+(z) > E^-(z)$, we take $\alpha_i = \alpha_2 = -3$ and $z = -0.0014$ m, and for the second case, $E^+(z) < E^-(z)$, we have $\alpha_i = \alpha_2 = 3$ and $z = -0.0006$ m, in which the values of z may refer to the cases (a) and (b) in Figure 4. Besides, the theoretical value of vibration frequency may be obtained via the expression $\omega^2 = a^4 E^* I^y / \overline{m}$ and the numerical result of frequency is also from the 15th order modes, which are also listed in Table 6.

Table 6. Equivalent modulus and frequency from theoretical and numerical results.

Cases	Equivalent Modulus E^* (GPa)	Vibration Frequency ω (Hz)	
		Theoretical Value	Numerical Value
$E^+(z) > E^-(z)$	11	17,506	28,006
$E^+(z) < E^-(z)$	227	28,019	34,599

Figures 7 and 8 show the two time-displacement curves from theoretical and numerical results, in which Figure 7 is for $E^+(z) > E^-(z)$ and Figure 8 is for $E^+(z) < E^-(z)$. It is easy to see that despite some differences, the theoretical curve basically agrees with the curve from numerical simulation, this validates the theoretical solution to some extent.

From Figures 7 and 8, we may also see that under two cases $E^+(z) > E^-(z)$ and $E^+(z) < E^-(z)$, the attenuation speed of the vibration from the numerical result is faster than that from theoretical solution, this is because the natural frequency from numerical simulation is greater than the one from theoretical solution, which may be easily seen from Table 6, in which for $E^+(z) > E^-(z)$, the value from numerical simulation is 28,006 Hz while the counterpart from theoretical solution is 17,506 Hz; for $E^+(z) < E^-(z)$, the two values from numerical simulation and theoretical solution are 34,599 Hz and 28,019 Hz, respectively. In addition, for $E^+(z) < E^-(z)$, the attenuation speed is obviously faster than that in the case $E^+(z) > E^-(z)$; this also may be easily seen from Table 6, in which the natural frequency from $E^+(z) < E^-(z)$ is greater than the one from $E^+(z) > E^-(z)$, for example, for theoretical solutions 28,019 Hz is greater than 17,506 Hz, while the numerical simulation 34,599 Hz is also greater than 28,006 Hz. It may be concluded that the relative magnitudes of the tensile and compressive moduli have influence on the attenuation speed of the vibration.

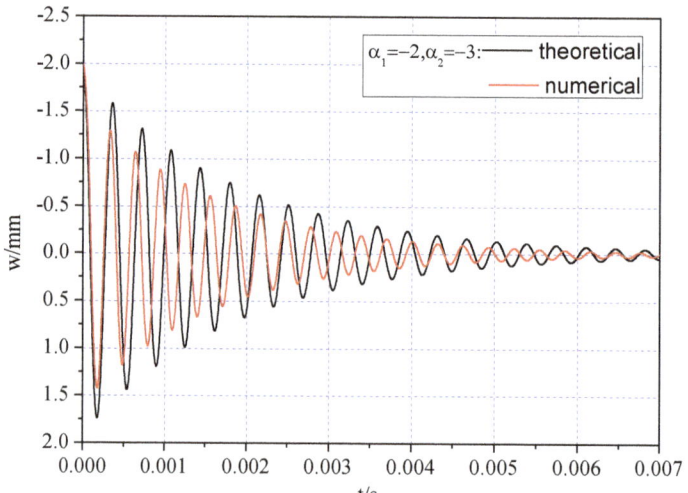

Figure 7. Comparison of two time-displacement curves for $E^+(z) > E^-(z)$

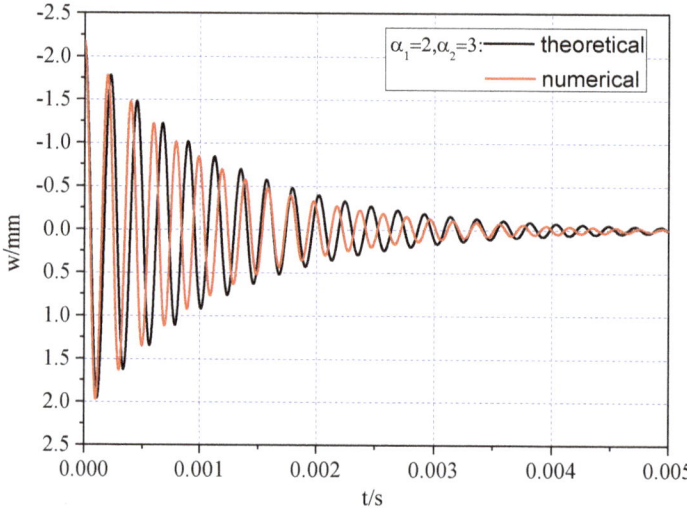

Figure 8. Comparison of two time-displacement curves for $E^+(z) < E^-(z)$

It should be noted here that there are some differences between the theoretical solution and the numerical simulation, mainly due to the different mechanical models on which the two methods are established. In the theoretical analysis, the introduction of the equivalent modulus of elasticity may greatly simplify the derivation process; on the other hand, this equivalent practice inevitably cause some errors, in other words, the so-called equivalence is actually a compromise. Based on this consideration, the numerical simulation seems to be more accurate than the theoretical analysis; that is to say, the theoretical analysis and numerical simulation have their own advantages and complement each other.

5.2. Comparison of Theoretical and Experimental Results

At present, the relevant vibration experiment for the functionally-graded piezoelectric cantilever beam has not been found, not to mention the consideration for bimodular effect of the materials.

The comparison should therefore be based on the experimental data available. Aiming at purely piezoelectric materials (which means there is no bimodular effect and also no functionally-graded characteristic), Yang et al. [33] performed an experiment of free damping vibration of a cantilever beam. In this section, the theoretical solution derived in this study with the experimental results from [33] will be compared.

Before comparison, it is necessary for us to reduce our theoretical solution to a purely piezoelectric case, to agree with the material model adopted in [33]. For this purpose, we need to presume $\alpha_1 = \alpha_2 = 0$ in Equation (8); thus, the equivalent modulus of elasticity E^* may be simplified as

$$E' = \frac{\lambda_{33}}{s_{11}\lambda_{33} - (d_{31})^2} \quad (27)$$

in which E' stands for the equivalent modulus of elasticity without bimodular and functionally-graded characteristic. Accordingly, Equation (18) may be changed as

$$\omega'_D = \omega' \sqrt{1 - \xi^2} \quad (28)$$

in which ω'_D is the natural damped frequency in the same case of materials.

In the comparison we should take the same parameters in accordance with the experimental model in [33], including the shape dimension of the beam, $h = 0.0001$ m, $b = 0.02$ m, $l = 0.05$ m, the damping ration $\xi = 0.03$ and the uniformly-distributed mass $\overline{m} = 0.015$ Kg/m, as well as the material parameters of PZT-5 (shown in Table 7). Besides, three different initial displacements in [33] are also adopted in our theoretical results; thus, Table 8 lists the vibration frequencies from experimental measurement and theoretical solution. It is easily seen that the differences between experimental results and theoretical results are small, which indicates the theoretical vibration frequency is reliable.

Table 7. Materials parameters of PZT-5 [33].

Elastic Constant (10^{-12} m^2·N^{-1})					Piezoelectric Constant (10^{-12} C·N^{-1})			Dielectric Constant (10^{-8} F·m^{-1})	
s_{11}	s_{12}	s_{13}	s_{33}	s_{44}	d_{31}	d_{33}	d_{15}	λ_{11}	λ_{33}
16.4	−5.74	−7.22	18.8	47.5	−172	374	584	1.505	1.531

Table 8. Frequencies of experimental and theoretical results.

Initial Displacements (mm)	Vibration Frequency (Hz)		Relative Errors (%)
	Experimental Results Reference [33]	Theoretical Results (This Paper)	
0.475	123.25	123.220	0.02
0.750	123.25	123.220	0.02
1.342	123.25	123.220	0.02

Figures 9–11 show time-displacement curves from theoretical and experimental results under the three different initial displacements. It is readily found that the theoretical results agree with the experimental ones, although there are some differences between them, which may be caused by some uncontrollable factors in the experimental operation.

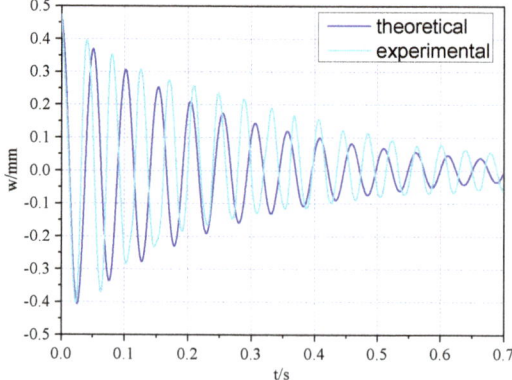

Figure 9. Comparison of two time-displacement curves under initial displacement 0.475 mm.

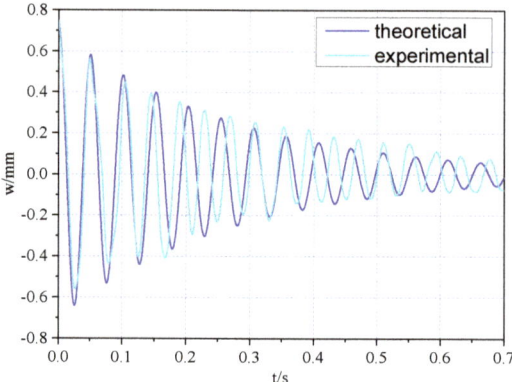

Figure 10. Comparison of two time-displacement curves under initial displacement 0.750 mm.

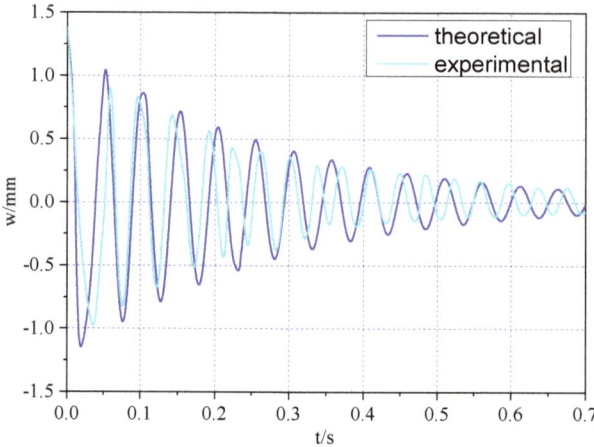

Figure 11. Comparison of two time-displacement curves under initial displacement 1.342 mm.

6. Conclusions

In this study, we investigate the free damping vibration problem of a bimodular FGPM cantilever beam by using analytical and numerical methods. By comparisons, the theoretical results basically agree with the numerical results, although there are some differences, mainly due to the different mechanical models. In addition, the theoretical solutions after regression agree well with the experimental results of the pure piezoelectric cantilever, which also proves, indirectly, the effectiveness of theoretical work. The following three conclusions can be drawn:

(i) Under two cases, $E^+(z) > E^-(z)$ and $E^+(z) < E^-(z)$, the attenuation speed of the vibration from numerical simulation is faster than that from theoretical solution; besides, the attenuation speed in the case $E^+(z) < E^-(z)$ is obviously faster than that in the case $E^+(z) > E^-(z)$.
(ii) The bimodular functionally-graded properties may change, to some extent, the dynamic response of the piezoelectric cantilever beam; however, the influence could be relatively small and unobvious.
(iii) In the frame of simple models, with analytical considerations, this work may be helpful for the analysis and design of flexible and lightweight cantilever-type elements composed of piezoelectric materials, especially when the bimodular functionally-graded properties of materials cannot be completely ignored.

Although the results in this study are obtained on the piezoelectric ceramics (PZT), this work will also be helpful for predicting the mechanical behaviors of a vibrational cantilever made of piezoelectric polymer films (PVDF); however, there may be a huge difference in their behavior and property about vibrational cantilever. Among the two main types of piezoelectric materials, due to the fact that the elastic modulus of PZT is far greater than that of PVDF; thus, the vibration frequency of PZT is far greater than that of PVDF. This should be give more attention in analyzing the vibration characteristic of cantilevers made of the two different piezoelectric materials.

Author Contributions: Conceptualization, X.-T.H. and J.-Y.S.; funding acquisition, X.-T.H. and J.-Y.S.; methodology, X.-T.H. and H.-X.J.; software, H.-X.J.; writing—original draft preparation, X.-T.H. and H.-X.J.; writing—review and editing, D.-D.P. and D.-W.D.; visualization, D.-D.P. and D.-W.D. All authors have read and agreed to the published version of the manuscript.

Funding: This research was funded by National Natural Science Foundation of China (Grant No. 11572061 and 11772072).

Conflicts of Interest: The authors declare no conflict of interest.

References

1. Yeo, H.G.; Ma, X.; Rahn, C.; Trolier-McKinstry, S. Efficient piezoelectric energy harvesters utilizing (001) textured bimorph PZT films on flexible metal foils. *Adv. Funct. Mater.* **2016**, *26*, 5940–5946. [CrossRef]
2. Won, S.S.; Seo, H.; Kawahara, M.; Glinsek, S.; Lee, J.; Kim, Y.; Jeong, C.K.; Kingon, A.I.; Kim, S.H. Flexible vibrational energy harvesting devices using strain-engineered perovskite piezoelectric thin films. *Nano Energy* **2019**, *55*, 182–192. [CrossRef]
3. Huang, D.J.; Ding, H.J.; Chen, W.Q. Piezoelasticity solutions for functionally-graded piezoelectric beams. *Smart Mater. Struct.* **2007**, *16*, 687–695. [CrossRef]
4. Bodaghi, M.; Damanpack, A.R.; Aghdam, M.M.; Shakeri, M. Geometrically non-linear transient thermo-elastic response of FG beams integrated with a pair of FG piezoelectric sensors. *Compos. Struct.* **2014**, *107*, 48–59. [CrossRef]
5. Kulikov, G.M.; Plotnikova, S.V. An analytical approach to three-dimensional coupled thermoelectroelastic analysis of functionally-graded piezoelectric plates. *J. Intell. Mater. Syst. Struct.* **2017**, *28*, 435–450. [CrossRef]
6. Alibeigloo, A. Thermo elasticity solution of functionally-graded, solid, circular, and annular plates integrated with piezoelectric layers using the differential quadrature method. *Mech. Adv. Mater. Struct.* **2018**, *25*, 766–784. [CrossRef]

7. Heydarpour, Y.; Malekzadeh, P.; Dimitri, R.; Tornabene, F. Thermoelastic analysis of functionally graded cylindrical panels with piezoelectric layers. *Appl. Sci.* **2020**, *10*, 1397. [CrossRef]
8. Arefi, M.; Bidgoli, E.M.R.; Dimitri, R.; Bacciocchi, M.; Tornabene, F. Application of sinusoidal shear deformation theory and physical neutral surface to analysis of functionally graded piezoelectric plate. *Compos. Part B Eng.* **2018**, *151*, 35–50. [CrossRef]
9. Mahinzare, M.; Ranjbarpur, H.; Ghadiri, M. Free vibration analysis of a rotary smart two directional functionally-graded piezoelectric material in axial symmetry circular nanoplate. *Mech. Syst. Signal Process.* **2018**, *100*, 188–207. [CrossRef]
10. Yao, R.X.; Shi, Z.F. Steady-State forced vibration of functionally-graded piezoelectric beams. *J. Intell. Mater. Syst. Struct.* **2011**, *22*, 769–779. [CrossRef]
11. Shakeri, M.; Mirzaeifar, R. Static and dynamic analysis of thick functionally-graded plates with piezoelectric layers using layerwise finite element model. *Mech. Adv. Mater. Struct.* **2009**, *16*, 561–575. [CrossRef]
12. Ebrahimi, F. Analytical investigation on vibrations and dynamic response of functionally-graded plate integrated with piezoelectric layers in thermal environment. *Mech. Adv. Mater. Struct.* **2013**, *20*, 854–870. [CrossRef]
13. Chen, M.F.; Chen, H.L.; Ma, X.L.; Jin, G.Y.; Ye, T.G.; Zhang, Y.T.; Liu, Z.G. The isogeometric free vibration and transient response of functionally-graded piezoelectric curved beam with elastic restraints. *Results Phys.* **2018**, *11*, 712–725. [CrossRef]
14. Li, S.R.; Su, H.D.; Cheng, C.J. Free vibration of functionally-graded material beams with surface-bonded piezoelectric layers in thermal environment. *Appl. Math. Mech.* **2009**, *30*, 969–982. [CrossRef]
15. Huang, X.L.; Shen, H.S. Vibration and dynamic response of functionally-graded plates with piezoelectric actuators in thermal environments. *J. Sound Vib.* **2006**, *289*, 25–53. [CrossRef]
16. Fu, Y.; Wang, J.; Mao, Y. Nonlinear analysis of buckling, free vibration and dynamic stability for the piezoelectric functionally-graded beams in thermal environment. *Appl. Math. Modell.* **2012**, *36*, 4324–4340. [CrossRef]
17. Li, Y.; Shi, Z. Free vibration of a functionally-graded piezoelectric beam via state-space based differential quadrature. *Compos. Struct.* **2009**, *87*, 257–264. [CrossRef]
18. Barak, M.M.; Currey, J.D.; Weiner, S.; Shahar, R. Are tensile and compressive Young's moduli of compact bone different. *J. Mech. Behav. Biomed. Mater.* **2009**, *2*, 51–60. [CrossRef]
19. Destrade, M.; Gilchrist, M.D.; Motherway, J.A.; Murphy, J.G. Bimodular rubber buckles early in bending. *Mech. Mater.* **2010**, *42*, 469–476. [CrossRef]
20. Jones, R.M. Stress–strain relations for materials with different moduli in tension and compression. *AIAA J.* **1977**, *15*, 16–23. [CrossRef]
21. Ambartsumyan, S.A. *Elasticity Theory of Different Moduli*; Wu, R.F.; Zhang, Y.Z., Translators; China Railway Publishing House: Beijing, China, 1986.
22. Zhang, Y.Z.; Wang, Z.F. Finite element method of elasticity problem with different tension and compression moduli. *Comput. Struct. Mech. Appl.* **1989**, *6*, 236–245.
23. Ye, Z.M.; Chen, T.; Yao, W.J. Progresses in elasticity theory with different moduli in tension and compression and related FEM. *Mech. Eng.* **2004**, *26*, 9–14.
24. Sun, J.Y.; Zhu, H.Q.; Qin, S.H.; Yang, D.L.; He, X.T. A review on the research of mechanical problems with different moduli in tension and compression. *J. Mech. Sci. Technol.* **2010**, *24*, 1845–1854. [CrossRef]
25. Du, Z.L.; Zhang, Y.P.; Zhang, W.S.; Guo, X. A new computational framework for materials with different mechanical responses in tension and compression and its applications. *Int. J. Solids Struct.* **2016**, *100*, 54–73. [CrossRef]
26. He, X.T.; Li, Y.H.; Liu, G.H.; Yang, Z.X.; Sun, J.Y. Non-linear bending of functionally graded thin plates with different moduli in tension and compression and its general perturbation solution. *Appl. Sci.* **2018**, *8*, 731.
27. He, X.T.; Wang, Y.Z.; Shi, S.J.; Sun, J.Y. An electroelastic solution for functionally-graded piezoelectric material beams with different moduli in tension and compression. *J. Intell. Mater. Syst. Struct.* **2018**, *29*, 1649–1669. [CrossRef]
28. He, X.T.; Yang, Z.X.; Jing, H.X.; Sun, J.Y. One-dimensional theoretical solution and two-dimensional numerical simulation for functionally-graded piezoelectric cantilever beams with different properties in tension and compression. *Polymers* **2019**, *11*, 1728.

29. Yu, T.; Zhong, Z. Bending analysis of a functionally-graded piezoelectric cantilever beam. *Sci. China Ser. G Phys. Mech. Astron.* **2007**, *50*, 97–108. [CrossRef]
30. Clough, R.W.; Penzien, J. *Dynamics of Structures*, 2nd ed.; Wang, G.Y., Translator; Higher Education Press: Beijing, China, 2006.
31. Martínez-Pañeda, E. On the finite element implementation of functionally graded materials. *Materials* **2019**, *12*, 287. [CrossRef]
32. Ruan, X.P.; Danforth, S.C.; Safari, A.; Chou, T.W. Saint-Venant end effects in piezoceramic materials. *Int. J. Solids Struct.* **2000**, *37*, 2625–2637. [CrossRef]
33. Yang, Z.X.; He, X.T.; Peng, D.D.; Sun, J.Y. Free damping vibration of piezoelectric cantilever beams: A biparametric perturbation solution and its experimental verification. *Appl. Sci.* **2020**, *10*, 215. [CrossRef]

© 2020 by the authors. Licensee MDPI, Basel, Switzerland. This article is an open access article distributed under the terms and conditions of the Creative Commons Attribution (CC BY) license (http://creativecommons.org/licenses/by/4.0/).

Article

Nonlinear Vibration of Functionally Graded Graphene Nanoplatelets Polymer Nanocomposite Sandwich Beams

Mohammad Sadegh Nematollahi [1], Hossein Mohammadi [1,*], Rossana Dimitri [2] and Francesco Tornabene [2,*]

[1] School of Mechanical Engineering, Shiraz University, Shiraz 71936, Iran; mo.nematollahi@shirazu.ac.ir
[2] Department of Innovation Engineering, University of Salento, 73100 Lecce, Italy; rossana.dimitri@unisalento.it
* Correspondence: h_mohammadi@shirazu.ac.ir (H.M.); francesco.tornabene@unisalento.it (F.T.)

Received: 18 July 2020; Accepted: 14 August 2020; Published: 15 August 2020

Abstract: We provide an analytical investigation of the nonlinear vibration behavior of thick sandwich nanocomposite beams reinforced by functionally graded (FG) graphene nanoplatelet (GPL) sheets, with a power-law-based distribution throughout the thickness. We assume the total amount of the reinforcement phase to remain constant in the beam, while defining a relationship between the GPL maximum weight fraction, the power-law parameter, and the thickness of the face sheets. The shear and rotation effects are here considered using a higher-order laminated beam model. The nonlinear partial differential equations (PDEs) of motion are derived from the Von Kármán strain-displacement relationships, here solved by applying an expansion of free vibration modes. The numerical results demonstrate the key role of the amplitudes on the vibration response of GPL-reinforced sandwich beams, whose nonlinear oscillation behavior is very important in the physical science, mechanical structures and other mathematical analyses. The sensitivity of the response to the total amount of GPLs is explored herein, along with the possible effects related to the power-law parameter, the structural geometry, and the environmental conditions. The results indicate that changing the nanofiller distribution patterns with the proposed model can remarkably increase or decrease the effective stiffness of laminated composite beams.

Keywords: functional reinforcement; graphene nanoplatelets; higher-order shear deformable laminated beams; nanocomposites; nonlinear free vibration; sandwich beams

1. Introduction

Sandwich structures, generally made of a soft core and two hard face sheets, are largely used in the aerospace, oil, gas, and petrochemical industries, due to their enhanced mechanical properties, namely, a high strength-to-weight ratio and a high resistance to heat, humidity, and noise [1–5]. Hence, in recent decades, much attention has been paid to the mechanical behavior of these structures [6–12]. Based on the available literature, it seems that the geometry of the layers, the mechanical properties of the constituents, and the geometrical properties of the whole structure can have a meaningful effect on the static and dynamic behavior of sandwich structures [13–20]. The presence of some reinforcing layers in sandwich structures represents one important issue to consider for a general improvement of their mechanical properties [21–24]. Nowadays, with the advancement of nanotechnology, carbon nanotubes (CNTs) and graphene sheets (GSs) are two alternative options for the reinforcement of structures, due to their extraordinary properties. This has led to an extensive research on the behavior of sandwich structures reinforced with nanocomposites [25–28]. Among different reinforcement possibilities, graphene nanoplatelets (GPLs) provide a uniform reinforced assembly, as well as the easiest manufacturing process, as discussed in [29–32]. Graphene is a monolayer structure of carbon atoms with extraordinary electrical, mechanical, thermal, and optical properties [33–35], which make it very attractive for high-tech device applications, such as micro/nano-electromechanical systems [36]. Graphene and its derivatives—namely, GPLs—are increasingly applied as a reinforcement material in many nanocomposite structures [37–43]. This justifies the large attention paid in the literature to the mechanical behavior of structures reinforced with graphenic materials [44–53].

Despite the extensive literature available on the behavior of composite structures reinforced with nanostructures, there is a general lack of works focusing on the nonlinear dynamic and vibration behavior of sandwich beams reinforced by GPLs. This is here investigated for thick polymer sandwich beams with face sheets reinforced by GPLs, in a context where the reduced weight of polymers and the high strength of GPLs can provide remarkable properties in the equivalent composite structure. A novel reinforcement model is proposed herein, which considers the functionality of the GPLs distribution throughout the thickness of the face sheets, and a constant total amount of the reinforced material. A higher-order laminated beam theory is applied to include the shear and rotation effects on the thick GPL-reinforced sandwich beam, where the nonlinear governing equations of the problem are solved in a straightforward manner by means of the multiple timescales method. The main advantage of the present method is that it can cover weak or strong nonlinearities with possible damping effects. The method is demonstrated to be very simple and accurate with respect to other existing predictions and theories from the literature.

The reinforcement phase varies along the thickness according to a power-law distribution, whereby the effective material properties of the nanocomposite beam are determined by means of the Halpin–Tsai micromechanics model and the rule of mixtures. The nonlinear partial equations of motion are derived by the Hamilton's principle, in accordance with the third-order shear deformation theory and the Von Kármán strain-displacement relationships. We then apply Galerkin's approach to discretize the nonlinear differential equations of motion, while determining the frequency equations by means of the multiple timescales method. Various numerical examples indicate the accuracy of the proposed model and check for the sensitivity of the vibration response of GPL-reinforced sandwich beams, of great interest for design and practical purposes.

The paper is organized as follows. In Section 2 the mechanical and geometrical properties of materials and their structure are briefly described. Section 3 presents the theoretical formulation of the problem, along with the numerical procedure. A number of illustrative applications and comparative evaluations with the available literature are proposed in Section 4. Finally, in Section 5 some concluding remarks are reported.

2. Material Properties and Geometry

A nanocomposite sandwich beam with length L, thickness h_t and width b is considered, as shown in Figure 1. The Cartesian coordinate system (x, z) is here used to derive the equations of motion, where the structural mid-plane is parallel to the x-axis. The beam is made of a homogeneous core and two face sheets with a symmetric GPL-based reinforcement, whose weight fraction satisfies the following power-law:

$$\Gamma(z) = \Gamma_{max}\left(\frac{2|z| - h_c}{h_t - h_c}\right)^\kappa, \qquad (1)$$

where Γ_{max} is the maximum value of the distribution function and κ is a power-law parameter, which defines the GPLs dispersion throughout the thickness of the face sheets.

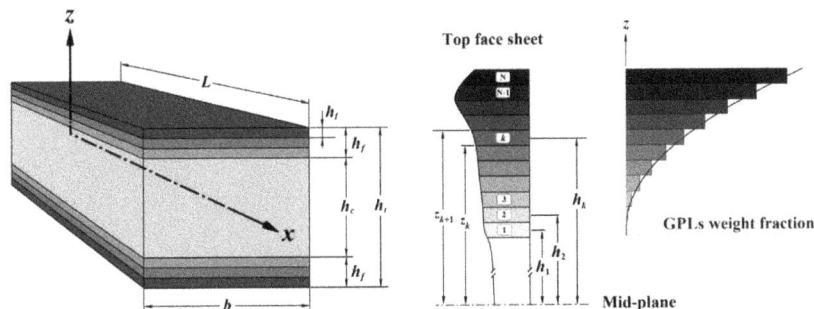

Figure 1. General configuration and graphene nanoplatelet (GPL) dispersion description in a laminated GPL-reinforced sandwich beam.

For a proper analysis, the total amount of GPLs in the beam remains constant independently of the distribution pattern (see Figure 2). This means that, if we keep constant the total amount of GPLs in the beam, Γ_b, the maximum value of the GPL weight fraction, Γ_{max}, increases by increasing the power-law parameter. Note that the total amount of reinforced GPLs decreases by increasing the power-law parameter if the maximum value of the GPL weight fraction is kept constant.

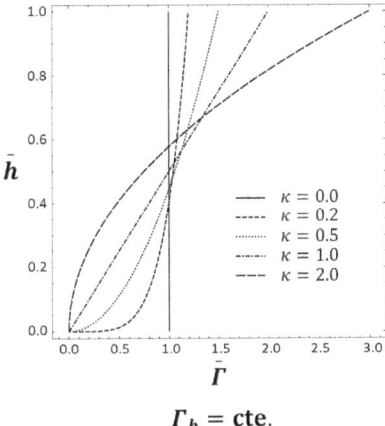

$\Gamma_b = $ cte.

Figure 2. Variation of the dimensionless GPL weight fraction ($\bar{\Gamma} = \Gamma(z)/\Gamma_b$) through the thickness of the top face sheet with respect to various power-law parameters ($\bar{h} = (2z - h_c)/2h_f$).

Due to possible difficulties during the manufacturing process of a functional reinforced lamina, each face sheet is considered to be made of N layers with equal thickness and the GPL reinforcement is assumed to be uniform within each layer (see Figure 1).

Therefore, the GPL weight fraction in the kth lamina can be defined as

$$\Gamma^{(k)} = \Gamma_{max}\left(\frac{2|h_k| - h_c}{h_t - h_c}\right)^{\kappa}, \qquad (2)$$

where h_k is the distance between the mid-plane of the beam and the mid-plane of the kth layer.

The volume fraction of the reinforced GPLs can be related to their weight fraction as

$$V_{GPL}^{(k)} = \frac{\Gamma^{(k)}}{\Gamma^{(k)} + (\rho_{GPL}/\rho_M)\left(1 - \Gamma^{(k)}\right)}, \qquad (3)$$

ρ_M and ρ_{GPL} being the mass density of the matrix and GPLs, respectively. Here, the Halpin–Tsai micromechanical model is adopted to define the effective elastic modulus of the reinforced face sheets [54]. Moreover, the GPL reinforcement is assumed to be randomly oriented in each lamina [27]. Therefore, the elastic modulus for the kth layer can be expressed as

$$E_C^{(k)} = \frac{3}{8}\left(\frac{1+\xi_L \eta_L V_{GPL}^{(k)}}{1-\eta_L V_{GPL}^{(k)}}\right)E_M + \frac{5}{8}\left(\frac{1+\xi_W \eta_W V_{GPL}^{(k)}}{1-\eta_W V_{GPL}^{(k)}}\right)E_M, \tag{4}$$

where

$$\eta_W = \frac{(E_{GPL}/E_M)-1}{(E_{GPL}/E_M)+\xi_W}, \quad \eta_L = \frac{(E_{GPL}/E_M)-1}{(E_{GPL}/E_M)+\xi_L}, \tag{5}$$

$$\xi_W = \frac{2w_{GPL}}{h_{GPL}}, \quad \xi_L = \frac{2l_{GPL}}{h_{GPL}}, \tag{6}$$

and h_{GPL}, l_{GPL}, w_{GPL} stand for the average thickness, length, and width of GPLs, respectively; E_M and E_{GPL} denote the Young modulus of the matrix and GPLs, respectively.

Using the rule of mixtures, the effective mass density and Poisson's ratio for the kth layer can be defined as

$$\rho_C^{(k)} = \rho_{GPL} V_{GPL}^{(k)} + \rho_M \left(1 - V_{GPL}^{(k)}\right), \tag{7}$$

$$\nu_C^{(k)} = \nu_{GPL} V_{GPL}^{(k)} + \nu_M \left(1 - V_{GPL}^{(k)}\right), \tag{8}$$

ν_M and ν_{GPL} being the Poisson's ratio of the matrix and GPLs, respectively.

3. Theoretical Formulations

In this section, the nonlinear governing equations of the problem for functionally graded (FG) GPL-reinforced sandwich beams are derived by Hamilton's principle, while using a higher-order shear deformation approach.

3.1. Displacement Field and Strains

In agreement with the third-order shear deformation theory [55,56], the displacement components $u_1(x,t)$ and $u_3(x,t)$ of an arbitrary point in the x and z directions for shear deformable sandwich beams can be expressed as

$$u_1(x,z,t) = u(x,t) + z\phi(x,t) - \frac{4z^3}{3h_t^2}\left(\phi + \frac{\partial w}{\partial x}\right), \tag{9}$$

$$u_3(x,z,t) = w(x,t), \tag{10}$$

where $u(x,t)$ and $w(x,t)$ are the displacement components of a point at the mid-plane of the beam in the x and z directions, respectively. Moreover, $\phi(x,t)$ denotes the slope of a transverse normal at $z = 0$. Based on the Von Kármán strain-displacement relationships, the nonlinear strain components associated with the displacement field (9)–(10) can be written as

$$\varepsilon_{xx} = \varepsilon_{xx}^{(0)} + z\varepsilon_{xx}^{(1)} + z^3\varepsilon_{xx}^{(3)}, \tag{11}$$

$$\gamma_{xz} = \gamma_{xz}^{(0)} + z^2\gamma_{xz}^{(2)}, \tag{12}$$

where

$$\varepsilon_{xx}^{(0)} = \frac{\partial u}{\partial x} + \frac{1}{2}\left(\frac{\partial w}{\partial x}\right)^2, \quad \varepsilon_{xx}^{(1)} = \frac{\partial \phi}{\partial x}, \quad \varepsilon_{xx}^{(3)} = -c_1\left(\frac{\partial \phi}{\partial x} + \frac{\partial^2 w}{\partial x^2}\right), \tag{13}$$

$$\gamma_{xz}^{(0)} = \phi + \frac{\partial w}{\partial x}, \quad \gamma_{xz}^{(2)} = -c_2\left(\phi + \frac{\partial w}{\partial x}\right), \tag{14}$$

and
$$c_1 = \frac{4}{3h_t^2}, \quad c_2 = 3c_1 = \frac{4}{h_t^2}. \tag{15}$$

3.2. Equations of Motion

The equations of motion of FG-GPL reinforced sandwich beams are derived from Hamilton's principle. Accordingly, we have

$$\int_{t_1}^{t_2} (\delta U - \delta W - \delta K) dt = 0, \tag{16}$$

where U is the strain energy, W is the work done by external forces, and K is the kinetic energy. The virtual strain energy δU for the third-order shear deformable sandwich beams reads as follows

$$\begin{aligned}
\delta U &= \int_A \int_0^L (\sigma_{xx} \delta \varepsilon_{xx} + \sigma_{xz} \delta \gamma_{xz}) dx dA \\
&= \int_A \int_0^L \left[\sigma_{xx} \left(\delta \varepsilon_{xx}^{(0)} + z \delta \varepsilon_{xx}^{(1)} + z^3 \delta \varepsilon_{xx}^{(3)} \right) + \sigma_{xz} \left(\delta \gamma_{xz}^{(0)} + z^2 \delta \gamma_{xz}^{(2)} \right) \right] dx dA \\
&= \int_0^L \left[-\frac{\partial N_{xx}}{\partial x} \delta u - \frac{\partial}{\partial x} \left(N_{xx} \frac{\partial w}{\partial x} \right) \delta w - \frac{\partial M_{xx}}{\partial x} \delta \phi + c_1 \frac{\partial P_{xx}}{\partial x} \delta \phi - c_1 \frac{\partial^2 P_{xx}}{\partial x^2} \delta w \right. \\
&\quad + Q_x \delta \phi - \frac{\partial Q_x}{\partial x} \delta w - c_2 R_x \delta \phi + c_2 \frac{\partial R_x}{\partial x} \delta w \bigg] dx \\
&\quad + \left[N_{xx} \delta u + N_{xx} \frac{\partial w}{\partial x} \delta w + M_{xx} \delta \phi - c_1 P_{xx} \delta \phi - c_1 P_{xx} \frac{\partial}{\partial x} \delta w + c_1 \frac{\partial P_{xx}}{\partial x} \delta w + Q_x \delta w - c_2 R_x \delta w \right]_0^L,
\end{aligned} \tag{17}$$

where

$$\begin{bmatrix} N_{xx} \\ M_{xx} \\ P_{xx} \end{bmatrix} = \int_A \sigma_{xx} \begin{bmatrix} 1 \\ z \\ z^3 \end{bmatrix} dA, \quad \begin{bmatrix} Q_x \\ R_x \end{bmatrix} = \int_A \sigma_{xz} \begin{bmatrix} 1 \\ z^2 \end{bmatrix} dA. \tag{18}$$

The virtual kinetic energy δK is defined as

$$\begin{aligned}
\delta K &= \int_A \int_0^L \rho(z) (\dot{u}_1 \delta \dot{u}_1 + \dot{u}_3 \delta \dot{u}_3) dA dx \\
&= \int_0^L \left[\left(m_0 \ddot{u} + m_1 \ddot{\phi} - c_1 m_3 \left(\ddot{\phi} + \frac{\partial \ddot{w}}{\partial x} \right) \right) \delta u \right. \\
&\quad + \left(m_1 \ddot{u} + m_2 \ddot{\phi} - c_1 m_3 \ddot{u} - c_1 m_4 \left(2\ddot{\phi} + \frac{\partial \ddot{w}}{\partial x} \right) + c_1^2 m_6 \left(\ddot{\phi} + \frac{\partial \ddot{w}}{\partial x} \right) \right) \delta \phi \\
&\quad + \left(m_0 \ddot{w} + c_1 m_3 \frac{\partial \ddot{u}}{\partial x} + c_1 m_4 \frac{\partial \ddot{\phi}}{\partial x} - c_1^2 m_6 \left(\frac{\partial \ddot{\phi}}{\partial x} + \frac{\partial^2 \ddot{w}}{\partial x^2} \right) \right) \delta w \bigg] dx,
\end{aligned} \tag{19}$$

with

$$m_i = \int_A \rho(z) z^i dA = \sum_{k=1}^{2N+1} b \int_{z_k}^{z_{k+1}} \rho_c^{(k)} z^i dz, \quad i = 0, 1, 2, 3, 4, 6. \tag{20}$$

In the total absence of external forces on the structure, it is $\delta W = 0$. Therefore, by substitution of Equations (17) and (19) into Equation (16), by integrating the result by parts, and equating the coefficients of δu, δw, and $\delta \phi$ to zero separately, we get the following nonlinear equations of motion:

$$\frac{\partial N_{xx}}{\partial x} = m_0 \ddot{u} + m_1 \ddot{\phi} - c_1 m_3 \left(\ddot{\phi} + \frac{\partial \ddot{w}}{\partial x} \right), \tag{21}$$

$$\frac{\partial M_{xx}}{\partial x} - c_1 \frac{\partial P_{xx}}{\partial x} - Q_x + c_2 R_x = m_1 \ddot{u} + m_2 \ddot{\phi} - c_1 m_3 \ddot{u} - c_1 m_4 \left(2\ddot{\phi} + \frac{\partial \ddot{w}}{\partial x} \right) + c_1^2 m_6 \left(\ddot{\phi} + \frac{\partial \ddot{w}}{\partial x} \right), \tag{22}$$

$$\frac{\partial}{\partial x} \left(N_{xx} \frac{\partial w}{\partial x} \right) + c_1 \frac{\partial^2 P_{xx}}{\partial x^2} + \frac{\partial Q_x}{\partial x} - c_2 \frac{\partial R_x}{\partial x} = m_0 \ddot{w} + c_1 m_3 \frac{\partial \ddot{u}}{\partial x} + c_1 m_4 \frac{\partial \ddot{\phi}}{\partial x} - c_1^2 m_6 \left(\frac{\partial \ddot{\phi}}{\partial x} + \frac{\partial^2 \ddot{w}}{\partial x^2} \right). \tag{23}$$

Thus, we define the stress resultants in terms of the displacement and rotation components of the sandwich beam as

$$N_{xx} = A_{11}\left(\frac{\partial u}{\partial x} + \frac{1}{2}\left(\frac{\partial w}{\partial x}\right)^2\right) + B_{11}\left(\frac{\partial \phi}{\partial x}\right) - c_1 E_{11}\left(\frac{\partial \phi}{\partial x} + \frac{\partial^2 w}{\partial x^2}\right), \tag{24}$$

$$M_{xx} = B_{11}\left(\frac{\partial u}{\partial x} + \frac{1}{2}\left(\frac{\partial w}{\partial x}\right)^2\right) + D_{11}\left(\frac{\partial \phi}{\partial x}\right) - c_1 F_{11}\left(\frac{\partial \phi}{\partial x} + \frac{\partial^2 w}{\partial x^2}\right), \tag{25}$$

$$P_{xx} = E_{11}\left(\frac{\partial u}{\partial x} + \frac{1}{2}\left(\frac{\partial w}{\partial x}\right)^2\right) + F_{11}\left(\frac{\partial \phi}{\partial x}\right) - c_1 H_{11}\left(\frac{\partial \phi}{\partial x} + \frac{\partial^2 w}{\partial x^2}\right), \tag{26}$$

$$Q_x = (A_{55} - c_2 D_{55})\left(\phi + \frac{\partial w}{\partial x}\right), \tag{27}$$

$$R_x = (D_{55} - c_2 F_{55})\left(\phi + \frac{\partial w}{\partial x}\right), \tag{28}$$

where

$$(A_{11}, B_{11}, D_{11}, E_{11}, F_{11}, H_{11}) = \sum_{k=1}^{N} b \int_{h_k}^{h_{k+1}} Q_{11}^{(k)}\left(1, z, z^2, z^3, z^4, z^6\right) dz, \tag{29}$$

$$(A_{55}, D_{55}, F_{55}) = \sum_{k=1}^{N} b \int_{h_k}^{h_{k+1}} Q_{55}^{(k)}\left(1, z^2, z^4\right) dz, \tag{30}$$

and

$$Q_{11}^{(k)} = \frac{E_C^{(k)}}{1 - \left(v_C^{(k)}\right)^2}, \quad Q_{55}^{(k)} = G_C^{(k)} = \frac{E_C^{(k)}}{2\left(1 + v_C^{(k)}\right)}, \tag{31}$$

In view of Equations (21)–(28), the nonlinear partial differential equations of motion of FG GPL-reinforced sandwich beams can be written as

$$\begin{aligned}A_{11}\left(\frac{\partial^2 u}{\partial x^2} + \frac{\partial^2 w}{\partial x^2}\frac{\partial w}{\partial x}\right) &+ (B_{11} - c_1 E_{11})\frac{\partial^2 \phi}{\partial x^2} - c_1 E_{11}\frac{\partial^3 w}{\partial x^3} \\ &= m_0 \frac{\partial^2 u}{\partial t^2} + (m_1 - c_1 m_3)\frac{\partial^2 \phi}{\partial t^2} - c_1 m_3 \frac{\partial^3 w}{\partial x \partial t^2},\end{aligned} \tag{32}$$

$$\begin{aligned}(B_{11} - c_1 E_{11})\left(\frac{\partial^2 u}{\partial x^2} + \frac{\partial^2 w}{\partial x^2}\frac{\partial w}{\partial x}\right) &+ \left(D_{11} - 2c_1 F_{11} + c_1^2 H_{11}\right)\frac{\partial^2 \phi}{\partial x^2} + \left(-c_1 F_{11} + c_1^2 H_{11}\right)\frac{\partial^3 w}{\partial x^3} \\ &+ \left(-A_{55} + 2c_2 D_{55} - c_2^2 F_{55}\right)\left(\phi + \frac{\partial w}{\partial x}\right) \\ &= (m_1 - c_1 m_3)\frac{\partial^2 u}{\partial t^2} + \left(m_2 - 2c_1 m_4 + c_1^2 m_6\right)\frac{\partial^2 \phi}{\partial t^2} + \left(-c_1 m_4 + c_1^2 m_6\right)\frac{\partial^3 w}{\partial x \partial t^2},\end{aligned} \tag{33}$$

$$\begin{aligned}c_1 E_{11}\left(\frac{\partial^3 u}{\partial x^3} - \left(\frac{\partial^2 w}{\partial x^2}\right)^2\right) &+ \left(c_1 F_{11} - c_1^2 H_{11}\right)\frac{\partial^3 \phi}{\partial x^3} - c_1^2 H_{11}\frac{\partial^4 w}{\partial x^4} \\ &+ \left(A_{55} - 2c_2 D_{55} + c_2^2 F_{55}\right)\left(\frac{\partial \phi}{\partial x} + \frac{\partial^2 w}{\partial x^2}\right) \\ &+ A_{11}\left(\frac{\partial^2 u}{\partial x^2}\frac{\partial w}{\partial x} + \frac{\partial u}{\partial x}\frac{\partial^2 w}{\partial x^2} + \frac{3}{2}\left(\frac{\partial^2 w}{\partial x^2}\left(\frac{\partial w}{\partial x}\right)^2\right)\right) \\ &+ (B_{11} - c_1 E_{11})\left(\frac{\partial^2 \phi}{\partial x^2}\frac{\partial w}{\partial x} + \frac{\partial \phi}{\partial x}\frac{\partial^2 w}{\partial x^2}\right) \\ &= m_0 \frac{\partial^2 w}{\partial t^2} + c_1 m_3 \frac{\partial^3 u}{\partial x \partial t^2} + \left(c_1 m_4 - c_1^2 m_6\right)\frac{\partial^3 \phi}{\partial x \partial t^2} - c_1^2 m_6 \frac{\partial^4 w}{\partial x^2 \partial t^2}.\end{aligned} \tag{34}$$

3.3. Solution Procedure

In this section, the nonlinear equations of motion are solved numerically, in order to obtain the linear and nonlinear frequency equations. In this regard, the nonlinear partial differential equations (PDEs) of motion (32)–(34) are discretized as ordinary differential equations by employing the Galerkin method. Afterwards, the multiple timescales approach is used to obtain the nonlinear frequency equation. Here, we assume that the GPL-reinforced sandwich beam is simply supported at both ends with movable supports. Based on these assumptions, the displacement and rotation field of the beam can be defined as expansions of the free vibration mode shapes, namely,

$$u(x,t) = \sum_{n=1}^{\infty} U_n(t)\chi_n(x), \quad \chi_n(x) = \cos\left(\frac{n\pi x}{L}\right), \tag{35}$$

$$\phi(x,t) = \sum_{n=1}^{\infty} \varphi_n(t)\psi_n(x), \quad \psi_n(x) = \cos\left(\frac{n\pi x}{L}\right), \tag{36}$$

$$w(x,t) = \sum_{n=1}^{\infty} W_n(t)\lambda_n(x), \quad \lambda_n(x) = \sin\left(\frac{n\pi x}{L}\right), \tag{37}$$

U_n, φ_n and W_n being the unknown generalized coordinates which stand for the amplitude of the vibration. Moreover, the following functions $\chi_n(x)$, $\psi_n(x)$, and $\lambda_n(x)$ are introduced to satisfy all boundary conditions of the system. By considering a single mode approximate solution and by substitution of Equations (35)–(37) into Equations (32)–(34), after multiplying the results by χ, ψ, and λ and after their integration over the domain of the system, the following nonlinear differential equations of motion are obtained

$$g_{11}U_n + g_{12}W_n + g_{13}W_n^2 + g_{14}\varphi_n = 0, \tag{38}$$

$$g_{21}U_n + g_{22}W_n + g_{23}W_n^2 + g_{24}\varphi_n = 0, \tag{39}$$

$$g_{31}U_n + g_{32}W_n + g_{33}U_n W_n + g_{34}W_n^2 + g_{35}W_n^3 + g_{36}\varphi_n + g_{37}W_n\varphi_n + g_{38}U_n'' + g_{39}W_n'' + g_{310}\varphi_n'' = 0, \tag{40}$$

where coefficients $g_{11}, g_{12}, \ldots, g_{310}$ are detailed in Appendix A. The ordinary differential equation of transverse motion can be obtained by solving U_n and φ_n in terms of W_n from Equations (38) and (39) and substituting the results in Equation (40). Thus, we get

$$W_n'' + \alpha_1 W_n + \alpha_2 W_n^3 = 0, \tag{41}$$

where

$$\alpha_1 = \omega_L^2 = \frac{g_{14}(g_{22}g_{31} - g_{21}g_{32}) + g_{12}(-g_{24}g_{31} + g_{21}g_{36}) + g_{11}(g_{24}g_{32} - g_{22}g_{36})}{g_{24}(-g_{12}g_{38} + g_{11}g_{39}) + g_{14}(g_{22}g_{38} - g_{21}g_{39}) + (g_{12}g_{21} - g_{11}g_{22})g_{310}}, \tag{42}$$

$$\alpha_2 = \frac{g_{14}(-g_{23}g_{33} + g_{21}g_{35}) + g_{13}(g_{24}g_{33} - g_{21}g_{37}) + g_{11}(-g_{24}g_{35} + g_{23}g_{37})}{g_{24}(g_{12}g_{38} - g_{11}g_{39}) + g_{14}(-g_{22}g_{38} + g_{21}g_{39}) + (-g_{12}g_{21} + g_{11}g_{22})g_{310}}, \tag{43}$$

and ω_L is the natural frequency of the nanocomposite sandwich beam. According to the multiple timescale approach [57,58], we approximate the solution of Equation (41) by means of the following expansion,

$$W(t,\epsilon) = \epsilon W_1(T_0, T_1, T_2, \ldots) + \epsilon^2 W_2(T_0, T_1, T_2, \ldots) + \epsilon^3 W_3(T_0, T_1, T_2, \ldots) + \ldots \tag{44}$$

where ϵ is a small perturbation parameter and $T_n = \epsilon^n t$ refers to the independent variables for $n = 0, 1, 2, \ldots$, whose derivatives with respect to t are defined as follows:

$$\begin{aligned}
\frac{d}{dt} &= \frac{dT_0}{dt}\frac{\partial}{\partial T_0} + \frac{dT_1}{dt}\frac{\partial}{\partial T_1} + \ldots = D_0 + \epsilon D_1 + \ldots \\
\frac{d^2}{dt^2} &= D_0^2 + 2\epsilon D_0 D_1 + \epsilon^2 (D_1^2 + 2D_0 D_2) + 2\epsilon^3 D_1 D_2 + \epsilon^4 D_2^2 + \ldots
\end{aligned} \tag{45}$$

where $D_n = \partial/\partial T_n$. In our case, we apply the expansion up to $O(\epsilon^3)$, such that we need T_0, T_1 and T_2. By substitution of Equations (44) and (45) into Equation (41), expanding and equating coefficients of ϵ, ϵ^2, and ϵ^3 to zero, we get the following relations:

$$\text{Order } \epsilon: \quad D_0^2 W_1 + \omega_L^2 W_1 = 0 \tag{46}$$

$$\text{Order } \epsilon^2: \quad D_0^2 W_2 + \omega_L^2 W_2 = -2 D_0 D_1 W_1 \tag{47}$$

$$\text{Order } \epsilon^3: \quad D_0^3 W_3 + \omega_L^2 W_3 = -2 D_0 D_2 W_2 - D_1^2 W_1 - 2 D_0 D_2 W_1 - \alpha_2 W_1^3 \tag{48}$$

The solution of Equation (46) takes the following form:

$$W_1 = A(T_1, T_2) \exp(i\omega_L T_0) + \overline{A} \exp(-i\omega_L T_0) \tag{49}$$

where A is an unknown complex function and $\overline{A}$ is its complex conjugate. By substitution of Equation (49) into Equation (46) we obtain the following relation:

$$D_0^2 W_2 + \omega_L^2 W_2 = -2 i \omega_L D_1 A \exp(i\omega_L T_0) + cc \tag{50}$$

cc being the complex conjugate of the previous term. Any particular solution of Equation (50) has a secular term containing the factor $T_0 \exp(i\omega_L T_0)$ unless $D_1 A = 0$. This means that A is independent of T_1, whereby the solution of Equation (50) is verified to be identically null.

By substitution of $W_2 = 0$, together with Equation (49), into Equation (48) we get the following expression

$$D_0^2 W_3 + \omega_L^2 W_3 = -\left[2 i \omega_L D_2 A - 3\alpha_2 A^2 \overline{A}\right] \exp(i\omega_L T_0) - \alpha_2 A^3 \exp(i\omega_L T_0) \tag{51}$$

In this last relation the secular terms containing $\exp(i\omega_L T_0)$ must be equal to zero to have a periodic solution, which corresponds to enforce the following relation:

$$2 i \omega_L D_2 A - 3\alpha_2 A^2 \overline{A} = 0 \tag{52}$$

whose solution can be found by defining A as

$$A = \frac{1}{2} a \exp(i\beta) \tag{53}$$

where a and β are real functions of T_2.

By substituting Equation (53) into Equation (52) and by equating the real and imaginary parts to zero, we obtain

$$\omega_L a' = 0 \text{ and } \omega_L a \beta' - 3/8 \alpha_2 a^3 = 0 \tag{54}$$

where the prime denotes the derivative with respect to T_2. Solving both relations in Equation (54), it follows that a is a constant and

$$\beta = 3/8 \frac{\alpha_2}{\omega_L} a^2 T_2 + \beta_0 \tag{55}$$

where β_0 is a constant. By combination of Equations (53) and (55) with Equation (49), we obtain the following closed-form solution for the nonlinear frequency of the transverse vibration of GPL-reinforced sandwich beams based on third-order shear deformation theory:

$$\omega_{NL} = \omega_L \left(1 + \frac{3}{8} \frac{\alpha_2}{\omega_L^2} \epsilon^2 a^2\right), \tag{56}$$

where a is the amplitude of the vibration.

4. Numerical Results

In this section, we present the numerical results from a large parametric investigation into the nonlinear vibration behavior of thick sandwich beams reinforced with GPLs. To check for the accuracy of the proposed model, our numerical results obtained for FG GPL-reinforced beams are compared with those available from the literature. In this regard, the thickness of the core layer in the present model is assumed to be zero ($h_c = 0$). In Tables 1 and 2, the dimensionless free linear ($\bar{\omega}_L = \omega_L \times L \times \sqrt{m_{10}/A_{10}}$) and nonlinear ($\bar{\omega}_{NL} = \omega_{NL} \times L \times \sqrt{m_{10}/A_{10}}$) frequencies are provided for a simply supported, GPL-reinforced beam and compared with numerical results reported by Feng et al. [30]. The reference model is based on a Timoshenko beam theory, whereby two different patterns are considered for validation purposes. The following properties are assumed for the beam: $E_M = 2.85$ GPa, $\rho_M = 1200$ Kg/m^3, $E_{GPL} = 1.01$ TPa, $\rho_{GPL} = 1062.5$ Kg/m^3, $w_{GPL} = 1.5$ μm, $l_{GPL} = 2.5$ μm, and $h_{GPL} = 1.5$ nm.

Table 1. First three dimensionless natural frequencies ($\bar{\omega}_L = \omega_L \times L \times \sqrt{m_{10}/A_{10}}$) of simply supported laminated beams reinforced with GPLs ($L/h_t = 20$).

Pattern	Reference	Mode		
		1	2	3
UD ($\kappa = 0$)	Feng et al. [30]	0.21542	0.85226	1.88292
	Present	0.23482	0.92903	2.05352
FG-X ($\kappa = 1$)	Feng et al. [30]	0.25853	1.01309	2.20666
	Present	0.26759	1.05199	2.30258

Table 2. First three dimensionless nonlinear frequencies ($\bar{\omega}_{NL} = \omega_{NL} \times L \times \sqrt{m_{10}/A_{10}}$) of simply supported laminated beams reinforced with GPLs ($L/h_t = 20$).

Pattern	Reference	Mode		
		1	2	3
UD ($\kappa = 0$)	Feng et al. [30]	0.27259	1.07270	2.33122
	Present	0.29020	1.08015	2.41741
FG-X ($\kappa = 1$)	Feng et al. [30]	0.31973	1.20509	2.54097
	Present	0.31619	1.18544	2.62712

In Table 3, a comparison has been attempted between results from the present formulation for the first four dimensionless frequencies ($\bar{\omega}_L = \omega_L \times (L^2/h) \sqrt{\rho/E_{11}}$) of simply supported orthotropic beams and those from the literature, based on different higher-order shear deformation theories [59–61]. The material properties are assumed to be $E_{11} = 144.9$ GPa, $E_{22} = 9.65$ GPa, $G_{12} = G_{13} = 4.14$ GPa, $G_{23} = 3.45$ GPa, $\rho = 1389.23$ Kg/m^3, and $\nu_{12} = \nu_{21} = 0.3$.

Table 3. Comparison of the first four dimensionless natural frequencies ($\bar{\omega}_L = \omega_L \times (L^2/h) \sqrt{\rho/E_{11}}$) of orthotropic thick beams based on different higher-order shear deformable beam theories ($L/h_t = 10$).

Reference	Mode			
	1	2	3	4
Shen et al. [61]	2.3100	6.9538	11.9707	17.0393
Li and Qiao [60]	2.3188	7.0204	12.0894	17.3139
Vo and Thai [59]	2.3198	7.0091	12.1250	17.2949
Present	2.4038	7.2110	12.3975	17.6279

As is clearly visible in Table 3, the numerical results based on the proposed model agree very well with predictions from the literature based on other shear deformable models. It seems that higher-order-models available in the literature [59–61] get more conservative results than our formulation. At the same time, the proposed multiple timescale approach proves to be an efficient analytical tool to solve nonlinear systems in a very easy and straightforward manner.

In the benchmark Tables 4 and 5, we report the numerical results in terms of the first-order nonlinear frequency and nonlinear-to-linear frequency ratio for a GPL-reinforced sandwich beam. The material and geometrical properties of the beam are considered to be $E_M = 2.85$ GPa, $\rho_M = 1200$ Kg/m^3, $E_{GPL} = 1.01$ TPa, $\rho_{GPL} = 1062.5$ Kg/m^3, $w_{GPL} = 1.5$ μm, $l_{GPL} = 2.5$ μm, and $h_{GPL} = 1.5$ nm. These properties are kept constant for the following examples. The numerical results are obtained for a different power-law parameter, length-to-total thickness of the beam ratio, as well as for a different total weight fraction of the GPLs reinforced in the beam (Γ_b). As mentioned before, Γ_b is defined such that the total amount of GPLs remains constant with respect to any change in the power-law parameter or thickness of the face sheets. According to Tables 4 and 5, an increased total weight fraction of GPLs (Γ_b) yields a meaningful increase of the nonlinear frequency of the beam, while decreasing the nonlinear-to-linear frequency ratio. In addition, an increased length-to-thickness ratio provides a decreasing effect on the nonlinear frequency of the system and its associated nonlinear-to-linear ratio.

Table 4. First-order nonlinear frequency of a simply supported GPL-reinforced sandwich beam ($h_c/h_t = 0.6, N = 10, a/h_t = 1$).

L/h_t	Γ_b	κ			
		0.5	1	2	5
10	0.5	1.52551	1.53220	1.54115	1.54952
	1	1.96610	1.97452	1.98470	1.99110
	2	2.62940	2.63683	2.64207	2.63204
20	0.5	0.38485	0.38708	0.39006	0.39311
	1	0.49821	0.50165	0.50612	0.51036
	2	0.66931	0.67405	0.67978	0.68391

Table 5. First-order nonlinear-to-linear frequency ratio of a simply supported GPL-reinforced sandwich beam ($h_c/h_t = 0.6, N = 10, a/h_t = 1$).

L/h_t	Γ_b	κ			
		0.5	1	2	5
10	0.5	1.59007	1.56735	1.54135	1.51106
	1	1.55652	1.53643	1.51484	1.49074
	2	1.54324	1.52830	1.51607	1.50875
20	0.5	1.55431	1.53007	1.50227	1.47047
	1	1.51103	1.48708	1.46023	1.42993
	2	1.48700	1.46431	1.43990	1.41399

On the other hand, an increased power-law parameter gets an increased nonlinear frequency and a decreased nonlinear-to-linear frequency ratio. A non-uniform behavior can be observed, sometimes, for an increasing power-law parameter and for a large amount of Γ_b. This aspect is illustrated in detail as follows.

4.1. Effect of the Amplitude of the Vibrations

Figures 3 and 4 show the effect of an increasing amplitude on the nonlinear frequency and the nonlinear-to-linear frequency ratio. According to both figures, the nonlinear frequency of a GPL-reinforced sandwich beam generally increases by increasing the amplitude of the vibrations as well as its nonlinear-to-linear frequency. However, an increased vibration amplitude significantly affects the nonlinear frequency and its rational form, for smaller values of the power-law parameter. It seems that for sandwich beams with a larger amount of GPL-reinforcement, the nonlinear frequency is increasingly affected by larger vibration amplitudes.

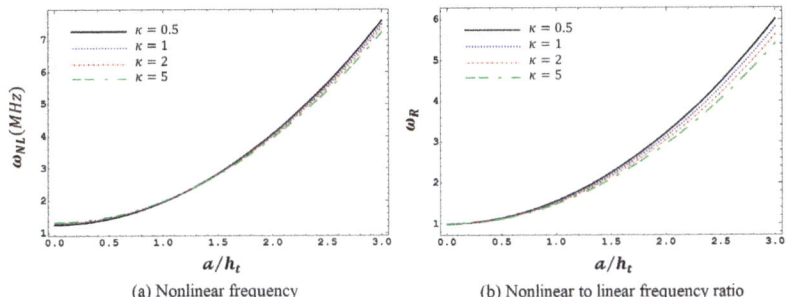

Figure 3. Variation of the first-order nonlinear frequency and nonlinear-to-linear frequency ratio of a GPL-reinforced sandwich beam with respect to an increasing amplitude of vibrations, and for different power-law parameters ($h_c/h_t = 0.6, N = 10, \Gamma_b = 1\%, L/h_t = 10$).

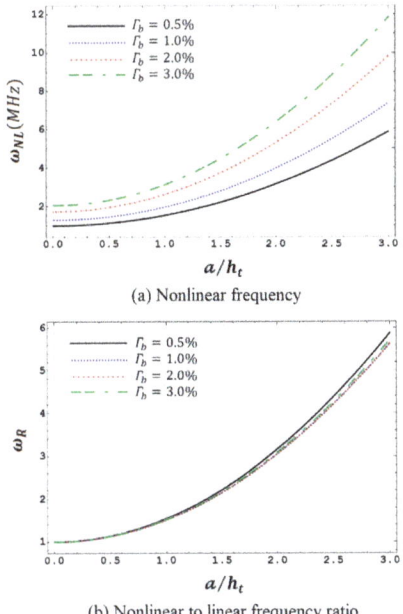

Figure 4. Variation of the first-order nonlinear frequency and nonlinear-to-linear frequency ratio of a GPL-reinforced sandwich beam with respect to an increasing amplitude of vibrations, and for a different total amount of GPLs in the beam ($h_c/h_t = 0.6, N = 10, a/h_t = 1, L/h_t = 10, \kappa = 2$).

4.2. Effect of the Power-Law Parameter

Here, we study the effect of the power-law parameter on the nonlinear vibration behavior of the nanocomposite structure. Figure 5 depicts the variation of the nonlinear frequency as a function of the power-law parameter for different thickness ratios, h_c/h_t, and vibration amplitudes a/h_t. Note that, for a small amplitude of vibration, the nonlinear frequency of the system increases by increasing the power-law parameter. The effect of an increasing power-law parameter on the vibration response of the system is different depending on the amount of the thickness ratio. For an increased amplitude of the vibration up to a threshold value, the results become non-uniform for an increased power-law parameter. An increased amplitude of vibration will completely change the behavior of the system; namely, by increasing the power-law parameter, the nonlinear frequency of the GPL-reinforced sandwich beam decreases for a large amplitude of vibration.

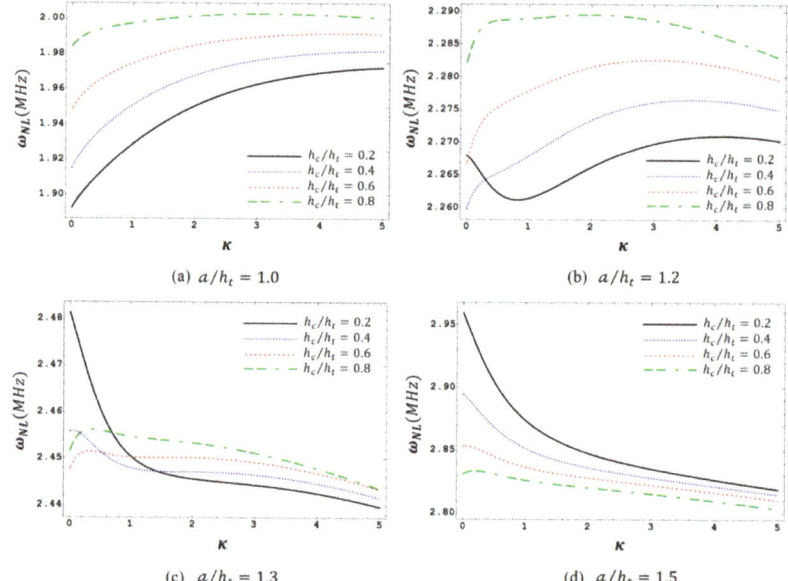

Figure 5. Variation of the first-order nonlinear frequency of a GPL-reinforced sandwich beam due to increasing power-law parameter for different core-to-beam thickness ratios and amplitudes of vibration ($N = 10, \Gamma_b = 1\%, L/h_t = 10$).

In Figure 6, we plot the variation of the nonlinear-to-linear frequency ratio of the nanocomposite structure vs. the power-law parameter. It is worth noting that an increased power-law parameter has a decreasing effect on the nonlinear-to-linear frequency ratio of the system. These effects become even more pronounced for larger vibration amplitudes, while leaving the overall behavior almost unaltered.

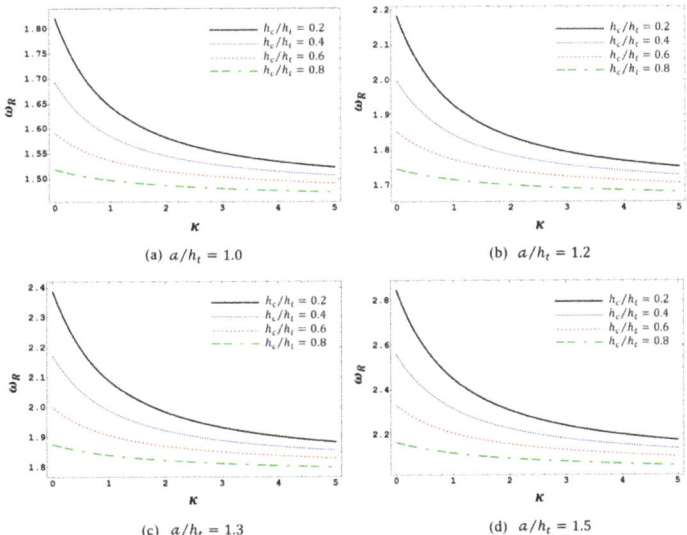

Figure 6. Variation of the first-order nonlinear-to-linear frequency ratio of a GPL-reinforced sandwich beam due to an increased power-law parameter for different core-to-beam thickness ratios and amplitudes of vibration ($N = 10, \Gamma_b = 1\%, L/h_t = 10$).

In Figures 7 and 8, the nonlinear frequency and the nonlinear-to-linear frequency ratio of a GPL-reinforced sandwich beam are plotted vs. the power-law parameter, while assuming different thickness ratios and total weight fractions of the GPL phase in the beam.

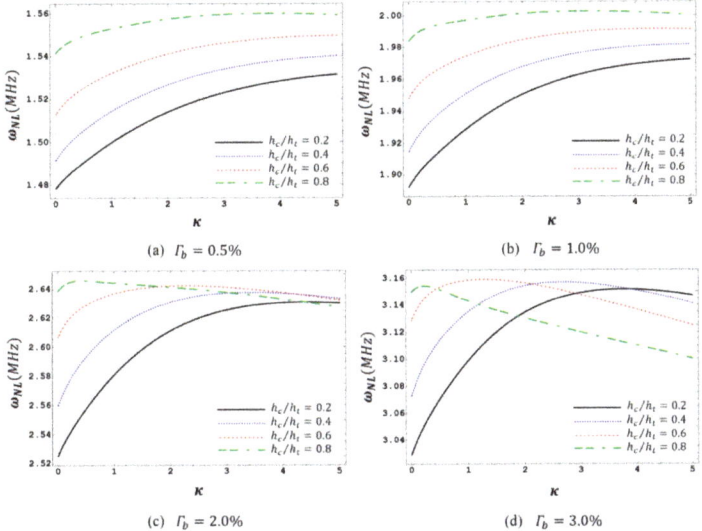

Figure 7. Variation of the first-order nonlinear frequency of a GPL-reinforced sandwich beam due to an increasing power-law parameter, for different core-to-beam thickness ratios and total amount of GPLs reinforcement in the beam ($N = 10, a/h_t = 1, L/h_t = 10$).

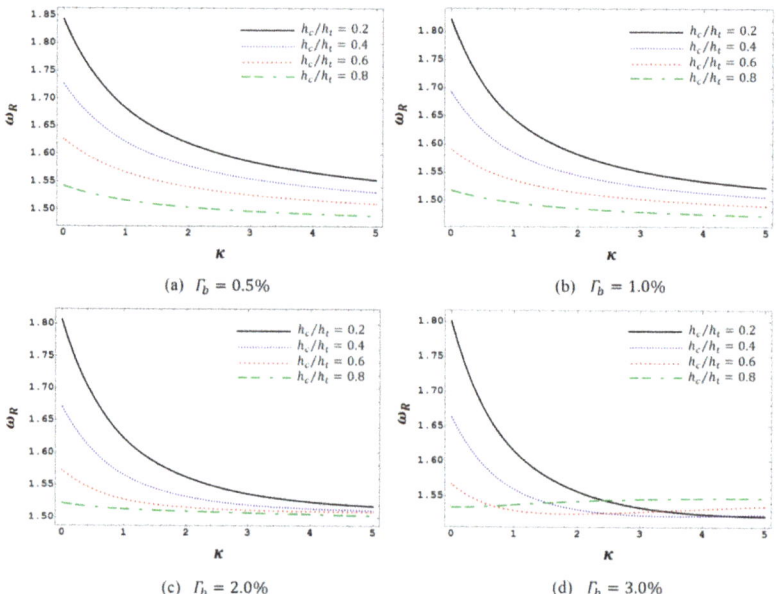

Figure 8. Variation of the first-order nonlinear-to-linear frequency ratio of a GPL-reinforced sandwich beam due to an increasing power-law parameter for different core-to-beam thickness ratios and total amounts of GPLs reinforcements in the beam ($N = 10, a/h_t = 1, L/h_t = 10$).

According to Figure 7, for a small amount of the GPLs weight fraction, the nonlinear frequency increases by increasing the power-law parameter. Moreover, the effect of an increasing power-law parameter becomes more pronounced for sandwich beams with thick face sheets. As the total weight fraction of GPLs increases, a different response is noticed, in terms of nonlinear frequency, by increasing the power-law parameter. More specifically, for a large GPL weight fraction, we notice a threshold value after which the nonlinear frequency decreases by increasing the power-law parameter. Of course, the value of the threshold point varies with the thickness of the face sheets. On the other hand, the results for the nonlinear-to-linear frequency ratio show some opposite effects for an increasing power-law parameter. An increased total amount of GPL reinforcement phase in the face sheets has a pronounced effect on the general behavior of the system, which has to be studied carefully.

4.3. Effect of the Thickness of the Face Sheets

Figures 9 and 10 show the effect of the core-to-face thickness ratio on the nonlinear vibration response of the structure. As visible in both figures, a decreasing thickness of the face sheets generally increases the nonlinear frequency of the system for a small vibration amplitude. By increasing the amplitude vibrations, the structural response changes due to an increased thickness ratio; namely, for large amplitude vibrations, the frequency decreases by increasing the thickness ratio, and the nonlinear-to-linear frequency ratio decreases accordingly. This behavior is almost unaffected by the amplitude vibrations.

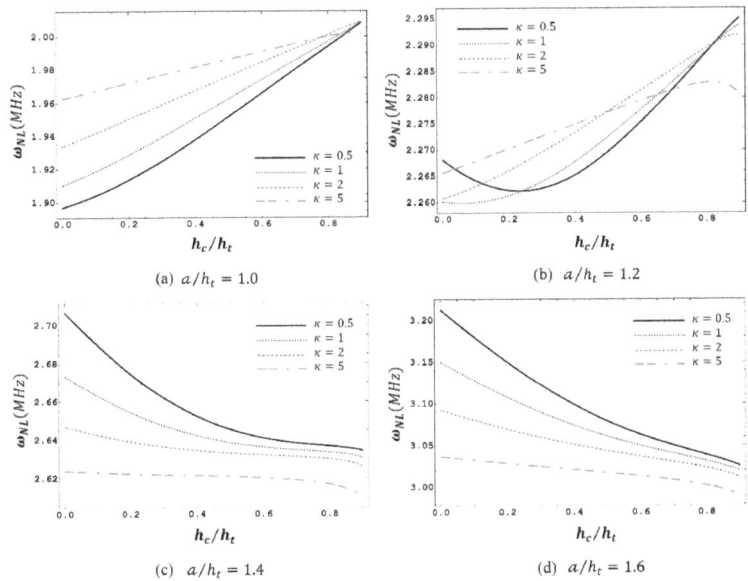

Figure 9. Variation of the first-order nonlinear frequency of a GPL-reinforced sandwich beam with respect to an increasing core-to-face thickness ratio for different power-law parameters and amplitudes of vibration ($N = 10, \Gamma_b = 1\%, L/h_t = 10$).

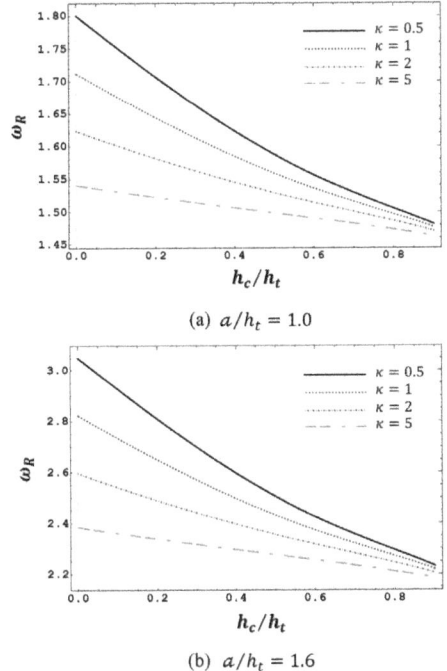

Figure 10. Variation of the first-order nonlinear-to-linear frequency ratio of a GPL-reinforced sandwich beam with respect to an increasing core-to-beam thickness ratio for different power-law parameters and amplitudes of vibration ($N = 10, \Gamma_b = 1\%, L/h_t = 10$).

4.4. Effects of the Total Weight Fraction of the GPLs

The last parametric investigation checks for the sensitivity of the response to an increased total amount of GPL phase in the face sheets of a sandwich beam, both in terms of nonlinear frequency and nonlinear-to-linear frequency. As shown in Figure 11, an increased total weight fraction of the GPLs will increase the nonlinear frequency of the system. Different effects on the nonlinear-to-linear frequency ratio are observable, depending on the value of the selected power-law parameter.

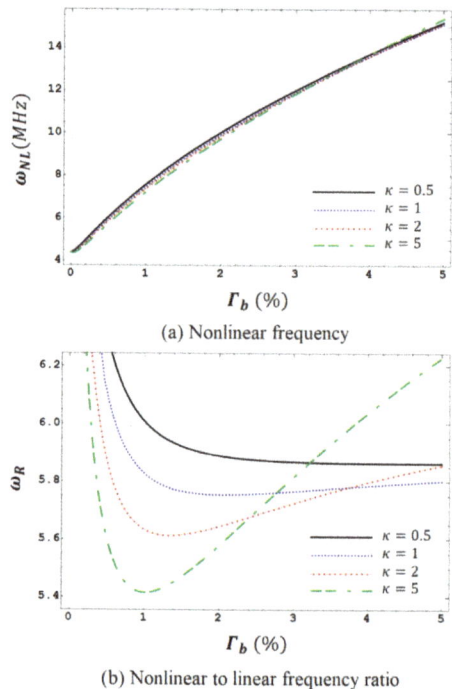

(a) Nonlinear frequency

(b) Nonlinear to linear frequency ratio

Figure 11. Variation of the first-order nonlinear frequency and nonlinear-to-linear frequency ratio of a GPL-reinforced sandwich beam with respect to an increasing total amount of GPLs in the beam, for different power-law parameters ($h_c/h_t = 0.6, N = 10, a/h_t = 3, L/h_t = 10$).

5. Concluding Remarks

In the present paper we analyze the nonlinear free vibration response of thick sandwich beams with FG GPL-reinforced face sheets, based on a novel dispersion model for the reinforcement phase. A higher-order laminated beam model is associated with the Von Kármán strain-displacement relationships to capture the shear and rotation effects on the structural behavior of the system. The nonlinear equations of motion are determined through Hamilton's principle, and they are discretized to ordinary differential equations by means of Galerkin's approach. An analytical solution procedure based on the multiple timescales method is then used to obtain the nonlinear frequency equation. A large numerical investigation analyzes the effect of the vibration amplitude, the thickness of the face sheets, and the GPL dispersion on the nonlinear vibration response of the reinforced sandwich structure, where the following conclusions can be summarized as follows:

- An increased amplitude of vibrations significantly increases the nonlinear frequency and its ratio to linear frequency. The sensitivity of the nonlinear response varies with the total amount of GPL reinforcement in the beam, as well as with the value of the power-law parameter and the thickness of the face sheets.

- An increased power-law parameter can have different effects on the stiffness of the GPL-reinforced sandwich beam, depending on the value of the vibration amplitude and the total weight fraction of the GPLs.
- There exists a threshold value for the vibration amplitude, after which the behavior of the system can change for an increased thickness of the face sheets and power-law parameter.
- For low amplitude vibrations, the nonlinear frequency increases by increasing the power-law parameter and by decreasing the thickness of the faces sheets. For large amplitude vibrations, the contrary occurs by increasing both the power-law parameter and core-to-face thickness ratio.
- The effect of an increasing power-law parameter and a decreasing thickness of the face sheets on the nonlinear-to-linear frequency ratio is independent of the vibration amplitude.
- An increasing total weight fraction of GPLs in the beam generally increases the nonlinear frequency of the system. The sensitivity of the nonlinear-to-linear frequency ratio can be more or less pronounced, depending on the GPL dispersion pattern in the face sheets.
- The proposed parametric study would be of great interest for optimization and design of materials and for an appropriate evaluation of stability for sandwich beams under different environmental conditions, which would prove useful in many space and aircraft applications.

Author Contributions: Formal analysis, M.S.N., H.M., R.D. and F.T.; investigation, M.S.N., H.M., R.D. and F.T.; methodology, H.M., R.D. and F.T.; software, H.M., F.T. and R.D.; supervision, F.T.; validation, M.S.N. and R.D.; writing—original draft, M.S.N. and H.M.; writing—review and editing, R.D. and F.T. All authors have read and agreed to the published version of the manuscript.

Funding: This research received no external funding.

Conflicts of Interest: The authors declare no conflict of interest.

Appendix A

In the following, we provide the extended relations for the coefficients $g_{11}, g_{12}, \ldots$ in Equations (38)–(40), i.e.,

$$g_{11} = \frac{n^2 \pi^2 A_{11}}{2L}, \tag{A1}$$

$$g_{12} = -\frac{n^3 \pi^3 c_1 E_{11}}{2L^2}, \tag{A2}$$

$$g_{13} = \frac{\left(1 + (-1)^{n+1}\right) n^2 \pi^2 A_{11}}{3L^2}, \tag{A3}$$

$$g_{14} = \frac{n^2 \pi^2 (B_{11} - c_1 E_{11})}{2L}, \tag{A4}$$

$$g_{21} = \frac{n^2 \pi^2 (B_{11} - c_1 E_{11})}{2L}, \tag{A5}$$

$$g_{22} = \frac{n^3 \pi^3 \left(-c_1 F_{11} + c_1^2 H_{11}\right)}{2L^2} - \frac{n\pi \left(-A_{55} + 2c_2 D_{55} - c_2^2 F_{55}\right)}{2}, \tag{A6}$$

$$g_{23} = \frac{\left(1 + (-1)^{n+1}\right) n^2 \pi^2 (B_{11} - c_1 E_{11})}{3L^2}, \tag{A7}$$

$$g_{24} = \frac{n^2 \pi^2 \left(D_{11} - 2c_1 F_{11} + c_1^2 H_{11}\right)}{2L} - \frac{L\left(-A_{55} + 2c_2 D_{55} - c_2^2 F_{55}\right)}{2}, \tag{A8}$$

$$g_{31} = \frac{n^3 \pi^3 c_1 E_{11}}{2L^2}, \tag{A9}$$

$$g_{32} = \frac{n^2\pi^2\left(-A_{55} + 2c_2 D_{55} - c_2^2 F_{55}\right)}{2L} - \frac{n^4\pi^4 c_1^2 H_{11}}{2L^3} \tag{A10}$$

$$g_{33} = \frac{n^2\pi^2\left(-1 + (-1)^n + 4(2 + (-1)^n)\mathrm{Sin}^4\left(\frac{n\pi}{2}\right)\right)A_{11}}{3L^2}, \tag{A11}$$

$$g_{34} = \frac{4\left(-2 + (-1)^{n+1}\right)n^3\pi^3 \mathrm{Sin}^4\left(\frac{n\pi}{2}\right)c_1 E_{11}}{3L^3}, \tag{A12}$$

$$g_{35} = -\frac{3n^4\pi^4 A_{11}}{16L^3}, \tag{A13}$$

$$g_{36} = \frac{n\pi\left(-A_{55} + 2c_2 D_{55} - c_2^2 F_{55}\right)}{2} - \frac{n^3\pi^3\left(-c_1 F_{11} + c_1^2 H_{11}\right)}{2L^2}, \tag{A14}$$

$$g_{37} = \frac{n^2\pi^2\left(-1 + (-1)^n + 4(2 + (-1)^n)\mathrm{Sin}^4\left(\frac{n\pi}{2}\right)\right)(B_{11} - c_1 E_{11})}{3L^2}, \tag{A15}$$

$$g_{38} = \frac{n\pi c_1 m_3}{2}, \tag{A16}$$

$$g_{39} = -\frac{n^2\pi^2 c_1^2 m_6}{2L} - \frac{L m_0}{2}, \tag{A17}$$

$$g_{310} = \frac{n\pi\left(-c_1 m_4 + c_1^2 m_6\right)}{2}. \tag{A18}$$

References

1. Birman, V.; Kardomateas, G.A. Review of current trends in research and applications of sandwich structures. *Compos. Part B Eng.* **2018**, *142*, 221–240. [CrossRef]
2. Vescovini, R.; D'Ottavio, M.; Dozio, L.; Polit, O. Buckling and wrinkling of anisotropic sandwich plates. *Int. J. Eng. Sci.* **2018**, *130*, 136–156. [CrossRef]
3. Safaei, B.; Dastjerdi, R.M.; Qin, Z.; Chu, F. Frequency-dependent forced vibration analysis of nanocomposite sandwich plate under thermo-mechanical loads. *Compos. Part B Eng.* **2019**, *161*, 44–54. [CrossRef]
4. Solyaev, Y.; Lurie, S.; Koshurina, A.; Dobryanskiy, V.; Kachanov, M. On a combined thermal/mechanical performance of a foam-filled sandwich panels. *Int. J. Eng. Sci.* **2019**, *134*, 66–76. [CrossRef]
5. Waddar, S.; Pitchaimani, J.; Doddamani, M.; Barbero, E. Buckling and vibration behaviour of syntactic foam core sandwich beam with natural fiber composite facings under axial compressive loads. *Compos. Part B Eng.* **2019**, *175*, 107133. [CrossRef]
6. Kahya, V.; Turan, M. Vibration and stability analysis of functionally graded sandwich beams by a multi-layer finite element. *Compos. Part B Eng.* **2018**, *146*, 198–212. [CrossRef]
7. Li, Y.H.; Dong, Y.H.; Qin, Y.; Lv, H.W. Nonlinear forced vibration and stability of an axially moving viscoelastic sandwich beam. *Int. J. Mech. Sci.* **2018**, *138*, 131–145. [CrossRef]
8. Safaei, B.; Dastjerdi, R.M.; Chu, F. Effect of thermal gradient load on thermo-elastic vibrational behavior of sandwich plates reinforced by carbon nanotube agglomerations. *Compos. Struct.* **2018**, *192*, 28–37. [CrossRef]
9. Aria, A.I.; Friswell, M.I. Computational hygro-thermal vibration and buckling analysis of functionally graded sandwich microbeams. *Compos. Part B Eng.* **2019**, *165*, 785–797. [CrossRef]
10. Huang, Q.; Choe, J.; Yang, J.; Hui, Y.; Xu, R.; Hu, H. An efficient approach for post-buckling analysis of sandwich structures with elastic-plastic material behavior. *Int. J. Eng. Sci.* **2019**, *142*, 20–35. [CrossRef]
11. Joubaneh, E.F.; Barry, O.R.; Oguamanam, D.C.D. Vibrations of sandwich beams with tip mass: Numerical and experimental investigations. *Compos. Struct.* **2019**, *210*, 628–640. [CrossRef]
12. Nguyen, N.D.; Nguyen, T.K.; Vo, T.P.; Nguyen, T.N.; Lee, S. Vibration and buckling behaviours of thin-walled composite and functionally graded sandwich I-beams. *Compos. Part B Eng.* **2019**, *166*, 414–427. [CrossRef]
13. Youzera, H.; Meftah, S.A. Nonlinear damping and forced vibration behaviour of sandwich beams with transverse normal stress. *Compos. Struct.* **2017**, *179*, 258–268. [CrossRef]

14. Wang, Z.X.; Shen, H.S. Nonlinear vibration of sandwich plates with FG-GRC face sheets in thermal environments. *Compos. Struct.* **2018**, *192*, 642–653. [CrossRef]
15. Karamanli, A.; Aydogdu, M. Buckling of laminated composite and sandwich beams due to axially varying in-plane loads. *Compos. Struct.* **2019**, *210*, 391–408. [CrossRef]
16. Nematollahi, M.S.; Mohammadi, H. Geometrically nonlinear vibration analysis of sandwich nanoplates based on higher-order nonlocal strain gradient theory. *Int. J. Mech. Sci.* **2019**, *156*, 31–45. [CrossRef]
17. Wang, X.; Tang, F.; Qi, X.; Lin, Z. Mechanical, electrochemical, and durability behavior of graphene nano-platelet loaded epoxy-resin composite coatings. *Compos. Part B Eng.* **2019**, *176*, 107103. [CrossRef]
18. Wang, Y.; Feng, C.; Santiuste, C.; Zhao, Z.; Yang, J. Buckling and postbuckling of dielectric composite beam reinforced with Graphene Platelets (GPLs). *Aerosp. Sci. Technol.* **2019**, *91*, 208–218. [CrossRef]
19. Wang, Y.Q.; Ye, C.; Zu, J.W. Nonlinear vibration of metal foam cylindrical shells reinforced with graphene platelets. *Aerosp. Sci. Technol.* **2019**, *85*, 359–370. [CrossRef]
20. Wang, Z.; Li, Z.; Xiong, W. Numerical study on three-point bending behavior of honeycomb sandwich with ceramic tile. *Compos. Part B Eng.* **2019**, *167*, 63–70. [CrossRef]
21. Sobhy, M.; Zenkour, A.M. Magnetic field effect on thermomechanical buckling and vibration of viscoelastic sandwich nanobeams with CNT reinforced face sheets on a viscoelastic substrate. *Compos. Part B Eng.* **2018**, *154*, 492–506. [CrossRef]
22. Wang, M.; Li, Z.M.; Qiao, P. Vibration analysis of sandwich plates with carbon nanotube-reinforced composite face-sheets. *Compos. Struct.* **2018**, *200*, 799–809. [CrossRef]
23. Mehar, K.; Panda, S.K. Theoretical deflection analysis of multi-walled carbon nanotube reinforced sandwich panel and experimental verification. *Compos. Part B Eng.* **2019**, *167*, 317–328. [CrossRef]
24. Naderi Beni, N. Free vibration analysis of annular sector sandwich plates with FG-CNT reinforced composite face-sheets based on the carrera's unified formulation. *Compos. Struct.* **2019**, *214*, 269–292. [CrossRef]
25. Zegeye, E.; Ghamsari, A.K.; Woldesenbet, E. Mechanical properties of graphene platelets reinforced syntactic foams. *Compos. Part B Eng.* **2014**, *60*, 268–273. [CrossRef]
26. Liang, J.Z.; Du, Q.; Tsui, G.C.P.; Tang, C.Y. Tensile properties of graphene nano-platelets reinforced polypropylene composites. *Compos. Part B Eng.* **2016**, *95*, 166–171. [CrossRef]
27. Nieto, A.; Bisht, A.; Lahiri, D.; Zhang, C.; Agarwal, A. Graphene reinforced metal and ceramic matrix composites: A review. *Int. Mater. Rev.* **2017**, *62*, 241–302. [CrossRef]
28. Nazarenko, L.; Chirkov, A.Y.; Stolarski, H.; Altenbach, H. On modeling of carbon nanotubes reinforced materials and on influence of carbon nanotubes spatial distribution on mechanical behavior of structural elements. *Int. J. Eng. Sci.* **2019**, *143*, 1–13. [CrossRef]
29. Aluko, O.; Gowtham, S.; Odegard, G.M. Multiscale modeling and analysis of graphene nanoplatelet/carbon fiber/epoxy hybrid composite. *Compos. Part B Eng.* **2017**, *131*, 82–90. [CrossRef]
30. Feng, C.; Kitipornchai, S.; Yang, J. Nonlinear free vibration of functionally graded polymer composite beams reinforced with graphene nanoplatelets (GPLs). *Eng. Struct.* **2017**, *140*, 110–119. [CrossRef]
31. Kitipornchai, S.; Chen, D.; Yang, J. Free vibration and elastic buckling of functionally graded porous beams reinforced by graphene platelets. *Mater. Des.* **2017**, *116*, 656–665. [CrossRef]
32. Karami, B.; Shahsavari, D.; Janghorban, M.; Tounsi, A. Resonance behavior of functionally graded polymer composite nanoplates reinforced with graphene nanoplatelets. *Int. J. Mech. Sci.* **2019**, *156*, 94–105. [CrossRef]
33. Soldano, C.; Mahmood, A.; Dujardin, E. Production, properties and potential of graphene. *Carbon* **2010**, *48*, 2127–2150. [CrossRef]
34. Singh, V.; Joung, D.; Zhai, L.; Das, S.; Khondaker, S.I.; Seal, S. Graphene based materials: Past, present and future. *Prog. Mater. Sci.* **2011**, *56*, 1178–1271. [CrossRef]
35. Nematollahi, M.S.; Mohammadi, H.; Nematollahi, M.A. Thermal vibration analysis of nanoplates based on the higher-order nonlocal strain gradient theory by an analytical approach. *Superlattices Microstruct.* **2017**, *111*, 944–959. [CrossRef]
36. Tian, F.; Lyu, J.; Shi, J.; Yang, M. Graphene and graphene-like two-denominational materials based fluorescence resonance energy transfer (FRET) assays for biological applications. *Biosens Bioelectr.* **2017**, *89*, 123–135. [CrossRef]
37. Wentzel, D.; Miller, S.; Sevostianov, I. Dependence of the electrical conductivity of graphene reinforced epoxy resin on the stress level. *Int. J. Eng. Sci.* **2017**, *120*, 63–70. [CrossRef]
38. Idowu, A.; Boesl, B.; Agarwal, A. 3D graphene foam-reinforced polymer composites—A review. *Carbon* **2018**, *135*, 52–71. [CrossRef]

39. Jiao, P.; Alavi, A.H. Buckling analysis of graphene-reinforced mechanical metamaterial beams with periodic webbing patterns. *Int. J. Eng. Sci.* **2018**, *131*, 1–18. [CrossRef]
40. Chiker, Y.; Bachene, M.; Guemana, M.; Attaf, B.; Rechak, S. Free vibration analysis of multilayer functionally graded polymer nanocomposite plates reinforced with nonlinearly distributed carbon-based nanofillers using a layer-wise formulation model. *Aerosp. Sci. Technol.* **2020**, *104*, 105913. [CrossRef]
41. Li, C.; Han, Q. Semi-analytical wave characteristics analysis of graphene-reinforced piezoelectric polymer nanocomposite cylindrical shells. *Int. J. Mech. Sci.* **2020**, *186*, 105890. [CrossRef]
42. Keshtegar, B.; Motezaker, M.; Kolahchi, R.; Trung, N.T. Wave propagation and vibration responses in porous smart nanocomposite sandwich beam resting on Kerr foundation considering structural damping. *Thin Wall. Struct.* **2020**, *154*, 106820. [CrossRef]
43. Do, V.N.V.; Lee, C.H. Static bending and free vibration analysis of multilayered composite cylindrical and spherical panels reinforced with graphene platelets by using isogeometric analysis method. *Eng. Struct.* **2020**, *215*, 110682. [CrossRef]
44. Shen, H.S.; Xiang, Y.; Fan, Y.; Hui, D. Nonlinear vibration of functionally graded graphene-reinforced composite laminated cylindrical panels resting on elastic foundations in thermal environments. *Compos. Part B Eng.* **2018**, *136*, 177–186. [CrossRef]
45. Yu, Y.; Shen, H.S.; Wang, H.; Hui, D. Postbuckling of sandwich plates with graphene-reinforced composite face sheets in thermal environments. *Compos. Part B Eng.* **2018**, *135*, 72–83. [CrossRef]
46. Arefi, M.; Rezaei, B.E.M.; Dimitri, R.; Bacciocchi, M.; Tornabene, F. Nonlocal bending analysis of curved nanobeams reinforced by graphene nanoplatelets. *Compos. Part B Eng.* **2019**, *166*, 1–12. [CrossRef]
47. Arefi, M.; Bidgoli, E.M.R.; Dimitri, R.; Tornabene, F.; Reddy, J.N. Size-dependent free vibrations of FG polymer composite curved nanobeams reinforced with graphene nanoplatelets resting on Pasternak foundations. *Appl. Sci.* **2019**, *9*, 1580. [CrossRef]
48. Jalaei, M.H.; Dimitri, R.; Tornabene, F. Dynamic stability of temperature-dependent graphene sheet embedded in an elastomeric medium. *Appl. Sci.* **2019**, *9*, 887. [CrossRef]
49. Liu, Z.; Yang, C.; Gao, W.; Wu, D.; Li, G. Nonlinear behaviour and stability of functionally graded porous arches with graphene platelets reinforcements. *Int. J. Eng. Sci.* **2019**, *137*, 37–56. [CrossRef]
50. Nguyen, N.V.; Lee, J.; Xuan, H.N. Active vibration control of GPLs-reinforced FG metal foam plates with piezoelectric sensor and actuator layers. *Compos. Part B Eng.* **2019**, *172*, 769–784. [CrossRef]
51. Polit, O.; Anant, C.; Anirudh, B.; Ganapathi, M. Functionally graded graphene reinforced porous nanocomposite curved beams: Bending and elastic stability using a higher-order model with thickness stretch effect. *Compos. Part B Eng.* **2019**, *166*, 310–327. [CrossRef]
52. Saidi, A.R.; Bahaadini, R.; Mozafari, K.M. On vibration and stability analysis of porous plates reinforced by graphene platelets under aerodynamical loading. *Compos. Part B Eng.* **2019**, *164*, 778–799. [CrossRef]
53. Heydarpour, Y.; Malekzadeh, P.; Dimitri, R.; Tornabene, F. Thermoelastic analysis of rotating multilayer FG-GPLRC truncated conical shells based on a coupled TDQM-NURBS scheme. *Compos. Struct.* **2020**, *235*, 111707. [CrossRef]
54. Affdl, J.C.H.; Kardos, J.L. The halpin-tsai equations: A review. *Polym. Eng. Sci.* **1976**, *16*, 344–352. [CrossRef]
55. Reddy, J.N. *Mechanics of Laminated Composite Plates and Shells: Theory and Analysis*, 2nd ed.; CRC Press: Boca Raton, FL, USA, 2004.
56. Reddy, J.N. *Theory and Analysis of Elastic Plates and Shells*, 2nd ed.; Taylor & Francis: Oxfordshire, UK, 2006.
57. Nayfeh, A.H. *Perturbation Methods*; Wiley: Hoboken, NJ, USA, 1973.
58. Nayfeh, A.H.; Mook, D.T. *Nonlinear Oscillations*; Wiley: Hoboken, NJ, USA, 1995.
59. Vo, T.P.; Thai, H.T. Free vibration of axially loaded rectangular composite beams using refined shear deformation theory. *Compos. Struct.* **2012**, *94*, 3379–3387. [CrossRef]
60. Li, Z.M.; Qiao, P. On an exact bending curvature model for nonlinear free vibration analysis shear deformable anisotropic laminated beams. *Compos. Struct.* **2014**, *108*, 243–258. [CrossRef]
61. Shen, H.S.; Lin, F.; Xiang, Y. Nonlinear vibration of functionally graded graphene-reinforced composite laminated beams resting on elastic foundations in thermal environments. *Nonlinear Dyn.* **2017**, *90*, 899–914. [CrossRef]

 © 2020 by the authors. Licensee MDPI, Basel, Switzerland. This article is an open access article distributed under the terms and conditions of the Creative Commons Attribution (CC BY) license (http://creativecommons.org/licenses/by/4.0/).

Article

Theoretical Analysis of Fractional Viscoelastic Flow in Circular Pipes: Parametric Study

Dmitry Gritsenko [1],[*] and Roberto Paoli [1],[2]

1. Department of Mechanical and Industrial Engineering, University of Illinois at Chicago, Chicago, IL 60607, USA; robpaoli@uic.edu
2. Computational Science Division and Leadership Computing Facility, Argonne National Laboratory, Lemont, IL 60439, USA
* Correspondence: dgrits2@uic.edu

Received: 29 October 2020; Accepted: 14 December 2020; Published: 18 December 2020

Abstract: Pipe flow is one of the most commonly used models to describe fluid dynamics. The concept of fractional derivative has been recently found very useful and much more accurate in predicting dynamics of viscoelastic fluids compared with classic models. In this paper, we capitalize on our previous study and consider space-time dynamics of flow velocity and stress for fractional Maxwell, Zener, and Burgers models. We demonstrate that the behavior of these quantities becomes much more complex (compared to integer-order classical models) when adjusting fractional order and elastic parameters. We investigate mutual influence of fractional orders and consider their limiting value combinations. Finally, we show that the models developed can be reduced to classical ones when appropriate fractional orders are set.

Keywords: fractional calculus; Riemann-Liouville fractional derivative; viscoelasticity; pipe flow; fractional Maxwell model; fractional Zener model; fractional Burgers model

1. Introduction

Fractional calculus in general and fractional derivative in particular has been gathering an increasing researchers interest recently. This mathematical tool provides the means for significant improvement of predictive power for numerous practical applications as summarized in a recent literature [1–18]. Heat conduction [5,18], anomalous diffusion [4,7], and viscoelastic properties of fluids and solids [3,9,10,13] are just a few examples of fundamental phenomena where fractional calculus finds immediate practical applications and provides promising results. While fractional models in general demonstrate better fit of experimental data compared to their classical counterparts, those involving variable-order fractional operators (VO-FC) are the handy tools for the cutting-edge research. VO-FC not only provides an effective tool to alter the system properties described by the fractional orders but also allows accounting for fractional orders variations with time, coordinate, and internal system-specific parameters.

One of the fields where fractional calculus has already proven its efficiency is the prediction of dynamics of viscoelastic flows in various constrained geometries, with circular pipe being one of the most popular ones. With theoretical foundations laid in [19], researchers considered numerous specific problems. The models studied here were addressed by Yin in [20] (Maxwell), by Shah in [21] (Burgers) to name a few. Fractional viscoelastic models, including Maxwell and Zener ones, were considered in close detail in series of works by Schiessel, Friedrich, and others [22,23]. Those authors provided in-depth physical analysis of mathematical concepts introduced via fractional calculus formalism, suggested, developed, and explained various ladder models. This brief literature overview is intended to mention the most recent achievements in a field of FC and to point out studies that specifically targeted fractional models studied here. A more detailed consideration that includes historic retrospective can be found in our previous study [1].

In this paper, we capitalize on the approaches developed in [1] to examine the behavior of fractional viscoelastic Maxwell, Zener, and Burgers models in application to pipe flow. We present an in-depth parametric analysis of these models and outline various operating regimes for each of them. Different from [1], here we consider 3D profiles of flow velocity and stress. It allows for the exploration of flow dynamics in spatial and frequency domains simultaneously and catches the specifics that cannot be easily visible in 2D projections. We consider various combinations of fractional orders, investigate their mutual influence, and establish an approach for viscoelastic flow optimization. The paper is organized as follows. We first provide domain definition. The governing equations for fluid flow and constitutive equations for the model considered are given in Sections 2.1 and 2.2, respectively. The main results of this study appear in Section 3. In particular, Section 3.1 describes the general approach for seeking the solution. Sections 3.2–3.7 present and discuss space-frequency dynamics of velocity and stress profiles. In particular, Section 3.2 presents velocity for fractional Maxwell model. Stress profiles for fractional Maxwell model are considered in Section 3.3. Zener model behavior is studied in Sections 3.4 and 3.5 (velocity and stress, respectively). Sections 3.6 and 3.7 are devoted to fractional Burgers model. The results of this study are summed up in Section 4. Hereafter we use the following notions. Under "classic" we understand either model or profile corresponding to integer values of fractional orders. "Fractional order" and "fractional parameter" is essentially the same. We also use "frequency domain" to refer to nondimensional frequency.

2. Problem Formulation

Here we provide brief problem formulation and key governing equations. All the details as well as derivations and overview of mathematical apparatus involved can be found in our previous study [1].

Let us introduce cylindrical coordinate system (r, θ, z) and consider laminar flow of incompressible viscoelastic fluid along z axis of infinitely long pipe of radius R with circular cross-section. The corresponding schematic is shown in Figure 1.

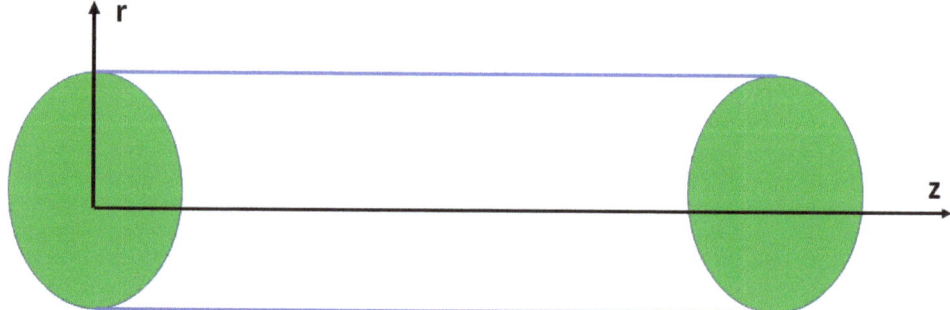

Figure 1. Domain definition for circular pipe.

2.1. Flow Dynamics

The flow dynamics is governed by continuity and momentum equations with no-slip boundary condition at the wall applied. Due to flow and geometry specifics, only z-component of momentum equation remains and reads as:

$$\rho \frac{\partial u_z}{\partial t} = -\frac{\partial p}{\partial z} + \frac{1}{r}\frac{\partial (r\sigma_{rz})}{\partial r} \qquad (1)$$

Here $\rho, u_z, p, \sigma_{rz}$ stand for fluid mass density, z-component of flow velocity, pressure, and rz-component of stress.

2.2. Constitutive Equations

First, we consider fractional Maxwell model (Figure 2a). Its constitutive equation reads as:

$$\sigma(t) + \tau^{\alpha-\beta} {}_aD_t^{\alpha-\beta}\sigma(t) = E\tau^\alpha {}_aD_t^\alpha \varepsilon(t), \quad 0 \leq \alpha \leq 1, \quad 0 \leq \beta \leq 1, \quad \alpha \geq \beta \qquad (2)$$

Here α and β are the fractional orders, τ, E are given by:

$$\tau = \left(\frac{E_1 \tau_1^\alpha}{E_2 \tau_2^\beta}\right)^{\frac{1}{\alpha-\beta}}, \quad E = E_1 \left(\frac{\tau_1}{\tau}\right)^\alpha, \quad E = E_2 \left(\frac{\tau_2}{\tau}\right)^\beta, \qquad (3)$$

where E_1, τ_1, E_2, τ_2 stand for the Young's moduli, relaxation times of elements "1" and "2", respectively. The ranges imposed for α and β have both physical and mathematical meaning. Physically, changing α and β from 0 to 1 allows balancing between viscous and elastic properties of fluid. Mathematically, all fractional orders should be non-negative, in order to ensure that we deal with fractional derivatives, not fractional integrals. An additional relation bounds α and β, so that the entire range for β can only be considered provided $\alpha = 1$. The limiting integer values of fractional orders deserve special attention, as they allow restoring various classical cases. In particular, $\alpha = 1$ and $\beta = 0$ correspond to the classical Maxwell fluid:

$$\sigma + \tau\dot\sigma = E\tau\dot\varepsilon, \qquad (4)$$

while $\alpha = 1$ and $\beta = 1$ results in a constitutive equation for the Newtonian fluid:

$$\sigma = \frac{E\tau}{2}\dot\varepsilon \qquad (5)$$

Here we omitted time dependence of stresses and strains for brevity and introduced dot as a time derivative operator. A closer look at the definition of τ reveals that the case $\alpha = \beta$ is somewhat special. Indeed, it corresponds to a critical gel. It means that τ blows up when $\alpha = \beta$. However, it happens when the base (expression in brackets) is greater than unity. If the opposite is true, then τ is bounded. However, the detailed analysis of all the specifics it brings is beyond the scope of this study.

Next, we consider fractional Zener model shown in Figure 2b. In this case constitutive equation is:

$$\sigma(t) + \tau^{\alpha-\beta}{}_aD_t^{\alpha-\beta}\sigma(t) = E\tau^\alpha {}_aD_t^\alpha \varepsilon(t) + E_0 \tau^\gamma {}_aD_t^\gamma \varepsilon(t) + E_0 \tau^{\gamma+\alpha-\beta} {}_aD_t^{\gamma+\alpha-\beta}\varepsilon(t),$$
$$0 \leq \alpha \leq 1, \quad 0 \leq \beta \leq 1, \quad 0 \leq \gamma \leq 1, \quad \alpha - \beta \geq 0, \quad \gamma + \alpha - \beta \geq 0 \qquad (6)$$

Here γ is the fractional order of element "3" and E_0 is defined as:

$$E_0 = E_3 \left(\frac{\tau_3}{\tau}\right)^\gamma, \qquad (7)$$

where E_3, τ_3 stand for Young's modulus and relaxation time of element "3" and all other notations are the same as above. At a glance, here we end up with two additional relations for the fractional orders $\alpha, \beta,$ and γ: $\alpha - \beta \geq 0$ and $\gamma + \alpha - \beta \geq 0$. A closer look to this inequalities reveals that if $\alpha - \beta \geq 0$ is held, the second one is satisfied automatically. That is, of course, only true for non-negative values of the fractional orders. To get classical Zener fluid, one should assume $\alpha = 1, \beta = 0,$ and $\gamma = 0$. Then the constitutive equation reduces to:

$$\sigma + \tau\dot\sigma = E_0\varepsilon + (E + E_0)\tau\dot\varepsilon \qquad (8)$$

Next, assuming $\alpha = 1$, $\beta = 1$, and $\gamma = 1$, one will end up with Newtonian fluid and constitutive equation of the form:

$$\sigma = \frac{E + 2E_0}{2}\tau\dot{\varepsilon} \qquad (9)$$

Finally, for fractional Burgers model, (Figure 2c) we have:

$$\sigma(t) + \frac{E}{E_0}\tau^{\alpha-\gamma}{}_aD_t^{\alpha-\gamma}\sigma(t) + \frac{E}{E_0}\tau^{\alpha-\delta}{}_aD_t^{\alpha-\delta}\sigma(t) + \frac{E}{E_0}\tau^{\beta-\gamma}{}_aD_t^{\beta-\gamma}\sigma(t) + \frac{E}{E_0}\tau^{\beta-\delta}{}_aD_t^{\beta-\delta}\sigma(t) =$$
$$= E\tau^{\alpha}{}_aD_t^{\alpha}\varepsilon(t) + E\tau^{\beta}{}_aD_t^{\beta}\varepsilon(t),$$
$$0 \leq \alpha \leq 1, \quad 1 \leq \beta \leq 2, \quad 0 \leq \gamma \leq 1, \quad 0 \leq \delta \leq 1, \quad \alpha - \gamma \geq 0, \quad \beta - \gamma \geq 0, \quad \alpha - \delta \geq 0, \quad \beta - \delta \geq 0 \qquad (10)$$

Here δ is the fractional order of element "4", (7) is still valid and additionally:

$$E_0 = E_4\left(\frac{\tau_4}{\tau}\right)^{\delta}, \qquad (11)$$

where E_4, τ_4 stand for Young's modulus and relaxation time of element "4" and all other notations being the same as above. Here we impose four additional conditions for fractional orders α, β, γ and δ. Different from previous models, the range of β has changed as the classical Burgers model should include second derivatives of stresses and strains with respect to time. Moreover, there are no relations for fractional orders α and β. At the same time, the whole range of values for fractional orders γ and δ can only be considered if $\alpha = 1$. To get classical Burgers model, one should set $\alpha = 1$, $\beta = 2$, $\gamma = 0$ and $\delta = 0$. Then the constitutive equation reads as:

$$\sigma + 2\frac{E}{E_0}\tau(\dot{\sigma} + \tau\ddot{\sigma}) = E\tau(\dot{\varepsilon} + \tau\ddot{\varepsilon}), \qquad (12)$$

where two dots stand for the second derivative with respect to time. Newtonian fluid can be obtained by setting $\alpha = \beta = \gamma = \delta = 1$. Corresponding constitutive equation reads as:

$$\sigma = \frac{2EE_0}{E_0 + 4E}\tau\dot{\varepsilon} \qquad (13)$$

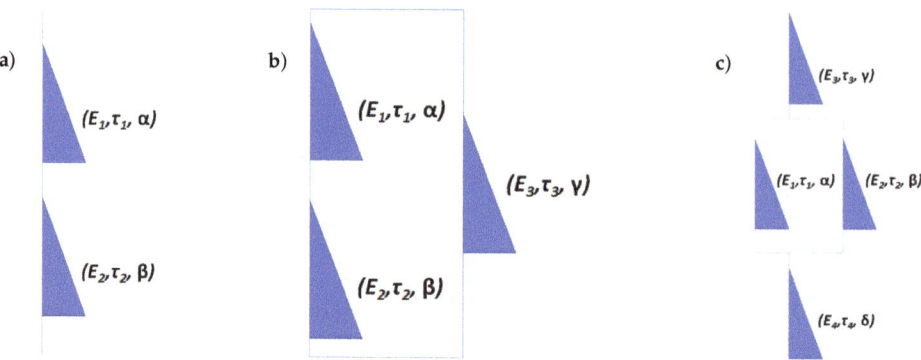

Figure 2. Fractional viscoelastic models: (**a**) Maxwell; (**b**) Zener; (**c**) Burgers.

3. Results and Discussion

3.1. General Solution

Here we follow the approach proposed earlier by Yin [20] for Maxwell model and then generalize it for all other models. Introduce pressure gradient of the form:

$$\frac{\partial p}{\partial z} = P_0 e^{i\omega t} \tag{14}$$

For all models considered in this study, nondimensional velocity and stress profiles are given by:

$$\tilde{u} = \left|\frac{u_z}{u_0}\right| = \left(1 - \frac{J_0(\sqrt{\bar{\omega}}\tilde{a}\phi)}{J_0(\sqrt{\bar{\omega}}\tilde{a})}\right), \quad \tilde{\sigma} = \left|\frac{\sigma_{rz}}{\sigma_0}\right| = \frac{J_1(\sqrt{\bar{\omega}}\tilde{a}\phi)}{\sqrt{\bar{\omega}}\tilde{a}J_0(\sqrt{\bar{\omega}}\tilde{a})}, \tag{15}$$

where we used the following quantities:

$$\tilde{r} = \frac{r}{\sqrt{\nu\tau}}, \quad \tilde{a} = \frac{R}{\sqrt{\nu\tau}}, \quad \bar{\omega} = \omega\tau, \quad \phi = \frac{\tilde{r}}{\tilde{a}}, \quad u_0 = \frac{1}{4\rho\omega}\frac{\partial p}{\partial z}, \quad r^2\zeta^2 = \bar{\omega}\tilde{r}^2, \tag{16}$$

and ν, ω stand for kinematic viscosity and angular frequency, $\sigma_0 = \frac{\partial p}{\partial z}R$, and $J_l(\cdot)$ is the Bessel function of first kind of order l. For all the models considered, we set $\tilde{a} = 0.1$. Table 1 provides expressions for parameter ζ^2 and ranges of fractional orders for all three models. The details of velocity and stress profiles derivation can be found in [1].

Table 1. Fractional models and their parameters.

Parameter Model	ζ^2	α	β	γ	δ
Fractional Maxwell	$(\rho\omega^2/E)\left[\frac{1}{(i\omega\tau)^\alpha} + \frac{1}{(i\omega\tau)^\beta}\right]$	$[0,1]$	$[0,1]$	–	–
Fractional Zener	$(\rho\omega^2/E)\left[\frac{1}{(E_0/E)(i\omega\tau)^\gamma + \frac{(i\omega\tau)^{\alpha+\beta}}{(i\omega\tau)^\alpha+(i\omega\tau)^\beta}}\right]$	$[0,1]$	$[0,1]$	$[0,1]$	–
Fractional Burgers	$(\rho\omega^2/E)\left[\frac{1}{(E_0/E)\left((i\omega\tau)^\gamma+(i\omega\tau)^\delta\right)} + \frac{1}{\left((i\omega\tau)^\alpha+(i\omega\tau)^\beta\right)}\right]$	$[0,1]$	$[1,2]$	$[0,1]$	$[0,1]$

3.2. Velocity Profiles for Fractional Maxwell Model

For fractional Maxwell model parameters that can vary are fractional orders α and β. We examine dynamics of normalized velocity along pipe diameter with respect to nondimensional frequency, $\bar{\omega}$. Both fractional orders α and β are varied considering $\alpha \geq \beta$. Let us first fix $\beta = 0.2$ and vary α: $\alpha = 0.2$, ..., 1 (See Supplementary Materials animation muz1 https://youtu.be/wNkKfdEOS4Q). Alternatively, $\alpha = 1$ and β is varied as: $\beta = 0.2$, ..., 1 (animation muz2 https://youtu.be/2dWZ6KNwL5I). These two velocity profiles have both similarities and distinct differences. Consider similarities first. Oscillatory behavior of the velocity profile is observed with respect to both spatial and frequency coordinates while for the latter these oscillations are also damped. Moreover, up to a certain nondimensional frequency, velocity increases rapidly until forming a nearly parabolic profile. Additionally, with α and β increasing, resonant peaks of velocity profile shift to higher values of non-dimensional frequency. Upon passing this profile, oscillations emerge along the pipe diameter. The dynamics of initial parabolic profile, however, differ when changing α and β. In particular, when β is fixed and α increases, peak amplitudes in initial parabolic profile oscillate. In contrast to it, when we fix α and vary β, initial parabolic profile peak amplitude gradually decreases with increasing β. What is even more noticeable, for $\alpha = 1$ and

$\beta = 0.2, ..., 1$, the velocity profile reduces to parabolic for the entire range of nondimensional frequency as β increases. This result is not surprising as for $\alpha = 1$ and $\beta = 1$ we end up with a Newtonian fluid.

3.3. Stress Profiles for Fractional Maxwell Model

Now let us have a look at stress profiles (animations msf1 https://youtu.be/3YnhhJUy1Wg and msf2 https://youtu.be/wTi8FdTpyok). Here again we either fix $\beta = 0.2$ and vary α: $\alpha = 0.2, ..., 1$ (animation msf1) or fix $\alpha = 1$ and vary β: $\beta = 0.2, ..., 1$ (animation msf2), while satisfying $\alpha \geq \beta$. Let us start from the similarities. The profiles are symmetrical with respect to pipe centerline. For both cases, stress exhibits aperiodic oscillations in a frequency domain and increases monotonically (nonlinear) from the pipe center up to the wall. Here come the differences. When β is fixed, stress peak amplitudes oscillate with increasing α in frequency domain with resonant peak shifting to higher values. More specifically, peak amplitude initially decay up to $\alpha \sim 0.5$. For $\alpha > 0.5$ trend reverses. Interestingly, the rate of amplitude change is higher for $\alpha < 0.5$. The situation changes dramatically when $\alpha = 1, \beta = 0.2, ..., 1$. While the initial behavior is very similar to the first case, the stress becomes independent of frequency and increases linearly from the pipe center up to the wall as β approaches unity (Newtonian fluid).

3.4. Velocity Profiles for Fractional Zener Model

Fractional Zener model is described by four parameters, namely three fractional orders α, β, γ and elastic parameter ψ ($\psi = E_0/E$). Let us first set $\psi = 1$ and examine contributions made by fractional orders α, β, γ. In the first pair of animations (zuz1 https://youtu.be/T8LLEwgiC4k and zuz2 https://youtu.be/dXQCrXTNLbQ) we investigate the influence of fractional order α ($\alpha = 0.2, ..., 1$) for two limiting cases of β and γ ($\beta = \gamma = 0.2$ and $\beta = \gamma = 1$, respectively). In both cases we observe parabolic profiles along the pipe diameter with the amplitude increasing when so does nondimensional frequency. However, in case of lower limits for fractional orders β and γ, the parabolic profile is followed by complex oscillatory pattern. Moreover, peak values of initial parabolic profile also oscillate. The situation is strikingly different if we consider the upper limit for fractional orders β and γ ($\beta = \gamma = 1$). Here parabolic profile occupies the entire range of nondimensional frequency considered. Moreover, it is independent from fractional order α.

Let us now have a look at the impact of the fractional order β. More specifically, we fix fractional orders $\alpha = 1, \gamma = 0.2$ (animation zuz3 https://youtu.be/bjLWJxFWy8I) and $\alpha = 1, \gamma = 1$ (animation zuz4 https://youtu.be/9VaIO3HkW8Y), while varying β: $\beta = 0.2, ..., 1$. Initial velocity profiles look somewhat similar to the previous case. That is, for lower limit value of γ profile is initially parabolic and switches to oscillatory in both space- and frequency domains. For the upper limit of γ, increasing parabolic profile along pipe diameter occupies entire frequency domain. However, it all changes when β start increasing. In particular, velocity profile gradually becomes parabolic along pipe diameter for the entire frequency range thus delivering a case of Newtonian fluid (animation zuz3 https://youtu.be/bjLWJxFWy8I). For the upper limit value of γ (animation zuz4 https://youtu.be/9VaIO3HkW8Y), the amplitudes of parabolic profile gradually decrease (up to roughly 60% from the maximum value) with increasing β.

Next, we investigate the effect of varying fractional order γ. More specifically, we fix $\alpha = 1$ for both lower limit of β ($\beta = 0.2$, animation zuz5 https://youtu.be/j8AGHtMGXV8) and upper limit of β ($\beta = 1$, animation zuz6 https://youtu.be/-wJNeA8hTXY), while varying γ: $\gamma = 0.2, ..., 1$. While the plots exhibit some similarities with the case of varying fractional order β, their dynamics is different. In particular, the rate of amplitudes decay for both parabolic and oscillating components is much higher. The same can be said when the profile turns to a parabolic for the entire frequency domain. As for the upper limit value of β, the amplitudes of parabolic profile also decay relatively faster (up to roughly 35% from the maximum value).

Finally, we consider the impact of elastic parameter ψ. More specifically, we fix $\alpha = 0.5$ and $\beta = 0.2$ for both lower (animation zuz7 https://youtu.be/GExLFiUC8Hk) and upper (animation zuz8 https://youtu.be/YbKXMIgcwSY) limits of fractional order γ ($\gamma = 0.2$ and $\gamma = 1$, respectively). At the same time, we vary ψ: $\psi = 0.01...100$. For the lower limit of γ, the peak amplitudes of the velocity profile increase until switching happens ($\psi \sim 20$) from parabolic+oscillatory behavior to purely parabolic. Upon reaching parabolic profile, its amplitudes start decaying while further increasing ψ. The rate of decay, however, decreases with increasing ψ (especially for $\psi > 30$). Different from all other cases with upper limit values of fractional order set, here for $\gamma = 1$ we observe a combination of parabolic and oscillatory profiles for lower values of ψ. It changes abruptly to a parabolic one ($\psi \sim 0.1$) and almost vanish for $\psi > 5$. This more complex structure can, however, be attributed to the fact that we set α not at the upper limit value ($\alpha = 1$) but at $\alpha = 0.5$.

Generally speaking, the influence of fractional orders and elastic parameter on a velocity profile is quite different. The strongest effect is achieved when varying elastic parameter ψ. Among fractional orders, the most powerful contribution in a profile dynamics is made by the fractional order γ, while influence of α is the weakest one. This trend is especially visible for the upper limit values of the fractional orders.

3.5. Stress Profiles for Fractional Zener Model

We further consider the dynamics of stress profiles for fractional Zener model. We first fix $\psi = 1$ and examine the influence of fractional parameters only. We consider the contribution made by the different parameters in the same order as for velocity profiles, i.e., varying α, β, γ and ψ. Let us start from the fractional order α. Corresponding plots are shown in animations zsf1 https://youtu.be/SOV4_XRiL1s and zsf2 https://youtu.be/ajgZpM0JQS8. For the first one we set $\beta = 0.2$, $\gamma = 0.2$ and vary α: $\alpha = 0.2, ..., 1$. As previously, profile is symmetric with respect to the pipe centerline. Moreover, it exhibits abruptly decaying aperiodic oscillations in a frequency domain. As α increases, peak amplitudes of the stress initially reduce, followed by increase for $\alpha > 0.5$. Different from fractional Maxwell model (see animation msf1 https://youtu.be/3YnhhJUy1Wg), the rate of peak amplitude change appears to be almost the same for $\alpha < 0.5$ and $\alpha > 0.5$. The stress profile for upper limit values of β and γ is shown in animation zsf2 https://youtu.be/ajgZpM0JQS8. It reveals that the stress is independent from the fractional order α and remains almost the same for the entire frequency domain.

Next, we examine the influence of fractional order β. More specifically, α is fixed ($\alpha = 1$) with β varying ($\beta = 0.2, ..., 1$) and limiting values of γ. Stress dynamics for $\gamma = 0.2$ is given in animation zsf3 https://youtu.be/jSkii6WKFlg. Different from the case with varying α, here we observe switching from oscillatory to constant behavior in a frequency domain as β increases. Stress profile for upper limit of γ ($\gamma = 1$) is shown in animation zsf4 https://youtu.be/zpyOc_J-Xnc. It turns out to be very slightly increasing at the pipe walls with increasing β in a frequency domain, while linearly increasing from the pipe center up to its walls.

Let us now vary fractional order γ. We set $\alpha = 1$; $\beta = 0.2$ (animation zsf5) https://youtu.be/M72Hy5s1D_U or $\beta = 1$ (animation zsf6 https://youtu.be/A-yh4HRlXz4) and vary γ: $\gamma = 0.2, ..., 1$. Here again we observe a switch from oscillatory to V-shape profile with increasing γ ($\beta = 0.2$). When $\beta = 1$, V-shape profile is present for the entire range of γ. It slightly decays at the pipe walls. However, this decay vanishes with increasing γ to to being completely independent from $\tilde{\omega}$.

Finally, we looked at the difference made by varying elastic parameter ψ. We fixed $\alpha = 0.5$, $\beta = 0.2$, vary ψ ($\psi = 0.01...100$). The lower limit case ($\gamma = 0.2$) is given in animation zsf7 https://youtu.be/qlLd7LQ6mtw. Switching between oscillatory and V-shape profiles is observed. However, its dynamics is different from the cases of varying fractional orders. More specifically, the peak stress values first increase with increasing ψ. This trend is observed until switching to V-shape profile, when stress amplitudes start gradually decrease. Next, we set $\gamma = 1$ (animation zsf8 https://youtu.be/WqIpXuk0tG8). Here too, a drastic difference from all the cases with varying fractional orders is observed. That is, complex symmetric oscillatory profile is present for the lowest values of ψ. As ψ starts increasing, stress amplitudes drop abruptly and switching to a V-shape profile occurs. Upon reaching this point, stress amplitude at the pipe walls starts growing.

3.6. Velocity Profiles for Fractional Burgers Model

Finally, let us examine the most complex model presented in this study, fractional Burgers model. Here five parameters should be considered, namely four fractional orders $\alpha, \beta, \gamma, \delta$ and elastic parameter ψ. Here we follow the pattern used above to maintain consistency. In particular, we first fix elastic parameter value and study the influence of fractional parameters on model dynamics and then include varying elastic parameter in consideration. To demonstrate the model to be universal, we picked a different value of elastic parameter, $\psi = 10$, when varying fractional orders α, β, γ and δ. It is worth remembering that for fractional Burgers model the range of fractional order β is different ($1 \leq \beta \leq 2$) from those for fractional Maxwell and Zener model ($0 \leq \beta \leq 1$).

As previously, let us start from a varying fractional order α ($\alpha = 0.2, ..., 1$). We fix $\beta = 1.2$ and consider lower ($\gamma = 0.2, \delta = 0.2$) and upper ($\gamma = 1, \delta = 1$) limits of the fractional orders γ and δ. Corresponding plots are shown in animations buz1 https://youtu.be/tLqceVl3ywk and buz2 https://youtu.be/JlIdYSZ-hbk. For lower limit values of fractional orders γ and δ, velocity profile appears to be independent from the fractional order α. Moreover, profile itself looks different from those fractional Maxwell and Zener models (for varying α). In particular, as nondimensional frequency increases, the velocity profile changes from parabolic to a resonant with plateaus. This difference, however, can be attributed to the fact that we picked a different value of elastic parameter ψ. When upper limit values of γ and δ are considered, velocity profile changes to parabolic with gradually increasing amplitude for the entire frequency domain. As α increases, no changes in profile shape are observed until $\alpha \sim 0.6$. Starting from it, peak profile value starts decreasing and reduces by approximately 40%.

Next, we fix $\alpha = 1$ and consider the impact of fractional order β. Velocity profile for lower limit of the remaining fractional orders ($\gamma = 0.2, \delta = 0.2$) is given in animation buz3 https://youtu.be/TUZE3HmaUg8. While the initial profile looks similar to the previous case, its dynamics with β differs. More specifically, profile consists of parabolic and resonant profiles. As β increases, peak value of parabolic profile slowly increases. Upon reaching $\beta \sim 1.8$ it remains almost unchanged. In contrast to it, as shown in animation buz4 https://youtu.be/Cp6h-9S-7hA, for $\gamma = \delta = 1$, there is only parabolic profile present for the entire frequency domain. Its peak amplitudes decrease gradually with increasing β dropping by approximately 70% from their maximum values. Moreover, as β surpasses a certain value ($\beta \sim 1.7$), profile remains almost unchanged up to upper limit ($\beta = 2$).

Now let us have a look at the influence made by varying fractional order γ. More specifically, we fix $\alpha = 1$, consider lower ($\beta = 1, \delta = 0.2$) and upper ($\beta = 2, \delta = 1$) limits for fractional orders β and δ. Finally, we vary γ in a range: $\gamma = 0.2, ..., 1$. For the lower limit (animation buz5) https://youtu.be/X2DWL3tJDV8 we start with a combination of parabolic and resonant profiles. As γ increases, peak value amplitudes for both components gradually decrease. Simultaneously, the profile degenerates to a parabolic for the entire frequency domain. These trends are observed up to $\gamma \sim 0.6$. Upon passing this value, profile remains parabolic. Its peak amplitude starts slowly increasing up to the limit value ($\gamma = 1$). When it comes to upper limit ((animation buz6) https://youtu.be/hs7pUv2nybo), dynamic parabolic profile is observed for the entire frequency domain. It, however, has some specifics that distinguishes it from previous cases. That is, for lower values of γ, peak values of parabolic profiles approach resonance at a certain $\tilde{\omega}$, then starts decaying, and finally reaches nearly constant value. The situation, however, changes when γ starts increasing. Peak values of parabolic profile start reducing until peak itself disappears. Then for the entire frequency domain parabolic profile is observed with amplitudes gradually increasing with $\tilde{\omega}$.

The last, but the least fractional order to vary is δ. Other fractional orders are as follows: $\alpha = 1$; $\beta = 1, \gamma = 0.2$ (lower limit) and $\beta = 2, \gamma = 1$ (upper limit). Finally, $\delta = 0.2, ..., 1$. Lower limit case is given in animation buz7 https://youtu.be/wmkBTIXXI0I. Similar to animation buz5 https://youtu.be/X2DWL3tJDV8, where we varied γ, a combination of parabolic and resonant profiles is observed. With δ increasing it degenerates to a parabolic profile for the entire frequency range with a resonance at a certain $\tilde{\omega}$. For the upper limit (animation buz8 https://youtu.be/0UQ6tFeV3yc) the profile is overall similar to animation buz6 https://youtu.be/hs7pUv2nybo, where varying γ was considered.

Finally, we consider the impact made by a varying elastic parameter ψ. Here ψ changes in a range $\psi = 0.1, ..., 100$. We start from $\alpha = 1$, $\beta = 1$, $\gamma = 0.2$ and $\delta = 0.2$ (animation buz9 https://youtu.be/t_BV2foEg5w). The profile dynamics appears to be the most complex considered so far. For the lowest values of ψ we have a combination of parabolic profile, inclined decaying oscillations that are symmetric with respect to pipe centerline. The latter enclose a plateau. As ψ increases, this complex combination transforms to parabolic+oscillatory profile observed for fractional Maxwell and Zener models. Next, it turns to a combination of parabolic and resonant profiles. With further increase in ψ, parabolic profile with resonance at certain $\tilde{\omega}$ emerges. Finally, it becomes a simple parabolic profile with the amplitude increasing with $\tilde{\omega}$. The next set of fractional parameters considered is: $\alpha = 1$, $\beta = 2$, $\gamma = 0.2$, and $\delta = 0.2$ (animation buz10 https://youtu.be/Fyel0IDrbtU). The variety of the profiles remains the same. However, in this case switching between them occurs slower compared to the previous one. Moreover, relative peak amplitudes turn out to be higher. Two final profiles correspond to parameter sets: $\alpha = 1$, $\beta = 1$, $\gamma = 1$, $\delta = 1$ and $\alpha = 1$, $\beta = 2$, $\gamma = 1$, $\delta = 1$ (animations buz11 https://youtu.be/MboUM5_YEM0 and buz12 https://youtu.be/U7uPEOWOc5s, respectively). This pair appears to be a way more simpler compared its immediate predecessors. More specifically, for both cases plots represent a combination of parabolic and M-shape profiles. Both reduce to simple parabolic profiles as ψ increases, with amplitudes increasing when so does $\tilde{\omega}$. In contrast to animation buz10 https://youtu.be/Fyel0IDrbtU, in animation buz12 relative peak amplitudes decay faster for the upper limit of β ($\beta = 2$). Relative simplicity of these two profiles can be attributed to the fact that all fractional orders as well as their differences are set to integers.

3.7. Stress Profiles for Fractional Burgers Model

Finally, we will have a look at the dynamics of stresses for fractional Burgers model. In picking the values for parameters affecting system behavior, we pursue the logic developed for the velocity profiles. In particular, we first fix the value of the elastic parameter at $\psi = 10$, set limiting values for three out of four fractional parameters and vary the fourth one. Then vary ψ for various limiting values of fractional orders. Once again, let us start from varying fractional order α ($\alpha = 0.2, ..., 1$). Simultaneously, we set $\beta = 1.2$, $\gamma = 0.2$, $\delta = 0.2$ for lower limit (animation bsf1 https://youtu.be/Pa3jHar4Ld8) and $\beta = 1.2$, $\gamma = 1$, $\delta = 1$ for the upper limit (animation bsf2 https://youtu.be/mTf29Vn1TSQ). When compared to animations zsf1 https://youtu.be/SOV4_XRiL1s and zsf2 https://youtu.be/ajgZpM0JQS8 for fractional Zener model, both similarities and differences can be outlined. For upper limit values, stresses are independent from α and do not change in frequency domain. For lower limit, plots have similar structure and are independent from α. At the same time, for Zener model major peak is sharper and higher, while for Burgers it has lower peak amplitude and is way more dispersive.

Next, we fix $\alpha = 1$ and consider the impact of fractional order β. Velocity profile for lower limit of the remaining fractional orders ($\gamma = 0.2$, $\delta = 0.2$) is given in animation bsf3 https://youtu.be/UdAFrW3Ut-8. It differs dramatically from a similar plot for fractional Zener model (animation zsf3 https://youtu.be/jSkii6WKFlg). In particular, while aperiodic oscillations are also observed in both frequency and space domains, no switching to V-shape profile occurs. Instead, amplitude of major peak slowly increases with increasing β. As for the upper limit ($\gamma = 1$, $\delta = 1$), static V-shape profile is observed.

Now let us have a look on the influence made by varying fractional order γ. More specifically, we fix $\alpha = 1$, consider lower ($\beta = 1$, $\delta = 0.2$) and upper ($\beta = 2$, $\delta = 1$) limits of fractional orders β and δ. Finally, we vary γ in a range: $\gamma = 0.2, ..., 1$. For lower limit, plot is shown in animation bsf5 https://youtu.be/eRX9wVxeW_Y. It exhibits aperiodic oscillations in both space- and frequency domains. As γ increases, the value of the major peak increases too. Moreover, the peak itself shifts to higher values of $\tilde{\omega}$. When compared to appropriate Zener model (animation zsf5 https://youtu.be/M72Hy5s1D_U), a striking difference can been seen. More specifically, for Zener model, increasing γ

results in switching profile from oscillating to V-shape. For the upper limit, plot is given in animation bsf6 https://youtu.be/owVOe5Nvc9Q. As γ increases, it demonstrates switching between oscillatory and V-shape profiles. When compared to appropriate Zener model (animation zsf6 https://youtu.be/A-yh4HRlXz4), the major difference is that for the latter only V-shape profile is present.

Next in line, we consider varying fractional order δ ($\delta = 0.2...1$). Corresponding plots for lower limit ($\alpha = 1, \beta = 1, \gamma = 0.2$) and upper limit ($\alpha = 1, \beta = 2, \gamma = 1$) are shown in animations bsf7 https://youtu.be/8H_w4KVlAp8 and bsf8 https://youtu.be/Djga0UAMybI, respectively. We start with the lower limit case. Here we observe aperiodic oscillatory behavior of the stress profile. As δ increases up to $\delta \sim 0.5$, amplitudes of the major peaks decrease. Upon passing this point, trend reverses and peak amplitudes start increasing up to the upper limit of δ ($\delta = 1$). For the upper limit case switching between aperiodic oscillatory and V-shape profiles occurs as δ increases. Interestingly, relative stress amplitudes keep decreasing for the entire δ range. At the same time, the rate of this decrease gradually becomes smaller and is almost not visible for $\delta > 0.8$.

Finally, we consider the impact made by a varying elastic parameter ψ. Here ψ changes in a range $\psi = 0.1, ..., 100$. We start from the pair $\alpha = 1, \beta = 1, \gamma = 0.2, \delta = 0.2$ (animation bsf9 https://youtu.be/a_TNASPHc8Y) and $\alpha = 1, \beta = 2, \gamma = 0.2, \delta = 0.2$ (animation bsf10 https://youtu.be/v35B5JzHzmE). Different from all the cases with varying fractional orders, here we observe very sharp major peaks in a frequency domain followed by abrupt decay and almost zero, flat plateaus. As ψ increases, aperiodic oscillatory profiles in space domain emerge. Moreover, major peaks become more dispersive and move to higher values of nondimensional frequency. For both cases profiles are on the track to V-shape but end up with nonlinear behavior in both space and frequency domains. Thus, a typical picture observed earlier with V-shape profile and stress independent from nondimensional frequency is not achieved. The difference between two is in the relative amplitudes of the major peaks. These are larger when the upper limit for β is set ($\beta = 2$, animation bsf10 https://youtu.be/v35B5JzHzmE). Animations bsf11 https://youtu.be/KKaSzIMAMDY and bsf12 https://youtu.be/d_qQx1P7fXw show the cases with the upper limits of fractional orders γ and δ. More specifically, animation bsf11 corresponds to $\alpha = 1, \beta = 1, \gamma = 1, \delta = 1$, while animation bsf12 shows the case of $\alpha = 1, \beta = 2, \gamma = 1, \delta = 1$. For both cases curved V-shaped profiles are observed starting from the lowest values of ψ. As ψ increases, both profiles tend V-shape and become independent from nondimensional frequency $\tilde{\omega}$. Both profiles remain unchanged for $\psi > 5$. The only difference is that for upper limit of β ($\beta = 2$, animation bsf12) the rate of transformation to V-shape profile is higher.

We would like to pay attention to a parameter ζ. It has been earlier introduced as a wave number but is also related to other physical quantities. A closer look at Table 1 reveals similarities in the form of these expressions. More specifically, ζ^2 for fractional Maxwell, Zener, and Burgers models differ by the terms in square brackets. What are these terms? In fact, divided by E, these are complex compliances, $J^*(\omega)$ ($\zeta^2 = \rho\omega^2 J^*(\omega)$), as they were defined in literature (see, for example, [22] for fractional Maxwell and fractional Zener models). As ζ enters both velocity and stress profiles, it delivers deeper physical meaning and establishes its role in fluid flow characterization. Moreover, with complex compliance introduced, other physical quantities can be readily obtained, including creep compliance, complex, and relaxation moduli. All these quantities are routinely measured in experiments for material properties characterization.

Fractional orders, their mutual influence, and underlying physical meaning should also be considered. Each fractional order (except for β in fractional Burgers model) varies in a range from 0 to 1. As outlined earlier, it allows to balance material properties between purely elastic and purely viscous. The same is true for all the differences of fractional orders. However, when both elastic and viscous behavior are present, damped oscillations of flow velocity are observed (in nondimensional frequency domain). Velocity profiles for models considered in this study are not symmetrical with respect to fractional orders. To better understand this phenomenon, let us look again at velocity profiles for fractional Maxwell model. Why does profile behavior in frequency domain differ in animations muz1 https://youtu.be/wNkKfdEOS4Q and muz2 https://youtu.be/2dWZ6KNwL5l? In animation muz1 an initial situation ($\alpha = \beta = 0.2$) results in $\alpha - \beta = 0$ in the left-hand side of the constitutive equation. At the same time, in the right-hand side $\alpha = 0.2$. Consequently, damped ($\alpha \neq 0$) oscillations are observed. When α increases, $\alpha - \beta \neq 0$. It starts contributing in oscillatory behavior of the flow. That is why damped oscillations are present for $\alpha = 1$. Now examine animation muz2. At the very beginning we have $\alpha = 1$ and $\alpha - \beta = 0.8$. When β starts increasing, oscillations in frequency domain gradually disappear. Why? Because we are on the track to pure dashpot that is reached for $\alpha = \beta = 1$ ($\alpha = 1$ in a right-hand side of (7) and $\alpha - \beta = 0$ in a left-hand side of (7)).

Finally, if we examine Figure 2, simple-to-complex approach in constructing fractional viscoelastic models can be restored. All the models are constructed via connecting fractional elements in series/parallel. For instance, fractional Zener model is readily obtained from Maxwell one via adding fractional element in parallel. More complex models can be built in a similar fashion. Fractional Burgers model, for example, can be obtained from fractional Kelvin-Voigt one, as outlined in [1]. Fractional viscoelastic models are principally different from their integer-order counterparts. An increase in the number of fractional elements is made to properly describe the specific class of materials. For integer-order models, adding more elements often serves to increase model accuracy. Thus, fractional viscoelastic models not only help to describe behaviors missed by integer-order ones but also simplify problem solving and material characterization.

4. Conclusions

This study has provided detailed parametric analysis of velocity and stress profiles for three fractional viscoelastic models, namely Maxwell model, Zener model, and Burgers model. These models were chosen to represent two-, three-, and four-element viscoelastic systems. Two types of parameters were considered: fractional and elastic. We have extended 2D projections of velocity and stress profiles studied in [1] and considered 3D dynamic velocity and stress surfaces in space and frequency domains. It allowed better visual representation of quantities studied. Surface plots simplified comparison of contributions made by varying individual fractional orders for each model considered. Dependent on researchers' needs, proper altering of fractional and elastic parameters can be made. Whether achieving of local/global minimum/maximum is required for practical applications, an optimal operating regime can found and visualized. In case the predictions of fractional models should be compared with classical or Newtonian ones, we have provided corresponding constitutive equations.

Fractional Maxwell model was found to be dependent on a pair (α, β) of fractional parameters only. The model has demonstrated dynamic behavior with fixed one and varying another fractional parameter, switching from a combination of parabolic and oscillatory behaviors (for lower values of fractional parameters) to purely classic parabolic behavior for the velocity profile (for higher values of fractional parameters). The latter took place only when both tended to unity that corresponded to Newtonian fluid. Fractional Zener model was found to be dependent on three fractional parameters and one elastic parameter. Among these four parameters, the strongest impact is made by the elastic one. As for varying fractional orders, the strongest effect (profile switching rate) was caused by γ, while the weakest by α. Moreover, for all combinations of upper limits for fractional orders, only parabolic profiles were observed. Fractional parameters were found to be proportional to the "rate" of switching between parabolic+oscillating and parabolic profile as well as determining the position of the dominant resonant peak in frequency domain. Fractional Burgers model was defined by four fractional parameters and one elastic parameter. As changing various parameter combinations, several distinct profile types have been obtained. When elastic parameter was fixed, combinations of three fractional parameters set to lower limit with fourth varying resulted in a combination of parabolic and complex resonant (with plateaus) components of varying amplitudes. No switching to purely parabolic profiles was observed. In contrast to it, when three out of four fractional parameters were set to their upper limits, velocity profiles appeared to be purely parabolic with amplitudes decreasing with fractional orders α and β increasing. At the same time, peak amplitudes grew monotonically in a frequency domain. When fractional parameters γ and δ varied, the shape of the parabolic profile changed. More specifically, in a frequency domain it demonstrated a presence of resonant peak followed by decay and finally reaching a nearly constant value. For changing elastic parameter, new types of velocity profiles emerged. In particular, setting fractional orders to lower limits resulted in a series of changing profiles. It started with a combination of parabolic profile, inclined decaying oscillations that were symmetric with respect to pipe centerline. The latter enclosed a plateau. Next, it transformed to parabolic+oscillatory profile. Further increase in ψ resulted in parabolic profile with resonance at certain $\tilde{\omega}$. We ended up with a simple parabolic profile with the amplitude increasing with $\tilde{\omega}$. For upper limit values of fractional parameters, a dynamic combination of parabolic and M-shape profile was obtained.

Stress profiles for all three models exhibited three types of behavior. These were aperiodic oscillations in spatial and frequency domains and V-shape profiles (curved or straight). Individual features of the stress profiles, however, varied from model to model. In particular, for fractional Maxwell model the stress profiles changed dynamically with the fractional parameters. More specifically, an increase in fractional parameters resulted in switching from aperiodic oscillations to V-shape profile. For fractional Zener model, two types of behavior were obtained: switching and complete independence from fractional parameters. Stress profiles independent from fractional parameters corresponded to "classic" case. Finally, for fractional Burgers model all three types of profiles were obtained. Different from fractional Zener model, however, here V-shape profiles were observed for both lower and upper limits of fractional orders.

The variety of profile shapes and its dynamics clearly demonstrates how powerful the approach developed is. It is applicable for completely different materials in a wide range of viscous and elastic properties. Dependent on the material used, an appropriate model can be chosen for more accurate predication of viscoelastic pipe flow dynamics. The beauty of the approach developed is that it provides analytical generalization of several most commonly used viscoelastic models as well as clearly demonstrates how rich and complex flow dynamics is when applying fractional calculus methods.

The methods developed in this study not only allow representation of different viscoelastic models in a similar functional form but also relate purely mathematical quantities with that measured experimentally. Thus, theoretical considerations appear to be on a venue of immediate practical applications.

Supplementary Materials: The following are available online at http://www.mdpi.com/2076-3417/10/24/9080/s1.

Author Contributions: Conceptualization, D.G. and R.P.; methodology, D.G.; software, D.G.; validation, R.P.; formal analysis, D.G.; investigation, D.G.; resources, R.P.; writing—original draft preparation, D.G.; writing—review and editing, R.P.; funding acquisition, R.P. All authors have read and agreed to the published version of the manuscript.

Funding: This research was funded by Argonne National Laboratory through grant #ANL 4J-303061-0030A "Multiscale modeling of complex flows".

Conflicts of Interest: The authors declare no conflict of interest.

References

1. Gritsenko, D.; Paoli, R. Theoretical analysis of fractional viscoelastic flow in circular pipes: General solutions. *Appl. Sci.* **2020**, *1*, 5, [CrossRef]
2. Sun, H.; Zhang, Y.; Baleanu, D.; Chen, W.; Chen, Y. A new collection of real world applications of fractional calculus in science and engineering. *Commun. Nonlinear Sci. Numer. Simul.* **2018**, *64*, 213–231. [CrossRef]
3. Atanacković, T.M.; Janev, M.; Pilipović, S. On the thermodynamical restrictions in isothermal deformations of fractional Burgers model. *Philos. Trans. R. Soc. A* **2020**, *378*, 20190278. [CrossRef]
4. Chen, R.; Wei, X.; Liu, F.; Anh, V.V. Multi-term time fractional diffusion equations and novel parameter estimation techniques for chloride ions sub-diffusion in reinforced concrete. *Philos. Trans. R. Soc. A* **2020**, *378*, 20190538. [CrossRef]

5. Li, S.N.; Cao, B.Y. Fractional-order heat conduction models from generalized Boltzmann transport equation. *Philos. Trans. R. Soc. A* **2020**, *378*, 20190280. [CrossRef] [PubMed]
6. Bologna, E.; Di Paola, M.; Dayal, K.; Deseri, L.; Zingales, M. Fractional-order nonlinear hereditariness of tendons and ligaments of the human knee. *Philos. Trans. R. Soc. A* **2020**, *378*, 20190294. [CrossRef] [PubMed]
7. Chugunov, V.; Fomin, S. Effect of adsorption, radioactive decay and fractal structure of matrix on solute transport in fracture. *Philos. Trans. R. Soc. A* **2020**, *378*, 20190283. [CrossRef] [PubMed]
8. Failla, G.; Zingales, M. Advanced Materials Modelling via Fractional Calculus: Challenges And Perspectives. *Philos. Trans. R. Soc. A* **2020**, *378*, 20200050. [CrossRef]
9. Fang, C.; Shen, X.; He, K.; Yin, C.; Li, S.; Chen, X.; Sun, H. Application of fractional calculus methods to viscoelastic behaviours of solid propellants. *Philos. Trans. R. Soc. A* **2020**, *378*, 20190291. [CrossRef]
10. Ionescu, C.M.; Birs, I.R.; Copot, D.; Muresan, C.; Caponetto, R. Mathematical modelling with experimental validation of viscoelastic properties in non-Newtonian fluids. *Philos. Trans. R. Soc. A* **2020**, *378*, 20190284. [CrossRef]
11. Li, J.; Ostoja-Starzewski, M. Thermo-poromechanics of fractal media. *Philos. Trans. R. Soc. A* **2020**, *378*, 20190288. [CrossRef] [PubMed]
12. Tenreiro Machado, J.; Lopes, A.M.; de Camposinhos, R. Fractional-order modelling of epoxy resin. *Philos. Trans. R. Soc. A* **2020**, *378*, 20190292. [CrossRef] [PubMed]
13. Di Paola, M.; Alotta, G.; Burlon, A.; Failla, G. A novel approach to nonlinear variable-order fractional viscoelasticity. *Philos. Trans. R. Soc. A* **2020**, *378*, 20190296. [CrossRef] [PubMed]
14. Patnaik, S.; Hollkamp, J.P.; Semperlotti, F. Applications of variable-order fractional operators: A review. *Proc. R. Soc. A* **2020**, *476*, 20190498. [CrossRef] [PubMed]
15. Patnaik, S.; Semperlotti, F. Variable-order particle dynamics: Formulation and application to the simulation of edge dislocations. *Philos. Trans. R. Soc. A* **2020**, *378*, 20190290. [CrossRef] [PubMed]
16. Povstenko, Y.; Kyrylych, T. Fractional thermoelasticity problem for an infinite solid with a penny-shaped crack under prescribed heat flux across its surfaces. *Philos. Trans. R. Soc. A* **2020**, *378*, 20190289. [CrossRef]
17. Zhang, X.; Ostoja-Starzewski, M. Impact force and moment problems on random mass density fields with fractal and Hurst effects. *Philos. Trans. R. Soc. A* **2020**, *378*, 20190591. [CrossRef]
18. Zorica, D.; Oparnica, L. Energy dissipation for hereditary and energy conservation for non-local fractional wave equations. *Philos. Trans. R. Soc. A* **2020**, *378*, 20190295. [CrossRef]
19. Bagley, R.L.; Torvik, P. A theoretical basis for the application of fractional calculus to viscoelasticity. *J. Rheol.* **1983**, *27*, 201–210. [CrossRef]
20. Yin, Y.; Zhu, K.Q. Oscillating flow of a viscoelastic fluid in a pipe with the fractional Maxwell model. *Appl. Math. Comput.* **2006**, *173*, 231–242. [CrossRef]
21. Shah, S.H.A.M.; Qi, H. Starting solutions for a viscoelastic fluid with fractional Burgers' model in an annular pipe. *Nonlinear Anal. Real World Appl.* **2010**, *11*, 547–554. [CrossRef]
22. Schiessel, H.; Metzler, R.; Blumen, A.; Nonnenmacher, T. Generalized viscoelastic models: Their fractional equations with solutions. *J. Phys. A Math. Gen.* **1995**, *28*, 6567. [CrossRef]
23. Friedrich, C.; Schiessel, H.; Blumen, A. Constitutive behavior modeling and fractional derivatives. In *Rheology Series*; Elsevier: 1999; Volume 8, pp. 429–466.

Publisher's Note: MDPI stays neutral with regard to jurisdictional claims in published maps and institutional affiliations.

© 2020 by the authors. Licensee MDPI, Basel, Switzerland. This article is an open access article distributed under the terms and conditions of the Creative Commons Attribution (CC BY) license (http://creativecommons.org/licenses/by/4.0/).

Article

Theoretical Analysis of Fractional Viscoelastic Flow in Circular Pipes: General Solutions

Dmitry Gritsenko [1,*] and Roberto Paoli [1,2]

1. Department of Mechanical and Industrial Engineering, University of Illinois at Chicago, Chicago, IL 60607, USA; robpaoli@uic.edu
2. Computational Science Division and Leadership Computing Facility, Argonne National Laboratory, Lemont, IL 60439, USA
* Correspondence: dgrits2@uic.edu

Received: 29 October 2020; Accepted: 2 December 2020; Published: 18 December 2020

Abstract: Fractional calculus is a relatively old yet emerging field of mathematics with the widest range of engineering and biomedical applications. Despite being an incredibly powerful tool, it, however, requires promotion in the engineering community. Rheology is undoubtedly one of the fields where fractional calculus has become an integral part of cutting-edge research. There exists extensive literature on the theoretical, experimental, and numerical treatment of various fractional viscoelastic flows in constraint geometries. However, the general theoretical approach that unites several most commonly used models is missing. Here we present exact analytical solutions for fractional viscoelastic flow in a circular pipe. We find velocity profiles and shear stresses for fractional Maxwell, Kelvin–Voigt, Zener, Poynting–Thomson, and Burgers models. The dynamics of these quantities are studied with respect to normalized pipe radius, fractional orders, and elastic moduli ratio. Three different types of behavior are identified: monotonic increase, resonant, and aperiodic oscillations. The models developed are applicable in the widest material range and allow for the alteration of the balance between viscous and elastic properties of the materials.

Keywords: Riemann–Liouville fractional derivative; viscoelasticity; pipe flow; fractional Maxwell model; fractional Kelvin–Voigt model; fractional Zener model; fractional Poynting–Thomson model; fractional Burgers model

1. Introduction

The very first consideration of the problem involving fractional differential equation is traced back to the very end of the seventeenth century. It was Leibniz who introduced the notation d^n/dx^n and L'Hospital who asked Leibniz: "What if n be $1/2$?". Leibniz response to this question has laid the foundations of what is today known as fractional calculus (FC) [1]. Despite these early attempts made by outstanding minds who were definitely ahead of their time, further development towards practical applications appeared to be somewhat slow. An interested reader can refer to an excellent review of Ross for details on early stages of FC development [2].

The fundamentals of FC along with a brief applications overview can be found in classical books by Miller and Ross [3], Podlubny [4], or more recent one by Mainardi [5]. In the following decades it became obvious that with the incredible potential of FC to tackle problems in absolutely different fields, each of them should be addressed separately. By these means the general approaches in solving fractional differential equations could be tailored to reflect the specifics of a given field, optimize solution process, and formulate reliable constraints and ranges for, sometimes, purely mathematical parameters. Along these lines, readers with a background in physics can refer to [6–9], while those from the engineering field can refer to [10,11].

The comprehensive overview summarizing state-of-the-art practical applications of FC has been recently published by The Royal Society Publishing. The sixteen-paper issue entitled "Advanced materials modeling via fractional calculus: challenges and perspectives" [12–27] covers applications of constant-order (CO) and variable-order (VO) fractional differential operators to several fundamental phenomena. These include anomalous diffusion, [13,16] heat conduction [14,27], fractional viscoelasticity of fluids [19], and materials [12,18,22]. The approach to model viscoelastic properties of materials with VO FC operators is undoubtedly among the most promising ones, as it allows for the consideration of fractional order dynamics with respect to time, space, and material variables [22].

Viscoelasticity has been in the scope of researchers' interest since the nineteenth century and has gone through gradual development. With mathematical apparatus available at the time, several mechanical models have been proposed, developed, and generalized to address the problem of a more accurate description of material properties observed. Those classical models named after researchers who made a significant contribution to the field include but are not limited to Maxwell, Kelvin, Voigt, Zener, and Burgers. Based on two principal elements, spring and dashpot, connected in series and/or in parallel, those models met the needs of adequate material response description towards stresses and strains applied. However, as it often happened, experimental results started to accumulate, which illustrated the behaviors beyond the above-mentioned models. This issue was addressed from scratch, namely via rethinking of a basic element by Scott-Blair in 1947 [28]. He came up with an idea of considering a single element capable of a simultaneous description of viscous and elastic properties of the material. This approach capitalized on a notion of fractional derivative and allowed for the alteration of the balance between viscous and elastic properties without any additional complexity, but covering a wide range of materials. The systematic study of fractional calculus applications to viscoelasticity was made by Bagley and Torvik [29], who laid theoretical foundations of this approach. Since that time the field had emerged, especially when biomedical applications were outlined. Historically, several approaches were developed to implement balancing between viscous and elastic properties of materials. Initially, ordinary first-order constitutive equations were straightforwardly replaced with fractional counterparts [30–34]. An obvious drawback of this approach was its purely phenomenological character. An alternative approach implied physical representation of fractional constitutive equations via hierarchical combinations of dashpots and springs [35–38]. This concept (so-called ladder models) has been successfully implemented in further studies by Schiessel et al. [39] and Friedrich et al. [40]. The authors have considered generalized viscoelastic models, replaced them with fractional ones, and obtained their analytical solutions in terms of relaxation modulus and creep compliance.

Numerous studies have demonstrated the superiority of fractional calculus approach compared to classical one in predicting viscoelastic properties of materials and flows in various geometries. Hernandez et al. [41] studied the behavior of relaxation modulus for polymethyl methacrylate (PMMA) and polytetrafluoroethelene (PTFE) and demonstrated much more accurate fitting of experimental data using fractional Maxwell model compared to integer-order one. Markis et al. [42] proposed and experimentally verified the generalized fractional Maxwell model in the design of damper systems for seismic and vibration isolation. Zhang and coworkers [43] have studied the stress-relaxation behavior of fabrics coated with PTFE under various temperatures. The authors demonstrated the superiority of fractional Maxwell model in predicting stress-relaxation behavior. This behavior was experimentally proven to be nonlinear, while predicted to be linear by the classic Maxwell model. Moreover, compared to the generalized Maxwell model, the fractional one appeared to be much easier, as it did not require a large number of structural units to increase accuracy. The similar results for fractional Maxwell and fractional Zener models were obtained for elastomers (carbon-black filled resins) [44–46], polymers and rocks [47], and biological materials [48]. Fractional Kelvin–Voigt model was found to be efficient in predicting viscoelastic behavior of sludge [49].

Fractional calculus approach to viscoelasticity found applications not only for solids, as shown above, but also for various fluids. A brief historic retrospective reveals that both classic and fractional viscoelastic fluid flows in constrained geometries, especially pipes and ducts, turned out to be of a particular researchers' interest for about a century. This can be attributed to the fact that such geometries are widely used for practical applications and also admit relatively easy analytical solutions. Oscillatory pipe flows were among thoroughly investigated. Its classical consideration was traced back to 1920–1930s [50,51]. These seminal studies were further extended and generalized by Wornersley, Uchida, and others [52–55]. Later researchers considered dynamics of oscillatory pipe flows at various Reynolds numbers [56,57], both experimentally and theoretically, and specifically investigated transition to turbulence [58,59]. The foundations of fractional derivatives towards fluids viscoelastic behavior were laid by Bagley and Torvik [29]. Wood [60] studied viscoelastic transient flows in cylindrical pipes and annulus. Yin et al. [61] provided a theoretical framework for oscillating viscoelastic pipe flow using fractional Maxwell model. The authors demonstrated a drastic difference in velocity profile compared to the integer-order Maxwell model. Viscoelastic start-up flow with the fractional Maxwell model was considered by Yang et al. [62]. A similar problem in the annular pipe using fractional Burgers model was solved by Shah et al. [63]. An unsteady viscoelastic flow in a cylinder [64] and rectangular duct [65] (using fractional Maxwell model) were also considered. The exact solutions of unsteady flow in cylindrical domains with Maxwell fractional model were derived by Khandelwal and Mathur [66,67]. Maqbool and coworkers [68] considered a flow of generalized fractional Burgers fluid in inclined tube. Tang et al. [69] studied nonlinear free vibrations of a pipe conveying fractional viscoelastic fluid. They demonstrated decreasing mode amplitudes with increasing fractional order. Wang and Chen [70] considered a similar problem for the pipeline conveying fractional fluid more accurately employing Legendre polynomials. Javadi et al. [71] investigated the effect of gravity on fractional viscoelastic fluid flow in a pipe and addressed the problem of stability for it. This selected list of fractional calculus applications is far from complete. An interested reader can refer to a recent review of Sun et al. [72]. A big picture of various applications of FC is also given in [17,23].

However, despite abundant theoretical, numerical, and experimental results on fractional viscoelastic flow in pipes, the general, unifying theoretical approach to tackle flow dynamics is still missing. To fill in this research gap, this paper provides exact analytical solutions for velocity profiles and shear stresses. We demonstrate that the same solution form is applicable for different viscoelastic models including Maxwell, Kelvin–Voigt, Zener, Poynting–Thomson, and Burgers. Velocity profiles and shear stresses are studied parametrically with respect to fractional order and elastic properties (Young's modulus ratio, starting from 3-element model). This paper is organized as follows. We first provide general problem formulation for fluid flow along with domain definition (Section 2.1). In Section 2.2 we introduce the notion of the fractional element and its governing equation. Sections 2.3–2.7 provide constitutive equations for the most common fractional viscoelastic models. The main results of this study appear in Section 3. In particular, Section 3.1 describes the general approach of seeking the solution along with brief revisiting of applicable transformations. Sections 3.2–3.6 present and discuss centerline velocity and shear stress profiles dynamics with varying fractional order and elastic parameters. Two-, three-, and four-component fractional models are considered. Finally, the results of this study are summarized in Section 4.

2. Problem Formulation

2.1. Domain Definition

Introduce cylindrical coordinate system (r, θ, z) and consider laminar flow of incompressible viscoelastic fluid along z axis of infinitely long pipe of radius R with circular cross-section. The domain of interest is shown in Figure 1. The governing equations are momentum and continuity that read as:

$$\rho \frac{d\mathbf{u}}{dt} = -\nabla p + \operatorname{div} \sigma \tag{1}$$

$$\text{div} \boldsymbol{u} = 0, \qquad (2)$$

where $\rho, \boldsymbol{u}, p, \sigma$ are fluid mass density, flow velocity, pressure, and stress, respectively and $\frac{d}{dt}$ stands for total derivative. Provided $\boldsymbol{u} = u_z(r,t)\boldsymbol{e}_z$, where $\boldsymbol{e}_z$ is a unit vector in z-direction and all but σ_{rz} stress tensor components are set to 0, the momentum equation reduces to:

$$\rho \frac{\partial u_z}{\partial t} = -\frac{\partial p}{\partial z} + \frac{1}{r}\frac{\partial (r\sigma_{rz})}{\partial r}, \qquad (3)$$

where u_z stands for z-component of flow velocity and all other quantities are defined above. Additionally, we apply nonslip boundary condition at the wall.

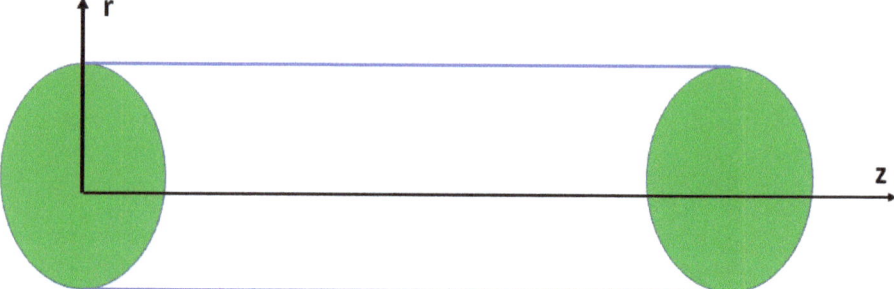

Figure 1. Domain definition for the circular pipe.

2.2. Fractional Element

The idea of introducing basic element accounting for both viscous and elastic properties of the material belongs to Blair [28]. Here we follow a conventional approach of constructing more complex mechanical models based on this element. For such an element, the stress-strain relation (constitutive equation) reads as:

$$\sigma(t) = E\tau^\alpha {}_aD_t^\alpha \varepsilon(t), \quad 0 \leq \alpha \leq 1, \qquad (4)$$

where $\sigma(t), \varepsilon(t), E, \tau, \alpha$ stand for stress, strain, Young's modulus, relaxation time and fractional order, respectively. The limiting cases of $\alpha = 0$ and $\alpha = 1$ represent spring and dashpot, respectively. The schematic of the element is given in Figure 2a. Here we use a conventional notation of Riemann–Liouville fractional derivative of a smooth function, $f(t)$, given by:

$${}_aD_t^\alpha f(t) = \frac{1}{\Gamma(k-\alpha)} \frac{d^k}{dt^k} \int_a^t (t-t_0)^{k-\alpha-1} f(t_0) dt_0, \qquad (5)$$

where k is the integer, α is the fractional order, and $\Gamma(\cdot)$ stands for Gamma-function defined as:

$$\Gamma(x) = \int_0^\infty e^{-t_0} t_0^{x-1} dt_0 \qquad (6)$$

In the following subsections we will provide constitutive equations for two-, three-, and four-component fractional viscoelastic models.

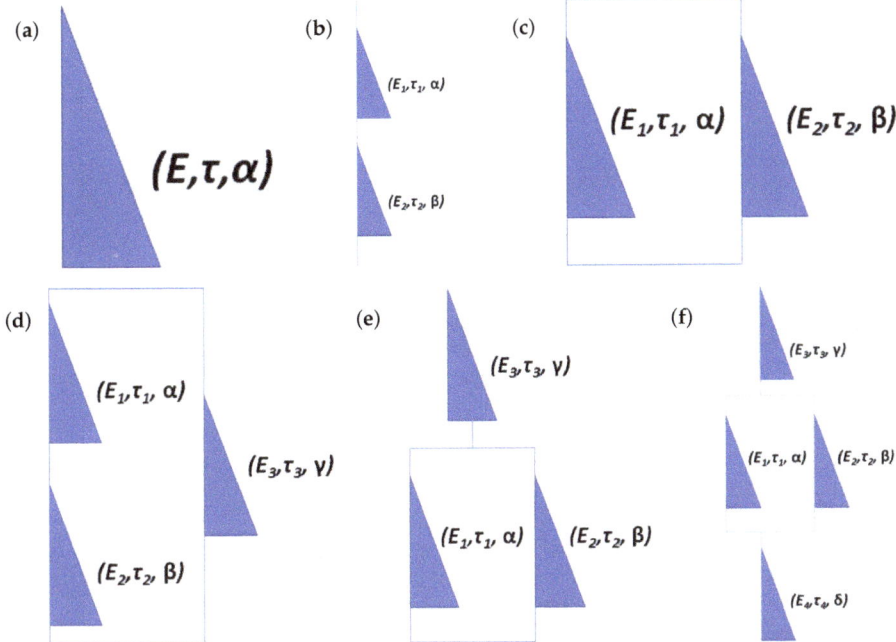

Figure 2. Fractional models: (**a**) basic element; (**b**) Maxwell; (**c**) Kelvin–Voigt; (**d**) Zener; (**e**) Poynting–Thomson; (**f**) Burgers.

2.3. Fractional Maxwell Model

We start our consideration from the simplest yet commonly used two-component fractional model with basic elements connected in series (fractional Maxwell model, Figure 2b). The constitutive equation for this model reads as [39]:

$$\sigma(t) + \tau^{\alpha-\beta}{_aD_t^{\alpha-\beta}}\sigma(t) = E\tau^{\alpha}{_aD_t^{\alpha}}\varepsilon(t), \quad 0 \leq \alpha \leq 1, \quad 0 \leq \beta \leq 1, \quad \alpha \geq \beta, \tag{7}$$

where α and β stand for the fractional orders, τ, E are defined as:

$$\tau = \left(\frac{E_1\tau_1^{\alpha}}{E_2\tau_2^{\beta}}\right)^{\frac{1}{\alpha-\beta}}, \quad E = E_1\left(\frac{\tau_1}{\tau}\right)^{\alpha}, \quad E = E_2\left(\frac{\tau_2}{\tau}\right)^{\beta}, \tag{8}$$

and E_1, τ_1, E_2, τ_2 are the Young's moduli, relaxation times for elements "1" and "2", respectively. Constitutive equation of fractional Maxwell model (7) can be reduced to classical one, if $\alpha = 1$ and $\beta = 0$. Moreover, if we set $\alpha = 1$ and $\beta = 1$, Newtonian fluid can be obtained. The ranges set for fractional orders α and β have both mathematical and physical meaning. Non-negative values of fractional orders and their difference reflect the fact that the dynamics of processes considered are described by fractional derivatives not fractional integrals. The upper limit of fractional orders range is set to get the corresponding classical viscoelastic model as a limiting case of a fractional one. For τ as it was defined above, $\alpha = \beta$ represents a special case. Indeed, it corresponds to relaxation time blowing up. Physically it means that a fluid becomes a critical gel. This special case is, however, beyond the scope of the current study. It is worth mentioning that relaxation time blows up only provided that the base of τ is greater than unity. If, however, that is not the case, i.e., τ base is less than unity, τ itself remains bound.

2.4. Fractional Kelvin–Voigt Model

Now let us connect two basic fractional elements in parallel. The dynamics of corresponding model (Kelvin–Voigt fractional model) are described as follows [39]:

$$\sigma(t) = E\tau^\alpha {}_aD_t^\alpha \varepsilon(t) + E\tau^\beta {}_aD_t^\beta \varepsilon(t), \quad E = E_2\left(\frac{\tau_2}{\tau}\right)^\beta, \tag{9}$$

where all the notations are similar to those for fractional Maxwell model. The schematic of the model is given in Figure 2c. If $\alpha = 1$ and $\beta = 0$, we appear at classical Kelvin–Voigt model.

2.5. Fractional Zener Model

Let us now consider 3-element model referred to as fractional Zener model (Figure 2d). The corresponding constitutive equation reads as [39]:

$$\sigma(t) + \tau^{\alpha-\beta} {}_aD_t^{\alpha-\beta}\sigma(t) = E\tau^\alpha {}_aD_t^\alpha \varepsilon(t) + E_0\tau^\gamma {}_aD_t^\gamma \varepsilon(t) + E_0\tau^{\gamma+\alpha-\beta} {}_aD_t^{\gamma+\alpha-\beta}\varepsilon(t),$$
$$0 \leq \alpha \leq 1, \quad 0 \leq \beta \leq 1, \quad 0 \leq \gamma \leq 1, \quad \alpha - \beta \geq 0, \quad \gamma + \alpha - \beta \geq 0, \tag{10}$$

where γ stands for the fractional order of element "3" and E_0 is given by:

$$E_0 = E_3\left(\frac{\tau_3}{\tau}\right)^\gamma, \tag{11}$$

with E_3, τ_3 being Young's modulus and relaxation time for element "3" and all other notations being the same as above. Assuming $\alpha = 1$, $\beta = 0$, and $\gamma = 0$, classical Zener model can be restored.

2.6. Fractional Poynting–Thomson Model

For fractional Poynting–Thomson model (Figure 2e) the constitutive equation is [39]:

$$\sigma(t) + \frac{E}{E_0}\tau^{\alpha-\gamma} {}_aD_t^{\alpha-\gamma}\sigma(t) + \frac{E}{E_0}\tau^{\beta-\gamma} {}_aD_t^{\beta-\gamma}\sigma(t) = E\tau^\alpha {}_aD_t^\alpha \varepsilon(t) + E\tau^\beta {}_aD_t^\beta \varepsilon(t),$$
$$0 \leq \alpha \leq 1, \quad 0 \leq \beta \leq 1, \quad 0 \leq \gamma \leq 1, \quad \alpha - \gamma \geq 0, \quad \beta - \gamma \geq 0, \tag{12}$$

with all the notations similar to above, but different limitations applied to fractional orders α, β and γ. Provided $\alpha = 1$, $\beta = 0$, and $\gamma = 0$, we get classical Poynting–Thomson model.

2.7. Fractional Burgers Model

Finally, for the most complex, 4-element model considered in this study (fractional Burgers model, Figure 2f) the constitutive equation reads as [5]:

$$\sigma(t) + \frac{E}{E_0}\tau_a^{\alpha-\gamma}D_t^{\alpha-\gamma}\sigma(t) + \frac{E}{E_0}\tau^{\alpha-\delta} {}_aD_t^{\alpha-\delta}\sigma(t) + \frac{E}{E_0}\tau_a^{\beta-\gamma}D_t^{\beta-\gamma}\sigma(t) + \frac{E}{E_0}\tau_a^{\beta-\delta}D_t^{\beta-\delta}\sigma(t) =$$
$$= E\tau_a^\alpha D_t^\alpha \varepsilon(t) + E\tau_a^\beta D_t^\beta \varepsilon(t), \tag{13}$$
$$0 \leq \alpha \leq 1, \quad 1 \leq \beta \leq 2, \quad 0 \leq \gamma \leq 1, \quad 0 \leq \delta \leq 1, \quad \alpha - \gamma \geq 0, \quad \beta - \gamma \geq 0, \quad \alpha - \delta \geq 0, \quad \beta - \delta \geq 0$$

where δ stands for the fractional order of element "4", (11) is still valid and additionally:

$$E_0 = E_4\left(\frac{\tau_4}{\tau}\right)^\delta, \tag{14}$$

with E_4, τ_4 being Young's modulus and relaxation time for element "4" and all other notations being the same as above. If we set $\alpha = 1$, $\beta = 2$, $\gamma = 0$, and $\delta = 0$, classical Burgers model can be restored. Here $\beta - \gamma \geq 0$ and $\beta - \delta \geq 0$ are satisfied automatically, given the range for β, γ, and δ.

3. Results and Discussion

3.1. General Solution

Here we follow the approach proposed earlier by Yin [61] for Maxwell model and then generalize it for all other models. Consider a specific type of pressure gradient given as follows:

$$\frac{\partial p}{\partial z} = P_0 e^{i\omega t} \tag{15}$$

Introduce Fourier and inverse Fourier transforms of flow velocity and pressure:

$$U_z(r,\omega) = \int_{-\infty}^{+\infty} u_z(r,t) e^{-i\omega t} dt, \quad u_z(r,t) = \frac{1}{2\pi} \int_{-\infty}^{+\infty} U_z(r,\omega) e^{i\omega t} d\omega \tag{16}$$

$$P(r,\omega) = \int_{-\infty}^{+\infty} p(r,t) e^{-i\omega t} dt, \quad p(r,t) = \frac{1}{2\pi} \int_{-\infty}^{+\infty} P(r,\omega) e^{i\omega t} d\omega \tag{17}$$

Recall Fourier transform rule for fractional derivative:

$$\mathcal{F}\{{}_aD_t^\alpha u_z\} = (i\omega)^\alpha U_z(r,\omega), \quad \mathcal{F}\{{}_aD_t^{\alpha-\beta}\frac{\partial p}{\partial z}\} = (i\omega)^{\alpha-\beta}\frac{\partial P(r,\omega)}{\partial z}, \tag{18}$$

where appropriate integration limits are implied. In particular, lower terminal value is set to: $a = -\infty$. Then the general algorithm to find flow velocity is the following: (1) express stress and its derivative from the constitutive eq-n; (2) plug stress and its derivative in momentum eq-n; (3) eliminate stress and its derivative and get modified momentum eq-n; (4) perform Fourier transform of modified momentum eq-n; (5) solve corresponding ODE for Fourier transform of velocity ($U_z(r,\omega)$); (6) change variables with ξ for simplicity; (7) perform inverse Fourier transform of $U_z(r,\omega)$ to get $u_z(r,t)$. Regardless of the model considered, z-component of the velocity reads as:

$$u_z(r,t) = \frac{i}{\rho\omega}\frac{\partial p(z,t)}{\partial z}\left[1 - \frac{J_0(\xi r)}{J_0(\xi R)}\right], \tag{19}$$

where $J_n(\cdot)$ is the Bessel function of the first kind of order n. The only difference that defines behavior of the system is hidden in ξ. At the same time, the form of the solution reproduces a well-known classical result (see, for example Reference [73]). Thus it represents a natural extension of integer-order models using apparatus of FC. As long as velocity profile was defined, we could also get expression for the stress. The direct integration of (3) accounting for (15) results in:

$$\sigma_{rz} = \frac{\partial p}{\partial z}\frac{J_1(\xi r)}{\xi J_0(\xi R)}, \tag{20}$$

where again the individual properties of the model are hidden in parameter ξ. Physical meaning of ξ is worth consideration. First, of all, $[\xi] = [1/m]$, so that it can be considered as an "effective" wave number. Moreover, as shown below, ξ is related to complex compliance. Introducing $\sigma_0 = \frac{\partial p}{\partial z}R$, nondimensional stress $\tilde{\sigma} = \sigma_{rz}/\sigma_0$ is given by:

$$\tilde{\sigma} = \frac{J_1(\xi r)}{\xi R J_0(\xi R)} \tag{21}$$

For practical purposes and analysis simplification, it is more convenient to introduce the following nondimensional quantities and relationships:

$$\tilde{r} = \frac{r}{\sqrt{\nu\tau}}, \quad \tilde{a} = \frac{R}{\sqrt{\nu\tau}}, \quad \tilde{\omega} = \omega\tau, \quad \phi = \frac{\tilde{r}}{\tilde{a}}, \quad u_0 = \frac{1}{4\rho\omega}\frac{\partial p}{\partial z}, \quad r^2\zeta^2 = \tilde{\omega}\tilde{r}^2, \qquad (22)$$

where ν is a kinematic viscosity and all other quantities were defined earlier. Then velocity and stress profiles are given by:

$$\tilde{u} = \left|\frac{u_z}{u_0}\right| = \left(1 - \frac{J_0(\sqrt{\tilde{\omega}}\tilde{a}\phi)}{J_0(\sqrt{\tilde{\omega}}\tilde{a})}\right), \quad \tilde{\sigma} = \left|\frac{\sigma_{rz}}{\sigma_0}\right| = \frac{J_1(\sqrt{\tilde{\omega}}\tilde{a}\phi)}{\sqrt{\tilde{\omega}}\tilde{a}J_0(\sqrt{\tilde{\omega}}\tilde{a})} \qquad (23)$$

As velocity profiles and shear stresses have been defined, we can now proceed with considering specific fractional viscoelastic models and outlining their specifics. Six parameters will be considered: fractional order ($\alpha, \beta, \gamma, \delta$), nondimensional radius, $\tilde{a}$, and elastic ratio, ψ.

3.2. Fractional Maxwell Model

Let us start from the simplest two-parameter fractional Maxwell model. As outlined above, the key quantity that defines the behavior of a model is ζ. For this model ζ is given by:

$$\zeta^2 = \frac{\rho\omega^2}{E}\left[\frac{1}{(i\omega\tau)^\alpha} + \frac{1}{(i\omega\tau)^\beta}\right], \qquad (24)$$

or alternatively in nondimensional form:

$$\tilde{\omega} = \tilde{\omega}^2\left[\frac{1}{(i\tilde{\omega})^\alpha} + \frac{1}{(i\tilde{\omega})^\beta}\right] \qquad (25)$$

Centerline velocity profiles for fractional Maxwell model are shown in Figure 3. For all the plots $\alpha = 0.5$. The value of $\tilde{a}$ ranges from $\tilde{a} = 0.01$ (Figure 3a) to $\tilde{a} = 0.1$ (Figure 3c). For this model an additional condition is imposed: $\alpha \geq \beta$, thus only a half of β range is considered given the value of α set. As can be seen from this figure, the behavior of centerline velocity changes dramatically from monotonically increasing to resonant or oscillatory. Centerline velocity profile also changes with increasing fractional order (β). For lower values of $\tilde{a}$, an increase of β results in decreasing of centerline velocity up to the constant value (as $\beta \to 0.5$). For higher values of $\tilde{a}$ (Figure 3b,c) the system exhibits aperiodic oscillations at low β and demonstrates a switch from resonant behavior to monotonically increasing as β increases. It is also worth mentioning that peak values decrease with increasing β. What is the reason for such a dramatic change of a velocity profile dynamics with varying $\tilde{a}$? In fact, $\tilde{a}^2$ turns out to be nothing else but a Reynolds number: $\tilde{a}^2 = Re$, one of the key flow parameters.

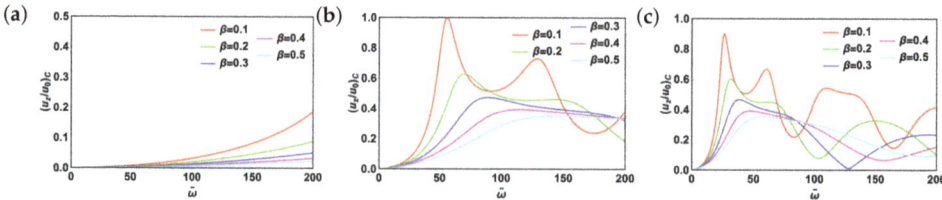

Figure 3. Centerline velocity profile for fractional Maxwell model: (a) $\tilde{a} = 0.01$; (b) $\tilde{a} = 0.05$; (c) $\tilde{a} = 0.1$. For all plots $\alpha = 0.5$. Subscript "c" stands for "centerline".

Now let us have a look at the shear stress dynamics (at the wall) shown in Figure 4. Here again we fix $\alpha = 0.5$ and vary β ($\alpha \geq \beta$). Similar to velocity profile, shear stress increases monotonically with $\tilde{\omega}$ for lower values of $\tilde{a}$. The trend, however, changes to almost constant with increasing fractional order. As $\tilde{a}$ increases, the system exhibits more complex behaviors. In particular, there are aperiodic oscillations observed for lower values of fractional order that switch to resonant ones and monotonic decay as the fractional order starts to increase. Moreover, peak values shift to higher values of $\tilde{\omega}$ with increasing fractional order and the peaks themselves become more dispersive. At the same time, an increase of $\tilde{a}$ results in resonant peaks becoming sharper and shifting to lower values of $\tilde{\omega}$.

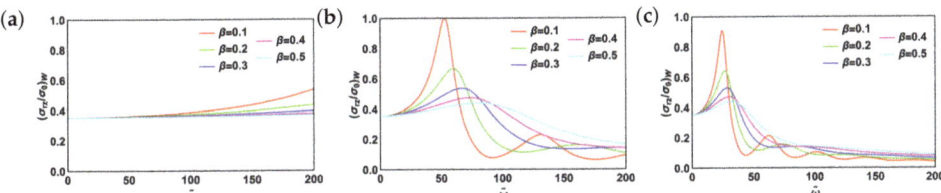

Figure 4. Shear stress profile at the pipe wall for fractional Maxwell model: (**a**) $\tilde{a} = 0.01$; (**b**) $\tilde{a} = 0.05$; (**c**) $\tilde{a} = 0.1$. For all plots $\alpha = 0.5$. Subscript "w" stands for "wall".

Finally, for relatively high values of $\tilde{\omega}$ and $\tilde{a}$, shear stress plots converge and become nearly independent from the fractional order β. To better understand physical meaning behind fractional orders α and β, we have examined its influence on centerline velocity and shear stress dynamics in a wider range of $\tilde{\omega}$ ($0 \leq \tilde{\omega} \leq 1000$) with the fixed value of $\tilde{a}$ ($\tilde{a} = 0.1$) as shown in Figure 5. We have first fixed $\alpha = 0.5$ and varied β: $0.1 \leq \beta \leq 0.5$. For both centerline velocity (Figure 5a) and shear stress (Figure 5b) low values of β resulted in oscillatory profiles that in turn reflects the fact that the fractional order β describes elastic properties of the fluid considered. In contrast, if we fix $\beta = 0.5$ and vary α: $0.6 \leq \alpha \leq 1$, the oscillations are damped much quicker for both centerline velocity and shear stress (Figure 5c,d). Finally, let us set $\alpha = 1$ and vary β: $0.25 \leq \beta \leq 1$. Corresponding plots for centerline velocity and shear stress are shown in Figure 5e,f, respectively. When $\alpha = 1$ and $\beta = 1$, Newtonian fluid is obtained with centerline velocity monotonically reaching constant value and shear stress monotonically decaying.

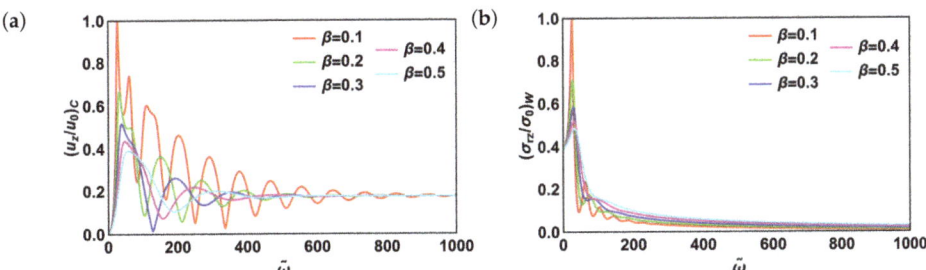

Figure 5. Cont.

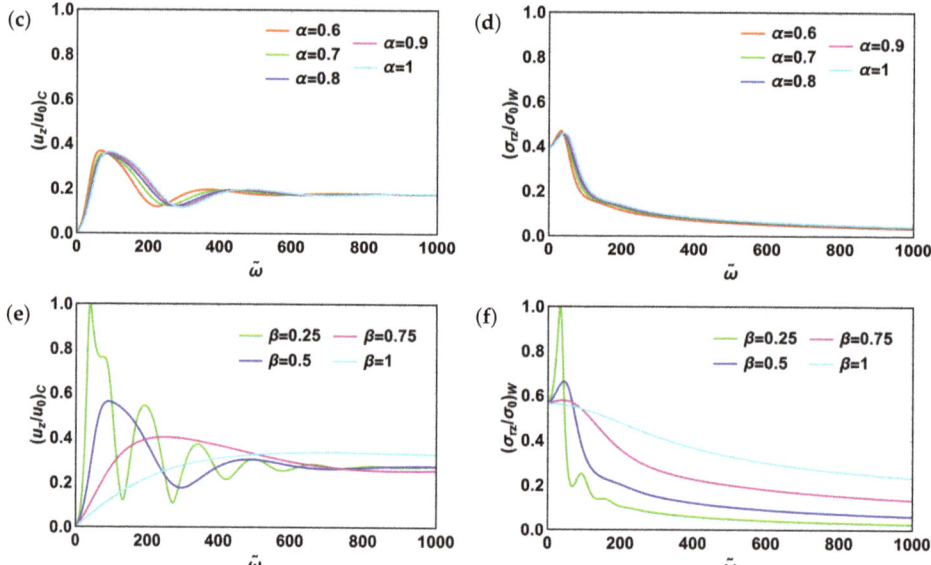

Figure 5. Fractional Maxwell model: (**a**) centerline velocity, $\alpha = 0.5$; (**b**) shear stress, $\alpha = 0.5$; (**c**) centerline velocity, $\beta = 0.5$; (**d**) shear stress, $\beta = 0.5$; (**e**) centerline velocity, $\alpha = 1$; (**f**) shear stress, $\alpha = 1$. For all plots $\tilde{a} = 0.1$. Subscripts "c" and "w" stand for "centerline" and "wall", respectively.

3.3. Fractional Kelvin–Voigt Model

Next, we consider another commonly used two-parameter fractional model referred to as Kelvin–Voigt. We start from parameter ζ, responsible for system behavior. It reads as:

$$\zeta^2 = \frac{\rho \omega^2}{E} \left[\frac{1}{(i\omega\tau)^\alpha + (i\omega\tau)^\beta} \right], \qquad (26)$$

or alternatively in nondimensional form:

$$\bar{\omega} = \frac{\tilde{\omega}^2}{\left[(i\tilde{\omega})^\alpha + (i\tilde{\omega})^\beta \right]} \qquad (27)$$

In contrast to fractional Maxwell model, no additional relations between α and β are implied here. Thus, we have more freedom to set the values of these fractional orders within the entire range.

Centerline velocity profiles and shear stresses for fractional Kelvin–Voigt model are given in Figure 6. Here again we fix $\alpha = 0.5$ and vary β. Different from fractional Maxwell model, centerline velocity profiles (Figure 6a,c,e) do not exhibit aperiodic oscillations with varying both fractional order and $\tilde{a}$. Centerline velocity amplitudes increase with increasing $\tilde{a}$ and decrease with increasing fractional order β. Shear stresses' dynamics for the same model are presented in Figure 6b,d,f. Here again all three types of behavior encountered for fractional Maxwell model are present. More specifically, as $\tilde{a}$ increases, shear stresses experience three different types of behavior: from monotonically increasing through resonant to oscillatory. Relative stress amplitudes decrease with increasing fractional order β.

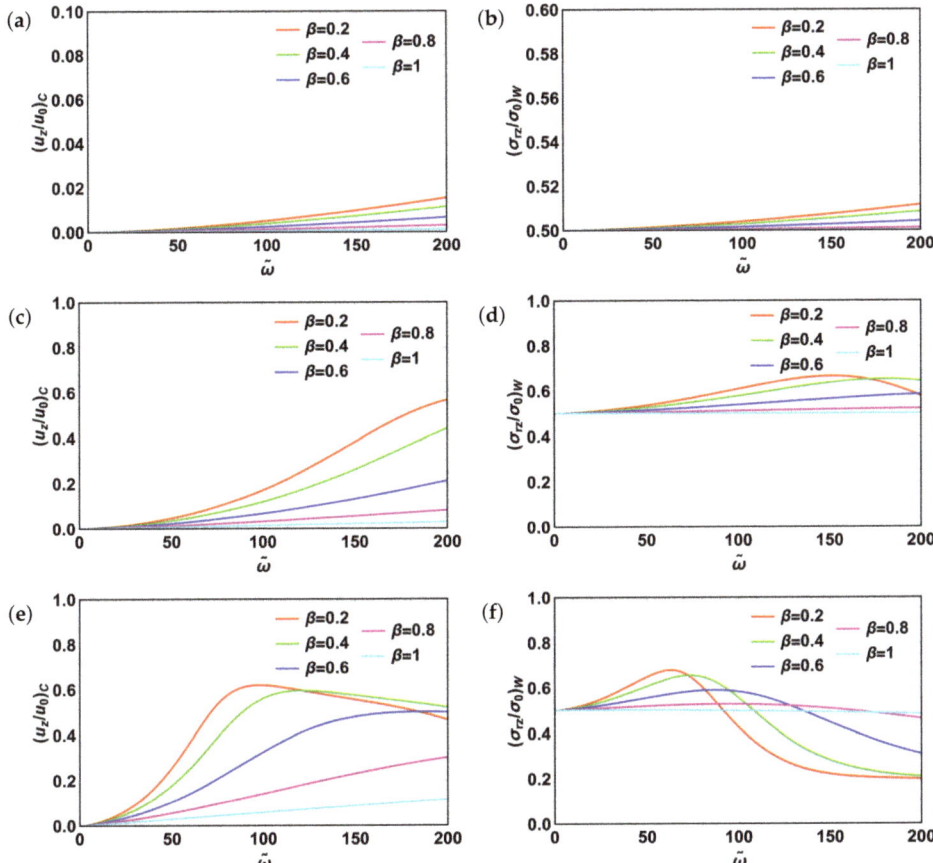

Figure 6. Fractional Kelvin–Voigt model: (**a**) centerline velocity, $\tilde{a} = 0.01$; (**b**) shear stress, $\tilde{a} = 0.01$; (**c**) centerline velocity, $\tilde{a} = 0.05$; (**d**) shear stress, $\tilde{a} = 0.05$; (**e**) centerline velocity, $\tilde{a} = 0.1$; (**f**) shear stress, $\tilde{a} = 0.1$. For all plots $\alpha = 0.5$. Subscripts "c" and "w" stand for "centerline" and "wall", respectively.

3.4. Fractional Zener Model

Now examine the behavior of three-element fractional Zener model. The key model variable, ξ, in this case becomes:

$$\xi^2 = \frac{\rho \omega^2}{E} \left[\frac{1}{(E_0/E)(i\omega\tau)^\gamma + \frac{(i\omega\tau)^{\alpha+\beta}}{(i\omega\tau)^\alpha + (i\omega\tau)^\beta}} \right], \tag{28}$$

or alternatively in nondimensional form:

$$\bar{\omega} = \frac{\tilde{\omega}^2}{\psi(i\tilde{\omega})^\gamma + \frac{(i\tilde{\omega})^{\alpha+\beta}}{(i\tilde{\omega})^\alpha + (i\tilde{\omega})^\beta}}, \tag{29}$$

where $\psi = \frac{E_0}{E}$ is an elastic parameter. By definition elastic parameter ψ can be rewritten as:

$$\psi = \frac{E_3 \tau_3^\gamma}{E_1 \tau_1^\alpha} \tau^{\gamma-\alpha} \tag{30}$$

For classical Zener model ($\alpha = 1$, $\beta = 0$, and $\gamma = 0$), ψ reduces to: $\psi = E_3/E_2$, i.e., The ratio of Young's moduli, thus justifying the notion introduced above. Similar to fractional Maxwell model, we impose additional conditions for fractional orders α, β, and γ. From two inequalities: $\alpha - \beta \geq 0$ and $\gamma + \alpha - \beta \geq 0$, we end up with the first one. The second one is satisfied automatically, provided the first one is and γ being non-negative.

We first examine the dynamics of centerline velocity. Different from two-element models, where we considered the influence of three parameters (α, β, and $\tilde{a}$) on the system behavior, here we need to account for five parameters ($\tilde{a}$, α, β, γ and ψ). Let us fix $\alpha = 0.5$ (as previously), $\gamma = 0.5$ and $\psi = 1$. Since $\alpha - \beta \geq 0$, we only consider $0 \leq \beta \leq 0.5$. Corresponding centerline velocity profiles are given in Figure 7a,d,g. Centerline velocity changes its behavior from monotonically increasing to resonant with increasing $\tilde{a}$. Moreover, relative velocity amplitudes increase with increasing $\tilde{a}$. In addition to it, an interesting phenomenon is observed in Figure 7g. Relative velocity amplitudes tend to decrease with increasing fractional-order β up to a certain value ($\tilde{\omega} \sim 150$). At this point, centerline velocities for all fractional orders become close, and then the trend inverses, namely velocity amplitudes start to increase with increasing fractional-order β.

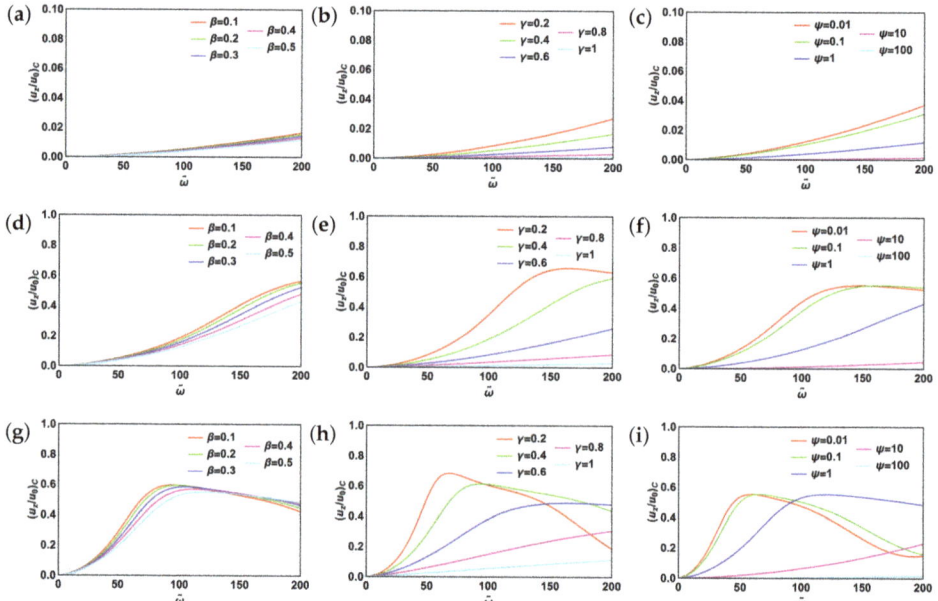

Figure 7. Centerline velocity for fractional Zener model: (**a**) $\tilde{a} = 0.01$, $\alpha = \gamma = 0.2$, $\psi = 1$; (**b**) $\tilde{a} = 0.01$, $\alpha = \beta = 0.5$, $\psi = 1$; (**c**) $\tilde{a} = 0.01$, $\alpha = \beta = \gamma = 0.5$; (**d**) $\tilde{a} = 0.05$, $\alpha = \gamma = 0.2$, $\psi = 1$; (**e**) $\tilde{a} = 0.05$, $\alpha = \beta = 0.5$, $\psi = 1$; (**f**) $\tilde{a} = 0.05$, $\alpha = \beta = \gamma = 0.5$; (**g**) $\tilde{a} = 0.1$, $\alpha = \gamma = 0.2$, $\psi = 1$; (**h**) $\tilde{a} = 0.1$, $\alpha = \beta = 0.5$, $\psi = 1$; (**i**) $\tilde{a} = 0.1$, $\alpha = \beta = \gamma = 0.5$. Subscript "c" stands for "centerline".

Next, let us fix $\alpha = \beta = 0.5$, $\psi = 1$ varying γ and $\tilde{a}$. The corresponding plots for centerline velocity are shown in Figure 7b,e,h. Here we can observe switching from monotonically increasing to resonant behavior of the system with $\tilde{a}$ increasing and relative velocity amplitude decrease with increasing fractional order γ. Moreover, for larger values of $\tilde{a}$ centerline velocity behavior changes from resonant to almost linearly increasing with increasing fractional order γ.

Finally, we fixed $\alpha = \beta = \gamma = 0.5$ and varied $\tilde{a}$ and ψ as shown in Figure 7c,f,i. As can be seen, the system behavior changes from monotonically increasing to resonant with increasing $\tilde{a}$. Moreover, resonant curves become less dispersive, with increasing $\tilde{a}$ and more dispersive with increasing ψ. Relative amplitudes of centerline velocity also decrease with increasing ψ.

The dynamics of shear stresses are given in Figure 8. The first column in Figure 8 (Figure 8a,d,g) corresponds to $\alpha = \gamma = 0.5$, $\psi = 1$ and varying $\tilde{a}$ and β. In Figure 8g there exist two values of $\tilde{\omega}$ where shear stresses become very close for different values of fractional order β. Upon reaching the one corresponding to lower value of $\tilde{\omega}$, the trend reverses with relative stress amplitude decreasing for increasing fractional order, β. For large values of $\tilde{\omega}$, shear stress becomes almost independent of the fractional order, β. For the second column of Figure 8 (Figure 8b,e,h), we set $\alpha = \beta = 0.5$, $\psi = 1$ and varied $\tilde{a}$ along with γ. The new phenomenon observed can be seen in Figure 8e,h. That is, the resonant behavior changes to monotonically increasing and finally to constant with increasing fractional order γ. The similar trend is observed in case of varying ψ as shown in Figure 8c,f,i ($\alpha = \beta = \gamma = 0.5$).

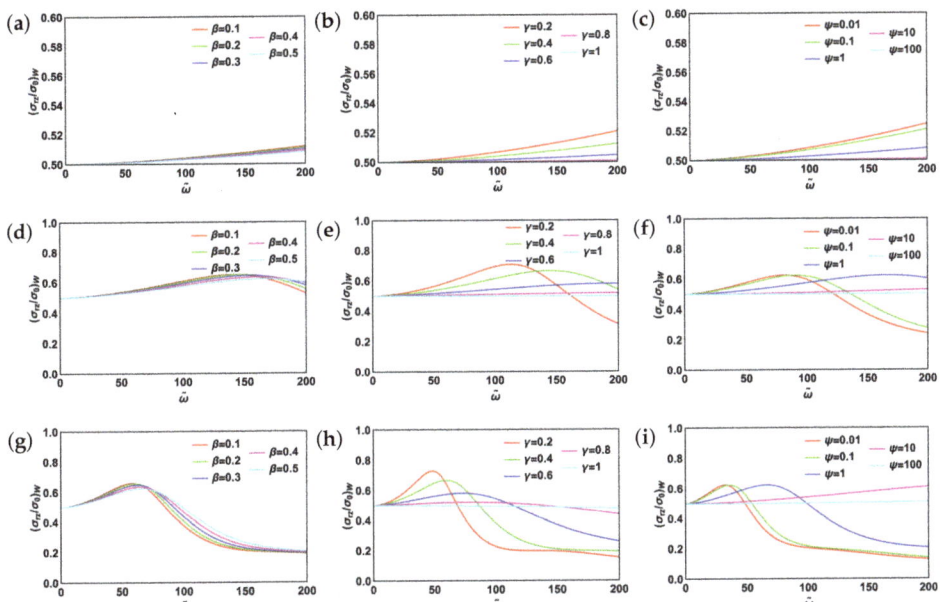

Figure 8. Shear stress at the pipe wall for fractional Zener model: (**a**) $\tilde{a} = 0.01$, $\alpha = \gamma = 0.2$, $\psi = 1$; (**b**) $\tilde{a} = 0.01$, $\alpha = \beta = 0.5$, $\psi = 1$; (**c**) $\tilde{a} = 0.01$, $\alpha = \beta = \gamma = 0.5$; (**d**) $\tilde{a} = 0.05$, $\alpha = \gamma = 0.2$, $\psi = 1$; (**e**) $\tilde{a} = 0.05$, $\alpha = \beta = 0.5$, $\psi = 1$; (**f**) $\tilde{a} = 0.05$, $\alpha = \beta = \gamma = 0.5$; (**g**) $\tilde{a} = 0.1$, $\alpha = \gamma = 0.2$, $\psi = 1$; (**h**) $\tilde{a} = 0.1$, $\alpha = \beta = 0.5$, $\psi = 1$; (**i**) $\tilde{a} = 0.1$, $\alpha = \beta = \gamma = 0.5$. Subscript "w" stands for "wall".

3.5. Fractional Poynting–Thomson Model

The next model to be considered is fractional Poynting–Thomson one. Here ζ reads as:

$$\zeta^2 = \frac{\rho \omega^2}{E}\left[\frac{1}{(E_0/E)(i\omega\tau)^\gamma} + \frac{1}{((i\omega\tau)^\alpha + (i\omega\tau)^\beta)}\right], \quad (31)$$

or alternatively in nondimensional form:

$$\tilde{\omega} = \tilde{\omega}^2\left(\frac{1}{\psi(i\tilde{\omega})^\gamma} + \frac{1}{(i\tilde{\omega})^\alpha + (i\tilde{\omega})^\beta}\right) \quad (32)$$

Here we impose two additional conditions for fractional orders α, β, and γ: $\alpha - \gamma \geq 0$ and $\beta - \gamma \geq 0$. As previously, we first examine the behavior of centerline velocity (Figure 9). Fixing $\alpha = \gamma = 0.2$, $\psi = 1$, we vary $\tilde{a}$ and β. The trends are somewhat repeatable compared to fractional Maxwell model (Figure 3a,d,g) and demonstrate elastic behavior for the entire range of β. Next, we fixed $\alpha = \beta = 0.2$, $\psi = 1$ and varied $\tilde{a}$ along with fractional order γ. The corresponding plots are presented in Figure 9b,e,h. Here aperiodic oscillations for lower values of fractional order γ, oscillatory and monotonically increasing velocity profiles with increasing γ are observed. Finally, we set $\alpha = \beta = \gamma = 0.5$ and varied $\tilde{a}$ and ψ (Figure 9c,f,i). Here, resonant behavior is observed even for lower values of $\tilde{a}$ (Figure 9c). Moreover, for $\psi < 1$ aperiodic oscillations of velocity are observed followed by it becoming independent from $\tilde{\omega}$. The trend, however, changes to resonant for $\psi > 1$. Overall, the velocity oscillations are damped slower with ψ increasing.

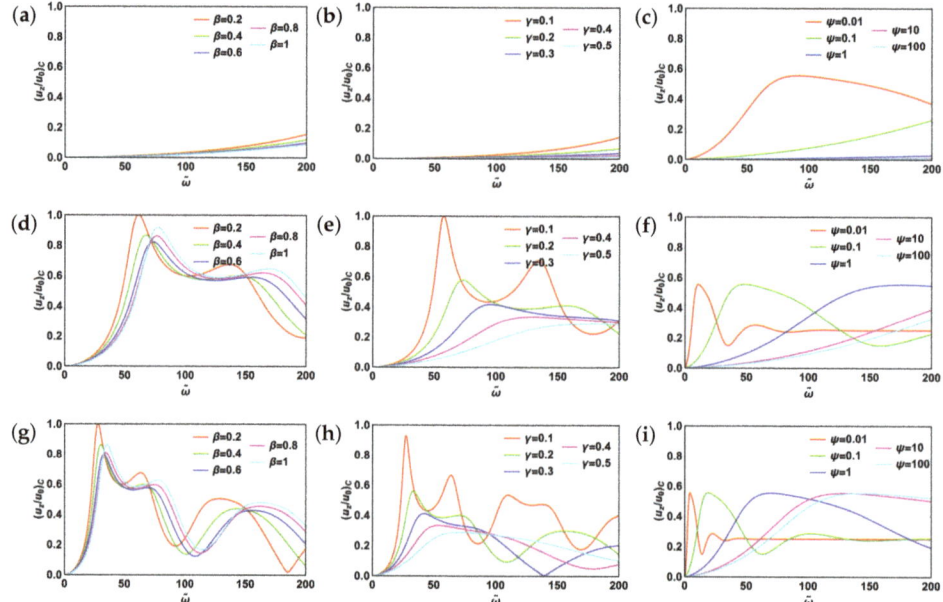

Figure 9. Centerline velocity for fractional Poynting-Thomson model: (**a**) $\tilde{a} = 0.01$, $\alpha = \gamma = 0.2$, $\psi = 1$; (**b**) $\tilde{a} = 0.01$, $\alpha = \beta = 0.5$, $\psi = 1$; (**c**) $\tilde{a} = 0.01$, $\alpha = \beta = \gamma = 0.5$; (**d**) $\tilde{a} = 0.05$, $\alpha = \gamma = 0.2$, $\psi = 1$; (**e**) $\tilde{a} = 0.05$, $\alpha = \beta = 0.5$, $\psi = 1$; (**f**) $\tilde{a} = 0.05$, $\alpha = \beta = \gamma = 0.5$; (**g**) $\tilde{a} = 0.1$, $\alpha = \gamma = 0.2$, $\psi = 1$; (**h**) $\tilde{a} = 0.1$, $\alpha = \beta = 0.5$, $\psi = 1$; (**i**) $\tilde{a} = 0.1$, $\alpha = \beta = \gamma = 0.5$. Subscript "c" stands for "centerline".

The shear stress dynamics at the wall are shown in Figure 10. As previously, we first fixed $\alpha = \gamma = 0.2$, $\psi = 1$ and varied $\tilde{a}$ and β (Figure 10a,d,g). After major peak, shear stress drops dramatically with oscillations being damped for the entire range of β, except for Figure 10a, where it increases monotonically. The value of the major peak itself oscillates with increasing β. Next, as shown in Figure 10b,e,h, system parameters were set at $\alpha = \beta = 0.2$, $\psi = 1$ with varying $\tilde{a}$ and γ. Monotonically increasing trend for lower values of $\tilde{a}$ changes to aperiodic oscillations and resonant for higher ones. Moreover, the value of the major peak decreases monotonically with increasing fractional order γ. Finally, we set $\alpha = \beta = \gamma = 0.5$ and varied $\tilde{a}$ and ψ. The corresponding plots are shown in Figure 10c,f,i. As can be seen from it, resonant behavior is present for all values of $\tilde{a}$. However, it switches to monotonically increasing with increasing ψ. Moreover, the switch happens at higher values of ψ as $\tilde{a}$ increases and is absent for $\tilde{a} = 0.1$ (Figure 10i).

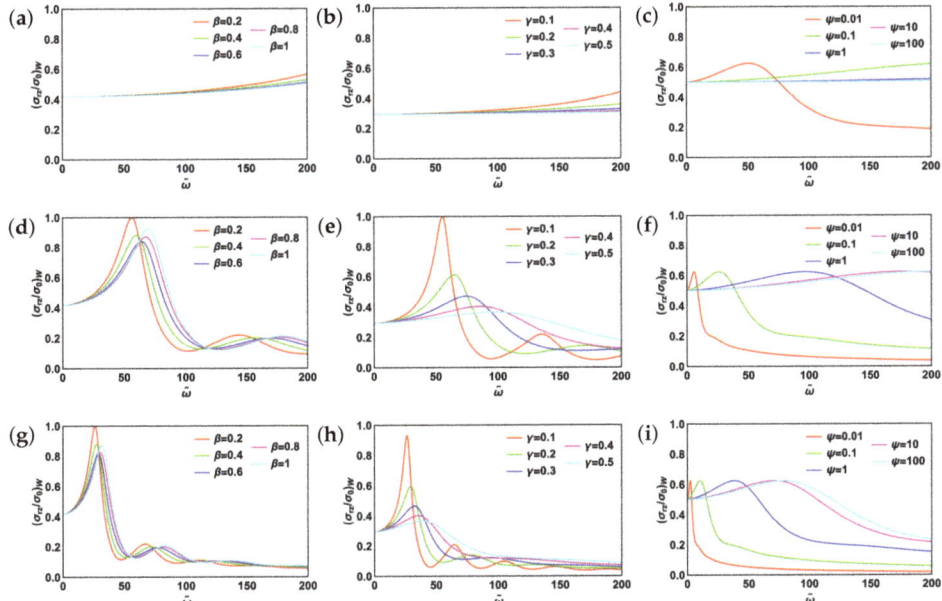

Figure 10. Shear stress at the pipe wall for fractional Poynting-Thomson model: (**a**) $\tilde{a} = 0.01$, $\alpha = \gamma = 0.2$, $\psi = 1$; (**b**) $\tilde{a} = 0.01$, $\alpha = \beta = 0.5$, $\psi = 1$; (**c**) $\tilde{a} = 0.01$, $\alpha = \beta = \gamma = 0.5$; (**d**) $\tilde{a} = 0.05$, $\alpha = \gamma = 0.2$, $\psi = 1$; (**e**) $\tilde{a} = 0.05$, $\alpha = \beta = 0.5$, $\psi = 1$; (**f**) $\tilde{a} = 0.05$, $\alpha = \beta = \gamma = 0.5$; (**g**) $\tilde{a} = 0.1$, $\alpha = \gamma = 0.2$, $\psi = 1$; (**h**) $\tilde{a} = 0.1$, $\alpha = \beta = 0.5$, $\psi = 1$; (**i**) $\tilde{a} = 0.1$, $\alpha = \beta = \gamma = 0.5$. Subscript "w" stands for "wall".

3.6. Fractional Burgers Model

Here we discuss the most complex fractional viscoelastic model, Burgers model. For this model ζ is given by:

$$\zeta^2 = \frac{\rho \omega^2}{E}\left[\frac{1}{(E_0/E)\left((i\omega\tau)^\gamma + (i\omega\tau)^\delta\right)} + \frac{1}{\left((i\omega\tau)^\alpha + (i\omega\tau)^\beta\right)}\right], \tag{33}$$

or alternatively in nondimensional form:

$$\tilde{\omega} = \tilde{\omega}^2\left(\frac{1}{\psi[(i\tilde{\omega})^\gamma + (i\tilde{\omega})^\delta]} + \frac{1}{(i\tilde{\omega})^\alpha + (i\tilde{\omega})^\beta}\right) \tag{34}$$

It is worth noting that for fractional Burgers model the range of fractional orders differs from all other models considered above. In particular, while the range for orders α, γ, and δ is still from 0 to 1, $1 \leq \beta \leq 2$. This change affects the dynamics of the centerline velocity profile dramatically. Centerline velocity profiles are shown in Figure 11. We first fixed $\alpha = \gamma = \delta = 0.5$, $\psi = 1$ and varied $\tilde{a}$ and β. The corresponding plots are shown in Figure 11a,d,g. The profile appears to be independent from the fractional order β. The situation changes when we set $\beta = 1.5$ and vary δ (Figure 11b,e,h). Velocity profile is very sensitive to the changes in fractional order δ when it has relatively low values. Aperiodic oscillations with reducing peak values for increasing values of δ are observed. For varying ψ (Figure 11 c,f,i), trends are overall similar to that for the fractional Poynting–Thomson model with the peaks for the fractional Burgers model being slightly sharper. This similarity can be attributed to the specific values of the fractional orders α, γ and δ set in Figure 11c,f,i ($\alpha = \gamma = \delta = 0.5$). This specific setting has consequences for constitutive equations in both models. That is, the terms containing fractional derivatives of the orders $\alpha - \gamma$ and $\alpha - \delta$ become of the order zero.

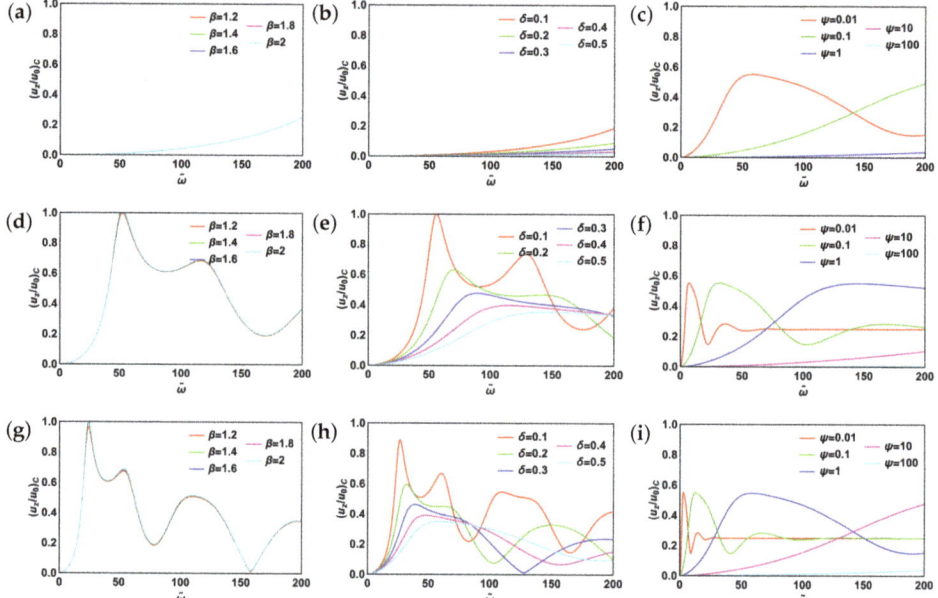

Figure 11. Centerline velocity for fractional Burgers model: (**a**) $\tilde{a} = 0.01$, $\alpha = \gamma = 0.5$, $\psi = 1$; (**b**) $\tilde{a} = 0.01$, $\alpha = \beta = 0.5$, $\psi = 1$; (**c**) $\tilde{a} = 0.01$, $\alpha = \beta = \gamma = 0.5$; (**d**) $\tilde{a} = 0.05$, $\alpha = \gamma = 0.5$, $\psi = 1$; (**e**) $\tilde{a} = 0.05$, $\alpha = \beta = 0.5$, $\psi = 1$; (**f**) $\tilde{a} = 0.05$, $\alpha = \beta = \gamma = 0.5$; (**g**) $\tilde{a} = 0.1$, $\alpha = \gamma = 0.5$, $\psi = 1$; (**h**) $\tilde{a} = 0.1$, $\alpha = \beta = 0.5$, $\psi = 1$; (**i**) $\tilde{a} = 0.1$, $\alpha = \beta = \gamma = 0.5$. Subscript "c" stands for "centerline".

Finally, let us have a look at the shear stresses shown in Figure 12. As can be seen from it, shear stress, too, is independent from the fractional order β (Figure 12a,d,g). Stress dynamics change in Figure 12b,e,h. Quickly decaying aperiodic oscillations are observed. In particular, major peak values decrease monotonically with increasing value of fractional order δ. Moreover, these peaks become sharper and shift to lower values of $\tilde{\omega}$ with increasing $\tilde{a}$. Next, we fixed $\alpha = \gamma = \delta = 0.5$ and $\beta = 1.5$. Here we observed resonant behavior for all values of $\tilde{a}$ and quickly damped oscillations for low values of ψ (Figure 12c,f,i).

A closer look at expressions for ξ corresponding to different models reveals their similarity. To illustrate this statement, let us introduce complex compliance, $J^*(\omega)$. For all the models considered, it has the following functional form: [...]$/E$, where an expression in square brackets is model-specific. Both ξ and $\tilde{\omega}$ can be expressed in terms of complex compliance:

$$\xi^2 = \rho \omega^2 J^*(\omega), \quad \tilde{\omega} = \tilde{\omega}^2 E J^*(\omega), \tag{35}$$

as it was defined in [39]. Thus, generally speaking, ξ is a complex quantity. Moreover, as we got complex compliance, other physical quantities including creep compliance, complex modulus, and relaxation modulus can be restored. For instance, $J^*(\omega) = 1/G^*(\omega)$, where $G^*(\omega)$ stands for complex modulus. Then for fractional Maxwell model the loss modulus, $G''(\omega) = \text{Im}(G^*(\omega))$, reads as [39]:

$$G''(\omega) = E \frac{(\omega \tau)^\alpha \sin(\pi \alpha/2) + (\omega \tau)^{2\alpha - \beta} \sin(\pi \beta/2)}{1 + (\omega \tau)^{2(\alpha - \beta)} + 2(\omega \tau)^{\alpha - \beta} \cos(\pi(\alpha - \beta)/2)} \tag{36}$$

Another insight can be obtained upon examination of Figure 2b–f and corresponding constitutive equations. The general observation is that more complex models can be obtained via consecutive combination of simpler ones with fractional elements. In particular, by adding fractional element to fractional Maxwell model in parallel, we arrive at fractional Zener model. Alternatively, starting from fractional Kelvin–Voigt model and adding fractional element in series, we get fractional Poynting–Thomson model. Adding one more fractional element in series, we end up with fractional Burgers model. The same "additive" behavior is observed for complex compliances.

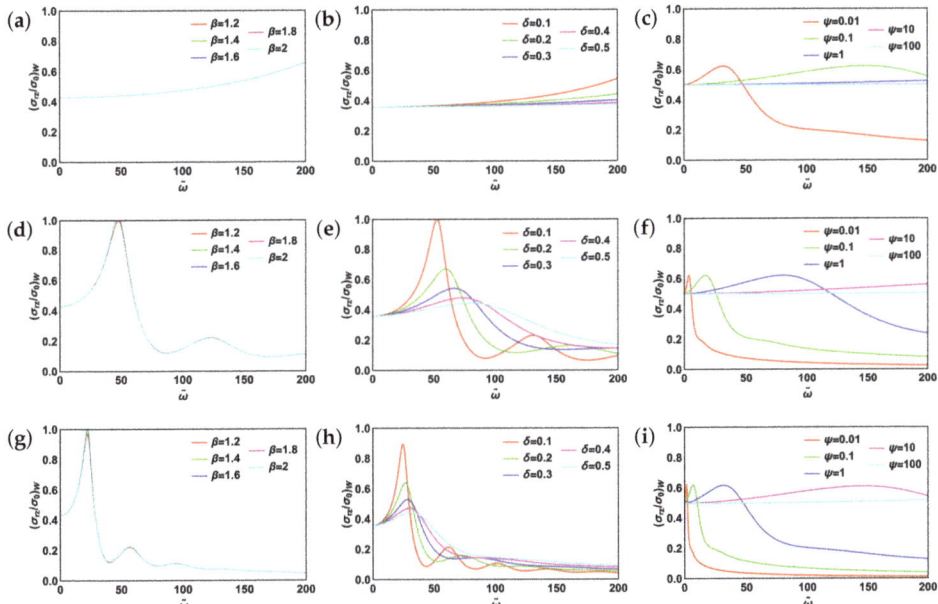

Figure 12. Shear stress at the pipe wall for fractional Burgers model: (**a**) $\tilde{a} = 0.01$, $\alpha = \gamma = 0.5$, $\psi = 1$; (**b**) $\tilde{a} = 0.01$, $\alpha = \beta = 0.5$, $\psi = 1$; (**c**) $\tilde{a} = 0.01$, $\alpha = \beta = \gamma = 0.5$; (**d**) $\tilde{a} = 0.05$, $\alpha = \gamma = 0.5$, $\psi = 1$; (**e**) $\tilde{a} = 0.05$, $\alpha = \beta = 0.5$, $\psi = 1$; (**f**) $\tilde{a} = 0.05$, $\alpha = \beta = \gamma = 0.5$; (**g**) $\tilde{a} = 0.1$, $\alpha = \gamma = 0.5$, $\psi = 1$; (**h**) $\tilde{a} = 0.1$, $\alpha = \beta = 0.5$, $\psi = 1$; (**i**) $\tilde{a} = 0.1$, $\alpha = \beta = \gamma = 0.5$. Subscript "w" stands for "wall".

It is worth emphasizing the generality of the approach developed. Provided the flow is axisymmetric, the pipe having a circular cross-section, and no-slip condition at the wall is satisfied, the solutions derived are suitable for various viscoelastic models commonly used for practical applications. For the same form of the solution of velocity and stress profiles, the prediction of flow dynamics for completely different fluids can be obtained via deriving a single model parameter, $\tilde{\zeta}$. Moreover, the functional form of $\tilde{\zeta}$ itself is also identical for all the models considered. What makes a given model specific is the expression for the complex compliance. This powerful approach simplifies significantly the implementation of other possible viscoelastic models within a given framework. The classical viscoelastic and Newtonian fluid models present the limiting cases of the theoretical approach proposed in this study. Velocity profiles' dynamics for fractional Maxwell and Newtonian fluids in a circular pipe are shown in Supplementary Materials 1. These clearly demonstrate the complexity of behavior for fractional Maxwell model.

4. Conclusions

We have obtained exact analytical solutions for velocity profiles and shear stresses. Fractional Maxwell, Kelvin-Voigt, Zener, Poynting–Thomson, and Burgers models were considered. We demonstrated that the same form of the solution is applicable to all the models considered, thus generalizing prior studies. For both centerline velocity and shear profiles, three types of behavior (monotonically increasing, resonant, oscillating aperiodic) have been identified. In addition to it, switching between trends of relative velocity amplitudes have been predicted. Monotonically decreasing trends were found to be typical for relatively low values of normalized pipe radius with more complex behaviors taking place for higher values of the normalized radius. In addition to this, centerline velocity profiles featured almost constant plateaus. The proposed models cover the widest range of viscoelastic materials both in terms of the balance between viscous and elastic properties (via fractional-order) and the ratio of elastic properties for complex materials.

The approach we developed features an advantage that is worth mentioning. It allows the wide range of fractional viscoelastic models to be represented and solved in the same functional form. In fact, the entire system behavior can be described by a single function ζ (or $\bar{\omega}$). It, in turn, is related to a complex compliance with the functional form identical for all five models considered in this study. Knowing complex compliance, complex and relaxation moduli and creep compliance are readily obtained. These quantities can be directly measured thus providing immediate practical applications for materials analysis. However, for some specific fields (e.g., polymers), experimental creep compliance data for short times has not been well-established so far. This problem was outlined in recent studies (see, for example, Reference [74]). It definitely deserves attention of the research community and will hopefully be addressed in future studies.

Supplementary Materials: The following are available online at http://www.mdpi.com/2076-3417/10/24/9093/s1.

Author Contributions: Conceptualization, D.G.; methodology, D.G.; software, D.G.; validation, R.P.; formal analysis, D.G.; investigation, D.G.; resources, R.P.; writing—original draft preparation, D.G.; writing—review and editing, R.P.; funding acquisition, R.P. All authors have read and agreed to the published version of the manuscript.

Funding: This research was funded by Argonne National Laboratory through grant #ANL 4J-303061-0030A "Multiscale modeling of complex flows".

Conflicts of Interest: The authors declare no conflict of interest.

References

1. Leibniz, G. Letter from Hanover, Germany, to GFA L'Hopital, September 30; 1695. *Math. Schriften* **1849**, *2*, 301–302.
2. Ross, B. The development of fractional calculus 1695–1900. *Hist. Math.* **1977**, *4*, 75–89. [CrossRef]
3. Miller, K.S.; Ross, B. *An Introduction to the Fractional Calculus and Fractional Differential Equations*; Wiley: Hoboken, NJ, USA, 1993.
4. Podlubny, I. *Fractional Differential Equations: An Introduction to Fractional Derivatives, Fractional Differential Equations, to Methods of their Solution and Some of their Applications*; Elsevier: Amsterdam, The Netherlands, 1998.
5. Mainardi, F. *Fractional Calculus: Theory and Applications*; MDPI: Basel, Switzerland, 2018.
6. Hilfer, R. (Ed.) *Applications of Fractional Calculus in Physics*; World Scientific: Singapore, 2000; Volume 35.
7. Herrmann, R. *Fractional Calculus: An Introduction for Physicists*; World Scientific: Singapore, 2014.
8. Tarasov, V. *Handbook of Fractional Calculus with Applications, Volume 4: Applications in Physics, Part A*; De Gruyter: Berlin, Germany, 2019.
9. Tarasov, V. *Handbook of Fractional Calculus with Applications, Volume 5: Applications in Physics, Part B*; De Gruyter: Berlin, Germany, 2019.
10. Baleanu, D.; Lopes, A.M. *Handbook of Fractional Calculus with Applications, Volume 7: Applications in Engineering, Life and Social Sciences, Part A*; De Gruyter: Berlin, Germany, 2019.
11. Baleanu, D.; Lopes, A.M. *Handbook of Fractional Calculus with Applications, Volume 8: Applications in Engineering, Life and Social Sciences, Part B*; De Gruyter: Berlin, Germany, 2019.
12. Atanacković, T.M.; Janev, M.; Pilipović, S. On the thermodynamical restrictions in isothermal deformations of fractional Burgers model. *Philos. Transact. R. Soc. A* **2020**, *378*, 20190278. [CrossRef]
13. Chen, R.; Wei, X.; Liu, F.; Anh, V.V. Multi-term time fractional diffusion equations and novel parameter estimation techniques for chloride ions sub-diffusion in reinforced concrete. *Philos. Transact. R. Soc. A* **2020**, *378*, 20190538. [CrossRef] [PubMed]
14. Li, S.N.; Cao, B.Y. Fractional-order heat conduction models from generalized Boltzmann transport equation. *Philos. Transact. R. Soc. A* **2020**, *378*, 20190280. [CrossRef] [PubMed]

15. Bologna, E.; Di Paola, M.; Dayal, K.; Deseri, L.; Zingales, M. Fractional-order nonlinear hereditariness of tendons and ligaments of the human knee. *Philos. Transact. R. Soc. A* **2020**, *378*, 20190294. [CrossRef] [PubMed]
16. Chugunov, V.; Fomin, S. Effect of adsorption, radioactive decay and fractal structure of matrix on solute transport in fracture. *Philos. Transact. R. Soc. A* **2020**, *378*, 20190283. [CrossRef] [PubMed]
17. Failla, G.; Zingales, M. Advanced materials modelling via fractional calculus: Challenges and perspectives. *Philos. Transact. R. Soc. A* **2020**, *378*, 20200050. [CrossRef]
18. Fang, C.; Shen, X.; He, K.; Yin, C.; Li, S.; Chen, X.; Sun, H. Application of fractional calculus methods to viscoelastic behaviours of solid propellants. *Philos. Transact. R. Soc. A* **2020**, *378*, 20190291. [CrossRef]
19. Ionescu, C.M.; Birs, I.R.; Copot, D.; Muresan, C.; Caponetto, R. Mathematical modelling with experimental validation of viscoelastic properties in non-Newtonian fluids. *Philos. Transact. R. Soc. A* **2020**, *378*, 20190284. [CrossRef]
20. Li, J.; Ostoja-Starzewski, M. Thermo-poromechanics of fractal media. *Philos. Transact. R. Soc. A* **2020**, *378*, 20190288. [CrossRef] [PubMed]
21. Tenreiro Machado, J.; Lopes, A.M.; de Camposinhos, R. Fractional-order modelling of epoxy resin. *Philos. Transact. R. Soc. A* **2020**, *378*, 20190292. [CrossRef] [PubMed]
22. Di Paola, M.; Alotta, G.; Burlon, A.; Failla, G. A novel approach to nonlinear variable-order fractional viscoelasticity. *Philos. Transact. R. Soc. A* **2020**, *378*, 20190296. [CrossRef]
23. Patnaik, S.; Hollkamp, J.P.; Semperlotti, F. Applications of variable-order fractional operators: A review. *Proc. R. Soc. A* **2020**, *476*, 20190498. [CrossRef]
24. Patnaik, S.; Semperlotti, F. Variable-order particle dynamics: Formulation and application to the simulation of edge dislocations. *Philos. Transact. R. Soc. A* **2020**, *378*, 20190290. [CrossRef]
25. Povstenko, Y.; Kyrylych, T. Fractional thermoelasticity problem for an infinite solid with a penny-shaped crack under prescribed heat flux across its surfaces. *Philos. Transact. R. Soc. A* **2020**, *378*, 20190289. [CrossRef] [PubMed]
26. Zhang, X.; Ostoja-Starzewski, M. Impact force and moment problems on random mass density fields with fractal and Hurst effects. *Philos. Transact. R. Soc. A* **2020**, *378*, 20190591. [CrossRef] [PubMed]
27. Zorica, D.; Oparnica, L. Energy dissipation for hereditary and energy conservation for non-local fractional wave equations. *Philos. Transact. R. Soc. A* **2020**, *378*, 20190295. [CrossRef] [PubMed]
28. Blair, G.S. The role of psychophysics in rheology. *J. Coll. Sci.* **1947**, *2*, 21–32. [CrossRef]
29. Bagley, R.L.; Torvik, P. A theoretical basis for the application of fractional calculus to viscoelasticity. *J. Rheol.* **1983**, *27*, 201–210. [CrossRef]
30. Gloeckle, W.G.; Nonnenmacher, T.F. Fractional integral operators and Fox functions in the theory of viscoelasticity. *Macromolecules* **1991**, *24*, 6426–6434. [CrossRef]
31. Metzler, R.; Schick, W.; Kilian, H.G.; Nonnenmacher, T.F. Relaxation in filled polymers: A fractional calculus approach. *J. Chem. Phys.* **1995**, *103*, 7180–7186. [CrossRef]
32. Friedrich, C.; Braun, H. Generalized Cole-Cole behavior and its rheological relevance. *Rheol. Acta* **1992**, *31*, 309–322. [CrossRef]
33. Friedrich, C. Relaxation and retardation functions of the Maxwell model with fractional derivatives. *Rheol. Acta* **1991**, *30*, 151–158. [CrossRef]
34. Nonnenmacher, T.; Glöckle, W. A fractional model for mechanical stress relaxation. *Philos. Mag. Lett.* **1991**, *64*, 89–93. [CrossRef]
35. Schiessel, H.; Blumen, A. Mesoscopic pictures of the sol-gel transition: Ladder models and fractal networks. *Macromolecules* **1995**, *28*, 4013–4019. [CrossRef]
36. Schiessel, H.; Blumen, A. Hierarchical analogues to fractional relaxation equations. *J. Phys. A Math. Gen.* **1993**, *26*, 5057. [CrossRef]
37. Schiessel, H.; Blumen, A.; Alemany, P. Dynamics in disordered systems. In *Transitions in Oligomer and Polymer Systems*; Springer: Berlin, Germany, 1994; pp. 16–21.
38. Heymans, N.; Bauwens, J.C. Fractal rheological models and fractional differential equations for viscoelastic behavior. *Rheol. Acta* **1994**, *33*, 210–219. [CrossRef]
39. Schiessel, H.; Metzler, R.; Blumen, A.; Nonnenmacher, T. Generalized viscoelastic models: Their fractional equations with solutions. *J. Phys. A Math. Gen.* **1995**, *28*, 6567. [CrossRef]

40. Friedrich, C.; Schiessel, H.; Blumen, A. Constitutive behavior modeling and fractional derivatives. In *Rheology Series*; Elsevier: Amsterdam, The Netherlands, 1999; Volume 8, pp. 429–466.
41. Hernández-Jiménez, A.; Hernández-Santiago, J.; Macias-Garcıa, A.; Sánchez-González, J. Relaxation modulus in PMMA and PTFE fitting by fractional Maxwell model. *Polym. Test.* **2002**, *21*, 325–331. [CrossRef]
42. Makris, N.; Constantinou, M. Spring-viscous damper systems for combined seismic and vibration isolation. *Earthq. Eng. Struct. Dyn.* **1992**, *21*, 649–664. [CrossRef]
43. Zhang, Y.; Xu, S.; Zhang, Q.; Zhou, Y. Experimental and theoretical research on the stress-relaxation behaviors of PTFE coated fabrics under different temperatures. *Adv. Mater. Sci. Eng.* **2015**, *2015*, 319473. [CrossRef] [PubMed]
44. Khajehsaeid, H. Application of fractional time derivatives in modeling the finite deformation viscoelastic behavior of carbon-black filled NR and SBR. *Polym. Test.* **2018**, *68*, 110–115. [CrossRef]
45. Khajehsaeid, H. A Comparison Between Fractional-Order and Integer-Order Differential Finite Deformation Viscoelastic Models: Effects of Filler Content and Loading Rate on Material Parameters. *Int. J. Appl. Mech.* **2018**, *10*, 1850099. [CrossRef]
46. Yin, B.; Hu, X.; Song, K. Evaluation of classic and fractional models as constitutive relations for carbon black–filled rubber. *J. Elastom. Plast.* **2018**, *50*, 463–477. [CrossRef]
47. Xu, H.; Jiang, X. Creep constitutive models for viscoelastic materials based on fractional derivatives. *Comp. Math. Appl.* **2017**, *73*, 1377–1384. [CrossRef]
48. Stankiewicz, A. Fractional Maxwell model of viscoelastic biological materials. In *BIO Web of Conferences*; EDP Sciences: Les Ulis, France, 2018; Volume 10, p. 02032.
49. Farno, E.; Baudez, J.C.; Eshtiaghi, N. Comparison between classical Kelvin-Voigt and fractional derivative Kelvin-Voigt models in prediction of linear viscoelastic behaviour of waste activated sludge. *Sci. Total Environ.* **2018**, *613*, 1031–1036. [CrossRef]
50. Grace, S. XCIV. Oscillatory motion of a viscous liquid in a long straight tube. *Lond. Edinb. Dublin Philos. Mag. J. Sci.* **1928**, *5*, 933–939. [CrossRef]
51. Sexl, T. Uber den von EG Richardson entdeckten "Annulareffekt". *Z. Phys.* **1930**, *61*, 349–362. [CrossRef]
52. Womersley, J.R. Method for the calculation of velocity, rate of flow and viscous drag in arteries when the pressure gradient is known. *J. Physiol.* **1955**, *127*, 553. [CrossRef]
53. Womersley, J.R. XXIV, Oscillatory motion of a viscous liquid in a thin-walled elastic tube—I: The linear approximation for long waves. *Lond. Edinb. Dublin Philos. Mag. J. Sci.* **1955**, *46*, 199–221. [CrossRef]
54. Uchida, S. The pulsating viscous flow superposed on the steady laminar motion of incompressible fluid in a circular pipe. *Z. Angew. Math. Phys. ZAMP* **1956**, *7*, 403–422. [CrossRef]
55. Taylor, M. An approach to an analysis of the arterial pulse wave II. Fluid oscillations in an elastic pipe. *Phys. Med. Biol.* **1957**, *1*, 321. [CrossRef] [PubMed]
56. Sergeev, S. Fluid oscillations in pipes at moderate Reynolds numbers. *Fluid Dyn.* **1966**, *1*, 121–122. [CrossRef]
57. Ramaprian, B.; Tu, S.W. An experimental study of oscillatory pipe flow at transitional Reynolds numbers. *J. Fluid Mech.* **1980**, *100*, 513–544. [CrossRef]
58. Harris, J.; Maheshwari, R. Oscillatory pipe flow: A comparison between predicted and observed displacement profiles. *Rheol. Acta* **1975**, *14*, 162–168. [CrossRef]
59. Hino, M.; Sawamoto, M.; Takasu, S. Experiments on transition to turbulence in an oscillatory pipe flow. *J. Fluid Mech.* **1976**, *75*, 193–207. [CrossRef]
60. Wood, W. Transient viscoelastic helical flows in pipes of circular and annular cross-section. *J. Non Newton. Fluid Mech.* **2001**, *100*, 115–126. [CrossRef]
61. Yin, Y.; Zhu, K.Q. Oscillating flow of a viscoelastic fluid in a pipe with the fractional Maxwell model. *Appl. Math. Comp.* **2006**, *173*, 231–242. [CrossRef]
62. Yang, D.; Zhu, K.Q. Start-up flow of a viscoelastic fluid in a pipe with a fractional Maxwell's model. *Comp. Math. Appl.* **2010**, *60*, 2231–2238. [CrossRef]
63. Shah, S.H.A.M.; Qi, H. Starting solutions for a viscoelastic fluid with fractional Burgers' model in an annular pipe. *Nonlinear Anal. Real World Appl.* **2010**, *11*, 547–554. [CrossRef]
64. Nazar, M.; Fetecau, C.; Awan, A. A note on the unsteady flow of a generalized second-grade fluid through a circular cylinder subject to a time dependent shear stress. *Nonlinear Anal. Real World Appl.* **2010**, *11*, 2207–2214. [CrossRef]

65. Qi, H.; Liu, J. Some duct flows of a fractional Maxwell fluid. *Eur. Phys. J. Spec. Top.* **2011**, *193*, 71–79. [CrossRef]
66. Khandelwal, K.; Mathur, V. Exact solutions for an unsteady flow of viscoelastic fluid in cylindrical domains using the fractional Maxwell model. *Int. J. Appl. Comput. Math.* **2015**, *1*, 143–156. [CrossRef]
67. Mathur, V.; Khandelwal, K. Exact Solutions for the Flow of Fractional Maxwell Fluid in Pipe-Like Domains. *Adv. Appl. Math. Mech.* **2016**, *8*, 784–794. [CrossRef]
68. Maqbool, K.; Mann, A.; Siddiqui, A.M.; Shaheen, S. Fractional generalized Burgers' fluid flow due to metachronal waves of cilia in an inclined tube. *Adv. Mech. Eng.* **2017**, *9*, 1687814017715565. [CrossRef]
69. Tang, Y.; Zhen, Y.; Fang, B. Nonlinear vibration analysis of a fractional dynamic model for the viscoelastic pipe conveying fluid. *Appl. Math. Model.* **2018**, *56*, 123–136. [CrossRef]
70. Wang, Y.; Chen, Y. Dynamic Analysis of the Viscoelastic Pipeline Conveying Fluid with an Improved Variable Fractional Order Model Based on Shifted Legendre Polynomials. *Fractal Fract.* **2019**, *3*, 52. [CrossRef]
71. Javadi, M.; Noorian, M.; Irani, S. Stability analysis of pipes conveying fluid with fractional viscoelastic model. *Meccanica* **2019**, *54*, 399–410. [CrossRef]
72. Sun, H.; Zhang, Y.; Baleanu, D.; Chen, W.; Chen, Y. A new collection of real world applications of fractional calculus in science and engineering. *Commun. Nonlinear Sci. Num. Simulat.* **2018**, *64*, 213–231. [CrossRef]
73. Crandall, I.B. *Theory of Vibrating Systems and Sound*; D. Van Nostrand Company: New York, NY, USA, 1926.
74. Urbanowicz, K.; Duan, H.F.; Bergant, A.; Urbanowicz, K.; Bergant, H.F.D.A. Transient Liquid Flow in Plastic Pipes. *STROJNISKI VESTNIK J. Mech. Eng.* **2020**, *66*, 77–90. [CrossRef]

Publisher's Note: MDPI stays neutral with regard to jurisdictional claims in published maps and institutional affiliations.

© 2020 by the authors. Licensee MDPI, Basel, Switzerland. This article is an open access article distributed under the terms and conditions of the Creative Commons Attribution (CC BY) license (http://creativecommons.org/licenses/by/4.0/).

Buckling Analysis of CNTRC Curved Sandwich Nanobeams in Thermal Environment

Ahmed Amine Daikh [1], Mohammed Sid Ahmed Houari [1], Behrouz Karami [2], Mohamed A. Eltaher [3,4], Rossana Dimitri [5] and Francesco Tornabene [5,*]

1. Laboratoire d'Etude des Structures et de Mécanique des Matériaux, Département de Génie Civil, Faculté des Sciences et de la Technologie, Université Mustapha Stambouli, B.P. 305, Mascara 29000, Algeria; ahmed.daikh@univ-sba.dz (A.A.D.); ms.houari@univ-mascara.dz (M.S.A.H.)
2. Department of Mechanical Engineering, Marvdasht Branch, Islamic Azad University, Marvdasht 15914, Iran; behrouz.karami@miau.ac.ir
3. Faculty of Engineering, Mechanical Engineering Department, King Abdulaziz University, P.O. Box 80204, Jeddah 21589, Saudi Arabia; meltaher@kau.edu.sa
4. Faculty of Engineering, Mechanical Design and Production Department, Zagazig University, P.O. Box 44519, Zagazig 44519, Egypt
5. Department of Innovation Engineering, Università del Salento, 73100 Lecce, Italy; rossana.dimitri@unisalento.it
* Correspondence: francesco.tornabene@unisalento.it

Abstract: This paper presents a mathematical continuum model to investigate the static stability buckling of cross-ply single-walled (SW) carbon nanotube reinforced composite (CNTRC) curved sandwich nanobeams in thermal environment, based on a novel quasi-3D higher-order shear deformation theory. The study considers possible nano-scale size effects in agreement with a nonlocal strain gradient theory, including a higher-order nonlocal parameter (material scale) and gradient length scale (size scale), to account for size-dependent properties. Several types of reinforcement material distributions are assumed, namely a uniform distribution (UD) as well as X- and O- functionally graded (FG) distributions. The material properties are also assumed to be temperature-dependent in agreement with the Touloukian principle. The problem is solved in closed form by applying the Galerkin method, where a numerical study is performed systematically to validate the proposed model, and check for the effects of several factors on the buckling response of CNTRC curved sandwich nanobeams, including the reinforcement material distributions, boundary conditions, length scale and nonlocal parameters, together with some geometry properties, such as the opening angle and slenderness ratio. The proposed model is verified to be an effective theoretical tool to treat the thermal buckling response of curved CNTRC sandwich nanobeams, ranging from macroscale to nanoscale, whose examples could be of great interest for the design of many nanostructural components in different engineering applications.

Keywords: curved sandwich nanobeams; nonlocal strain gradient theory; quasi-3D higher-order shear theory; thermal-buckling

1. Introduction

Multilayered composites are widely used in various engineering structures, ranging from macroscale (i.e., aircraft, submarines, space-station structures, etc.) to nanoscale (nano-sensors, nano-actuators, nano-gears, and micro/nano-electro-mechanical systems (MEMS/NEMS)), due to the high stiffness and strength-to-weight ratios caused by fiber reinforcements. In the recent literature, reinforcements based on carbon nanotubes (CNTs) have been largely applied in lieu of conventional fibers due to their excellent properties in order to improve the mechanical, electrical, and thermal properties of composite structures. In [1,2], for example, different molecular dynamic simulations have been successfully applied by the authors to exploit the elastic moduli of polymer–CNT composites embedded in polymeric matrices. Fidelus et al. [3] examined the thermo-mechanical properties of different epoxy-based nanocomposites with randomly oriented single-walled (SW) and multi-walled (MW) CNTs. Moreover, Shen [4] investigated the nonlinear bending behavior of FG nanocomposite plates reinforced by SWCNTs subjected to a transverse uniform or sinusoidal load in a thermal environment using two different distribution functions. A nonlocal strain gradient theory was also proposed by Lim et al. [5] to study a wave propagation in macro and nanobeam structures for the first time. Wu and Kitipornchai [6] investigated the free vibration and elastic buckling of sandwich beams with a stiff core and functionally graded (FG)-CNTRC face sheets in a Timoshenko beam theoretical framework. Among coupled thermo-mechanical problems, Eltaher et al. [7] investigated the influence of a thermal loading and shear force on the nonlocal buckling response of nanobeams via higher-order shear deformation Eringen beam theories. Similarly, Ebrahimi and Farazmandnia [8] investigated the thermo-mechanical vibration of sandwich FG-CNTRC beams within a Timoshenko-based beam approach; Sobhy and Zenkour [9] illustrated the influence of a magnetic field on the thermo-mechanical buckling and vibration response of FG-CNTRC nanobeams with a viscoelastic substrate. In line with the previous works, Daikh and Megueni [10] studied the thermal buckling of FG sandwich higher-order plates with material temperature-dependent properties under a nonlinear temperature rise; Arefi and Arani [11] combined a third-order shear deformation approach together with the nonlocal elasticity to study the static deflection of FG nanobeams under a coupled thermo-electro-magneto-mechanical environment. A novel refined shear theory was recently proposed by Bekhadda et al. [12] for the study of a gradation influence on the vibration and buckling behavior of FG beams with a power-law function by means of Fourier series. Medani et al. [13], instead, applied the first order shear deformation and energy principle to study the static and dynamic behavior of FG-CNT-reinforced porous sandwich plates. Arani et al. [14] later performed a thermo-electro-mechanical buckling study of FG-CNTRC sandwich nanobeams based on a nonlocal strain gradient elasticity theory and differential quadrature numerical procedure. More complicated double-curved sandwich panels were accounted by Nejati et al. [15], who analyzed the thermal vibration in presence of pre-strained shape memory alloy wires. Chaht et al. [16] analyzed the size-dependent static behavior of FG nanobeams, including the thickness stretching effect; whereas a nonlocal trigonometric shear deformation theory and nonlocal quasi-3D theory were proposed in [17,18], respectively, to treat FG nanobeams. An efficient alternative tool to handle non-localities within nanostructures is represented by the strain gradient theory, as successfully applied in [19,20] for the thermal snap-buckling and bending analysis of FG curved porous and non-porous nanobeams and in [21,22] for the buckling study of porous FG sandwich nanoplates resting on a Kerr foundation due to a heat conduction. A theoretical formulation based on a Reddy shear deformation theory, has been also proposed in the recent work by Daikh et al. [23] to study the buckling and vibration of FG-CNTRC-laminated nanoplates in thermal environment, with promising results for engineering applications. Furthermore, Daikh et al. [24] investigated the thermal buckling response of FG sandwich beams under a power-law (P-FGM) or sigmoid (SFGM) variation. Further attempts of combining higher order theories and nonlocal approaches in a unified context, can be found in [25–28] to predict the influence of an axial in-plane load function on the critical buckling load and mode shape of composite beam members, also in presence of porosities. During fabrication, structural members can exhibit an initial curved shape as possible imperfection related to iterative heating and cooling processes. Many MEMS devices employ curved structures as well [29]. The initial curvature of a beam structure can be a source of difficulty in developing the constitutive relations, as verified by Emam et al. [30], who illustrated the possible effects of curvatures and imperfections on the post-buckling and free vibration response of multilayer nonlocal prestressed nanobeams. Shi et al. [31] also studied the effect of nanotube waviness and agglomeration on the elastic property of CNT-reinforced composites. A further systematic study was performed by Khater et al. [32], who investigated the impact of the surface energy and thermal loading on the static stability of curved nanowires, modeled as curved Euler-Bernoulli beams, accounting for both the von Karman and axial strain field. Among more sophisticated shell models, a valuable comparison between different higher-order formulations was proposed in [33–35] for the static analysis of multilayered composite and sandwich plates and shells, both from a theoretical and computational perspective. Mohamed et al. [36] later proposed a differential quadrature method to study the nonlinear free and forced vibrations of buckled curved beams resting on nonlinear elastic foundations. A further attempt of combining the nonlocal strain gradient and higher-order shell theories was conducted by Karami et al. [37] for a wave dispersion study in anisotropic doubly-curved nanoshells, as well as in [38–41] for FG-CNTRC curved nanobeams also in coupled piezoelectric conditions. In another work, Arefi et al. [42] predicted the static deflection and stress field of curved FG-CNTRC nonlocal Timoshenko nanobeams resting on an elastic foundation under four different distribution patterns of CNTs throughout the thickness direction. Eltaher et al. [43] also presented the influence of periodic and/or nonperiodic imperfections on the buckling, post-buckling and dynamic response of curved beams resting on nonlinear elastic foundations by means of high-performing numerical differential-integral quadrature methods (DIQMs). Malikan et al. [44] developed a theoretical model to study the dynamics of non-cylindrical curved viscoelastic SWCNTs by applying a second gradient theory of stress-strain, whereas Mohamed et al. [45] used an energy equivalent model to study the post-buckling response of imperfect CNTs resting on a nonlinear elastic foundation, including mid-plane stretching and nanoscale effects. Among the most recent works on the topic, Van Tham et al. [46] developed a novel four-variable refined shell theory to study the free vibration of multi-layered FG-CNTRC doubly curved shallow shell panels; Dindarloo et al. [47] exploited the strain-driven nonlocal integral theory to study the bending response of isotropic doubly curved high-order shear deformation nanoshells under a combined assumption of exponential and trigonometric shape functions. Furthermore, Eltaher and Mohamed [48] exploited the nonlinear stability and vibration of imperfect CNTs modeled as Euler-Bernoulli beams with a mid-plane stretching, while in [49–51], the authors studied the free and forced vibration and the dispersion behavior of elastic waves of doubly-curved nonlocal strain gradient theory nanoshells in conjunction with a higher-order shear deformation shell theory. Based on the available literature, however, the influence of a material scale, size scale, and graduation distribution functions on the thermal static stability of curved sandwich nanobeams with temperature-dependent material seems to be generally lacking. To this end, the present paper aims at providing a closed-form solution to the problem, for different boundary conditions, that could be useful as theoretical benchmark for different computational studies and engineering design applications. The paper is organized as follows. In Section 2, the theoretical formulation of curved sandwich CNTRC nanobeams is reviewed, including the kinematic field, relations and constitutive equations. Section 3 illustrates the governing equilibrium equation of curved sandwich beams in a classical and nonclassical domain, while discussing about different thermal field distributions and temperature-dependent properties of materials. Section 4 presents the analytical solutions of the problem for different boundary conditions, whose comparative study is performed systematically and discussed in Section 5. Finally, in Section 6, conclusions are drawn together with possible future research directions.

2. Theoretical Formulation
2.1. Geometric and Mechanical Properties

A symmetric cross-ply single-walled carbon nanotube reinforced composite (CNTRC) curved sandwich beam of length L, thickness h, and radius of curvature R is considered, as shown in Figure 1. Different volume fraction distributions of CNTs are here assumed throughout the thickness (see Figure 2), in agreement with the following relations [22]:

- UD (Uniformly-Distributed) CNTRC multilayered nanobeam:

$$V_{cnt} = V_{cnt}^* \qquad (1)$$

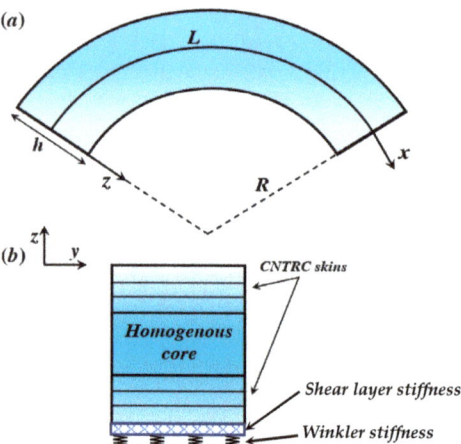

Figure 1. Geometry of a carbon nanotube reinforced composite (CNTRC) curved sandwich beam: (**a**) geometric parameters of the curved beam, (**b**) cross-section of the curved beam.

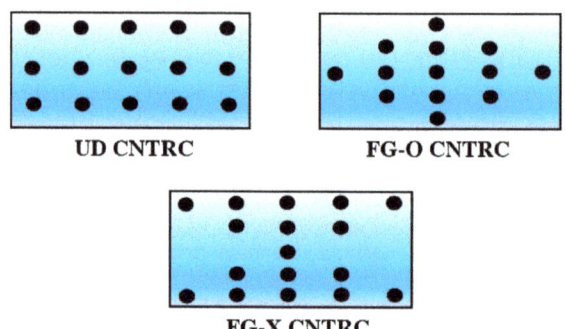

Figure 2. Cross-sections of various (CNTRC) carbon nanotube reinforced composite sandwich beams.

- FG-O CNTRC multilayered nanobeam:

$$V_{cnt} = 2\left(1 - \frac{\left|2|z| - \left|z_{(k-1)} + z_{(k)}\right|\right|}{z_{(k)} - z_{(k-1)}}\right) V_{cnt}^* \tag{2}$$

- FG-X CNTRC multilayered nanobeam:

$$V_{cnt} = 2\frac{\left|2|z| - \left|z_{(k-1)} + z_{(k)}\right|\right|}{z_{(k)} - z_{(k-1)}} V_{cnt}^* \tag{3}$$

More specifically, UD CNTRC refers to a uniform distribution of CNTs, whereas FG-V CNTRC, FG-O CNTRC and FG-X CNTRC account for different non-uniform FG distributions. Moreover, $z_{(k)}$ and $z_{(k-1)}$ refer to the thickness coordinates at the bottom and top sides of the kth layer within the laminated nanobeam; V_{cnt}^* is the total volume fraction of CNTs, defined as

$$V_{cnt}^* = \frac{W_{cnt}}{W_{cnt} + (\rho_{cnt}/\rho_m)(1 - W_{cnt})} \tag{4}$$

where W_{cnt} is the CNTs mass fraction; ρ_{cnt}, ρ_m refer to the CNTs and polymer mass density, respectively.

The Mori–Tanaka scheme [31] is here applied together with the rule of mixtures and molecular dynamics, as suggested in [1,2]. Thus, the effective Young's modulus and shear modulus for each CNTRC sheet is described as

$$E_{11}^k = \eta_1 V_{cnt}^k E_{11}^{cnt} + V_p^k E_p$$
$$\frac{\eta_2}{E_{22}^k} = \frac{V_{cnt}^k}{E_{22}^{cnt}} + \frac{V_p^k}{E_p} \quad (5)$$
$$\frac{\eta_3}{G_{12}^k} = \frac{V_{cnt}^k}{G_{12}^{cnt}} + \frac{V_p^k}{G_p}$$

where E_{11}^k and E_{22}^k are the elasticity modulus along the in-plane directions (x, z) for the kth layer and G_{12}^k is its shear modulus. The subscripts p and cnt refer to the polymer and SWCNT properties, respectively, assuming the CNT efficiency parameters η_1, η_2, η_3 as proposed in [6] and summarized in Table 1.

Table 1. CNT efficiency parameters.

V_{cnt}^*	η_1	η_2	η_3
0.12	0.137	1.022	0.715
0.17	0.142	1.626	1.138
0.28	0.141	1.585	1.109

The Poisson's ratio ν_{12}^k, the density ρ^k, and the thermal expansion coefficients in the longitudinal and transverse directions α_{11}^k, α_{22}^k, for each sheet are given as follows:

$$\nu_{12}^k = V_{cnt}^k \nu_{12}^{cnt} + V_p^k \nu_p \quad (6)$$

$$\rho^k = V_{cnt}^k \rho_{cnt} + V_p^k \rho_p \quad (7)$$

$$\alpha_{11}^k = V_{cnt}^k \alpha_{11}^{cnt} + V_p^k \alpha_p \quad (8)$$

$$\alpha_{22}^k = \left(1 + \nu_{12}^{cnt}\right) V_{cnt}^k \alpha_{22}^{cnt} + \left(1 + \nu_p\right) V_p^k \alpha_p - \nu_{12}^k \alpha_{11}^k \quad (9)$$

2.2. Kinematic Field

In the present work, a quasi-3D higher-order-shear deformation theory (HSDT) is used to define the governing equations for the buckling problem of CNTRC curved sandwich beams, whose displacements components are expressed in terms of the midline displacements and cross-section rotations as

$$u(x,z,t) = \left(1 + \frac{z}{R}\right) u_0 - z\frac{\partial w_0}{\partial x} + \Phi(z)\varphi_x$$
$$w(x,z,t) = w_0 + \Phi(z)' \varphi_z \quad (10)$$

A novel hyperbolic shape function $\Phi(z)$ is proposed herein to determine the distribution of the transverse shear strain and stress field along the thickness direction, namely

$$\Phi(z) = \frac{h\left(\pi \cosh\left(\frac{\pi}{2}\right)\tanh\left(\frac{z}{h}\right) - \sinh\left(\frac{\pi z}{h}\right)\left(1 - \tanh\left(\frac{1}{2}\right)^2\right)\right)}{\pi\left(\tanh\left(\frac{1}{2}\right)^2 + \cosh\left(\frac{\pi}{2}\right) - 1\right)} \quad (11)$$

Based on a quasi-3D theory, the strain fields of the curved sandwich beam have the following form:

$$\varepsilon_{xx} = \left[\frac{\partial u}{\partial x} + \frac{w}{R}\right] = \frac{\partial u_0}{\partial x} - z\frac{\partial^2 w_0}{\partial x^2} + \Phi(z)\frac{\partial \varphi_x}{\partial x} + \frac{w_0}{R} + \Phi(z)'\frac{\varphi_z}{R}$$
$$\varepsilon_{zz} = \left[\frac{\partial w}{\partial x}\right] = \Phi(z)'' \varphi_z \quad (12)$$
$$\gamma_{xz} = \left[\frac{\partial u}{\partial x} + \frac{\partial w}{\partial x} - \frac{u_0}{R}\right] = \Phi(z)'\left(\partial \varphi_x + \frac{\partial \varphi_z}{\partial x}\right)$$

2.3. Constitutive Equations

The stress field is governed by the following constitutive relations:

$$\left\{\begin{array}{c} \sigma_{xx} \\ \sigma_{zz} \\ \tau_{xz} \end{array}\right\}^{(k)} = \left[\begin{array}{ccc} \overline{Q}_{11}^k & \overline{Q}_{13}^k & 0 \\ \overline{Q}_{13}^k & \overline{Q}_{33}^k & 0 \\ 0 & 0 & \overline{Q}_{55}^k \end{array}\right]\left\{\begin{array}{c} \varepsilon_{xx} \\ \varepsilon_{zz} \\ \gamma_{xz} \end{array}\right\}^{(k)} \quad (13)$$

with $\overline{Q}_{ij}^k$ being the transformed material constants, defined by means of the lamination angle θ_k for the kth layer, as follows:

$$\overline{Q}_{11}^k = Q_{11}\cos^4\theta_k + 2(Q_{12} + 2Q_{66})\sin^2\theta_k \cos^2\theta_k + Q_{22}\sin^4\theta_k$$
$$\overline{Q}_{13}^k = Q_{13}\cos^2\theta_k + Q_{23}\sin^2\theta_k \quad (14)$$
$$\overline{Q}_{55}^k = Q_{55}\cos^2\theta_k + Q_{44}\sin^2\theta_k$$

and

$$Q_{11} = \frac{E_{11}}{1 - v_{12}v_{21}}$$
$$Q_{12} = Q_{13} = \frac{v_{12}E_{11}}{1 - v_{12}v_{21}} \quad (15)$$
$$Q_{23} = \frac{v_{21}E_{22}}{1 - v_{12}v_{21}}$$
$$Q_{22} = Q_{33} = \frac{E_{22}}{1 - v_{12}v_{21}}$$

$$E_{33} = E_{22}, \ G_{12} = G_{13} = G_{23}, \ v_{21} = \frac{E_{22}}{E_{11}}v_{12}, \ v_{13} = v_{12}, \ v_{31} = v_{21}, \ v_{32} = v_{23} = v_{21} \quad (16)$$

3. Equilibrium Governing Equations

3.1. Classical Formulation of Curved Sandwich Beams

Based on a classical formulation, the equilibrium equations of the problem are determined by means of the potential energy principle. In detail, the strain energy variation is defined as

$$\int_{-h/2}^{h/2}\int_0^L\left[\sigma_{xx}^{(k)}\delta\varepsilon_{xx} + \sigma_{zz}^k\varepsilon_{zz} + \tau_{xz}^{(k)(k)}\gamma_{xz}\right]dxdz - \int_0^L N_x^0\frac{\partial w_0}{\partial x}\frac{\partial \delta w_0}{\partial x}dx - \int_0^L\left[k_w w_0 \delta w_0 + k_g\frac{\partial w_0}{\partial x}\frac{\partial \delta w_0}{\partial x} + k_{NL}w_0^3\delta w_0\right]dx \quad (17)$$

in agreement with a quasi-3D theory, where k_w and k_g are the linear Winkler stiffness and the shear layer stiffness, respectively, and k_{NL} refers to the non-linear stiffness. The strain energy variation can be rewritten in terms of stress resultants as

$$\int_0^L\left[N_{xx}\frac{\partial \delta u_0}{\partial x} - M_{xx}\frac{\partial^2 \delta w_0}{\partial x^2} + P_{xx}\frac{\partial \delta \varphi_x}{\partial x} + N_{xx}\frac{\delta w_0}{R} + Q_x\frac{\delta \varphi_z}{R} + R_z\delta\varphi_z + Q_{xz}\delta\varphi_x + Q_{xz}\frac{\delta\delta\varphi_z}{\partial x}\right]dx \quad (18)$$

where

$$N_{xx} = \sum_{k=1}^{N} \int_{h_k}^{h_{k+1}} \sigma_{xx}^{(k)} dz = A_{11}\frac{\partial u_0}{\partial x} - B_{11}\frac{\partial^2 w_0}{\partial x^2} + B_{11}^s\frac{\partial \varphi_x}{\partial x} + A_{11}\frac{w_0}{R} + D_{11}\frac{\varphi_z}{R} + E_{12}\varphi_z$$

$$M_{xx} = \sum_{k=1}^{N} \int_{h_k}^{h_{k+1}} \sigma_{xx}^{(k)} z dz = B_{11}\frac{\partial u_0}{\partial x} - F_{11}\frac{\partial^2 w_0}{\partial x^2} + F_{11}^s\frac{\partial \varphi_x}{\partial x} + B_{11}\frac{w_0}{R} + D_{11}^s\frac{\varphi_z}{R} + J_{12}^s\varphi_z$$

$$P_{xx} = \sum_{k=1}^{N} \int_{h_k}^{h_{k+1}} \sigma_{xx}^{(k)} \Phi(z) dz = B_{11}^s\frac{\partial u_0}{\partial x} - F_{11}^s\frac{\partial^2 w_0}{\partial x^2} + G_{11}^s\frac{\partial \varphi_x}{\partial x} + B_{11}^s\frac{w_0}{R} + H_{11}^s\frac{\varphi_z}{R} + E_{12}^s\varphi_z$$

$$Q_x = \sum_{k=1}^{N} \int_{h_k}^{h_{k+1}} \sigma_{xx}^{(k)} \Phi(z)' dz = D_{11}\frac{\partial u_0}{\partial x} - D_{11}^s\frac{\partial^2 w_0}{\partial x^2} + H_{11}^s\frac{\partial \varphi_x}{\partial x} + D_{11}\frac{w_0}{R} + K_{11}^s\frac{\varphi_z}{R} + L_{12}^s\varphi_z \quad (19)$$

$$Q_{xz} = \sum_{k=1}^{N} \int_{h_k}^{h_{k+1}} \tau_{xz}^{(k)} \Phi(z)' dz = K_{33}^s\left(\varphi_x + \frac{\partial \varphi_z}{\partial x}\right)$$

$$R_z = \sum_{k=1}^{N} \int_{h_k}^{h_{k+1}} \sigma_{zz}^{(k)} \Phi(z)'' dz = E_{12}\frac{\partial u_0}{\partial x} - E_{12}^s\frac{\partial^2 w_0}{\partial x^2} + J_{12}^s\frac{\partial \varphi_x}{\partial x} + E_{12}\frac{w_0}{R} + L_{12}^s\frac{\varphi_z}{R} + L_{22}^s\varphi_z$$

and

$$\{A_{11}, B_{11}, F_{11}, B_{11}^s, F_{11}^s, G_{11}^s\} = \sum_{k=1}^{N} \int_{h_k}^{h_{k+1}} \overline{Q}_{11}^k\{1, z, z^2, \Phi(z), z\Phi(z), \Phi(z)^2\} dz$$

$$\{D_{11}, D_{11}^s, H_{11}^s, K_{11}^s\} = \sum_{k=1}^{N} \int_{h_k}^{h_{k+1}} \overline{Q}_{11}^k\{\Phi(z)', z\Phi(z)', \Phi(z)\Phi(z)', \Phi(z)'^2\} dz$$

$$\{E_{12}, E_{12}^s, J_{12}^s, L_{12}^s\} = \sum_{k=1}^{N} \int_{h_k}^{h_{k+1}} \overline{Q}_{12}^k\{\Phi(z)'', z\Phi(z)'', \Phi(z)\Phi(z)'', \Phi(z)'\Phi(z)''\} dz \quad (20)$$

$$L_{22}^s = \sum_{k=1}^{N} \int_{h_k}^{h_{k+1}} \overline{Q}_{22}^k \Phi(z)''^2 dz$$

$$K_{33}^s = \sum_{k=1}^{N} \int_{h_k}^{h_{k+1}} \overline{Q}_{33}^k \Phi(z)'^2 dz$$

Integrating by parts and setting the coefficients of δu_0, δw_0, $\delta \varphi_x$, and $\delta \varphi_z$ equal to zero, the equilibrium equations of the problem are as follows:

$$\begin{aligned} \delta u_0 &: \frac{\partial N_{xx}}{\partial x} = 0 \\ \delta w_0 &: \frac{\partial^2 M_{xx}}{\partial x^2} - \frac{N_{xx}}{R} - N_x^0\frac{\partial^2 w_0}{\partial x^2} - k_w w_0 + k_g\frac{\partial^2 w_0}{\partial x^2} - k_{NL} w_0^3 = 0 \\ \delta \varphi_x &: \frac{\partial P_{xx}}{\partial x} - Q_{xz} = 0 \\ \delta \varphi_z &: \frac{\partial Q_{xz}}{\partial x} - R_z - \frac{Q_x}{R} = 0 \end{aligned} \quad (21)$$

3.2. Nonlocal Strain Gradient Approach

We account for possible effects related to the strain gradient stress and nonlocal elastic stress fields, in line with [5], as follows:

$$\sigma_{ij} = \sigma_{ij}^{(0)} - \frac{d\sigma_{ij}^{(1)}}{dx} \quad (22)$$

where $\sigma_{ij}^{(0)}$ refers to the classical stress components corresponding to the strain field ε_{kl} and the higher-order stress $\sigma_{ij}^{(1)}$ corresponds to strain gradient $\varepsilon_{kl,x}$. The classical and higher-order stress components are described as

$$\begin{aligned} \sigma_{ij}^{(0)} &= \int_0^L C_{ijkl}\alpha_0(x, x', e_0 a)\varepsilon_{kl,x}(x') dx' \\ \sigma_{ij}^{(1)} &= l^2 \int_0^L C_{ijkl}\alpha_1(x, x', e_1 a)\varepsilon_{kl,x}(x') dx' \end{aligned} \quad (23)$$

where C_{ijkl} is an elastic constant and l is the material length scale parameter, here introduced to account for the strain gradient stress field; $e_0 a$ and $e_1 a$ are the nonlocal parameters defining the nonlocal elastic stress field.

The nonlocal kernel functions $\alpha_0(x, x', e_0 a)$ and $\alpha_1(x, x', e_1 a)$ satisfy the conditions developed by Eringen [52], whereby the general constitutive relations can be defined as

$$\left[1 - (e_1 a)^2 \nabla^2\right]\left[1 - (e_0 a)^2 \nabla^2\right] \sigma_{ij} = C_{ijkl}\left[1 - (e_1 a)^2 \nabla^2\right] \varepsilon_{kl} - C_{ijkl} l^2 \left[1 - (e_0 a)^2 \nabla^2\right] \nabla^2 \varepsilon_{kl} \quad (24)$$

$$\left[1 - \mu \nabla^2\right] \sigma_{ij} = C_{ijkl}\left[1 - \lambda \nabla^2\right] \varepsilon_{kl} \quad (25)$$

where $\mu = (ea)^2$ and $\lambda = l^2$.

In addition, the constitutive relations for a nonlocal shear deformable CNTRC curved sandwich nanobeam can be written as

$$\sigma_{xx} - \mu \frac{\partial^2 \sigma_{xx}}{\partial x^2} = \overline{Q}^k_{11}\left(\varepsilon_{xx} - \lambda \frac{\partial^2 \varepsilon_{xx}}{\partial x^2}\right) \quad (26)$$

$$\sigma_{xz} - \mu \frac{\partial^2 \sigma_{xx}}{\partial x^2} = \overline{Q}^k_{55}\left(\gamma_{xz} - \lambda \frac{\partial^2 \gamma_{xz}}{\partial x^2}\right) \quad (27)$$

Based on a nonlocal strain gradient theory, the following equilibrium equations are obtained in terms of the displacement components by substitution of Equation (19) into Equation (21).

$$\begin{aligned}
&\left(1 - \lambda \frac{\partial^2}{\partial x^2}\right)\left(A_{11} \frac{\partial^2 u_0}{\partial x^2} - B_{11} \frac{\partial^3 w_0}{\partial x^3} + \frac{A_{11}}{R} \frac{\partial w_0}{\partial x} + B^s_{11} \frac{\partial^2 \varphi_x}{\partial x^2} + \left(\frac{D_{11}}{R} + E_{12}\right) \frac{\partial \varphi_z}{\partial x}\right) = 0 \\
&\left(1 - \lambda \frac{\partial^2}{\partial x^2}\right)\left(\begin{array}{c} B_{11} \frac{\partial^3 u_0}{\partial x^3} - \frac{A_{11}}{R} \frac{\partial u_0}{\partial x} - F_{11} \frac{\partial^4 w_0}{\partial x^4} + \frac{2B_{11}}{R} \frac{\partial^2 w_0}{\partial x^2} - \frac{A_{11}}{R^2} w_0 \\ + F^s_{11} \frac{\partial^3 \varphi_x}{\partial x^3} - \frac{B^s_{11}}{R} \frac{\partial \varphi_x}{\partial x} + \left(\frac{D^s_{11}}{R} + J^s_{12}\right) \frac{\partial^2 \varphi_z}{\partial x^2} - \left(\frac{D_{11}}{R^2} + \frac{E_{12}}{R}\right) \varphi_z \end{array}\right) \\
&- \left(1 - \mu \frac{\partial^2}{\partial x^2}\right)\left(N^0_x \frac{\partial^2 w_0}{\partial x^2} - k_w w_0 - k_g \frac{\partial^2 w_0}{\partial x^2} - k_w w_0^3\right) = 0 \\
&\left(1 - \lambda \frac{\partial^2}{\partial x^2}\right)\left(B^s_{11} \frac{\partial^2 u_0}{\partial x^2} - F^s_{11} \frac{\partial^3 w_0}{\partial x^3} + \frac{B^s_{11}}{R} \frac{\partial w_0}{\partial x} + G^s_{11} \frac{\partial^2 \varphi_x}{\partial x^2} - K^s_{33} \varphi_x + \left(\frac{H^s_{11}}{R} + J^s_{12} - K^s_{33}\right) \frac{\partial \varphi_z}{\partial x}\right) = 0 \\
&\left(1 - \lambda \frac{\partial^2}{\partial x^2}\right)\left(\begin{array}{c} -\left(\frac{D_{11}}{R} - E_{12}\right) \frac{\partial u_0}{\partial x} + \left(\frac{D^s_{11}}{R} + E^s_{12}\right) \frac{\partial^2 w_0}{\partial x^2} - \left(\frac{D_{11}}{R^2} + \frac{E_{12}}{R}\right) w_0 \\ -\left(\frac{H^s_{11}}{R} + J^s_{12} - K^s_{33}\right) \frac{\partial \varphi_x}{\partial x} - \left(2\frac{L^s_{12}}{R} + \frac{K^s_{11}}{R^2} + L^s_{22}\right) \varphi_z + K^s_{33} \frac{\partial^2 \varphi_z}{\partial x^2} \end{array}\right) = 0
\end{aligned} \quad (28)$$

3.3. Temperature Field

In the present work we assume a uniform temperature field distribution on the CNTRC surfaces, labeled as T_m and T_p, on the bottom and top sandwich surfaces, respectively. A (10,10) SWCNT-based reinforcement is selected within the numerical investigation, with the same mechanical properties as assumed by Shen [4] and summarized in Table 2.

Table 2. Thermo-mechanical properties of SWCNTs.

T[K]	E^{cnt}_{11}[TPa]	E^{cnt}_{22}[TPa]	G^{cnt}_{12}[TPa]	v^{cnt}_{11}	$\alpha^{cnt}_{11}[10^{-6}]$/K	$\alpha^{cnt}_{22}[10^{-6}]$/K
300	5.6466	7.0800	1.9445	0.175	3.4584	5.1682
400	5.5679	6.9814	1.9703	0.175	4.1496	5.0905
500	5.5308	6.9348	1.9643	0.175	4.5361	5.0189
700	5.4744	6.8641	1.9644	0.175	4.6677	4.8943
1000	5.2814	6.6220	1.9451	0.175	4.2800	4.7532

To analyze the thermal effect on the buckling response of CNTRC curved sandwich nanobeams, we assume the following temperature-dependent material properties, in line with [53].

$$P = P_0\left(P_{-1}T^{-1} + 1 + P_1T + P_2T^2 + P_3T^3 + P_4T^4\right) \quad (29)$$

where $T = T_0 + \Delta T$, T_0 is the ambient temperature ($T_0 = 300$ K), ΔT is the temperature difference, and P_0, P_1, P_2, P_3 and P_4 are thermal coefficients listed in Table 3.

Table 3. Temperature-dependent coefficients of CNT material properties [22].

	P_0	P_{-1}	P_1	P_2	P_3	P_4
E_{11}^{cnt} [TPa]	6.5653	0	-8.9437×10^{-4}	1.9182×10^{-6}	-1.8198×10^{-9}	6.0043×10^{-13}
E_{22}^{cnt} [TPa]	8.2271	0	-8.9024×10^{-4}	1.9066×10^{-6}	-1.8063×10^{-9}	5.9486×10^{-13}
G_{12}^{cnt} [TPa]	1.1056	0	5.6727×10^{-3}	-1.4815×10^{-5}	1.6402×10^{-8}	-6.5007×10^{-12}
α_{11} [$10^{-6}/^\circ$C]	-1.1279	0	-2.0340×10^{-2}	2.5672×10^{-5}	-1.0186×10^{-8}	5.9455×10^{-14}
α_{22} [$10^{-6}/^\circ$C]	5.4359	0	-1.7906×10^{-4}	4.6367×10^{-8}	1.2424×10^{-11}	-5.3290×10^{-14}
ν_{12}^{cnt}	0.175	0	0	0	0	0

The polymeric matrix (PmPV) features temperature-dependent elastic properties, as follows:

$$E_m = (3.51 - 0.0047T) \text{ GPa} \quad (30)$$

$$\alpha_m = 45(1 + 0.0005\Delta T)10^{-6} \text{ GPa} \quad (31)$$

where the Poisson's ratio and mass density are set as $\nu_m = 0.34$ and $\rho_m = 1150$ kg/m^3, respectively.

4. Analytical Solution

In this section, the equilibrium equations are solved analytically using the Galerkin method for simply-supported (SS), clamped-clamped (CC) and clamped-hinged (CS) boundary conditions. The following displacement functions are thus assumed:

$$\left\{\begin{array}{c} u_0 \\ w_0 \\ \varphi_x \\ \varphi_z \end{array}\right\} = \sum_{m=1}^{\infty} \left\{\begin{array}{c} U_m \frac{\partial X_m}{\partial x} \\ W_m X_m \\ \psi_{xm} \frac{\partial X_m}{\partial x} \\ \psi_{zm} X_m \end{array}\right\} \quad (32)$$

with U_m, W_m, ψ_{xm} and ψ_{zm} being arbitrary parameters. The functions $X_m(x)$ that satisfy the selected boundary conditions are defined as

- For SS beam

$$X_m = \sin(\beta x), \quad \beta = \frac{m\pi}{L} \quad (33)$$

- For CC beam

$$X_m = 1 - \cos(\beta x), \quad \beta = \frac{2m\pi}{L} \quad (34)$$

- For CS beam

$$X_m = \sin(\beta x)[\cos(\beta x) - 1], \quad \beta = \frac{m\pi}{L} \quad (35)$$

By substituting Equation (32) in Equation (28), we get

$$[K_{ij}]\left\{\begin{array}{c} U_m \\ W_m \\ \psi_{xm} \\ \psi_{zm} \end{array}\right\} = 0, \quad i,j = 1:4 \tag{36}$$

where

$$K_{11} = A_{11}\left(\int_0^L \frac{\partial^3 X_m}{\partial x^3}\frac{\partial X_m}{\partial x}dx - \lambda\int_0^L \frac{\partial^5 X_m}{\partial x^5}\frac{\partial X_m}{\partial x}dx\right)$$

$$K_{12} = -B_{11}\left(\int_0^L \frac{\partial^3 X_m}{\partial x^3}\frac{\partial X_m}{\partial x}dx - \lambda\int_0^L \frac{\partial^5 X_m}{\partial x^5}\frac{\partial X_m}{\partial x}dx\right) + \frac{A_{11}}{R}\left(\int_0^L \left(\frac{\partial X_m}{\partial x}\right)^2 dx - \lambda\int_0^L \frac{\partial^3 X_m}{\partial x^3}\frac{\partial X_m}{\partial x}dx\right)$$

$$K_{13} = B_{11}^s\left(\int_0^L \frac{\partial^3 X_m}{\partial x^3}\frac{\partial X_m}{\partial x}dx - \lambda\int_0^L \frac{\partial^5 X_m}{\partial x^5}\frac{\partial X_m}{\partial x}dx\right)$$

$$K_{14} = \left(\frac{D_{11}}{R} + E_{12}\right)\left(\int_0^L \left(\frac{\partial X_m}{\partial x}\right)^2 dx - \lambda\int_0^L \frac{\partial^3 X_m}{\partial x^3}\frac{\partial X_m}{\partial x}dx\right)$$

$$K_{21} = B_{11}\left(\int_0^L \frac{\partial^4 X_m}{\partial x^4}X_m dx - \lambda\int_0^L \frac{\partial^6 X_m}{\partial x^6}X_m dx\right) - \frac{A_{11}}{R}\left(\int_0^L \frac{\partial^2 X_m}{\partial x^2}X_m dx - \lambda\int_0^L \frac{\partial^4 X_m}{\partial x^4}X_m dx\right)$$

$$K_{22} = -F_{11}\left(\int_0^L \frac{\partial^4 X_m}{\partial x^4}X_m dx - \lambda\int_0^L \frac{\partial^6 X_m}{\partial x^6}X_m dx\right) + 2\frac{B_{11}}{R}\left(\int_0^L \frac{\partial^2 X_m}{\partial x^2}X_m dx - \lambda\int_0^L \frac{\partial^4 X_m}{\partial x^4}X_m dx\right) - (N_x^0 - k_g)\left(\int_0^L \frac{\partial^2 X_m}{\partial x^2}X_m dx - \mu\int_0^L \frac{\partial^4 X_m}{\partial x^4}X_m dx\right) - k_w\left(\int_0^L X_m^2 dx - \mu\int_0^L \frac{\partial^2 X_m}{\partial x^2}X_m dx\right) - k_{NL}\left(\int_0^L X_m^4 dx - \mu\int_0^L \frac{\partial^2 X_m^3}{\partial x^2}X_m dx\right) \tag{37}$$

$$K_{23} = F_{11}^s\left(\int_0^L \frac{\partial^4 X_m}{\partial x^4}X_m dx - \lambda\int_0^L \frac{\partial^6 X_m}{\partial x^6}X_m dx\right) - \frac{B_{11}^s}{R}\left(\int_0^L \frac{\partial^2 X_m}{\partial x^2}X_m dx - \lambda\int_0^L \frac{\partial^4 X_m}{\partial x^4}X_m dx\right)$$

$$K_{24} = \left(\frac{D_{11}^s}{R} + E_{12}^s\right)\left(\int_0^L \frac{\partial^2 X_m}{\partial x^2}X_m dx - \lambda\int_0^L \frac{\partial^4 X_m}{\partial x^4}X_m dx\right) - \left(\frac{D_{11}}{R^2} + \frac{E_{12}}{R}\right)\left(\int_0^L X_m^2 dx - \lambda\int_0^L \frac{\partial^2 X_m}{\partial x^2}X_m dx\right)$$

$$K_{31} = B_{11}^s\left(\int_0^L \frac{\partial^3 X_m}{\partial x^3}\frac{\partial X_m}{\partial x}dx - \lambda\int_0^L \frac{\partial^5 X_m}{\partial x^5}\frac{\partial X_m}{\partial x}dx\right)$$

$$K_{32} = -F_{11}^s\left(\int_0^L \frac{\partial^3 X_m}{\partial x^3}\frac{\partial X_m}{\partial x}dx - \lambda\int_0^L \frac{\partial^5 X_m}{\partial x^5}\frac{\partial X_m}{\partial x}dx\right) - \frac{B_{11}^s}{R}\left(\int_0^L \frac{\partial X_m}{\partial x}^2 dx - \lambda\int_0^L \frac{\partial^3 X_m}{\partial x^3}\frac{\partial X_m}{\partial x}dx\right)$$

$$K_{33} = G_{11}^s\left(\int_0^L \frac{\partial^3 X_m}{\partial x^3}\frac{\partial X_m}{\partial x}dx - \lambda\int_0^L \frac{\partial^5 X_m}{\partial x^5}\frac{\partial X_m}{\partial x}dx\right) - K_{33}^s\left(\int_0^L \left(\frac{\partial X_m}{\partial x}\right)^2 dx - \lambda\int_0^L \frac{\partial^3 X_m}{\partial x^3}\frac{\partial X_m}{\partial x}dx\right)$$

$$K_{34} = \left(\frac{H_{11}^s}{R} + J_{12}^s - K_{33}^s\right)\left(\int_0^L \left(\frac{\partial X_m}{\partial x}\right)^2 dx - \lambda\int_0^L \frac{\partial^3 X_m}{\partial x^3}\frac{\partial X_m}{\partial x}dx\right)$$

$$K_{41} = -\left(\frac{D_{11}}{R} + E_{12}\right)\left(\int_0^L \frac{\partial^2 X_m}{\partial x^2}X_m dx - \lambda\int_0^L \frac{\partial^4 X_m}{\partial x^4}X_m dx\right)$$

$$K_{42} = \left(\frac{D_{11}^s}{R} + E_{12}^s\right)\left(\int_0^L \frac{\partial^2 X_m}{\partial x^2}X_m dx - \lambda\int_0^L \frac{\partial^4 X_m}{\partial x^4}X_m dx\right) - \left(\frac{D_{11}}{R^2} + \frac{E_{12}}{R}\right)\left(\int_0^L X_m^2 dx - \lambda\int_0^L \frac{\partial^2 X_m}{\partial x^2}X_m dx\right)$$

$$K_{43} = -\left(\frac{H_{11}^s}{R} + J_{12}^s - K_{33}^s\right)\left(\int_0^L \frac{\partial^2 X_m}{\partial x^2}X_m dx - \lambda\int_0^L \frac{\partial^4 X_m}{\partial x^4}X_m dx\right)$$

$$K_{44} = -\left(2\frac{L_{12}^s}{R} + \frac{K_{11}^s}{R^2} + L_{22}^s\right)\left(\int_0^L X_m^2 dx - \lambda\int_0^L \frac{\partial^2 X_m}{\partial x^2}X_m dx\right) + K_{33}^s\left(\int_0^L \frac{\partial^2 X_m}{\partial x^2}X_m dx - \lambda\int_0^L \frac{\partial^4 X_m}{\partial x^4}X_m dx\right)$$

The accuracy of the proposed theoretical solution is explored in the next section, within a large systematic investigation aimed at determining the sensitivity of the buckling response. The proposed model is limited to uniform cross-sectional curved FG-CNTRC nanobeams with SS, SC, and CC boundary conditions and linear variation of temperature across the beam thickness; a further expansion should include more complicated cross-sectional geometries and thermal variations.

5. Results and Discussion

In this section, various numerical applications are presented to determine the accuracy of a quasi-3D HSDT, to solve the buckling problem of FG-CNTRC straight sandwich beams, compared to some existing solutions from the literature. Then, we investigate the effect of curvature on the structural response of CNTRC sandwich beams, which could be of great interest for design purposes, among different engineering applications. In what follows, the critical buckling load and elastic foundation parameters are presented in dimensionless form, as follows:

$$\overline{N} = R^2 \frac{N_x^0}{A_{110}}, \qquad K_w = \frac{k_w L^2}{A_{110}}, \qquad K_g = \frac{k_g}{A_{110}}, \qquad K_{NL} = \frac{k_{NL} L^2}{A_{110}} \qquad (38)$$

where the coefficient A_{110} refers to a beam made of pure matrix material at room temperature $T = 300$ K. The length of the curved sandwich beam is kept equal to $L = 20$ for all the numerical examples.

5.1. Comparison Study

We start the numerical analysis by a comparative evaluation of our results with predictions from the open literature, while including possible thickness stretching effects. In Table 4, we summarize the results in terms of dimensionless critical buckling load for SS- and CC-CNTRC sandwich beams with and without thickness stretching effects and compare their accuracy against the numerical predictions by Wu et al. [6], based on a differential quadrature method (DQM). The face sheets are made of poly methyl methacrylate (PMMA) as matrix, with $E_m = 2.5$ GPa and $\nu_m = 0.3$, and armchair (10, 10) SWCNTs as reinforcement phase, with $E_{11}^{cnt} = 5.6466$ TPa, $E_{22}^{cnt} = 7.08$ TPa, $G_{12}^{cnt} = 1.9445$ TPa and $\nu_{cnt} = 0.175$ (in 300 K). Titanium alloy (Ti-6Al-4V) is used as core, with $E_m = 113.8$ GPa and $\nu_m = 0.342$. It is worth noticing the good correlation between our results (see Table 4) and the findings of [6] when the thickness stretching effect is neglected.

5.2. Parametric Study

The parametric study in this section assumes a PmPV as core material and as matrix phase for the face sheets of the sandwich structure, with mechanical properties as specified in Equations (30) and (31); (10,10) SWCNTs are considered as the reinforcement phase (Table 3). The mechanical properties of materials depend on the temperature. Table 5 presents the effect of the dimensionless thickness ratio L/h on the buckling load of a single layer CNTRC curved beam with various CNT volume fractions in the presence (or absence) of a thickness stretching effect ε_{zz}, while keeping the opening angle $\alpha = L/R$ equal to $\pi/3$. Note that increased values of L/h result in lower values of the buckling load, under the same assumptions for the reinforcement distribution, volume fraction and possible stretching effects. In any case, the worst buckling response is observed for an FG-O reinforcement distribution within the material, whereas a FG-X distribution seems to yield the highest buckling loads for fixed values of L/h, ε_{zz}, V_{cnt}^*. The stability of the curved beam increases significantly for higher values of V_{cnt}^*, with a small variation in the buckling load, depending on whether ε_{zz} is assumed (or not) equal to zero.

Table 4. Comparisons of dimensionless critical buckling loads for FG-CNTRC straight beams h_c/h_f, $V^*_{cnt} = 0.12$.

	L/h		CC			SS		
			V^*_{cnt}=12	V^*_{cnt}=17	V^*_{cnt}=28	V^*_{cnt}=12	V^*_{cnt}=17	V^*_{cnt}=28
UD	10	Wu [6]	0.0254	0.0296	0.0373	0.0070	0.0082	0.0107
		Present $\varepsilon_{zz}=0$	0.0271	0.0319	0.0413	0.0071	0.0084	0.0110
		Present $\varepsilon_{zz}\neq 0$	0.0267	0.0316	0.0410	0.0066	0.0080	0.0106
	20	Wu [6]	0.0070	0.0082	0.0107	0.0018	0.0021	0.0028
		Present $\varepsilon_{zz}=0$	0.0071	0.0084	0.0110	0.0018	0.0021	0.0028
		Present $\varepsilon_{zz}\neq 0$	0.0069	0.0082	0.0108	0.0017	0.0020	0.0027
	30	Wu [6]	0.0031	0.0037	0.0049	0.0008	0.0009	0.0012
		Present $\varepsilon_{zz}=0$	0.0032	0.0038	0.0049	0.0008	0.0009	0.0012
		Present $\varepsilon_{zz}\neq 0$	0.0031	0.0037	0.0049	0.0007	0.0008	0.0012
FG	10	Wu [6]	0.0261	0.0305	0.0387	0.0072	0.0085	0.0111
		Present $\varepsilon_{zz}=0$	0.0271	0.0319	0.0413	0.0071	0.0084	0.0110
		Present $\varepsilon_{zz}\neq 0$	0.0267	0.0316	0.0410	0.0066	0.0079	0.0106
	20	Wu [6]	0.0072	0.0085	0.0111	0.0018	0.0022	0.0029
		Present $\varepsilon_{zz}=0$	0.0071	0.0084	0.0110	0.0018	0.0021	0.0028
		Present $\varepsilon_{zz}\neq 0$	0.0069	0.0082	0.0108	0.0017	0.0020	0.0027
	30	Wu [6]	0.0032	0.0039	0.0051	0.0008	0.0010	0.0013
		Present $\varepsilon_{zz}=0$	0.0032	0.0038	0.0049	0.0008	0.0010	0.0012
		Present $\varepsilon_{zz}\neq 0$	0.0031	0.0037	0.0049	0.0007	0.0009	0.0012

Table 5. Effect of thickness ratio on the buckling load of a single layer CNTRC curved beam $\alpha = \frac{\pi}{3}$, T = 300 K.

	L/h	V^*_{cnt}=12		V^*_{cnt}=17		V^*_{cnt}=28	
		$\varepsilon_{zz}=0$	$\varepsilon_{zz}\neq 0$	$\varepsilon_{zz}=0$	$\varepsilon_{zz}\neq 0$	$\varepsilon_{zz}=0$	$\varepsilon_{zz}\neq 0$
UD	5	73.7930	73.4424	120.6917	120.1610	146.3642	145.5590
	10	49.0266	49.0250	77.6401	77.6399	101.4712	101.4484
	20	21.9451	21.9242	33.2103	33.1708	48.7346	48.7136
	30	11.4565	11.4369	17.0329	16.9992	26.2191	26.1931
FG-X	5	79.5433	79.1094	128.5687	127.9463	149.0114	148.3479
	10	57.1285	57.1134	90.5184	90.5055	111.0156	110.9778
	20	28.9721	28.9611	44.1395	44.1160	61.6479	61.6372
	30	15.9804	15.9653	23.8972	23.8700	35.6810	35.6608
FG-O	5	58.0980	57.9593	96.1446	95.9410	128.1600	127.5412
	10	33.6793	33.6650	52.7577	52.7221	75.9965	75.9952
	20	12.7261	12.6870	18.9830	18.9144	29.3401	29.2905
	30	6.2518	6.2261	9.1882	9.1452	14.5124	14.4755

In Table 6, we account for the influence of opening angles α, boundary conditions, and CNT reinforcement patterns on the dimensionless critical buckling load of $(0°/90°/c/90°/0°)$

sandwich beams. Note that the critical buckling load increases significantly for a decreased opening angle and increased CNT volume fraction. As summarized in Table 7, the dimensionless critical buckling load of curved sandwich $(0°/90°/0°/c/0°/90°/0°)$ nanobeams could be affected by nonlocal and length scale parameters as well as by the core-to-face sheet thickness ratio, h_c/h_f, and thermal condition. A meaningful reduction of the critical buckling load is observed for higher temperatures for a fixed geometry and nonlocal parameters μ, λ. An increased value of μ and a reduced value of λ reduce the critical buckling load of the nanostructure under the same thermal and geometric assumptions. Moreover, Table 8 summarizes the sensitivity of the buckling response of CNTRC sandwich $(0°/c/0°)$ beams to different elastic foundation parameters and boundary conditions, with an increased stability of the structure for more rigid boundary conditions and foundation.

Table 6. Effect of opening angle on the dimensionless buckling load of curved sandwich beam $(0°/90°/c/90°/0°)$ ($h_c/h_f = 4$, $h = L/10$, $T = 300$ K).

	α	SS			CC			CS		
		$V^*_{cnt}=12$	$V^*_{cnt}=17$	$V^*_{cnt}=28$	$V^*_{cnt}=12$	$V^*_{cnt}=17$	$V^*_{cnt}=28$	$V^*_{cnt}=12$	$V^*_{cnt}=17$	$V^*_{cnt}=28$
UD	$\pi/4$	74.6585	100.3288	139.9486	339.9366	442.8754	611.3111	212.5558	278.2913	381.2890
	$\pi/3$	41.9954	56.4350	78.7211	257.7059	343.7374	492.5488	149.4863	199.1205	281.3863
	$\pi/2$	18.6646	25.0822	34.9871	198.9635	272.9172	407.7090	104.4352	142.5681	210.0248
	$2\pi/3$	10.4989	14.1087	19.6803	178.3950	248.1198	378.0018	88.6652	122.7722	185.0451
FG-X	$\pi/4$	74.8276	100.6190	140.6169	340.6800	444.0969	613.9162	213.0218	279.0764	383.0227
	$\pi/3$	42.0905	56.5982	79.0970	258.1393	344.4732	494.1862	149.7553	199.5840	282.4389
	$\pi/2$	18.7069	25.1548	35.1542	199.1755	273.3061	408.6553	104.5634	142.8020	210.5910
	$2\pi/3$	10.5226	14.1496	19.7743	178.5296	248.3875	378.7063	88.7442	122.9258	185.4411
FG-O	$\pi/4$	74.5146	100.1307	139.7518	339.3283	442.1373	610.8484	212.1712	277.8003	380.9295
	$\pi/3$	41.9144	56.3235	78.6104	257.3788	343.3707	492.4602	149.2768	198.8661	281.2614
	$\pi/2$	18.6286	25.0327	34.9379	198.8371	272.8157	407.8875	104.3507	142.4828	210.0675
	$2\pi/3$	10.4786	14.0809	19.6526	178.3388	248.1110	378.2735	88.6244	122.7461	185.1463

Table 7. Effect of nonlocal and length scale parameter on the dimensionless buckling load of simply supported UD-CNTRC curved sandwich nanobeam $(0°/90°/0°/c/0°/90°/0°)$ ($\alpha = \pi/3$, $h = L/10$, $V^*_{cnt} = 28$).

			h_c/h_f							
μ		T = 300 K			T = 500 K			T = 700 K		
	λ	4	6	8	4	6	8	4	6	8
0	0	81.8686	73.5399	66.3319	61.1267	56.6032	52.3027	19.0516	18.8260	18.5363
	1	83.8886	75.3545	67.9686	62.6350	57.9999	53.5932	19.5217	19.2906	18.9937
	2	85.9086	77.1690	69.6053	64.1432	59.3965	54.8837	19.9918	19.7551	19.4511
	3	87.9287	78.9835	71.2419	65.6515	60.7931	56.1743	20.4618	20.2196	19.9084
1	0	79.8972	71.7691	64.7347	59.6548	55.2402	51.0433	18.5928	18.3727	18.0900
	1	81.8686	73.5399	66.3319	61.1267	56.6032	52.3027	19.0516	18.8260	18.5363
	2	83.8400	75.3108	67.9292	62.5987	57.9662	53.5622	19.5104	19.2794	18.9827
	3	85.8114	77.0816	69.5265	64.0706	59.3292	54.8216	19.9691	19.7327	19.4290
2	0	78.0185	70.0816	63.2125	58.2521	53.9413	49.8431	18.1557	17.9407	17.6646
	1	79.9436	71.8107	64.7722	59.6894	55.2723	51.0729	18.6036	18.3834	18.1005
	2	81.8686	73.5399	66.3319	61.1267	56.6032	52.3027	19.0516	18.8260	18.5363
	3	83.7936	75.2691	67.8916	62.5640	57.9342	53.5325	19.4996	19.2687	18.9722
3	0	76.2262	68.4715	61.7603	56.9139	52.7021	48.6980	17.7386	17.5285	17.2588
	1	78.1070	70.1610	63.2842	58.3181	54.0025	49.8996	18.1762	17.9610	17.6847
	2	79.9878	71.8505	64.8081	59.7224	55.3029	51.1011	18.6139	18.3935	18.1105
	3	81.8686	73.5399	66.3319	61.1267	56.6032	52.3027	19.0516	18.8260	18.5363

Table 8. Effect of hardening nonlinear parameters on the dimensionless buckling load of CNTRC curved sandwich beams $(0°/c/0°)$ ($\alpha = \pi/3$, $h = L/10$, $h_c/h_f = 4$, $V_{cnt}^* = 0.12$, $T = 300$ K).

K_w	K_g	K_{nl}	SS			CC			CS		
			UD	FG-X	FG-O	UD	FG-X	FG-O	UD	FG-X	FG-O
0	0	0	59.3208	59.9622	58.6879	406.4881	408.7038	404.3375	229.0790	230.5969	227.5994
		0.05	60.7067	61.3482	60.0738	410.5303	412.7460	408.3797	230.5082	232.0261	229.0286
		0.1	62.0926	62.7341	61.4597	414.5725	416.7882	412.4220	231.9374	233.4554	230.4578
	0.05	0	77.5586	78.2001	76.9257	424.7259	426.9416	422.5753	247.3168	248.8347	245.8372
		0.05	78.9446	79.5860	78.3117	428.7681	430.9838	426.6175	248.7460	250.2640	247.2664
		0.1	80.3305	80.9719	79.6976	432.8103	435.0261	430.6598	250.1752	251.6932	248.6957
	0.1	0	95.7965	96.4379	95.1636	442.9637	445.1794	440.8131	265.5546	267.0725	264.0750
		0.05	97.1824	97.8238	96.5495	447.0059	449.2216	444.8554	266.9838	268.5018	265.5042
		0.1	98.5683	99.2097	97.9354	451.0482	453.2639	448.8976	268.4130	269.9310	266.9335
0.05	0	0	61.1687	61.8101	60.5358	407.8740	410.0897	405.7234	230.2339	231.7518	228.7543
		0.05	62.5546	63.1960	61.9217	411.9162	414.1319	409.7656	231.6631	233.1811	230.1835
		0.1	63.9405	64.5819	63.3076	415.9584	418.1742	413.8079	233.0923	234.6103	231.6128
	0.05	0	79.4065	80.0479	78.7736	426.1118	428.3275	423.9612	248.4717	249.9897	246.9921
		0.05	80.7924	81.4338	80.1595	430.1540	432.3697	428.0034	249.9009	251.4189	248.4214
		0.1	82.1783	82.8197	81.5454	434.1962	436.4120	432.0457	251.3301	252.8481	249.8506
	0.1	0	97.6443	98.2857	97.0114	444.3496	446.5653	442.1990	266.7095	268.2275	265.2300
		0.05	99.0302	99.6717	98.3973	448.3918	450.6076	446.2413	268.1387	269.6567	266.6592
		0.1	100.4162	101.0576	99.7833	452.4341	454.6498	450.2835	269.5680	271.0859	268.0884
0.1	0	0	63.0166	63.6580	62.3837	409.2599	411.4756	407.1093	231.3888	232.9068	229.9092
		0.05	64.4025	65.0439	63.7696	413.3021	415.5178	411.1515	232.8180	234.3360	231.3385
		0.1	65.7884	66.4298	65.1555	417.3443	419.5601	415.1938	234.2473	235.7652	232.7677
	0.05	0	81.2544	81.8958	80.6215	427.4977	429.7134	425.3471	249.6266	251.1446	248.1471
		0.05	82.6403	83.2817	82.0074	431.5399	433.7556	429.3894	251.0559	252.5738	249.5763
		0.1	84.0262	84.6676	83.3933	435.5822	437.7979	433.4316	252.4851	254.0030	251.0055
	0.1	0	99.4922	100.1336	98.8593	445.7355	447.9512	443.5849	267.8645	269.3824	266.3849
		0.05	100.8781	101.5195	100.2452	449.7777	451.9935	447.6272	269.2937	270.8116	267.8141
		0.1	102.2640	102.9054	101.6311	453.8200	456.0357	451.6694	270.7229	272.2408	269.2433

Figure 3 also depicts the buckling response for a SS $(0°/90°/c/90°/0°)$ beam versus the thickness ratio, L/h, while varying the opening angles. All the plots in Figure 3 feature a monotone decreasing behavior for increasing values of L/h, reaching a plateau for $L/h \geq 30$. Note also that an increased opening angle value decreases significantly the buckling load of the structure for each fixed value of L/h.

In Figure 4 the critical buckling load versus the opening angle is illustrated, taking into account the core-to-face sheet thickness ratio variation. A clear reduction of the beam stiffness with an increased core layer can be observed for each fixed opening angle, which is even more pronounced for lower values of the opening angles.

Figure 5 also shows the double effect of the core-to-face sheet thickness ratio and CNT volume fraction on the dimensionless buckling load, with a clear shift of the curve upwards for increasing values of V_{cnt}. The highest critical buckling load is reached for a volume fraction $V_{cnt} = 28$, where the lowest stability is observed for $V_{cnt} = 12$. The impact of the thermal environment on critical buckling load is visible in Figure 6, where an increased temperature value leads to a clear reduction in the buckling load for all the selected boundary conditions because of the thermal dependence of the mechanical properties of the materials. As also expected, the highest stability is reached by CC sandwich beams, independently of the thermal environment. The further effect of nonlocal μ and length scale λ parameters on the critical buckling load is also plotted in Figures 7 and 8, respectively. One can easily note that the buckling load increases by decreasing the nonlocal parameter and by increasing the length scale parameter, in line with the information in Table 7. Unlike the length scale parameter λ, an increased nonlocal parameter μ leads to a stiffness reduction of CNTRC laminated nanobeams. The critical buckling load versus the thickness ratio L/h is finally illustrated in Figure 9 by assuming different elastic foundation parameters. An increased thickness ratio L/h leads to a monotone reduction of the buckling load, with a meaningful effect of the shear foundation parameter K_g on the buckling results.

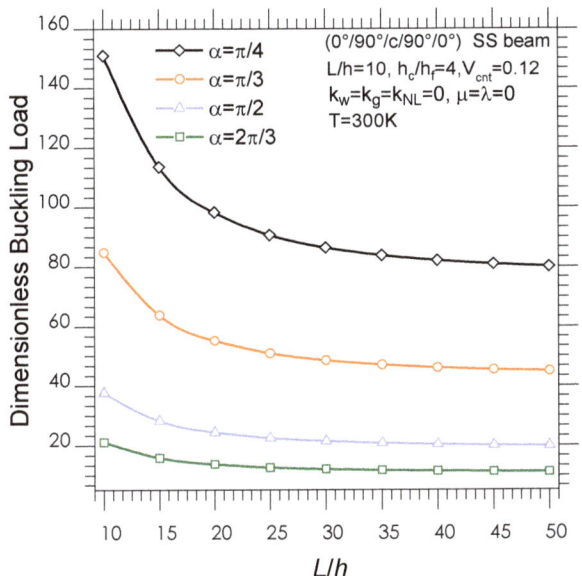

Figure 3. Dimensionless buckling load versus thickness ratio.

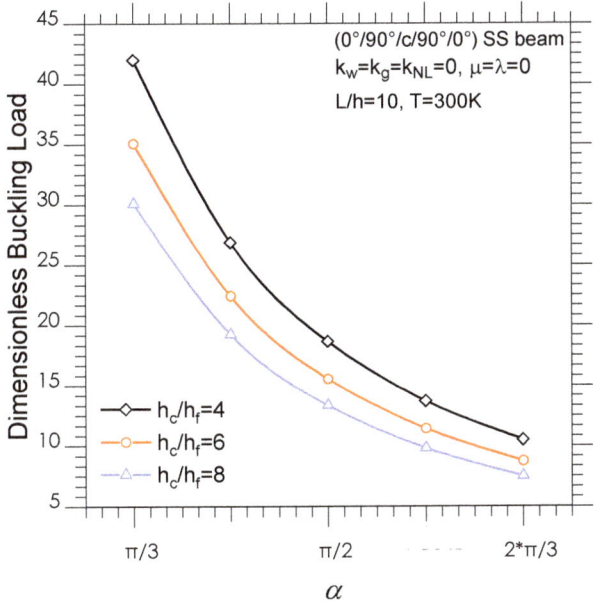

Figure 4. Dimensionless buckling load versus opening angle.

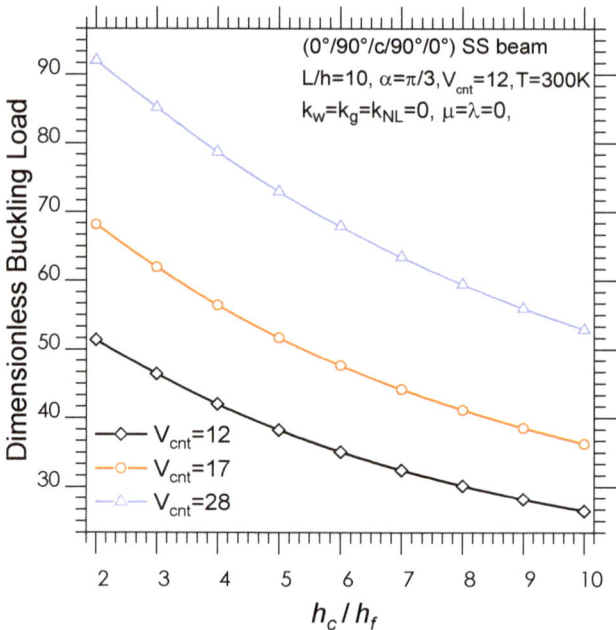

Figure 5. Dimensionless buckling load versus the core-to-face sheet thickness ratio.

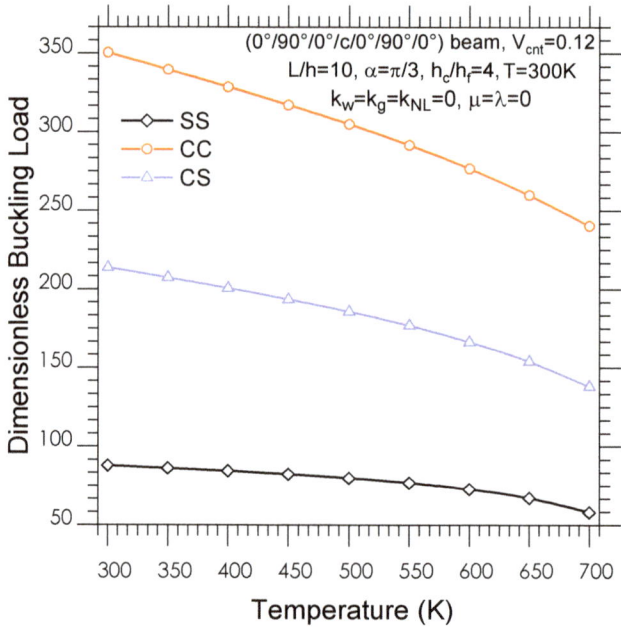

Figure 6. Dimensionless buckling load versus temperature.

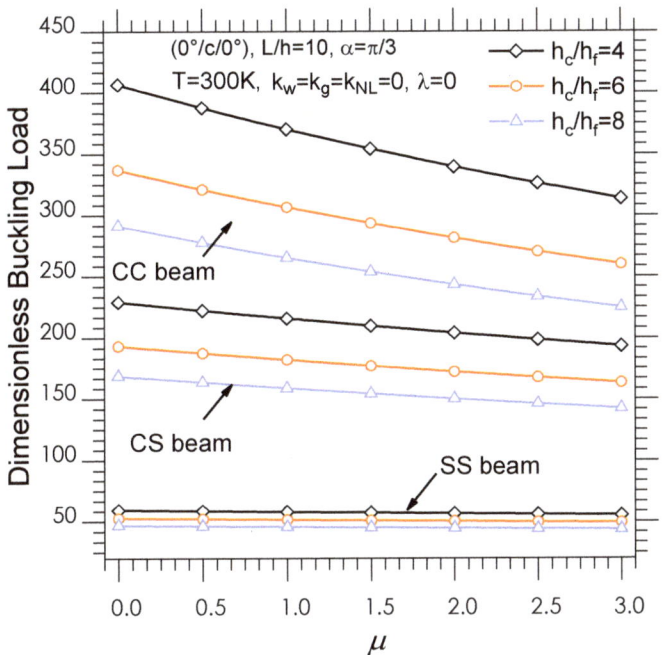

Figure 7. Dimensionless buckling load versus the nonlocal parameter.

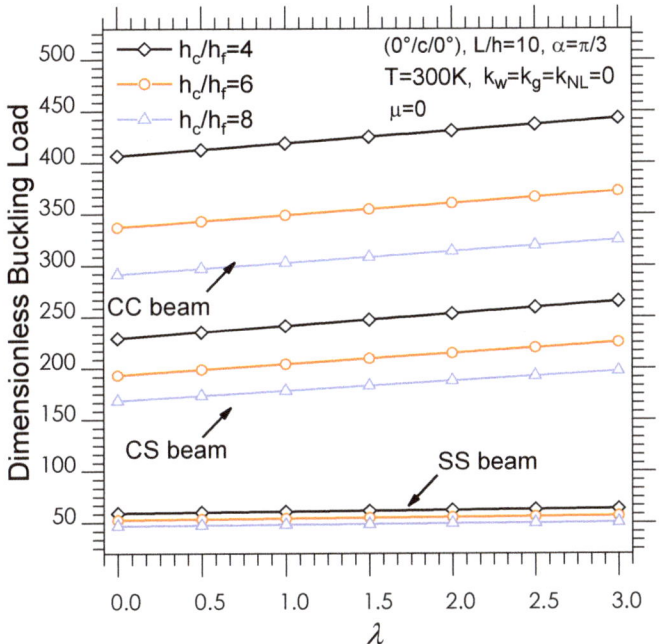

Figure 8. Dimensionless buckling load versus the length scale parameter.

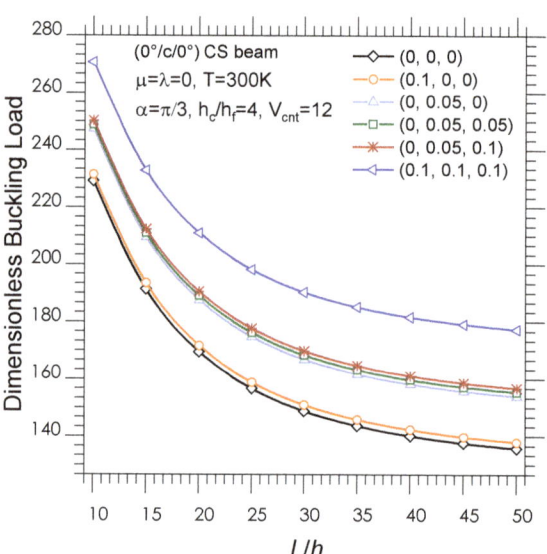

Figure 9. Effect of thickness ratio and elastic foundation on the dimensionless buckling load.

6. Conclusions

A novel quasi-3D higher-order shear deformation theory was proposed in this work to study the buckling response of CNTRC curved sandwich nanobeams for the first time. The problem was tackled theoretically, based on a Galerkin procedure, accounting for different boundary conditions and size-dependent effects. The material properties of CNTRC sheets were here assumed to be temperature-dependent, in agreement with the Touloukian principle.

A parametric study was performed systematically, to check for the influence of some significant parameters on the buckling response of CNTRC curved sandwich nanobeams, namely the CNTs reinforcement patterns and the nonlocal and length scale parameter, together with the geometric parameters. Based on the parametric investigation, it seems that the critical buckling load decreases for an increased temperature because of a global reduction in the stiffness of CNTRC curved sandwich nanobeams. Possible size effects can reduce the overall stiffness of CNTRC curved sandwich nanobeams, whereby the dimensionless critical buckling load decreases for an increased nonlocal parameter μ. Unlike the nonlocality effect, an increased length scale parameter λ leads to an increased buckling stability. More flexible elastic foundations and boundary conditions can reduce significantly the overall structural stability, which is also largely affected by a varying core-to-face sheet thickness ratio h_c/h_f, opening angle α, and CNT volume fractions. The results obtained by neglecting the effect of thickness stretching ($\varepsilon = 0$) are perfectly in line with predictions from the literature, thus confirming the good accuracy of the proposed method to handle similar problems. The results obtained in this work, could represent valid benchmarks for engineers and researchers to validate different numerical methods as well as for practical design purposes of nanostructures.

Author Contributions: Conceptualization, A.A.D., M.S.A.H., B.K., R.D. and F.T.; Data curation, A.A.D. and B.K.; Formal analysis, M.S.A.H., B.K., R.D. and F.T.; Investigation, A.A.D., R.D. and F.T.; Methodology, M.S.A.H., M.A.E., R.D. and F.T.; Supervision, R.D. and F.T.; Validation, A.A.D., B.K., M.A.E., R.D. and F.T.; Writing—original draft, A.A.D., M.S.A.H., B.K. and M.A.E.; Writing—review & editing, R.D. and F.T. All authors have read and agreed to the published version of the manuscript.

Funding: This research receive no funding.

Conflicts of Interest: The authors declare no conflict of interest.

References

1. Griebel, M.; Hamaekers, J. Molecular dynamics simulations of the elastic moduli of polymer–carbon nanotube composites. *Comput. Methods Appl. Mech. Eng.* **2004**, *193*, 1773–1788. [CrossRef]
2. Han, Y.; Elliott, J. Molecular dynamics simulations of the elastic properties of polymer/carbon nanotube composites. *Comput. Mater. Sci.* **2007**, *39*, 315–323. [CrossRef]
3. Fidelus, J.; Wiesel, E.; Gojny, F.; Schulte, K.; Wagner, H. Thermo-mechanical properties of randomly oriented carbon/epoxy nanocomposites. *Compos. Part A Appl. Sci. Manuf.* **2005**, *36*, 1555–1561. [CrossRef]
4. Shen, H.-S. Nonlinear bending of functionally graded carbon nanotube-reinforced composite plates in thermal environments. *Compos. Struct.* **2009**, *91*, 9–19. [CrossRef]
5. Lim, C.; Zhang, G.; Reddy, J. A higher-order nonlocal elasticity and strain gradient theory and its applications in wave propagation. *J. Mech. Phys. Solids* **2015**, *78*, 298–313. [CrossRef]
6. Wu, H.; Kitipornchai, S.; Yang, J. Free Vibration and Buckling Analysis of Sandwich Beams with Functionally Graded Carbon Nanotube-Reinforced Composite Face Sheets. *Int. J. Struct. Stab. Dyn.* **2015**, *15*, 1540011. [CrossRef]
7. Eltaher, M.; Khater, M.; Park, S.; Abdel-Rahman, E.; Yavuz, M. On the static stability of nonlocal nanobeams using higher-order beam theories. *Adv. Nano Res.* **2016**, *4*, 51–64. [CrossRef]
8. Ebrahimi, F.; Farazmandnia, N. Vibration analysis of functionally graded carbon nanotube-reinforced composite sandwich beams in thermal environment. *Adv. Aircr. Spacecr. Sci.* **2018**, *5*, 107. [CrossRef]
9. Sobhy, M.; Zenkour, A.M. Magnetic field effect on thermomechanical buckling and vibration of viscoelastic sandwich nanobeams with CNT reinforced face sheets on a viscoelastic substrate. *Compos. Part B Eng.* **2018**, *154*, 492–506. [CrossRef]
10. Daikh, A.A.; Megueni, A. Thermal buckling analysis of functionally graded sandwich plates. *J. Therm. Stress.* **2018**, *41*, 139–159. [CrossRef]
11. Arefi, M.; Arani, A.H.S. Higher order shear deformation bending results of a magnetoelectrothermoelastic functionally graded nanobeam in thermal, mechanical, electrical, and magnetic environments. *Mech. Based Des. Struct. Mach.* **2018**, *46*, 669–692. [CrossRef]
12. Bekhadda, A.; Cheikh, A.; Bensaid, I.; Hadjoui, A.; Daikh, A.A. A novel first order refined shear-deformation beam theory for vibration and buckling analysis of continuously graded beams. *Adv. Aircr. Spacecr. Sci.* **2019**, *6*, 189–206. [CrossRef]
13. Medani, M.; Benahmed, A.; Zidour, M.; Heireche, H.; Tounsi, A.; Bousahla, A.A.; Tounsi, A.; Mahmoud, S.R. Static and dynamic behavior of (FG-CNT) reinforced porous sandwich plate using energy principle. *Steel Compos. Struct.* **2019**, *32*, 595–610. [CrossRef]
14. Arani, A.G.; Pourjamshidian, M.; Arefi, M.; Arani, M.R. Thermal, electrical and mechanical buckling loads of sandwich nano-beams made of FG-CNTRC resting on Pasternak's foundation based on higher order shear deformation theory. *Struct. Eng. Mech.* **2019**, *69*, 439–455. [CrossRef]
15. Nejati, M.; Ghasemi-Ghalebahman, A.; Soltanimaleki, A.; Dimitri, R.; Tornabene, F. Thermal vibration analysis of SMA hybrid composite double curved sandwich panels. *Compos. Struct.* **2019**, *224*, 111035. [CrossRef]
16. Chaht, F.L.; Kaci, A.; Houari, M.S.A.; Tounsi, A.; Beg, O.A.; Mahmoud, S. Bending and buckling analyses of functionally graded material (FGM) size-dependent nanoscale beams including the thickness stretching effect. *Steel Compos. Struct.* **2015**, *18*, 425–442. [CrossRef]
17. Ahouel, M.; Houari, M.S.A.; Bedia, E.A.; Tounsi, A. Size-dependent mechanical behavior of functionally graded trigonometric shear deformable nanobeams including neutral surface position concept. *Steel Compos. Struct.* **2016**, *20*, 963–981. [CrossRef]
18. Bouafia, K.; Kaci, A.; Houari, M.S.A.; Benzair, A.; Tounsi, A. A nonlocal quasi-3D theory for bending and free flexural vibration behaviors of functionally graded nanobeams. *Smart Struct. Syst.* **2017**, *19*, 115–126. [CrossRef]
19. She, G.-L.; Jiang, X.Y.; Karami, B. On thermal snap-buckling of FG curved nanobeams. *Mater. Res. Express* **2019**, *6*, 115008. [CrossRef]
20. She, G.-L.; Yuan, F.-G.; Karami, B.; Ren, Y.-R.; Xiao, W.-S. On nonlinear bending behavior of FG porous curved nanotubes. *Int. J. Eng. Sci.* **2019**, *135*, 58–74. [CrossRef]
21. Daikh, A.A.; Houari, M.S.A.; Tounsi, A. Buckling analysis of porous FGM sandwich nanoplates due to heat conduction via nonlocal strain gradient theory. *Eng. Res. Express* **2019**, *1*, 015022. [CrossRef]
22. Daikh, A.A.; Bachiri, A.; Houari, M.S.A.; Tounsi, A. Size dependent free vibration and buckling of multilayered carbon nanotubes reinforced composite nanoplates in thermal environment. *Mech. Based Des. Struct. Mach.* **2020**, 1–29, in press. [CrossRef]
23. Daikh, A.A.; Drai, A.; Bensaid, I.; Houari, M.S.A.; Tounsi, A. On vibration of functionally graded sandwich nanoplates in the thermal environment. *J. Sandw. Struct. Mater.* **2020**, 1–28, in press. [CrossRef]
24. Daikh, A.A.; Guerroudj, M.; El Adjrami, M.; Megueni, A. Thermal Buckling of Functionally Graded Sandwich Beams. *Adv. Mater. Res.* **2019**, *1156*, 43–59. [CrossRef]
25. Eltaher, M.; Mohamed, S.; Melaibari, A. Static stability of a unified composite beams under varying axial loads. *Thin-Walled Struct.* **2020**, *147*, 106488. [CrossRef]
26. Hamed, M.A.; Abo-Bakr, R.M.; Mohamed, S.A.; Eltaher, M.A. Influence of axial load function and optimization on static stability of sandwich functionally graded beams with porous core. *Eng. Comput.* **2020**, *36*, 1929–1946. [CrossRef]

27. Melaibari, A.; Abo-Bakr, R.M.; Mohamed, S.; Eltaher, M. Static stability of higher order functionally graded beam under variable axial load. *Alex. Eng. J.* **2020**, *59*, 1661–1675. [CrossRef]
28. Zenkour, A.M.; Daikh, A.A. Bending of functionally graded sandwich nanoplates resting on Pasternak foundation under different boundary conditions. *J. Appl. Comput. Mech.* **2020**, *6*, 1245–1259. [CrossRef]
29. Senturia, S.D. *Microsystem Design*; Springer Science & Business Media: Berlin, Germany, 2007.
30. Emam, S.A.; Eltaher, M.A.; Khater, M.E.; Abdalla, W.S. Postbuckling and Free Vibration of Multilayer Imperfect Nanobeams under a Pre-Stress Load. *Appl. Sci.* **2018**, *8*, 2238. [CrossRef]
31. Shi, D.-L.; Feng, X.-Q.; Huang, Y.Y.; Hwang, K.-C.; Gao, H. The Effect of Nanotube Waviness and Agglomeration on the Elastic Property of Carbon Nanotube-Reinforced Composites. *J. Eng. Mater. Technol.* **2004**, *126*, 250–257. [CrossRef]
32. Khater, M.; Eltaher, M.; Abdel-Rahman, E.; Yavuz, M. Surface and thermal load effects on the buckling of curved nanowires. *Eng. Sci. Technol. Int. J.* **2014**, *17*, 279–283. [CrossRef]
33. Brischetto, S.; Tornabene, F. Advanced GDQ models and 3D stress recovery in multilayered plates, spherical and double-curved panels subjected to transverse shear loads. *Compos. Part B Eng.* **2018**, *146*, 244–269. [CrossRef]
34. Tornabene, F.; Bacciocchi, M. Dynamic stability of doubly-curved multilayered shells subjected to arbitrarily oriented angular velocities: Numerical evaluation of the critical speed. *Compos. Struct.* **2018**, *201*, 1031–1055. [CrossRef]
35. Tornabene, F. On the critical speed evaluation of arbitrarily oriented rotating doubly-curved shells made of functionally graded materials. *Thin-Walled Struct.* **2019**, *140*, 85–98. [CrossRef]
36. Mohamed, N.; Eltaher, M.; Mohamed, S.; Seddek, L. Numerical analysis of nonlinear free and forced vibrations of buckled curved beams resting on nonlinear elastic foundations. *Int. J. Non-Linear Mech.* **2018**, *101*, 157–173. [CrossRef]
37. Karami, B.; Janghorban, M.; Tounsi, A. Variational approach for wave dispersion in anisotropic doubly-curved nanoshells based on a new nonlocal strain gradient higher order shell theory. *Thin-Walled Struct.* **2018**, *129*, 251–264. [CrossRef]
38. Arefi, M.; Pourjamshidian, M.; Arani, A.G. Free vibration analysis of a piezoelectric curved sandwich nano-beam with FG-CNTRCs face-sheets based on various high-order shear deformation and nonlocal elasticity theories. *Eur. Phys. J. Plus* **2018**, *133*, 193. [CrossRef]
39. Arefi, M.; Bidgoli, E.M.-R.; Dimitri, R.; Tornabene, F.; Reddy, J.N. Size-Dependent Free Vibrations of FG Polymer Composite Curved Nanobeams Reinforced with Graphene Nanoplatelets Resting on Pasternak Foundations. *Appl. Sci.* **2019**, *9*, 1580. [CrossRef]
40. Karami, B.; Janghorban, M.; Shahsavari, D.; Dimitri, R.; Tornabene, F. Nonlocal Buckling Analysis of Composite Curved Beams Reinforced with Functionally Graded Carbon Nanotubes. *Molecules* **2019**, *24*, 2750. [CrossRef]
41. Karami, B.; Shahsavari, D.; Janghorban, M.; Li, L. Influence of homogenization schemes on vibration of functionally graded curved microbeams. *Compos. Struct.* **2019**, *216*, 67–79. [CrossRef]
42. Arefi, M.; Bidgoli, E.M.-R.; Dimitri, R.; Bacciocchi, M.; Tornabene, F. Nonlocal bending analysis of curved nanobeams reinforced by graphene nanoplatelets. *Compos. Part B Eng.* **2019**, *166*, 1–12. [CrossRef]
43. Eltaher, M.; Mohamed, N.; Mohamed, S.; Seddek, L. Periodic and nonperiodic modes of postbuckling and nonlinear vibration of beams attached to nonlinear foundations. *Appl. Math. Model.* **2019**, *75*, 414–445. [CrossRef]
44. Malikan, M.; Nguyen, V.B.; Dimitri, R.; Tornabene, F. Dynamic modeling of non-cylindrical curved viscoelastic single-walled carbon nanotubes based on the second gradient theory. *Mater. Res. Express* **2019**, *6*, 075041. [CrossRef]
45. Mohamed, N.; Eltaher, M.A.; Mohamed, S.A.; Seddek, L.F. Energy equivalent model in analysis of postbuckling of imperfect carbon nanotubes resting on nonlinear elastic foundation. *Struct. Eng. Mech.* **2019**, *70*, 737–750. [CrossRef]
46. Van Tham, V.; Quoc, T.H.; Tu, T.M. Free Vibration Analysis of Laminated Functionally Graded Carbon Nanotube-Reinforced Composite Doubly Curved Shallow Shell Panels Using a New Four-Variable Refined Theory. *J. Compos. Sci.* **2019**, *3*, 104. [CrossRef]
47. Dindarloo, M.H.; Li, L.; Dimitri, R.; Tornabene, F. Nonlocal Elasticity Response of Doubly-Curved Nanoshells. *Symmetry* **2020**, *12*, 466. [CrossRef]
48. Eltaher, M.A.; Mohamed, N. Nonlinear stability and vibration of imperfect CNTs by Doublet mechanics. *Appl. Math. Comput.* **2020**, *382*, 125311. [CrossRef]
49. Karami, B.; Shahsavari, D. On the forced resonant vibration analysis of functionally graded polymer composite doubly-curved nanoshells reinforced with graphene-nanoplatelets. *Comput. Methods Appl. Mech. Eng.* **2020**, *359*, 112767. [CrossRef]
50. Karami, B.; Janghorban, M.; Tounsi, A. Novel study on functionally graded anisotropic doubly curved nanoshells. *Eur. Phys. J. Plus* **2020**, *135*, 103. [CrossRef]
51. Mohamed, N.; Mohamed, S.A.; Eltaher, M.A. Buckling and post-buckling behaviors of higher order carbon nanotubes using energy-equivalent model. *Eng. Comput.* **2020**, 1–14, in press. [CrossRef]
52. Eringen, A.C. On differential equations of nonlocal elasticity and solutions of screw dislocation and surface waves. *J. Appl. Phys.* **1983**, *54*, 4703–4710. [CrossRef]
53. Touloukian, Y.S. *Thermophysical Properties of High Temperature Solid Materials*; MacMillan: New York, NY, USA, 1967.

Article

Theoretical and Numerical Solution for the Bending and Frequency Response of Graphene Reinforced Nanocomposite Rectangular Plates

Mehran Safarpour [1], Ali Forooghi [1], Rossana Dimitri [2] and Francesco Tornabene [2,*]

[1] Department of Mechanical Engineering, Tarbiat Modares University, Tehran 14115-336, Iran; m_safarpour@modares.ac.ir (M.S.); aliforooghi@modares.ac.ir (A.F.)
[2] Department of Innovation Engineering, University of Salento, 73100 Lecce, Italy; rossana.dimitri@unisalento.it
* Correspondence: francesco.tornabene@unisalento.it

Abstract: In this work, we study the vibration and bending response of functionally graded graphene platelets reinforced composite (FG-GPLRC) rectangular plates embedded on different substrates and thermal conditions. The governing equations of the problem along with boundary conditions are determined by employing the minimum total potential energy and Hamilton's principle, within a higher-order shear deformation theoretical setting. The problem is solved both theoretically and numerically by means of a Navier-type exact solution and a generalized differential quadrature (GDQ) method, respectively, whose results are successfully validated against the finite element predictions performed in the commercial COMSOL code, and similar outcomes available in the literature. A large parametric study is developed to check for the sensitivity of the response to different foundation properties, graphene platelets (GPL) distribution patterns, volume fractions of the reinforcing phase, as well as the surrounding environment and boundary conditions, with very interesting insights from a scientific and design standpoint.

Keywords: FG-GPL; GDQ; heat transfer equation; higher-order shear deformation theory

1. Introduction

Due to their outstanding thermal and mechanical properties, carbon-based nano-filler reinforced composites are widely applied in many engineering fields, such as civil, biomedical and automotive engineering [1–6]. In more detail, graphene platelets (GPLs) are increasingly introduced as carbon nano-fillers because of their relevant potentials in terms of high surface area, elasticity modulus, thermal conductivity, etc. GPLs, as one of novel nanosize reinforcements, have special properties, and their two-dimensional geometry enables them to be scattered in the matrix with less agglomeration, unlike the one-dimensional anisotropic ones. Due to their excellent mechanical, chemical, and physical properties, graphene-based composites demonstrate a wide range of applications in an engineering field, such as sensors, fuel cells, supercapacitors, and batteries. The addition of graphene as reinforcing agent in a polymer matrix, indeed, improves the overall performances and properties of composite materials, as largely demonstrated in the literature from researchers working in this area [7,8]. The primary interest of using graphene materials stems from their excellent mechanical, thermal, electrical and physicochemical properties with prosing results in all fields of technologies. For example, graphene represents one of the stiffest and most grounded materials, with an elastic modulus of ~ 1 TPa and quality of ~ 100 GPa [9–11]. By introducing 1 volume percent of graphene in a polymer matrix, the nanocomposite material reaches a conductivity of about 0.1 Sm^{-1} with adequate consequences for electrical applications, along with significant changes in quality and strength [12]. In such a context, several theories and computational models have been developed in the last decades in the field of GPL-reinforced media. Anamagh and Bediz [13], for example, studied the buckling and vibration response of GPL-reinforced rectangular plates with different boundary conditions, based on a spectral-Tchebychev model. Reddy et al. [14] surveyed the vibrational frequencies of the composite plates reinforced by GPLs, and investigated the effect of various parameters, primarily, boundary conditions, distribution patterns, geometry and weight fractions of GPLs, on the natural frequencies of the system. In addition, Qaderi and Ebrahimi [15] focused on the frequency response of GPL reinforced rectangular plates embedded on viscoelastic substrates, and their sensitivity to different damping coefficients. In line with the previous works, Song et al. [16] studied the free and forced vibration behavior of FG-GPL-reinforced (FG-GPLR) plates by applying a first-order shear deformation theory (FSDT), with a clear enhancement of the vibration performances even with the addition of small quantities of GPLs. Based on the Chebyshev—Ritz procedure, Yang et al. [17] investigated the natural frequencies and critical buckling loads of FG-GPLR nanocomposite plates in presence of different porosities levels. Among the recent literature, different continuum-based nonlocal models have been considered as effective methods to treat plate-like nanostructures and to avoid possible difficulties encountered during experimental characterizations or time-consuming computational atomistic simulations of nanotubes. In this context, some theoretical studies of the free vibration response of graphene sheets can be found in the recent works [18–24], based on different nonlocal theoretical assumptions, accounting for different small-scale parameters, geometrical properties, boundary and environmental conditions. It is also well-known that different substrates can surround a structural member, thus affecting its mechanical behavior and stability. Numerous engineering problems (e.g., heavy machines, pavement of roads, etc.), indeed, are modeled as structural members resting on an elastic medium [25]. The elastic substrates are commonly modeled as Winkler or Pasternak foundations by means of one or two parameters [26,27]. The effect of visco-Pasternak substrate on the nonlinear dynamic response of the FG-GPLRC rectangular plates can be found in the seminal works by Fan et al. [28], and by Liu et al. [29] along with a sensitivity study of the mechanical behavior to different foundation parameters and porosity distributions. Among further works, Gao et al. [30] analyzed the nonlinear vibrational frequencies of FG-GPLR porous plates embedded on a two-parameter-type elastic medium, where an increased porosity coefficient was found to reduce the overall stiffness of structures. The vibrational properties of FG rectangular plates resting on a two-parameter elastic substrate were also surveyed by Thai and Choi [31]. They demonstrated that an increased quantity of metal components can significantly increase the deformability in a structural system. Similarly, Zhou et al. [32] studied the frequency response of thick plates on elastic media, while checking for the effect of different parameters, namely, the foundation coefficients, boundary conditions and aspect ratios, on the structural stiffness. A FSDT was also proposed in [33] to assess the nonlinear vibrational frequency and dynamic behavior of FG-GPLR plates resting on a viscoelastic-Pasternak foundation, with a clear reduction of the structural capacity for increased compressive loads.

Starting with the available literature on the topic, the present work aims at determining a general thermo-elasticity solution to treat both the static and frequency problems of GPLRC rectangular plates under different boundary conditions and embedding foundations, as typically applied in many lightweight mechanical and biomedical components, as well as in membranes and flexible wearable sensors and actuators. Despite the available literature on plate-like nanostructures, usually based on nonclassical approaches, the proposed work explores the capability of a higher-order shear deformation plate formulation combined with a modified Halpin and Tsai model to handle the problem, and checks for the potentials of the generalized differential quadrature (GDQ) approach as high-performance numerical tool to solve the equations even with a reduced computational effort, in lieu of the most common continuum finite element methods from the literature. The governing equations are here derived by means of the Hamilton's principle, accounting for a modified Halpin–Tsai model for the definition of the material properties and the effect of the dispersion in nanocomposites. The GDQ-based solution is here compared to the analytical once based on a Navier-type expansion, and numerically. An extensive study is performed systematically to analyze the impact of different parameters such as the distribution patterns and weight fractions of the reinforcement phase, complex environments, Winkler–Pasternak foundation coefficients, and Kerr substrate constants on the overall response of FG-GPLRC rectangular plates. Results of the present study would be useful for the design of advanced lightweight composite members in civil and mechanical engineering, due to the importance of nanofillers dispersion and the application of foundation structures. The proposed GDQ method represents an innovative computational tool for design purposes, due to its great capability to solve challenging problems, with high simplicity and accuracy. A further extension of the formulation accounts for the thermal buckling of nanocomposite members within a unified setting, as useful for coupled problems for which theoretical predictions are usually cumbersome to obtain.

2. Theoretical Formulation

Here, we consider a FG-GPLRC rectangular plate resting on an elastic Winkler–Pasternak and Kerr medium, whose geometry and dimensions are depicted in Figure 1. The GPLs reinforcement is assumed to be distributed either uniformly (GPL-UD) or in a functionally graded way throughout the thickness, with two symmetric patterns, GPL-X, and GPL-O, respectively.

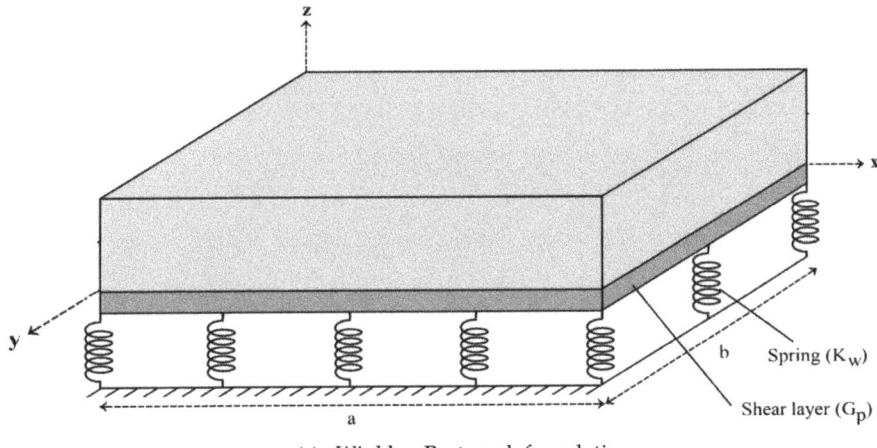

(a) Winkler–Pasternak foundation

Figure 1. *Cont.*

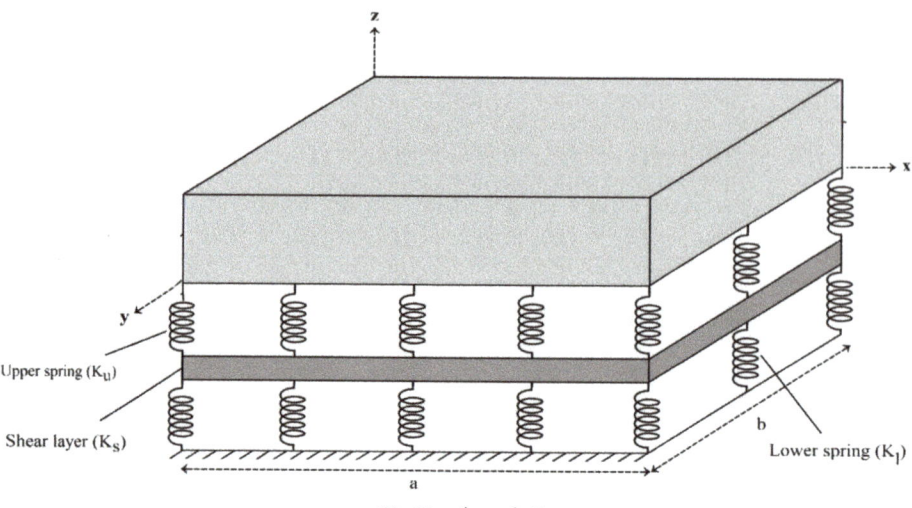

(b) Kerr foundation

Figure 1. Rectangular plate embedded on and elastic foundation.

2.1. Effective Material Properties

The material properties are here defined according to a modified Halpin–Tsai model, such that the effective Young's modulus of the GPL/polymer composite $\overline{E}$ reads as follows [34]:

$$\overline{E} = \frac{3}{8}\frac{(1+\xi_L \eta_L V_{GPL})}{(1-\eta_L V_{GPL})}E_M + \frac{5}{8}\frac{(1+\xi_W \eta_W V_{GPL})}{(1-\eta_W V_{GPL})}E_M \quad (1)$$

where

$$\xi_L = 2\frac{L_{GPL}}{t_{GPL}},\ \xi_W = 2\frac{W_{GPL}}{t_{GPL}},\ \eta_W = -\frac{1-\left(\frac{E_{GPL}}{E_M}\right)}{\xi_W + \left(\frac{E_{GPL}}{E_M}\right)},\ \eta_L = \frac{\left(\frac{E_{GPL}}{E_M}\right)-1}{\xi_L + \left(\frac{E_{GPL}}{E_M}\right)} \quad (2)$$

with E_M and E_{GPL} are the Young's moduli of the polymer matrix and GPLs, respectively; V_{GPL} is the GPL volume fraction, ξ_L and ξ_W are the parameters characterizing both the geometry and size of GPL nanofillers; L_{GPL}, W_{GPL} and t_{GPL} are the average length, width, and thickness of GPLs, respectively.

In line with findings by Rafiee et al. [35], the effective Young's modulus of GPL/polymer nanocomposites is well-approximated by the modified Halpin–Tsai model. The result determined by Equation (1), indeed, is only 2.7% higher than the experimental predictions. Based on the same rule of mixtures, the effective Poisson's ratio and mass density read as follows:

$$\overline{\rho} = \rho_{GPL}V_{GPL} + \rho_M(1-V_{GPL}),\quad \overline{\nu} = \nu_{GPL}V_{GPL} + \nu_M(1-V_{GPL}) \quad (3)$$

while, the effective shear modulus is defined as:

$$\overline{G} = \frac{\overline{E}}{2(1+\overline{\nu})} \quad (4)$$

As also depicted in Figure 2, we select three different distribution patterns of GPLs along the thickness direction of the structure, whose analytical expressions take the following form [36]:

$$V_{GPL}(z_j) = \begin{cases} \begin{cases} 2V_{GPL}^*\left(1-2\frac{Z}{h}\right) & 0 \leq Z \leq \frac{h}{2} \\ 2V_{GPL}^*\left(1-2\frac{(h-Z)}{h}\right) & \frac{h}{2} \leq Z \leq h \end{cases} & GPL-X \\ \begin{cases} 2V_{GPL}^*\left(1-2\frac{\left(\frac{h}{2}-Z\right)}{h}\right) & 0 \leq Z \leq \frac{h}{2} \\ 2V_{GPL}^*\left(1+2\frac{\left(\frac{h}{2}-Z\right)}{h}\right) & \frac{h}{2} \leq Z \leq h \end{cases} & GPL-O \\ \begin{cases} V_{GPL}^* & 0 \leq Z \leq \frac{h}{2} \\ V_{GPL}^* & \frac{h}{2} \leq Z \leq h \end{cases} & UD \end{cases} \quad (5)$$

being $V_{GPL}^* = \frac{\Lambda_{GPL}}{\left(\frac{\rho_{GPL}}{\rho_M}\right)(1-\Lambda_{GPL})+\Lambda_{GPL}}$ and $z_j = \left(\frac{1}{2}+\frac{1}{2n}-\frac{j}{N_L}\right)h$, $j=1,2,3,\ldots,N_L$.

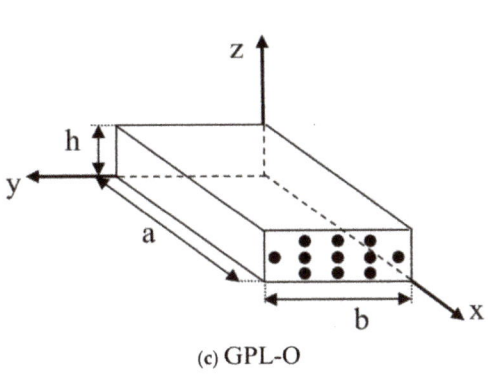

Figure 2. Distribution patterns of GPLs: (**a**) GPL-UD distribution, (**b**) GPL-X distribution, (**c**) GPL-O distribution.

2.2. Displacement Field

As already mentioned in the introduction, we follow a higher order shear deformation theory (HSDT) to define the kinematic field of the structure, i.e., [37].

$$\begin{aligned} u(x,y,z,t) &= u_0(x,y,t) + z\,u_1(x,y,t) + z^2 u_2(x,y,t) + z^3 u_3(x,y,t) \\ v(x,y,z,t) &= v_0(x,y,t) + z\,v_1(x,y,t) + z^2 v_2(x,y,t) + z^3 v_3(x,y,t) \\ w(x,y,z,t) &= w_0(x,y,t) \end{aligned} \quad (6)$$

where (u,v,w) refer to the axial displacement components of an arbitrary point (x,y,z) within the domain; (u_0,v_0,w_0) stand for the related components at the reference midplane; (u_1,v_1,w_1) are the rotations of the normal about the y-, x-, and z-axis respectively;

u_2, v_2, u_3, v_3 define the higher-order terms in the Taylor's series expansion. Also, the non-null strain components are defined in Appendix A.

The constitutive relations for the elastic problem are expressed as:

$$\left\{\begin{array}{c} \sigma_x \\ \sigma_y \\ \tau_{yz} \\ \tau_{xz} \\ \tau_{xy} \end{array}\right\}^{(K)} = \left[\begin{array}{ccccc} Q_{11} & Q_{12} & 0 & 0 & 0 \\ Q_{21} & Q_{22} & 0 & 0 & 0 \\ 0 & 0 & Q_{44} & 0 & 0 \\ 0 & 0 & 0 & Q_{55} & 0 \\ 0 & 0 & 0 & 0 & Q_{66} \end{array}\right]^{(K)} \left\{\begin{array}{c} \varepsilon_x \\ \varepsilon_y \\ \gamma_{yz} \\ \gamma_{xz} \\ \gamma_{xy} \end{array}\right\}^{(K)} \tag{7}$$

where the elastic constants are defined in Appendix A.

2.3. Hamilton's Principle and Governing Equations

The fundamental equations of the problem are determined by applying the Hamilton's principle, in the following variational energy form [38]:

$$\int_{t_1}^{t_2} (\delta\Phi_k - \delta\Phi_e - (\delta\Phi_{w1} + \delta\Phi_{w2} + \delta\Phi_{w3} + \delta\Phi_{w4}))dt = 0 \tag{8}$$

where Φ_k and Φ_e stand for the kinetic and elastic energy, respectively, and the external work Φ_w is split as Φ_{w1}, Φ_{w2}, Φ_{w3}, and Φ_{w4} whose definition depends on the elastic Winkler–Pasternak and Kerr substrates, as well as on the mechanical loading, respectively. The above-mentioned quantities are defined in a variational form as:

$$\delta\Phi_k = \int_V \rho \left(\frac{\partial U}{\partial t} \frac{\partial \delta U}{\partial t} + \frac{\partial V}{\partial t} \frac{\partial \delta V}{\partial t} + \frac{\partial W}{\partial t} \frac{\partial \delta W}{\partial t} \right) dV \tag{9}$$

$$\delta\Phi_e = \int_V \left(\sigma_{xx}\delta\varepsilon_{xx} + \sigma_{yy}\delta\varepsilon_{yy} + \sigma_{zz}\delta\varepsilon_{zz} + \tau_{xy}\delta\gamma_{xy} + \tau_{yz}\delta\gamma_{yz} + \tau_{xz}\delta\gamma_{xz} \right) dV \tag{10}$$

In addition

$$\delta\Phi_{w1} = \int_A \left(-k_w w_o + k_p \left(\frac{\partial^2 w_o}{\partial x^2} + \frac{\partial^2 w_o}{\partial y^2} \right) \right) \delta w_o dA \tag{11}$$

k_w and k_p being the Winkler and Pasternak constants.
While

$$\delta\Phi_{w2} = \int_A \left(-\frac{k_l k_u}{k_l + k_u} w_o + \frac{k_s k_u}{k_l + k_u} \left(\frac{\partial^2 w_o}{\partial x^2} + \frac{\partial^2 w_o}{\partial y^2} \right) \right) \delta w_o dA \tag{12}$$

where k_s, k_u, k_l, refer to the shear layer, upper, and lower spring layers, respectively [39]. The last energy contribution related to the external load P acting on the top surface of the plate reads as follows [40]:

$$\delta\Phi_{w3} = -\int_A P \delta w_o dA \tag{13}$$

In addition, the conductive layer reinforced with GPLs satisfies the following Fourier heat conduction relation:

$$\nabla^2 T + R = \rho c \frac{\partial T}{\partial t} \tag{14}$$

In absence of a thermal generation, in steady-state conditions, it is:

$$\nabla^2 T = 0 \tag{15}$$

For a conductive layer reinforced with a UD or FG distribution of GPLs, we get the following relations:

$$\frac{\partial^2 T}{\partial x^2} + \frac{\partial^2 T}{\partial y^2} + \frac{\partial^2 T}{\partial z^2} = 0 \quad \text{for} \quad UD \tag{16a}$$

$$\frac{\partial}{\partial x}\left(K_c \frac{\partial T}{\partial x}\right) + \frac{\partial}{\partial y}\left(K_c \frac{\partial T}{\partial y}\right) + \frac{\partial}{\partial z}\left(K_c \frac{\partial T}{\partial z}\right) = 0 \quad \text{for} \quad \text{FG} \quad (16b)$$

whose thermal boundary conditions read as follows:

$$T(0,y,z) = 0, \ T(a,y,z) = 0, \ T(x,0,z) = 0, \ T(x,b,z) = 0, \ T(x,y,0) = T_1, \ T(x,y,h) = T_2 \quad (17)$$

The last energy contribution related to the thermal load can be obtained as:

$$\delta \Phi_{w4} = \int_A \left((N_1^T) \frac{\partial w_0}{\partial x} \frac{\partial \delta w_0}{\partial x} + (N_2^T) \frac{\partial w_0}{\partial y} \frac{\partial \delta w_0}{\partial y} \right) dA \quad (18)$$

with

$$N_1^T = \int_{-\frac{h}{2}}^{\frac{h}{2}} (Q_{11} + Q_{12}) \alpha_C (T - T_0) dz, \quad N_2^T = \int_{-\frac{h}{2}}^{\frac{h}{2}} (Q_{21} + Q_{22}) \alpha_C (T - T_0) \, dz \quad (19)$$

and T_0 being the ambient temperature.

By substitution of Equations (9)–(13), and (18) into Equation (8), after a mathematical manipulation we get the following equations as presented in Appendix A (Equations (A4)–(A12)).

3. Thermal Field

To satisfy the thermal boundary conditions in Equation (17), we introduce a Fourier-type solution as follows:

$$T = \sum_{m=1}^{\infty} \sum_{n=1}^{\infty} T_{mn}(z) \sin(P_m x) \sin(P_n y) . \quad (20)$$

where $P_m = m\pi/a$ and $P_n = n\pi/b$. Moreover, the thermal conductivity coefficients related to the GPLs distribution pattern are determined as:

$$UD: \frac{K_c}{K_m} = 1 + D \quad (21a)$$

$$GPL - X : \begin{cases} \frac{K_c}{K_m} = 1 + 2D(1 - 2\frac{Z}{h}) & 0 \leq Z \leq \frac{h}{2} \\ \frac{K_c}{K_m} = 1 + 2D(-1 + 2\frac{Z}{h}) & \frac{h}{2} \leq Z \leq h \end{cases} \quad (21b)$$

$$GPL - O : \begin{cases} \frac{K_c}{K_m} = 1 + 4D(\frac{Z}{h}) & 0 \leq Z \leq \frac{h}{2} \\ \frac{K_c}{K_m} = 1 + 4D(1 - \frac{Z}{h}) & \frac{h}{2} \leq Z \leq h \end{cases} \quad (21c)$$

Inserting Equations (20) and (21a) into Equation (16a) and solving the equation analytically in its final form, we obtain the following expression of temperature gradient for GPLRC rectangular plates with a uniform distribution of GPLs.

$$T_{mn} = C_{11} e^{\sqrt{A_{11}} Z} + C_{22} e^{-\sqrt{A_{11}} Z} \quad (22)$$

where C_{11}, C_{22}, and A_{11} are arbitrary constants determined with appropriate enforcement of the thermal surface boundary conditions (see more details in Appendix B).

At the same time, by combining Equations (20), (21b) and (21c), the heat conduction differential Equation (16b) reduces to the following hypergeometric equation:

$$(A_1 Z + A_2) \frac{\partial^2 T_{mn}(z)}{\partial Z^2} + (A_3 Z + A_4) \frac{\partial T_{mn}(z)}{\partial Z} + (A_5 Z + A_6) T_{mn}(z) = 0 \quad (23)$$

with $A_1, A_2, \ldots, A_6$ being constant coefficients depending on the pattern of GPLs distribution (see Appendix B). The analytical solution of Equation (23) takes the following form:

$$T_{mn} = C_1 e^{\alpha_1} \text{KummerM}(\beta_1, \beta_2, \beta_3)(A_2 + A_1 Z)^{\alpha_2} + C_2 e^{\alpha_1} \text{KummerU}(\beta_1, \beta_2, \beta_3)(A_2 + A_1 Z)^{\alpha_2} \quad (24)$$

where the Kummer's function, also known as the confluent hypergeometric function of the first kind, is a solution to a Kummer's differential equation. In addition, KummerU and KummerM functions represent two special types of Kummer function.

As the thermal behavior of a structure depends on its thermo-mechanical properties, it is worth noticing in Figure 3 that the temperature distribution in thermoelastic solutions is completely different from a uniform or harmonic distribution.

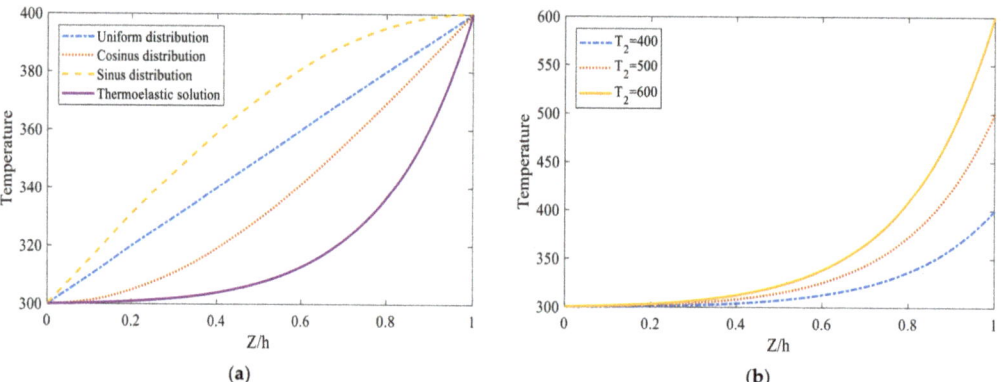

Figure 3. Different temperature distributions, when b = 10 h, a = b, $\Lambda_{GPL} = 0.3$ (wt%), GPL-X, $T_1 = 300$, $T_2 = 400$. (**a**) Representation of the effect of different distribution, (**b**) temperature variation for different temperature boundary T2.

4. Solution Procedure

4.1. Analytical Solution

Following a Navier-type procedure, we now introduce the analytical solution to the above-mentioned governing equations for simply supported FG-GNPRC plates, namely [37]:

$$(u_0, u_1, u_2, u_3) = \sum_{n=1}^{\infty} \sum_{m=1}^{\infty} (U_0, U_1, U_2, U_3) \cos(p_m x) \sin(p_n y) \exp(i\omega t) \quad (25a)$$

$$(v_0, v_1, v_2, v_3) = \sum_{n=1}^{\infty} \sum_{m=1}^{\infty} (V_0, V_1, V_2, V_3) \sin(p_m x) \cos(p_n y) \exp(i\omega t) \quad (25b)$$

$$w_0 = \sum_{n=1}^{\infty} \sum_{m=1}^{\infty} W_0 \sin(p_m x) \sin(p_n y) \exp(i\omega t) \quad (25c)$$

Substituting Equations (25a)–(25c) into Equations (A4)–(A12) (see Appendix A), it is possible to derive the following relations in matrix form, under the assumption $p = 0$

$$\left([K] - [M]\omega^2\right)\{\delta\} = \{0\} \quad (26)$$

with $[K]$ and $[M]$ being the stiffness and mass matrix, respectively, and $\{\delta\}^T = \{U_0, V_0, W_0, U_1, V_1, U_2, V_2, U_3, V_3\}$. Thus, the natural frequencies are determined by means of the following eigenvalue relation:

$$\left|[K] - [M]\omega^2\right| = 0 \quad (27)$$

4.2. Numerical Solution

The same problem is also solved numerically by means of the GDQ method, due to its capability to yield accurate solutions even with a reduced computational effort [41–46] while maintaining a certain flexibility when involving any kind of boundary condition along the structural edges. The proposed method allows solving of the problem in a strong form, by discretizing the derivatives of a function in the following form [41]:

$$\left.\frac{\partial f}{\partial x}\right|_{x=x_i,y=y_j} = \sum_{m=1}^{N_x} \sum_{n=1}^{N_y} A_{im}^x I_{jn}^y f_{mn} \tag{28a}$$

$$\left.\frac{\partial f}{\partial y}\right|_{x=x_i,y=y_j} = \sum_{m=1}^{N_x} \sum_{n=1}^{N_y} I_{im}^x A_{jn}^y f_{mn} \tag{28b}$$

$$\frac{\partial}{\partial r}\left(\left.\frac{\partial f}{\partial \theta}\right|_{x=x_i,y=y_j}\right) = \sum_{m=1}^{N_x} \sum_{n=1}^{N_y} A_{im}^x A_{jn}^y f_{mn} \tag{28c}$$

$$\left.\frac{\partial^2 f}{\partial x^2}\right|_{x=x_i,y=y_j} = \sum_{m=1}^{N_x} \sum_{n=1}^{N_y} B_{im}^x I_{jn}^y f_{mn} \tag{28d}$$

$$\left.\frac{\partial^2 f}{\partial y^2}\right|_{x=x_i,y=y_j} = \sum_{m=1}^{N_x} \sum_{n=1}^{N_y} I_{im}^x B_{jn}^y f_{mn} \tag{28e}$$

where I_{im}^x and I_{jn}^y are equal to one when $i = m$ and $j = n$, or equal to zero, otherwise. In addition, A_{im}^x, A_{jn}^y, B_{im}^x and B_{jn}^y are the weighting coefficients of the first and second-order derivatives along the x and y-directions, respectively, defined as:

$$A_{im}^{(1)} = \begin{cases} \frac{\zeta(x_i)}{(x_i - x_m)\zeta(x_m)} & \text{when } i \neq m \\ -\sum_{k=1,k\neq i}^{N_x} A_{ik}^{(1)} & \text{when } i = m \end{cases} \quad i,m = 1,2,\ldots,N_x \tag{29a}$$

$$A_{jn}^{(1)} = \begin{cases} \frac{\zeta(y_j)}{(y_j - y_n)\zeta(y_n)} & \text{when } j \neq n \\ -\sum_{k=1,k\neq j}^{N_y} A_{jk}^{(1)} & \text{when } j = n \end{cases} \quad j,n = 1,2,\ldots,N_y \tag{29b}$$

with

$$\zeta(x_i) = \prod_{k=1,k\neq i}^{N_x} (x_i - x_k) \tag{30a}$$

$$\zeta(y_j) = \prod_{k=1,k\neq j}^{N_y} (y_j - y_k) \tag{30b}$$

and

$$B_{im}^{(2)} = 2\left(A_{ii}^{(1)} A_{im}^{(1)} - \frac{A_{im}^{(1)}}{(x_i - x_m)}\right) \quad i,m = 1,2,\ldots,N_x,\ i \neq m \tag{31a}$$

$$B_{jn}^{(2)} = 2\left(A_{jj}^{(1)} A_{jn}^{(1)} - \frac{A_{jn}^{(1)}}{(y_j - y_n)}\right) \quad j,n = 1,2,\ldots,N_y,\ j \neq n \tag{31b}$$

$$B_{ii}^{(2)} = -\sum_{k=1,k\neq i}^{N_x} B_{ik}^{(2)}, \quad i = 1,2,\ldots,N_x,\ i = m \tag{31c}$$

$$B_{jj}^{(2)} = -\sum_{k=1, k\neq j}^{N_y} B_{jk}^{(2)}, \quad j = 1, 2, \ldots, N_y, \ j = n \tag{31d}$$

In what follows, we select a Chebyshev distribution of grid points within the domain defined as:

$$x_i = \frac{a}{2}\left(1 - \cos\left(\frac{(i-1)}{(N_x - 1)}\pi\right)\right) \quad i = 1, 2, 3, \ldots, N_x \tag{32a}$$

$$y_j = \frac{b}{2}\left(1 - \cos\left(\frac{(j-1)}{(N_y - 1)}\pi\right)\right) \quad j = 1, 2, 3, \ldots, N_y \tag{32b}$$

Thus, the algebraic eigenvalue problem can be redefined in matrix form as:

$$\left\{ \begin{bmatrix} [M_{dd}] & [M_{db}] \\ [M_{bd}] & [M_{bb}] \end{bmatrix} \omega_{mn}^2 + \begin{bmatrix} [K_{dd}] & [K_{db}] \\ [K_{bd}] & [K_{bb}] \end{bmatrix} \right\} \begin{Bmatrix} \delta_d \\ \delta_b \end{Bmatrix} = 0 \tag{33}$$

where we distinguish among inner and boundary grid-points, by means of subscripts d and b, respectively, whereas δ stands for the displacement vector. The natural frequencies of the problem are derived as solutions of Equation (33).

5. Minimum Total Potential Energy Principle

Now we apply the minimization procedure of the total energy associated to the structural system to study its static response [47], namely:

$$\delta(\Phi_e + \Phi_{w1} + \Phi_{w2} + \Phi_{w3} + \Phi_{w4}) = 0 \tag{34}$$

which is combined with the energy quantities in Equations (10)–(13) and (18) to yield the following governing equations of GPL reinforced composite rectangular plates (see Equations (A34)–(A42) in Appendix B).

Bending Analysis

The analytical solution for a static problem stems from a Fourier-type series discretization of the mechanical force, as follows [48]:

$$P_{mn} = \sum_{n=1}^{\infty} \sum_{m=1}^{\infty} q_{mn} \sin(p_m x) \cos(p_n y) \tag{35}$$

in which $q_{mn} = \frac{4p_0}{mn\pi^2}(1 - (-1)^n)(1 - (-1)^m)$, and $p_0 = 0.1$ MPa.

By substitution of Equations (35) and (25a)–(25c) into Equations (A34)–(A42), with the assumption $\omega = 0$, we get the following relation:

$$\{[F] + [K]\}\{\delta\} = 0 \tag{36}$$

which is solved in terms of the kinematic unknowns.

At the same time, based on a GDQ definition of the problem, the substitution of Equations (28a)–(28e) into Equations (A34)–(A42) leads to the following relation in matrix form:

$$\left\{ \begin{bmatrix} [F_{dd}] & [F_{db}] \\ [F_{bd}] & [F_{bb}] \end{bmatrix} + \begin{bmatrix} [K_{dd}] & [K_{db}] \\ [K_{bd}] & [K_{bb}] \end{bmatrix} \right\} \begin{Bmatrix} \delta_d \\ \delta_b \end{Bmatrix} = 0 \tag{37}$$

depending on the kinematic unknowns δ_d and δ_b.

6. Results and Discussion

6.1. Validation

We now present the results from a large numerical investigation aimed at studying the static and vibrational response of GPLRC multilayer rectangular plates, with material properties as summarized in Table 1. After a preliminary convergence study, we test the performances of our proposed formulation with a comparative evaluation against the open literature or further numerical methods.

Table 1. Material properties of the system (see Ref. [3]).

Polymer Epoxy (Matrix)	Graphene Platelets
$\nu_m = 0.34$	$\nu_{GPL} = 0.186$
$\rho_m \, [\text{kg/m}^3] = 1.2 \times 10^3$	$\rho_{GPL} \, [\text{kg/m}^3] = 1.06 \times 10^3$
$E_m \, [\text{GPa}] = 2.85$	$E_{GPL} \, [\text{TPa}] = 1.01$
	$l_{GPL} \, [\mu\text{m}] = 2.5$
	$w_{GPL} \, [\mu\text{m}] = 1.5$
	$t_{GPL} \, [\text{nm}] = 1.5$

The GDQ numerical study starts considering the effect of an increased grid point distribution on the structural response in terms of dimensionless fundamental frequency and bending deflection, for two different boundary conditions (completely clamped and simply-supported), as visible in Figures 4 and 5, respectively. Based on the plots in these figures it is worth noticing the very fast stabilization of results even with a reduced number of sampling points, whose rate of convergence maintains almost constant independently of the selected boundary conditions. As a further step we perform a parametric evaluation of the frequency response for GPL reinforced thick rectangular plates in terms of natural frequencies for different reinforcement distributions, accounting for different longitudinal and transverse modes, as summarized in Table 2. A FSDT is adopted in this case for comparative purposes with predictions by Song et al. [16], with a perfect matching for all the selected graphene distributions and mode shapes. Among the different GPL distributions, it seems that a GPL-X distribution predicts the highest vibrational frequencies of the system, whereas a pure epoxy material yields the lowest vibrational values.

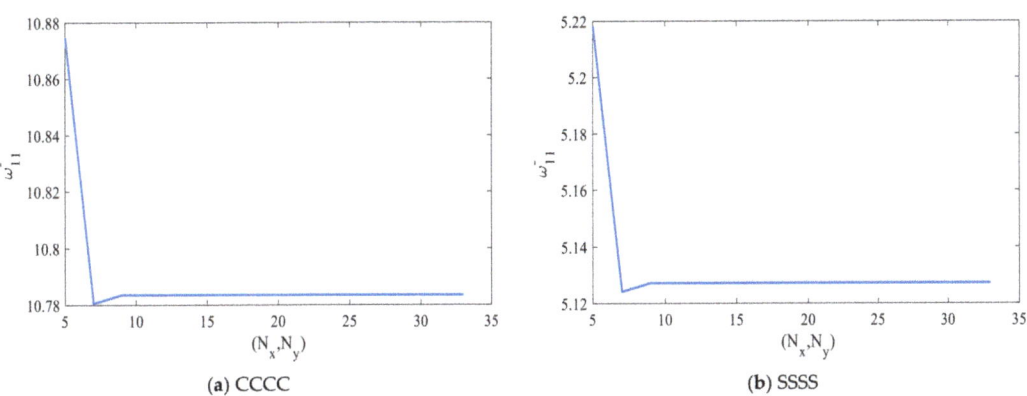

Figure 4. Convergence study of the first fundamental frequency, when assuming a GPL-UD, $b/a = 5$, $a/h = 10$, and $\Lambda_{GPL} = 0.5\%$: (**a**) CCCC boundary conditions, (**b**) SSSS boundary conditions.

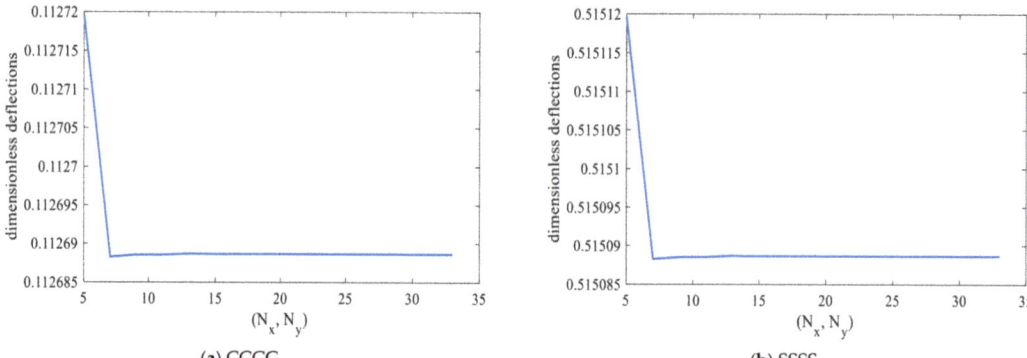

Figure 5. Convergence study of the bending deflection, when assuming a GPL-UD, $p_0 = 10^5$, $b/a = 5$, $a/h = 10$, and $\Lambda_{GPL} = 0.5\%$: (**a**) CCCC boundary conditions, (**b**) SSSS boundary conditions.

Table 2. Dimensionless natural frequency of a rectangular plate made by pure epoxy and different types of graphene distribution with various longitudinal and transverse mode shapes for $(a \times b \times h = 0.45 \text{ m} \times 0.45 \text{ m} \times 0.045 \text{ m})$, $\overline{\omega}_{mn} = \omega_{mn} h \sqrt{\rho_m/E_m}$, and $\Lambda_{GPL} = 1\%$.

GPL-X	GPL-O	GPL-UD	Pure Epoxy	$\overline{\omega}_{mn}$	(m, n)
0.1378	0.102	0.1216	0.0584	Ref. [16]	
0.1378	0.102	0.1216	0.0584	FSDT	(1, 1)
0%	0%	0%	0%	Error%	
0.3249	0.2456	0.2895	0.1391	Ref. [16]	
0.3249	0.2456	0.2895	0.1391	FSDT	(2, 1)
0%	0%	0%	0%	Error%	
0.4939	0.3796	0.4436	0.2132	Ref. [16]	
0.4939	0.3796	0.4436	0.2132	FSDT	(2, 2)
0%	0%	0%	0%	Error%	
0.5984	0.4645	0.54	0.2595	Ref. [16]	
0.5984	0.4645	0.54	0.2595	FSDT	(3, 1)
0%	0%	0%	0%	Error%	
0.7454	0.586	0.6767	0.3251	Ref. [16]	
0.7454	0.586	0.6767	0.3251	FSDT	(3, 2)
0%	0%	0%	0%	Error%	
0.969	0.7755	0.8869	0.4261	Ref. [16]	
0.969	0.7755	0.8869	0.4261	FSDT	(3, 3)
0%	0%	0%	0%	Error%	

We focus, now, on the statics of FG square plates, with the upper and lower surfaces made by a pure ceramic and metal, respectively. The mechanical properties of the system are assumed to vary along the thickness direction based on the following relation [49]:

$$E(z) = (E_c - E_m)\left(\frac{z}{h} + \frac{1}{2}\right)^p + E_m$$
$$\rho(z) = (\rho_c - \rho_m)\left(\frac{z}{h} + \frac{1}{2}\right)^p + \rho_m \quad (38)$$
$$v(z) = (v_c - v_m)\left(\frac{z}{h} + \frac{1}{2}\right)^p + v_m$$

where subscripts c and m refer to the ceramic and metal phases, respectively, and p is the power-law index of the FG material. The mechanical properties of pure ceramic and metal phases are summarized in Table 3, where the ceramic phase features a meaningful higher stiffness than a pure metal. This means that an increased quantity of metal components in the structure would increase its flexibility, as specified in the deflection response of

Table 4, for different exponents p, in line with findings by References [48,49]. As further validation, we compare our vibration results with finite element predictions as performed in COMSOL, for a square plate with constant thickness $h = 10$ mm, length 250 mm, and different GPL distribution profiles. As the GPLRC plate features an inhomogeneous behavior, the material properties are assumed to be graded in the thickness direction. In Tables 5–7, we compare the vibration results for the first six modes, and for three different reinforcement distributions, namely, a GPL–UD, –X, and –O profile, respectively, It is worth noticing the very good matching of results compared to finite elements, with a limited percentage error (lower than 1.04% in each case), despite the reduced computational effort. For a GPL-O distribution we also represent the related mode shapes in Figure 6, while keeping $\Lambda_{GPL} = 0.1$ (wt%), with a reasonable kinematic response with the selected simply-supported boundary condition.

Table 3. Material properties of the system.

Property	Value
E_c	380 GPa
E_m	70 GPa
v_c	0.3
v_m	0.3

Table 4. Effect of the volume fraction exponent of the FG material on the dimensionless deflections of SSSS FG square plates.

Present	Ref. [49]	Ref. [48]	p
0.4621	0.4666	0.4665	$p = 0$ (Ceramic)
0.9416	-	0.9287	1
1.2002	1.1908	1.194	2
1.3204	-	1.32	3
1.3869	1.3769	1.389	4
1.4342	-	1.4356	5
1.4741	1.4554	1.4727	6
1.5107	-	1.5049	7
1.5455	1.5157	1.5343	8
1.579	-	1.5617	9
1.6112	1.5695	1.5876	10
2.5085	-	2.5327	$p = \infty$ (Metal)

Table 5. Comparison with FEM (GPL-UD).

S. No	Mode	ω (Hz)	FEM (Hz)	Relative Error%
S1	1	494.9	495.9	0.19%
S2	2	996.8	997.3	0.05%
S3	3	1453.0	1451.9	0.07%
S4	4	1766.1	1763.9	0.12%
S5	5	2180.2	2175.1	0.23%
S6	6	2752.5	2743.3	0.33%

Table 6. Comparison between FEM-based predictions and results from our formulation (GPL-X).

S. No	Mode	ω (Hz)	FEM (Hz)	Relative Error
S1	1	520.2	520.51	0.06%
S2	2	1046.4	1044.0	0.23%
S3	3	1523.8	1516.3	0.49%
S4	4	1840.6	1840.1	0.027%
S5	5	2283.2	2264.3	0.83%
S6	6	2879.9	2850.2	1.04%

Table 7. Comparison between FEM-based predictions and results from our formulation (GPL-O).

S. No	Mode	ω (Hz)	FEM (Hz)	Relative Error
S1	1	468.1	470.1	0.42%
S2	2	944.0	947.3	0.34%
S3	3	1377.3	1381.4	0.29%
S4	4	1665.5	1670.2	0.28%
S5	5	2069.4	2074.0	0.22%
S6	6	2615.0	2620.4	0.20%

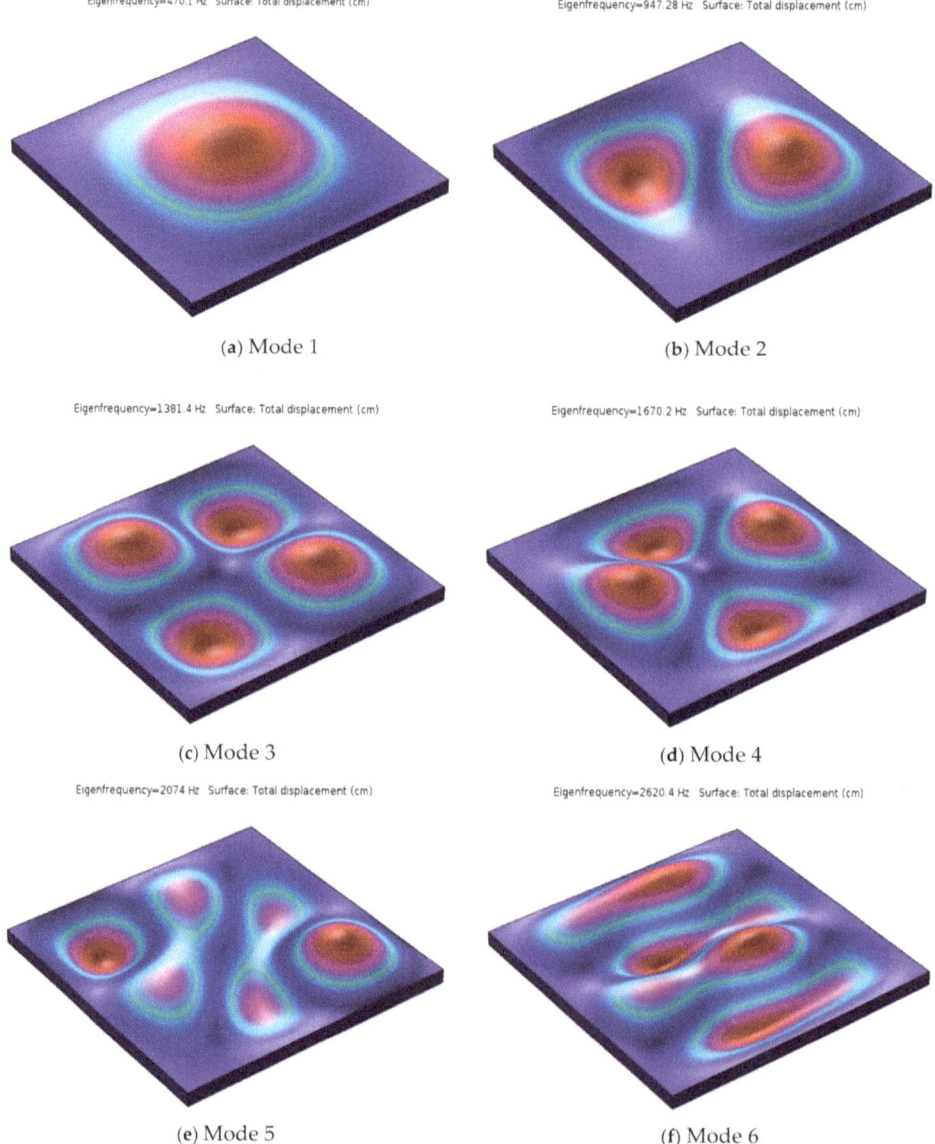

(a) Mode 1 (b) Mode 2
(c) Mode 3 (d) Mode 4
(e) Mode 5 (f) Mode 6

Figure 6. The first six mode shapes of a GPLRC rectangular plate with a GPL-O distribution pattern.

6.2. Parametric Study

We perform, now, a parametric study focusing on the sensitivity of the frequency response $\Delta\omega\% = (\omega_c - \omega_{epoxy})/\omega_{epoxy} \times 100$ against the weight fraction of GPLs for each selected distribution pattern (Figure 7), with a clear beneficial effect on the structural stiffness and stability as Λ_{GPL} increases within the material. Moreover, due to the presence of high normal stresses in the upper and lower sides of multilayered plates, a GPL-X distribution with an increased quantity of GPLs causes a hardening effect on the system, together with higher natural frequencies. On the other hand, a higher concentration of the reinforcing phase in the middle surface of the plate increases the structural deformability monotonically, whose variation rate is plotted in Figure 7, with a perfect agreement with predictions by Song et al. [16]. A further parameter affecting the frequency response is the number of layers N_L in the structure, as plotted in Figure 8 for three different reinforcement distributions. Except for the uniform reinforcement case for which the response is unaffected by N_L, a small increase of this parameter can cause some hardening or softening effects on the structural stiffness, with a sharp increase or decrease of the frequency, for a GPL-X or GPL-O distribution, respectively. Even for these two distributions, the solutions stabilize for a number of layers equal or higher than 10, as also predicted by Song et al. [16]. The sensitivity of the response to the number of layers in the thickness direction is plotted in Figure 9 in terms of dimensionless vibrational frequency, for a square plate with $a/h = 25$ and $\Lambda_{GPL} = 0.3\%$ under a completely-clamped (CCCC) and simply-supported (SSSS) boundary condition. Based on a comparative evaluation of the plots in these two figures, the best stability response seems to be reached for a CCCC multilayered structure with $N_L = 10$ and a GPL-X reinforcement distribution, in terms of vibration frequency (see Figure 9). At the same time, an increased number of layers more than 10 becomes deleterious for the overall stability of plates with a GPL-O type distribution, both for CCCC and SSSS boundary conditions. Figure 10 depicts the variation of the first vibration frequency against the thermal gradient ΔT of moderately thick square plates for different GPL weight fractions. Due to the variation of the thermoelastic properties of the system, together with the presence of initial internal stresses and strains in the structure by thermal attacks, this complex environment causes a combined hardening-softening impact on the system. More specifically, an increased thermal variation decreases the fundamental frequency of the system up to a certain value of ΔT, for which the fundamental frequency becomes zero, and the structure undergoes a static instability phenomenon. This critical temperature moves towards higher values for increased weight fractions of GPLs (from 0.2% up to 0.8%). It is worth also noticing that in the pre-divergence zone, by approaching the static instability phenomenon, an enhanced temperature causes a very fast reduction of the vibrational frequency of the system. In order to survey the effect of the Kerr foundation on the vibrational behavior and static instability of GPL-reinforced plates, we plot the variation of dimensionless first fundamental frequency versus the thermal variation for different substrate coefficients while selecting an SSSS and CCCC boundary condition, respectively, in Figure 11a,b. As the presence of an elastic foundation can vary the bending stiffness of a structure, we note that an enhanced value of the foundation constants improves the vibrational behavior of the system. Meanwhile, the divergence instability can be delayed by increasing the substrate coefficients values. In other words, the presence of an elastic medium gets higher critical values at which the static instability phenomenon takes place. In addition, clamped boundary constraints reduce the positive effect of foundations, with a less pronounced variation in the critical temperature corresponding to the static instability and natural vibrational frequencies of the system.

A further goal of the systematic analysis is also the evaluation of the maximum deflection of the structure, hereafter reported in dimensionless form. In Figure 12 we show the variation of this kinematic quantity versus the number of layers within a SSSS (Figure 12a) and CCCC (Figure 12b) laminated structure, accounting for the three different GPLs patterns. Unlike the UD of GPLs, the kinematic response seems to be clearly sensitive to the number of layers N_L within a multilayered structure, with a monotone increase or decrease depending on whether a GPL–O or –X distribution is selected as reinforcing phase, with a plateau obtained in correspondence of $N_L = 10$. With regard to the sensitivity of the deflection response to the reinforcement weight fraction, we plot the results in Figure 13a,b for a SSSS and CCCC boundary condition, respectively. Based on the plots in this figure, an increased amount of GPL nanofillers decreases monotonically the overall deformability of the structure. For example, the introduction of a small percentage of GPLs (equal to 0.5%) is able to reduce the deformability of the system up to a percentage of 360%, for a GPL-X symmetric distribution. This last reinforcement dispersion provides the highest stiffness in multilayered structures, for both a SSSS and CCCC boundary conditions, whereas the highest deformability is obtained for GPL-O distributions of the reinforcing phase under the same weight fraction assumptions. From a design standpoint, a GPL-X symmetric dispersion is desirable as the best reinforcing distribution among others, due to its capability to limit the structural deformability. At the same time, a further reduction in deformability can be obtained, accounting for the elastic properties of the surrounding medium, as plotted in Figures 14 and 15 for a Winkler–Pasternak or Kerr elastic substrate, respectively. In both cases, indeed, elastic foundations with increased stiffness properties get lower deflections, while keeping fixed the GPL weight fraction within the structure. This reduction is even more pronounced for more relaxing boundary conditions as simply-supports, while assuming a Kerr medium in lieu of a Winkler–Pasternak-type foundation (compare the plots of Figures 14 and 15).

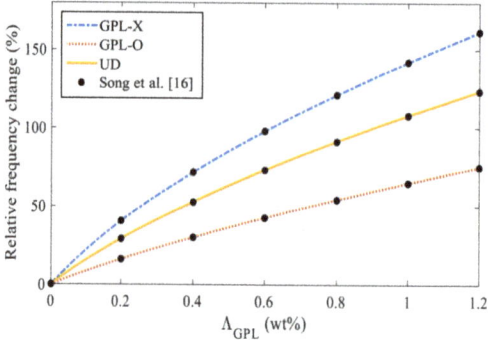

Figure 7. Relative frequency variation vs. the GPL weight fraction, for different GPL distribution patterns ($b/a = 1$, $a/h = 10$, and SSSS boundary condition).

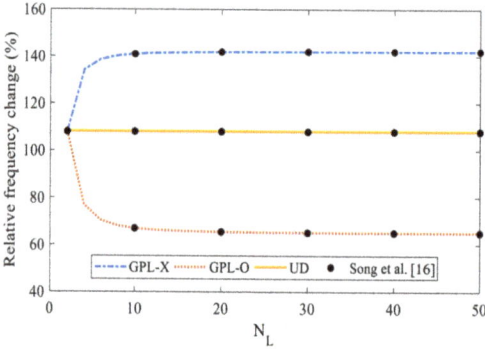

Figure 8. Relative frequency variation vs. the number of layers, for different GPL distribution patterns ($b/a = 1$, $a/h = 10$, and SSSS boundary condition).

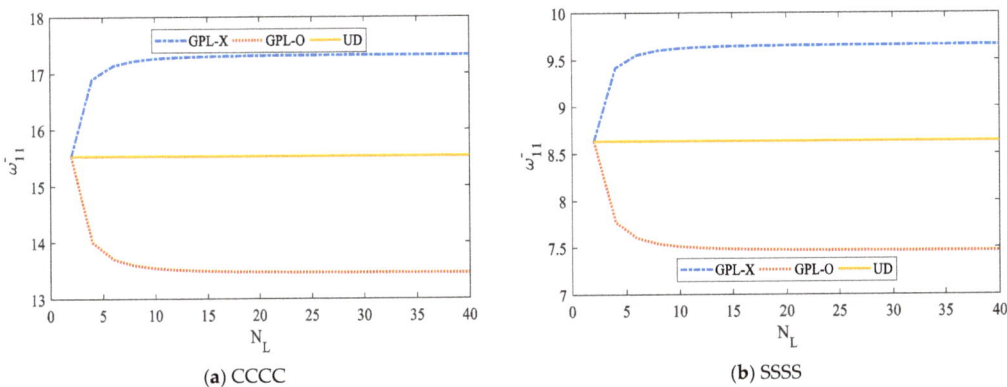

Figure 9. Dimensionless fundamental frequency vs. the number of layers, for different GPL distribution patterns ($b/a = 1$, $a/h = 25$, $\Delta_{GPL} = 0.3\%$): (**a**) CCCC boundary conditions, (**b**) SSSS boundary conditions.

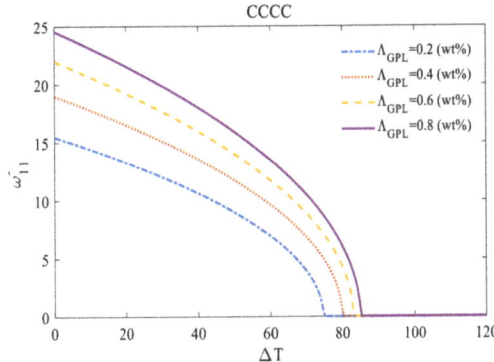

Figure 10. Dimensionless fundamental frequency vs. thermal variation, for different GPL weight fractions ($a = b$, $a/h = 25$, and GPL-X).

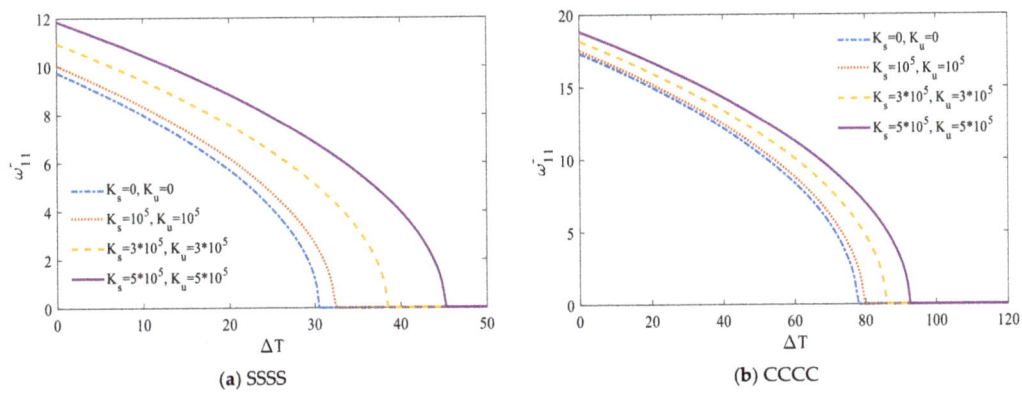

Figure 11. Dimensionless fundamental frequency vs. thermal variation, for different Kerr substrate coefficients ($a = b$, $a/h = 25$, $\Lambda_{GPL} = 0.3\%$, $K_l = 10^5$, and GPL-X): (**a**) SSSS boundary conditions, (**b**) CCCC boundary conditions.

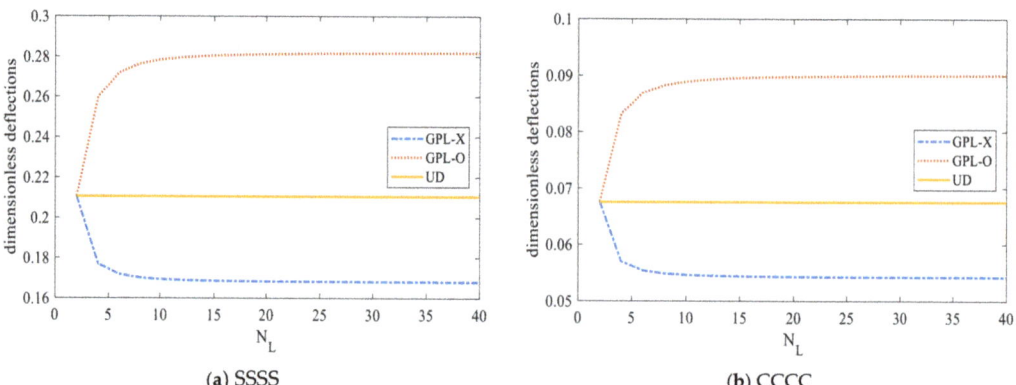

Figure 12. Dimensionless deflection vs. number of layers, for different distribution patterns of reinforcement ($b/a = 1$, $a/h = 25$, $\Delta_{GPL} = 0.3\%$, $p_0 = 10^5$): (**a**) SSSS boundary conditions, (**b**) CCCC boundary conditions.

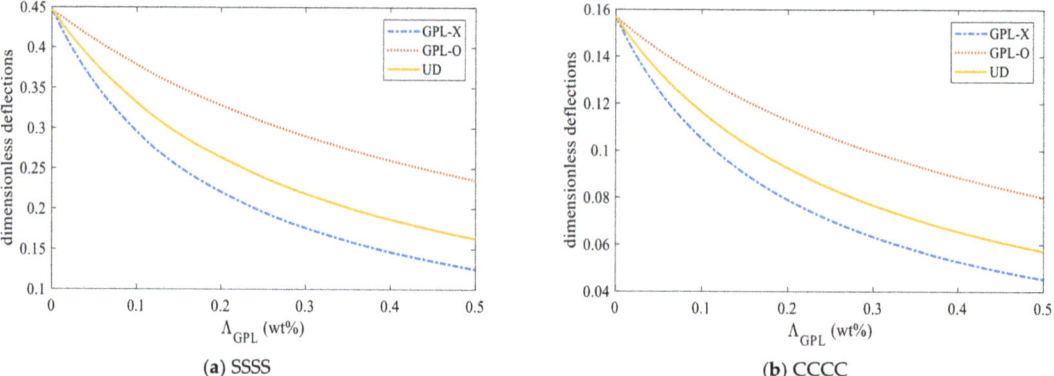

Figure 13. Dimensionless deflection changes vs. GPL weight fraction, for different distribution patterns of reinforcement ($b/a = 1$, $a/h = 10$, $p_0 = 10^5$): (**a**) SSSS boundary conditions, (**b**) CCCC boundary conditions.

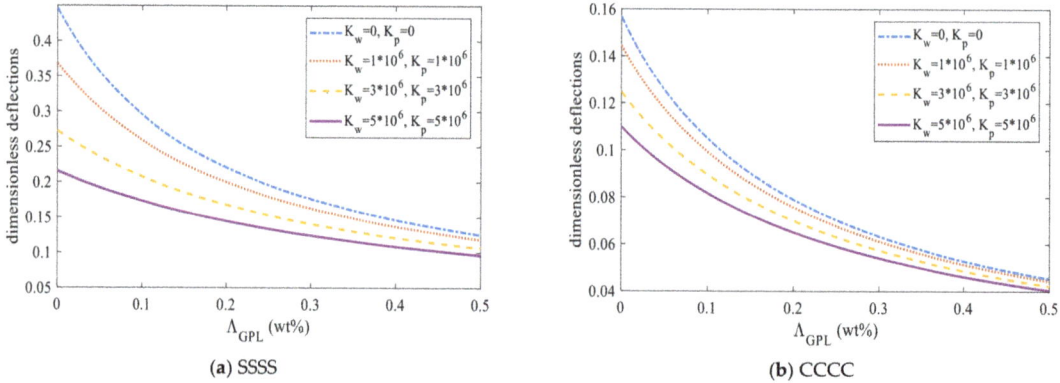

Figure 14. Dimensionless deflection vs. GPL weight fraction, for different elastic foundation coefficients ($b/a = 1$, $a/h = 10$, $GPL - X$, $p_0 = 10^5$): (**a**) SSSS boundary conditions, (**b**) CCCC boundary conditions.

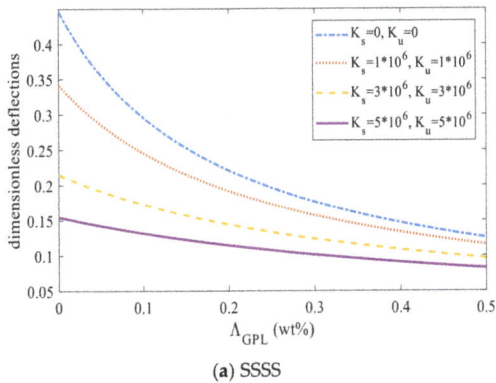
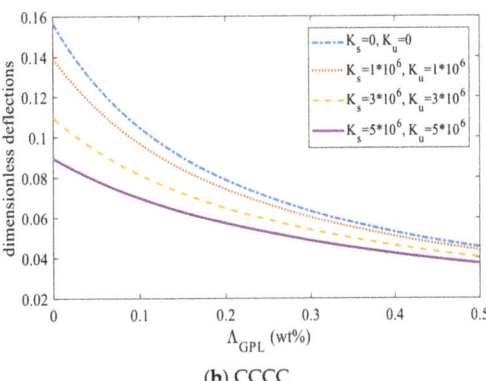

Figure 15. Dimensionless deflection vs. GPL weight fraction, for different Kerr foundation coefficients ($b/a = 1$, $a/h = 10$, $GPL - X$, $p_0 = 10^5$ and $K_I = 10^5$): (**a**) SSSS boundary conditions, (**b**) CCCC boundary conditions.

7. Conclusions

In this paper we focused on the vibrational and static response of FG-GPLRC multi-layer rectangular plates by combining a higher order formulation of shells together with a modified Halpin and Tsai model to account the effect of the dispersion in nanocomposites. The problem is here solved both theoretically with a Navier-type solution, and computationally, by means of the GDQ approach, as high-performance numerical tool. The proposed model is successfully validated in its accuracy against predictions from the literature and results from finite element formulations in the first part. Based on a parametric study, it seems that A GPL-X pattern in a multilayered member provides the highest fundamental frequencies and stiffness of the structure. These mechanical properties increase for an increased GPL weight fraction within the material. In addition, the vibration and kinematic results based on a uniform distribution of GPLs are always unaffected by the reinforcement weight fraction and number of layers within the structure. At the same time, the elastic foundations with increased stiffness properties reduce the overall deformability of multilayered GPL-reinforced structures, which confirms the importance of considering the correct mechanical performances of different substrates around a structural member for design purposes. It is also observed that the presence of a thermal environment reduces the structural efficiency and stiffness due to the introduction of an additional stress and strain field in the system. Meanwhile, elastic foundations with increased stiffness properties raise the critical temperature of multilayered structures while reducing their deformability. This study would provide useful scientific insights and an enhanced tool to engineers and designers for the development of novel and efficient composite structures and components, such as electronic circuits, sensors, or flexible electrodes for displays and solar cells.

Author Contributions: Conceptualization, M.S., A.F., R.D. and F.T.; methodology, M.S., A.F., R.D. and F.T.; software, M.S. and A.F.; validation, M.S., A.F., R.D. and F.T.; formal analysis, M.S. and A.F.; investigation, M.S., A.F., R.D. and F.T.; writing—original draft preparation, M.S. and A.F.; writing—review and editing, R.D. and F.T.; visualization, M.S., A.F., R.D. and F.T.; supervision, R.D. and F.T. All authors have read and agreed to the published version of the manuscript.

Funding: This research received no funding.

Conflicts of Interest: The authors declare no conflict of interest.

Appendix A

The non-null strain components are defined as

$$\left\{\begin{array}{c}\varepsilon_{xx}\\ \varepsilon_{yy}\\ \gamma_{yz}\\ \gamma_{xz}\\ \gamma_{xy}\end{array}\right\}=\left\{\begin{array}{c}\varepsilon_{xx}^0\\ \varepsilon_{yy}^0\\ \gamma_{yz}^0\\ \gamma_{xz}^0\\ \gamma_{xy}^0\end{array}\right\}+z\left\{\begin{array}{c}k_{xx}^0\\ k_{yy}^0\\ k_{yz}^0\\ k_{xz}^0\\ k_{xy}^0\end{array}\right\}+z^2\left\{\begin{array}{c}\varepsilon_{xx}^*\\ \varepsilon_{yy}^*\\ \gamma_{yz}^*\\ \gamma_{xz}^*\\ \gamma_{xy}^*\end{array}\right\}+z^3\left\{\begin{array}{c}k_{xx}^*\\ k_{yy}^*\\ k_{yz}^*\\ k_{xz}^*\\ k_{xy}^*\end{array}\right\} \quad \text{(A1)}$$

where

$$\left\{\begin{array}{c}\varepsilon_{xx}^0\\ \varepsilon_{yy}^0\\ \gamma_{yz}^0\\ \gamma_{xz}^0\\ \gamma_{xy}^0\end{array}\right\}=\left\{\begin{array}{c}\frac{\partial u_0}{\partial x}\\ \frac{\partial v_0}{\partial y}\\ v_1+\frac{\partial w_0}{\partial y}\\ u_1+\frac{\partial w_0}{\partial x}\\ \frac{\partial u_0}{\partial y}+\frac{\partial v_0}{\partial x}\end{array}\right\}, \left\{\begin{array}{c}k_{xx}^0\\ k_{yy}^0\\ k_{yz}^0\\ k_{xz}^0\\ k_{xy}^0\end{array}\right\}=\left\{\begin{array}{c}\frac{\partial u_1}{\partial x}\\ \frac{\partial v_1}{\partial y}\\ 2v_2\\ 2u_2\\ \frac{\partial u_1}{\partial y}+\frac{\partial v_1}{\partial x}\end{array}\right\}, \left\{\begin{array}{c}\varepsilon_{xx}^*\\ \varepsilon_{yy}^*\\ \gamma_{yz}^*\\ \gamma_{xz}^*\\ \gamma_{xy}^*\end{array}\right\}=\left\{\begin{array}{c}\frac{\partial u_2}{\partial x}\\ \frac{\partial v_2}{\partial y}\\ 3v_3\\ 3u_3\\ \frac{\partial u_2}{\partial y}+\frac{\partial v_2}{\partial x}\end{array}\right\}, \left\{\begin{array}{c}k_{xx}^*\\ k_{yy}^*\\ k_{yz}^*\\ k_{xz}^*\\ k_{xy}^*\end{array}\right\}=\left\{\begin{array}{c}\frac{\partial u_3}{\partial x}\\ \frac{\partial v_3}{\partial y}\\ 0\\ 0\\ \frac{\partial u_3}{\partial y}+\frac{\partial v_3}{\partial x}\end{array}\right\}. \quad \text{(A2)}$$

The elastic constants used in Equation (7) are introduced as

$$Q_{11}=\frac{E_C}{1-\nu_C^2}, Q_{12}=\frac{\nu_C E_C}{1-\nu_C^2}, Q_{13}=Q_{23}=Q_{12}, Q_{33}=Q_{22}=Q_{11}, Q_{44}=Q_{55}=Q_{66}=G_C \quad \text{(A3)}$$

The set of governing associated to the Hamilton's principle takes the following form

$$\delta u_0: \frac{\partial N_{xx}}{\partial x}+\frac{\partial N_{xy}}{\partial y}=I_0\ddot{u}_0+I_1\ddot{u}_1+I_2\ddot{u}_2+I_3\ddot{u}_3, \quad \text{(A4)}$$

$$\delta v_0: \frac{\partial N_{yy}}{\partial y}+\frac{\partial N_{xy}}{\partial x}=I_0\ddot{v}_0+I_1\ddot{v}_1+I_2\ddot{v}_2+I_3\ddot{v}_3, \quad \text{(A5)}$$

$$\delta w_0: \begin{array}{c}\frac{\partial Q_x}{\partial x}+\frac{\partial Q_y}{\partial y}-N_1^T\frac{\partial^2 w_0}{\partial x^2}-N_2^T\frac{\partial^2 w_0}{\partial y^2}+k_p\frac{\partial^2 w_0}{\partial x^2}+k_p\frac{\partial^2 w_0}{\partial y^2}-k_w w_0\\ -\frac{k_l k_u}{k_l+k_u}w_0+\frac{k_s k_u}{k_l+k_u}\frac{\partial^2 w_0}{\partial x^2}+\frac{k_s k_u}{k_l+k_u}\frac{\partial^2 w_0}{\partial y^2}-Pw_0=I_0\ddot{w}_0,\end{array} \quad \text{(A6)}$$

$$\delta u_1: \frac{\partial M_x}{\partial x}+\frac{\partial M_{xy}}{\partial y}-Q_x=I_1\ddot{u}_0+I_2\ddot{u}_1+I_3\ddot{u}_2+I_4\ddot{u}_3, \quad \text{(A7)}$$

$$\delta v_1: \frac{\partial M_y}{\partial y}+\frac{\partial M_{xy}}{\partial x}-Q_y=I_1\ddot{v}_0+I_2\ddot{v}_1+I_3\ddot{v}_2+I_4\ddot{v}_3, \quad \text{(A8)}$$

$$\delta u_2: \frac{\partial N_x^*}{\partial x}+\frac{\partial N_{xy}^*}{\partial y}-2S_x=I_2\ddot{u}_0+I_3\ddot{u}_1+I_4\ddot{u}_2+I_5\ddot{u}_3, \quad \text{(A9)}$$

$$\delta v_2: \frac{\partial N_y^*}{\partial y}+\frac{\partial N_{xy}^*}{\partial x}-2S_y=I_2\ddot{v}_0+I_3\ddot{v}_1+I_4\ddot{v}_2+I_5\ddot{v}_3, \quad \text{(A10)}$$

$$\delta u_3: \frac{\partial M_x^*}{\partial x}+\frac{\partial M_{xy}^*}{\partial y}-3Q_x^*=I_3\ddot{u}_0+I_4\ddot{u}_1+I_5\ddot{u}_2+I_6\ddot{u}_3, \quad \text{(A11)}$$

$$\delta v_3: \frac{\partial M_y^*}{\partial y}+\frac{\partial M_{xy}^*}{\partial x}-3Q_y^*=I_3\ddot{v}_0+I_4\ddot{v}_1+I_5\ddot{v}_2+I_6\ddot{v}_3, \quad \text{(A12)}$$

where $(\ddot{\circ})$ refers to the acceleration field, and I_i are the mass inertias. The corresponding boundary conditions are defined as

$$\begin{array}{lll}\delta u_0=0 & \text{or} & N_x\, n_x+N_{xy}n_y=0,\\ \delta v_0=0 & \text{or} & N_y\, n_y+N_{xy}n_x=0,\\ \delta w_0=0 & \text{or} & N_x\, n_x+Q_y n_y+Q_x n_x+N_1^T\frac{\partial w_0}{\partial x}n_x+N_2^T\frac{\partial w_0}{\partial y}n_y\\ & & -k_p\frac{\partial w_0}{\partial x}n_x-k_p\frac{\partial w_0}{\partial y}n_y-\frac{k_s k_u}{k_l+k_u}\frac{\partial w_0}{\partial x}n_x-\frac{k_s k_u}{k_l+k_u}\frac{\partial w_0}{\partial y}n_y=0,\end{array} \quad \text{(A13)}$$

$$\begin{array}{lll}\delta u_1=0 & \text{or} & M_x\, n_x+M_{xy}n_y=0,\\ \delta v_1=0 & \text{or} & M_y\, n_y n_x+M_{xy}n_x=0,\end{array} \quad \text{(A14)}$$

$$\begin{aligned}\delta u_2 = 0 \quad &\text{or} \quad N_x^* n_x + N_{xy}^* n_y = 0, \\ \delta v_2 = 0 \quad &\text{or} \quad N_y^* n_y + N_{xy}^* n_x = 0,\end{aligned} \qquad (A15)$$

$$\begin{aligned}\delta u_3 = 0 \quad &\text{or} \quad M_x^* n_x + M_{xy}^* n_y = 0, \\ \delta v_3 = 0 \quad &\text{or} \quad M_y^* n_y n_x + M_{xy}^* n_x = 0,\end{aligned} \qquad (A16)$$

where

$$\begin{bmatrix} N_x & N_x^* \\ N_y & N_y^* \\ N_{xy} & N_{xy}^* \end{bmatrix} = \int_{-\frac{h}{2}}^{\frac{h}{2}} \begin{Bmatrix} \sigma_x \\ \sigma_y \\ \sigma_{xy} \end{Bmatrix} \{1 \quad z^2\} dz, \begin{Bmatrix} M_x \\ M_y \\ M_{xy} \end{Bmatrix} = \int_{-\frac{h}{2}}^{\frac{h}{2}} \begin{Bmatrix} \sigma_x \\ \sigma_y \\ \sigma_{xy} \end{Bmatrix} z dz, \begin{Bmatrix} M_x^* \\ M_y^* \\ M_{xy}^* \end{Bmatrix} = \int_{-\frac{h}{2}}^{\frac{h}{2}} \begin{Bmatrix} \sigma_x \\ \sigma_y \\ \sigma_{xy} \end{Bmatrix} z^3 dz,$$

$$\begin{bmatrix} Q_x \\ Q_y \end{bmatrix} = \int_{-\frac{h}{2}}^{\frac{h}{2}} \begin{Bmatrix} \sigma_{xz} \\ \sigma_{yz} \end{Bmatrix} dz, \begin{bmatrix} S_x \\ S_y \end{bmatrix} = \int_{-\frac{h}{2}}^{\frac{h}{2}} \begin{Bmatrix} \sigma_{xz} \\ \sigma_{yz} \end{Bmatrix} z \, dz,$$

$$(I_0, I_1, I_2, I_3, I_4, I_5, I_6) = \int_z \rho(1, z^1, z^2, z^3, z^4, z^5, z^6) dz. \qquad (A17)$$

and

$$\begin{aligned}
N_{xx} &= A_{11}\tfrac{\partial u_0}{\partial x} + B_{11}\tfrac{\partial u_1}{\partial x} + C_{11}\tfrac{\partial u_2}{\partial x} + D_{11}\tfrac{\partial u_3}{\partial x} + A_{12}\tfrac{\partial v_0}{\partial y} + B_{12}\tfrac{\partial v_1}{\partial y} + C_{12}\tfrac{\partial v_2}{\partial y} + D_{12}\tfrac{\partial v_3}{\partial y} \\
N_{xy} &= A_{44}\left(\tfrac{\partial u_0}{\partial y} + \tfrac{\partial v_0}{\partial x}\right) + B_{44}\left(\tfrac{\partial u_1}{\partial y} + \tfrac{\partial v_1}{\partial x}\right) + C_{44}\left(\tfrac{\partial u_2}{\partial y} + \tfrac{\partial v_2}{\partial x}\right) + D_{44}\left(\tfrac{\partial u_3}{\partial y} + \tfrac{\partial v_3}{\partial x}\right) \\
N_{yy} &= A_{21}\tfrac{\partial u_0}{\partial x} + B_{21}\tfrac{\partial u_1}{\partial x} + C_{21}\tfrac{\partial u_2}{\partial x} + D_{21}\tfrac{\partial u_3}{\partial x} + A_{22}\tfrac{\partial v_0}{\partial y} + B_{22}\tfrac{\partial v_1}{\partial y} + C_{22}\tfrac{\partial v_2}{\partial y} + D_{22}\tfrac{\partial v_3}{\partial y} \\
Q_y &= A_{55}\left(\tfrac{\partial w_0}{\partial y} + v_1\right) + B_{55}(2v_2) + C_{55}(3v_3) \\
Q_x &= A_{66}\left(\tfrac{\partial w_0}{\partial x} + u_1\right) + B_{66}(2u_2) + C_{66}(3u_3) \\
M_x &= B_{11}\tfrac{\partial u_0}{\partial x} + C_{11}\tfrac{\partial u_1}{\partial x} + D_{11}\tfrac{\partial u_2}{\partial x} + E_{11}\tfrac{\partial u_3}{\partial x} + B_{12}\tfrac{\partial v_0}{\partial y} + C_{12}\tfrac{\partial v_1}{\partial y} + D_{12}\tfrac{\partial v_2}{\partial y} + E_{12}\tfrac{\partial v_3}{\partial y} \\
M_{xy} &= B_{44}\left(\tfrac{\partial u_0}{\partial y} + \tfrac{\partial v_0}{\partial x}\right) + C_{44}\left(\tfrac{\partial u_1}{\partial y} + \tfrac{\partial v_1}{\partial x}\right) + D_{44}\left(\tfrac{\partial u_2}{\partial y} + \tfrac{\partial v_2}{\partial x}\right) + E_{44}\left(\tfrac{\partial u_3}{\partial y} + \tfrac{\partial v_3}{\partial x}\right) \\
M_y &= B_{21}\tfrac{\partial u_0}{\partial x} + C_{21}\tfrac{\partial u_1}{\partial x} + D_{21}\tfrac{\partial u_2}{\partial x} + E_{21}\tfrac{\partial u_3}{\partial x} + B_{22}\tfrac{\partial v_0}{\partial y} + C_{22}\tfrac{\partial v_1}{\partial y} + D_{22}\tfrac{\partial v_2}{\partial y} + E_{22}\tfrac{\partial v_3}{\partial y} \\
N_x^* &= C_{11}\tfrac{\partial u_0}{\partial x} + D_{11}\tfrac{\partial u_1}{\partial x} + E_{11}\tfrac{\partial u_2}{\partial x} + F_{11}\tfrac{\partial u_3}{\partial x} + C_{12}\tfrac{\partial v_0}{\partial y} + D_{12}\tfrac{\partial v_1}{\partial y} + E_{12}\tfrac{\partial v_2}{\partial y} + F_{12}\tfrac{\partial v_3}{\partial y} \\
N_{xy}^* &= C_{44}\left(\tfrac{\partial u_0}{\partial y} + \tfrac{\partial v_0}{\partial x}\right) + D_{44}\left(\tfrac{\partial u_1}{\partial y} + \tfrac{\partial v_1}{\partial x}\right) + E_{44}\left(\tfrac{\partial u_2}{\partial y} + \tfrac{\partial v_2}{\partial x}\right) + F_{44}\left(\tfrac{\partial u_3}{\partial y} + \tfrac{\partial v_3}{\partial x}\right) \\
N_y^* &= C_{21}\tfrac{\partial u_0}{\partial x} + D_{21}\tfrac{\partial u_1}{\partial x} + E_{21}\tfrac{\partial u_2}{\partial x} + F_{21}\tfrac{\partial u_3}{\partial x} + C_{22}\tfrac{\partial v_0}{\partial y} + D_{22}\tfrac{\partial v_1}{\partial y} + E_{22}\tfrac{\partial v_2}{\partial y} + F_{22}\tfrac{\partial v_3}{\partial y} \\
M_x^* &= D_{11}\tfrac{\partial u_0}{\partial x} + E_{11}\tfrac{\partial u_1}{\partial x} + F_{11}\tfrac{\partial u_2}{\partial x} + Y_{11}\tfrac{\partial u_3}{\partial x} + D_{12}\tfrac{\partial v_0}{\partial y} + E_{12}\tfrac{\partial v_1}{\partial y} + F_{12}\tfrac{\partial v_2}{\partial y} + Y_{12}\tfrac{\partial v_3}{\partial y} \\
M_{xy}^* &= D_{44}\left(\tfrac{\partial u_0}{\partial y} + \tfrac{\partial v_0}{\partial x}\right) + E_{44}\left(\tfrac{\partial u_1}{\partial y} + \tfrac{\partial v_1}{\partial x}\right) + F_{44}\left(\tfrac{\partial u_2}{\partial y} + \tfrac{\partial v_2}{\partial x}\right) + Y_{44}\left(\tfrac{\partial u_3}{\partial y} + \tfrac{\partial v_3}{\partial x}\right) \\
M_y^* &= D_{21}\tfrac{\partial u_0}{\partial x} + E_{21}\tfrac{\partial u_1}{\partial x} + F_{21}\tfrac{\partial u_2}{\partial x} + Y_{21}\tfrac{\partial u_3}{\partial x} + D_{22}\tfrac{\partial v_0}{\partial y} + E_{22}\tfrac{\partial v_1}{\partial y} + F_{22}\tfrac{\partial v_2}{\partial y} + Y_{22}\tfrac{\partial v_3}{\partial y}
\end{aligned} \qquad (A18)$$

Based on relations (A13)–(A16), different boundary conditions can be set as follows

Clamped (C) edges:

$$\begin{cases} x = 0, a \\ y = 0, b \end{cases} \rightarrow \begin{cases} u_0 = 0 \quad u_1 = 0 \quad u_2 = 0 \quad u_3 = 0 \\ v_0 = 0 \quad v_1 = 0 \quad v_2 = 0 \quad v_3 = 0, \\ w_0 = 0 \end{cases} \qquad (A19)$$

Simply (S) edges: $\begin{cases} x = 0, a \rightarrow \{N_{xx} = M_{xx} = N_{xx}^* = M_{xx}^* = 0\} \\ y = 0, b \rightarrow \{N_{yy} = M_{yy} = N_{yy}^* = M_{yy}^* = 0\} \end{cases}. \qquad (A20)$

Appendix B

The parameters in Equations (21a)–(21c) are defined as follows

$$D = \frac{1}{3} V_{GPL}^* \left\{ \frac{2}{H + \frac{1}{\frac{K_x}{K_m} - 1}} + \frac{1}{\frac{1}{2}(1 - H) + \frac{1}{\frac{K_z}{K_m} - 1}} \right\}, \qquad (A21)$$

$$H = \frac{Ln\left[\rho\left(\rho + \sqrt{\rho^2 - 1}\right)\right]}{\sqrt{(\rho^2 - 1)^3}} - \frac{1}{\rho^2 - 1}, \quad \text{(A22)}$$

$$K_x = \frac{K_g}{\frac{2R_k K_g}{L} + 1}, \quad \text{(A23)}$$

$$K_z = \frac{K_g}{\frac{2R_k K_g}{t} + 1}, \quad \text{(A24)}$$

$$\rho = \frac{l_{GPL}}{t_{GPL}}. \quad \text{(A25)}$$

The constants in Equation (23) are determined as

$$C_{11} = \frac{\begin{vmatrix} T_1 & T_2 \\ 1 & e^{-\sqrt{A_1}h} \end{vmatrix}}{\begin{vmatrix} 1 & 1 \\ e^{\sqrt{A_1}h} & e^{-\sqrt{A_1}h} \end{vmatrix}}, \quad C_{22} = \frac{\begin{vmatrix} T_2 & T_1 \\ e^{\sqrt{A_1}h} & 1 \end{vmatrix}}{\begin{vmatrix} 1 & 1 \\ e^{\sqrt{A_1}h} & e^{-\sqrt{A_1}h} \end{vmatrix}}, \quad A_{11} = P_m^2 + P_n^2, \quad \text{(A26)}$$

where

$FG - X$:
$$\begin{aligned} Z < \tfrac{h}{2} &\Rightarrow A_1 = -\tfrac{4D}{h}\,;\ A_2 = 1 + 2D\,;\ A_3 = 0\,;\ A_4 = -\tfrac{4D}{h}\,;\ A_5 = -A_{11}\tfrac{4D}{h};\ A_6 = (1 + 2D)A_{11} \\ Z > \tfrac{h}{2} &\Rightarrow A_1 = \tfrac{4D}{h}\,;\ A_2 = 1 - 2D\,;\ A_3 = 0\,;\ A_4 = \tfrac{4D}{h}\,;\ A_5 = A_{11}\tfrac{4D}{h};\ A_6 = (1 - 2D)A_{11} \end{aligned} \quad \text{(A27)}$$

$FG - O$:
$$\begin{aligned} Z < \tfrac{h}{2} &\Rightarrow A_1 = \tfrac{4D}{h}\,;\ A_2 = 1;\ A_3 = 0\,;\ A_4 = \tfrac{4D}{h}\,;\ A_5 = A_{11}\tfrac{4D}{h};\ A_6 = A_{11} \\ Z > \tfrac{h}{2} &\Rightarrow A_1 = -\tfrac{4D}{h}\,;\ A_2 = 1 + 4D;\ A_3 = 0\,;\ A_4 = -\tfrac{4D}{h}\,;\ A_5 = -\tfrac{4D}{h}A_{11};\ A_6 = (1 + 4D)A_{11} \end{aligned} \quad \text{(A28)}$$

More details about coefficients in Equations (25a)–(25c) are defined in the following

$$\alpha_1 = -\frac{\left(A_3 + \sqrt{-4A_5 A_1 + A_3^2}\right)}{2A_1} Z \quad \text{(A29)}$$

$$\alpha_2 = \frac{A_1^2 - A_1 A_4 + A_2 A_3}{A_1^2} \quad \text{(A30)}$$

$$\beta_1 = \frac{-2A_6 A_1^2 - 2A_1^2 \sqrt{-4A_5 A_1 + A_3^2} - 2A_2 A_5 A_1}{2A_1^2 \sqrt{-4A_5 A_1 + A_3^2}} + \frac{-A_1 A_3 A_4 + A_1 A_4 \sqrt{-4A_5 A_1 + A_3^2} + A_2 A_3^2 - A_2 A_3 \sqrt{-4A_5 A_1 + A_3^2}}{2A_1^2 \sqrt{-4A_5 A_1 + A_3^2}} \quad \text{(A31)}$$

$$\beta_2 = \frac{2A_1^2 - A_1 A_4 + A_2 A_3}{A_1^2} \quad \text{(A32)}$$

$$\beta_3 = \frac{\sqrt{-4A_5 A_1 + A_3^2}(A_2 + A_1 Z)}{A_1^2} \quad \text{(A33)}$$

The governing equations of GPLRC plates determined by means of minimum total potential energy principle in Section 5 are defined as

$$\delta u_0 : \frac{\partial N_{xx}}{\partial x} + \frac{\partial N_{xy}}{\partial y} = 0, \quad \text{(A34)}$$

$$\delta v_0 : \frac{\partial N_{yy}}{\partial y} + \frac{\partial N_{xy}}{\partial x} = 0, \quad \text{(A35)}$$

$$\delta w_0 : \frac{\partial Q_x}{\partial x} + \frac{\partial Q_y}{\partial y} - N_1^T \frac{\partial^2 w_0}{\partial x^2} - N_2^T \frac{\partial^2 w_0}{\partial y^2} + k_p \frac{\partial^2 w_0}{\partial x^2} + k_p \frac{\partial^2 w_0}{\partial y^2} - k_w w_0 \\ - \frac{k_l k_u}{k_l + k_u} w_0 + \frac{k_s k_u}{k_l + k_u} \frac{\partial^2 w_0}{\partial x^2} + \frac{k_s k_u}{k_l + k_u} \frac{\partial^2 w_0}{\partial y^2} - P w_0 = 0, \quad \text{(A36)}$$

$$\delta u_1 : \frac{\partial M_x}{\partial x} + \frac{\partial M_{xy}}{\partial y} - Q_x = 0, \quad \text{(A37)}$$

$$\delta v_1 : \frac{\partial M_y}{\partial y} + \frac{\partial M_{xy}}{\partial x} - Q_y = 0, \quad \text{(A38)}$$

$$\delta u_2 : \frac{\partial N_x^*}{\partial x} + \frac{\partial N_{xy}^*}{\partial y} - 2S_x = 0, \quad \text{(A39)}$$

$$\delta v_2 : \frac{\partial N_y^*}{\partial y} + \frac{\partial N_{xy}^*}{\partial x} - 2S_y = 0, \quad \text{(A40)}$$

$$\delta u_3 : \frac{\partial M_x^*}{\partial x} + \frac{\partial M_{xy}^*}{\partial y} - 3Q_x^* = 0, \quad \text{(A41)}$$

$$\delta v_3 : \frac{\partial M_y^*}{\partial y} + \frac{\partial M_{xy}^*}{\partial x} - 3Q_y^* = 0, \quad \text{(A42)}$$

The boundary conditions and static quantities associated with the problem are the same as defined in Equations (A13)–(A16).

References

1. Thai, C.H.; Ferreira, A.J.M.; Tran, F.T.D.; Phung-Van, P. Free vibration, buckling and bending analyses of multilayer functionally graded graphene nanoplatelets reinforced composite plates using the NURBS formulation. *Compos. Struct.* **2019**, *220*, 749–759. [CrossRef]
2. Song, M.; Yang, J.; Kitipornchai, S. Bending and buckling analyses of functionally graded polymer composite plates reinforced with graphene nanoplatelets. *Compos. Part B Eng.* **2018**, *134*, 106–113. [CrossRef]
3. Yang, B.; Kitipornchai, S.; Yang, Y.-F.; Yang, J. 3D thermo-mechanical bending solution of functionally graded graphene reinforced circular and annular plates. *Appl. Math. Model.* **2017**, *49*, 69–86. [CrossRef]
4. Żur, K.K. Quasi-Green's function approach to free vibration analysis of elastically supported functionally graded circular plates. *Compos. Struct.* **2018**, *183*, 600–610. [CrossRef]
5. Jalaei, M.H.; Civalek, Ö. On dynamic instability of magnetically embedded viscoelastic porous FG nanobeam. *Int. J. Eng. Sci.* **2019**, *143*, 14–32. [CrossRef]
6. Civalek, Ö.; Dastjerdi, S.; Akbaş, Ş.D.; Akgöz, B. Vibration analysis of carbon nanotube-reinforced composite microbeams. *Math. Methods Appl. Sci.* **2021**. [CrossRef]
7. Moghadam, A.D.; Omrani, E.; Menezes, P.L.; Rohatgi, P.K. Mechanical and tribological properties of self-lubricating metal matrix nanocomposites reinforced by carbon nanotubes (CNTs) and graphene—A review. *Compos. Part B Eng.* **2015**, *77*, 402–420. [CrossRef]
8. Mohan, V.B.; Lau, K.-T.; Hui, D.; Bhattacharyya, D. Graphene-based materials and their composites: A review on product applications and product limitations. *Compos. Part B Eng.* **2018**, *142*, 200–220. [CrossRef]
9. Geim, A.K. Graphene: Status and prospects. *Science* **2009**, *324*, 1530–1534. [CrossRef]
10. Bonaccorso, F.; Sun, Z.; Hasan, T.; Ferrari, A. Graphene photonics and optoelectronics. *Nat. Photonics* **2010**, *4*, 611–622. [CrossRef]
11. Ferrari, A.C.; Bonaccorso, F.; Fal'Ko, V.; Novoselov, K.; Roche, S.; Bøggild, P.; Borini, S.; Koppens, F.; Palermo, V.; Pugno, N.; et al. Science and technology roadmap for graphene, related two-dimensional crystals, and hybrid systems. *Nanoscale* **2015**, *7*, 4598–4810. [CrossRef]
12. Sreenivasulu, B.; Ramji, B.R.; Nagaral, M. A Review on Graphene Reinforced Polymer Matrix Composites. *Mater. Today Proc.* **2018**, *5*, 2419–2428. [CrossRef]
13. Anamagh, M.R.; Bediz, B. Free vibration and buckling behavior of functionally graded porous plates reinforced by graphene platelets using spectral Chebyshev approach. *Compos. Struct.* **2020**, *253*, 37–55. [CrossRef]
14. Reddy, R.M.R.; Karunasena, W.; Lokuge, W. Free vibration of functionally graded-GPL reinforced composite plates with different boundary conditions. *Aerosp. Sci. Technol.* **2018**, *78*, 147–156. [CrossRef]
15. Qaderi, S.; Ebrahimi, F. Vibration Analysis of Polymer Composite Plates Reinforced with Graphene Platelets Resting on Two-Parameter Viscoelastic Foundation. *Eng. Comput.* **2020**. [CrossRef]
16. Song, M.; Kitipornchai, S.; Yang, J. Free and forced vibrations of functionally graded polymer composite plates reinforced with graphene nanoplatelets. *Compos. Struct.* **2017**, *159*, 579–588. [CrossRef]
17. Yang, J.; Chen, D.; Kitipornchai, S. Buckling and free vibration analyses of functionally graded graphene reinforced porous nanocomposite plates based on Chebyshev-Ritz method. *Compos. Struct.* **2018**, *193*, 281–294. [CrossRef]

18. Kiani, K. Revisiting the free transverse vibration of embedded single-layer graphene sheets acted upon by an in-plane magnetic field. *J. Mech. Sci. Technol.* **2014**, *28*, 3511–3516. [CrossRef]
19. Shen, L.; Shen, H.-S.; Zhang, C.-L. Nonlocal plate model for nonlinear vibration of single layer graphene sheets in thermal environments. *Comput. Mater. Sci.* **2010**, *48*, 680–685. [CrossRef]
20. Zhou, S.-M.; Sheng, L.-P.; Shen, Z.-B. Transverse vibration of circular graphene sheet-based mass sensor via nonlocal Kirchhoff plate theory. *Comput. Mater. Sci.* **2014**, *86*, 73–78. [CrossRef]
21. Kiani, K. Nonlocal continuum-based modeling of a nanoplate subjected to a moving nanoparticle. Part. I: Theoretical formulations. *Phys. E Low-Dimens. Syst. Nanostruct.* **2011**, *44*, 229–248.
22. Kiani, K. Nonlocal continuum-based modeling of a nanoplate subjected to a moving nanoparticle. Part. II: Parametric studies. *Phys. E Low-Dimens. Syst. Nanostruct.* **2011**, *44*, 249–269.
23. Kiani, K. Free vibration of conducting nanoplates exposed to unidirectional in-plane magnetic fields using nonlocal shear deformable plate theories. *Phys. E Low-Dimens. Syst. Nanostruct.* **2011**, *57*, 179–192. [CrossRef]
24. Pradhan, S.C.; Kumar, A. Vibration analysis of orthotropic graphene sheets embedded in Pasternak elastic medium using nonlocal elasticity theory and differential quadrature method. *Comput. Mater. Sci.* **2010**, *50*, 239–245. [CrossRef]
25. Chakraverty, S.; Pradhan, K. Free vibration of functionally graded thin rectangular plates resting on Winkler elastic foundation with general boundary conditions using Rayleigh–Ritz method. *Int. J. Appl. Mech.* **2014**, *6*, 1450043. [CrossRef]
26. Duc, N.D.; Lee, J.; Nguyen-Thoi, T.; Thang, P.T. Static response and free vibration of functionally graded carbon nanotube-reinforced composite rectangular plates resting on Winkler–Pasternak elastic foundations. *Aerosp. Sci. Technol.* **2017**, *68*, 391–402. [CrossRef]
27. Jena, S.K.; Chakraverty, S.; Malikan, M. Application of shifted Chebyshev polynomial-based Rayleigh–Ritz method and Navier's technique for vibration analysis of a functionally graded porous beam embedded in Kerr foundation. *Eng. Comput.* **2020**, 1–21. [CrossRef]
28. Fan, Y.; Xiang, Y.; Shen, H.-S. Nonlinear forced vibration of FG-GRC laminated plates resting on visco-Pasternak foundations. *Compos. Struct.* **2019**, *209*, 443–452. [CrossRef]
29. Liu, J.; Deng, X.; Wang, Q.; Zhong, R.; Xiong, R.; Zhao, J. A unified modeling method for dynamic analysis of GPL-reinforced FGP plate resting on Winkler-Pasternak foundation with elastic boundary conditions. *Compos. Struct.* **2020**, *244*, 112217. [CrossRef]
30. Gao, K.; Gao, W.; Chen, D.; Yang, J. Nonlinear free vibration of functionally graded graphene platelets reinforced porous nanocomposite plates resting on elastic foundation. *Compos. Struct.* **2018**, *204*, 831–846. [CrossRef]
31. Thai, H.-T.; Choi, D.-H. A refined shear deformation theory for free vibration of functionally graded plates on elastic foundation. *Compos. Part B Eng.* **2012**, *43*, 2335–2347. [CrossRef]
32. Zhou, D.; Cheung, Y.K.; Lo, S.H.; Au, F.T.K. Three-dimensional vibration analysis of rectangular thick plates on Pasternak foundation. *Int. J. Numer. Methods Eng.* **2004**, *59*, 1313–1334. [CrossRef]
33. Cong, P.H.; Duc, N.D. New approach to investigate the nonlinear dynamic response and vibration of a functionally graded multilayer graphene nanocomposite plate on a viscoelastic Pasternak medium in a thermal environment. *Acta Mech.* **2018**, *229*, 3651–3670. [CrossRef]
34. Zhao, Z.; Feng, C.; Wang, Y.; Yang, J. Bending and vibration analysis of functionally graded trapezoidal nanocomposite plates reinforced with graphene nanoplatelets (GPLs). *Compos. Struct.* **2017**, *180*, 799–808. [CrossRef]
35. Rafiee, M.; Rafiee, J.; Wang, Z.; Song, H.; Yu, Z.-Z.; Koratkar, N. Enhanced mechanical properties of nanocomposites at low graphene content. *ACS Nano* **2009**, *3*, 3884–3890.
36. Di Sciuva, M.; Sorrenti, M. Bending, free vibration and buckling of functionally graded carbon nanotube-reinforced sandwich plates, using the extended Refined Zigzag Theory. *Compos. Struct.* **2019**, *227*, 111324.
37. Tu, T.M.; Quoc, T.H.; Van Long, N. Vibration analysis of functionally graded plates using the eight-unknown higher order shear deformation theory in thermal environments. *Aerosp. Sci. Technol.* **2019**, *84*, 698–711.
38. Benahmed, A.; Houari, M.S.A.; Benyoucef, S.; Belakhdar, K.; Tounsi, A. A novel quasi-3D hyperbolic shear deformation theory for functionally graded thick rectangular plates on elastic foundation. *Geomech. Eng.* **2017**, *12*, 9–34. [CrossRef]
39. Shahraki, H.; Riahi, H.T.; Izadinia, M.; Talaeitaba, S.B.; Moghaddam, T.V.; Nikravesh, S.K.Y.; Khosravi, M.A. Buckling and vibration analysis of FG-CNT-reinforced composite rectangular thick nanoplates resting on Kerr foundation based on nonlocal strain gradient theory. *J. Vib. Control* **2020**, *26*, 277–305.
40. Al-Furjan, M.; Habibi, M.; Ghabussi, A.; Safarpour, H.; Safarpour, M.; Tounsi, A. Non-polynomial framework for stress and strain response of the FG-GPLRC disk using three-dimensional refined higher-order theory. *Eng. Struct.* **2020**, 111496.
41. Shu, C. *Differential Quadrature and Its Application in Engineering*; Springer Science & Business Media: Berlin, Germany, 2012.
42. Tornabene, F.; Fantuzzi, N.; Ubertini, F.; Viola, E. Strong formulation finite element method based on differential quadrature: A survey. *Appl. Mech. Rev.* **2015**, *67*, 20801. [CrossRef]
43. Tornabene, F.; Bacciocchi, M. *Anisotropic Doubly-Curved Shells. Higher-Order Strong and Weak Formulations for Arbitrarily Shaped Shell Structures*; Esculapio: Bologna, Italy, 2018.
44. Tornabene, F.; Brischetto, S. 3D capability of refined GDQ models for the bending analysis of composite and sandwich plates, spherical and doubly-curved shells. *Thin-Walled Struct.* **2018**, *129*, 94–124. [CrossRef]
45. Tornabene, F. On the critical speed evaluation of arbitrarily oriented rotating doubly-curved shells made of functionally graded materials. *Thin-Walled Struct.* **2019**, *140*, 85–98. [CrossRef]

46. Tornabene, F.; Dimitri, R.; Viola, E. Transient dynamic response of generally-shaped arches based on a GDQ-time-stepping method. *Int. J. Mech. Sci.* **2016**, *114*, 277–314. [CrossRef]
47. Zhang, L.; Song, Z.; Liew, K. Nonlinear bending analysis of FG-CNT reinforced composite thick plates resting on Pasternak foundations using the element-free IMLS-Ritz method. *Compos. Struct.* **2015**, *128*, 165–175. [CrossRef]
48. Zenkour, A.M. Generalized shear deformation theory for bending analysis of functionally graded plates. *Appl. Math. Model.* **2006**, *30*, 67–84. [CrossRef]
49. Chen, D.; Yang, J.; Kitipornchai, S. Buckling and bending analyses of a novel functionally graded porous plate using Chebyshev-Ritz method. *Arch. Civ. Mech. Eng.* **2019**, *19*, 157–170. [CrossRef]

Article

Three-Dimensional Buckling Analysis of Functionally Graded Saturated Porous Rectangular Plates under Combined Loading Conditions

Faraz Kiarasi [1], Masoud Babaei [1], Kamran Asemi [2,*], Rossana Dimitri [3] and Francesco Tornabene [3,*]

1. Department of Mechanical Engineering, University of Eyvanekey, Eyvanekey 99888-35918, Iran; f.kiarasi@eyc.ac.ir (F.K.); masoudbabaei@eyc.ac.ir (M.B.)
2. Department of Mechanical Engineering, Faculty of Engineering, North Tehran Branch, Islamic Azad University, Tehran 158474-3311, Iran
3. Department of Innovation Engineering, Faculty of Engineering, University of Salento, 73100 Lecce, Italy; rossana.dimitri@unisalento.it
* Correspondence: k.asemi@iau-tnb.ac.ir (K.A.); francesco.tornabene@unisalento.it (F.T.)

Abstract: The present work studies the buckling behavior of functionally graded (FG) porous rectangular plates subjected to different loading conditions. Three different porosity distributions are assumed throughout the thickness, namely, a nonlinear symmetric, a nonlinear asymmetric and a uniform distribution. A novel approach is proposed here based on a combination of the generalized differential quadrature (GDQ) method and finite elements (FEs), labeled here as the FE-GDQ method, while assuming a Biot's constitutive law in lieu of the classical elasticity relations. A parametric study is performed systematically to study the sensitivity of the buckling response of porous structures, to different input parameters, such as the aspect ratio, porosity and Skempton coefficients, along with different boundary conditions (BCs) and porosity distributions, with promising and useful conclusions for design purposes of many engineering structural porous members.

Keywords: buckling; FE-GDQ; functionally graded materials; porosity; 3D elasticity

1. Introduction

In the last decades, an increased interest in porous materials has arisen among scientists and designers regarding engineering materials and structures due to their remarkable mechanical properties, electrical conductivity and high permeability. Besides, porous materials can be used in the aerospace industry and sea structures because of their very low density, but also in submarines, reformers and catalysts owing to their high specific surfaces. Thus, many investigations on the mechanical behavior of functionally graded (FG) porous plate and shell structures have been increasingly conducted in the literature from a theoretical, experimental and computational standpoint. Biot [1] was one of the pioneers who investigated the buckling response of a fluid-saturated porous slab under an axial compression, and checked for the sensitivity of the buckling load to pore compressibility. Similarly, Magnucki and Stasiewicz [2] suggested an analytical determination of the critical buckling load of a compressed porous beam based on a broken-line hypothesis and the principle of stationary action for the total potential energy. A shear deformation theory was applied in [3] for the buckling study of porous beams with varying material properties, and in [4] for the bending and buckling of rectangular plates made of a foam material with a nonlinear symmetric porosity distribution. In the further work by Chen et al. [5], the elastic buckling behavior of shear deformable FG porous beams was studied systematically to check for the effect of different porosity distributions on the mechanical response. A multiple analytical, numerical and experimental approach was proposed by Jasion et al. [6] for the buckling study of plates and beams, with a foam core and external layers of perfect material. In the last decades, different higher-order assumptions have been integrated with high-performance computational methods to treat buckling problems of perfect and/or porous composite structures. A nonlinear dynamic buckling of FG porous beams was performed by Kang Gao et al. [7]. The Galerkin method was applied by the authors to determine the governing equations of the problem, which was then solved numerically by means of a fourth-order Runge–Kutta method. A different approach based on a generalized differential quadrature (GDQ) method was applied by Tang et al. [8] to analyze the nonlinear and linear buckling behavior of FG porous Euler–Bernoulli beams. A refined theory was also proposed by Ebrahimi and Jafari [9] to treat the buckling problem of smart magneto-electro-elastic-FG porous plates, accounting for two different FG distributions. Hyperbolic higher-order shear deformation theory (HSDT) was combined with a mesh-free approach in [10] to investigate the buckling and free vibration behavior of porous FG plates resting on an elastic foundation. Among coupled problems, Cong et al. [11] focused on the nonlinear thermomechanical buckling and post-buckling of porous FG plates with two poro/nonlinear symmetric and non-symmetric distributions, by using the Reddy's HSDT and Galerkin method. Further recent contributions on the buckling and free vibration response of perfect and porous FG plates applied first-order shear deformation theory (FSDT) combined with the Chebyshev Polynomials-Ritz method [12–15], even for graphene-reinforced nanocomposites. Tu et al. [16], instead, proposed a Galerkin-based solution for the nonlinear buckling and post-buckling study of imperfect porous plates subjected to different mechanical loads while applying classical shell theory–Von Karman nonlinearity. In addition, Sekkal et al. [17] focused on a novel quasi-3D HSDT to assess the buckling and vibration response of FG plates, whose solution was determined analytically. Shahsavari et al. [18] investigated the shear buckling of porous nanoplates with even, uneven and logarithmic-uneven distribution templates, by means of the Galerkin method, a novel size-dependent quasi-3D shear deformation theory and Eringen's nonlocal elasticity. Another successful application of mesh-free methods can be found in [19] for the thermal buckling response of porous sandwich plates with CNT-reinforced nanocomposite layers. At the same time, Li et al. [20] studied the nonlinear vibration and dynamic buckling of a sandwich FG porous plate reinforced by graphene platelets and resting on a Winkler–Pasternak elastic foundation, where the Galerkin method was proposed together with the fourth-order Runge–Kutta approach as theoretical and numerical tools. A conventional FSDT approach was also employed by Shahgholian et al. [21] for the study of the buckling behavior of FG graphene-reinforced porous cylindrical shells combined with the Rayleigh–Ritz numerical method, whereas Zhao et al. [22] applied the classical Euler–Bernoulli theory and the Galerkin method to check for the dynamic instability of FG porous arches reinforced by graphene platelets.

Based on the current literature on the buckling of FG porous structures, however, most studies rely on the use of simple elastic Hooke's laws, with limited attention to the effect of pore fluid pressures stemming from poroelastic constitutive Biot's laws. In such a context, Jabbari et al. [23,24] proposed a closed-form solution for the axial buckling of FG-saturated, porous, rectangular, simply supported Kirchhoff plates, immersed in a piezoelectric [23] or thermal field [24], respectively. In [25], the same authors studied the axisymmetric buckling of a saturated circular porous-cellular plate as provided by FSDT. In the further work by Jabbari et al. [26,27], classical plate theory (CPT) or HSDT was implemented for the analysis of the buckling capacity of circular porous plates under a radial compressive load, and its sensitivity to some important poroelastic material properties. In another work, Jabbari et al. [28] performed a buckling study of thin circular FG plates made of saturated porous-soft ferromagnetic materials in transverse magnetic fields, whereas in [29,30], a FSDT closed solution was proposed to the buckling problem of transversely graded saturated porous plates with piezoelectric layers, and the axisymmetric post-buckling study of saturated porous circular plates under a uniform radial compression. Among moderately thick plates, Rezaei and Saidi [31] assessed the buckling behavior of fluid-infiltrated porous annular sector plates, as provided by Mindlin plate theory involving fluid-saturated and fluid-free conditions. Additional buckling studies for structural members made of metals or composite materials can be found in [32–35]. The available literature, however, shows the potential application of buckling issues in many porous structures, although at the present state, a proper study of the buckling response of saturated porous rectangular plates subjected to normal and shear loads is still lacking. Based on the literature overview, it seems that the analysis of porous structures is usually based on FSDT and HSDT. Moreover, in most studies, the Hooke's law or drained condition is commonly assumed to model the porous behavior of structures. Based on the above-mentioned lacking aspects of the problem, in this work, the buckling behavior is investigated for FG-saturated porous rectangular plates subjected to a double normal and shear loads. To this end, 3D elasticity theory and Biot's constitutive law are applied, while proposing a mixed FE-DQM based on a Rayleigh–Ritz energy formulation as an efficient computational tool to solve the problem. The application of Biot's constitutive law in lieu of the simple Hooke's law provides more realistic results and conclusions, even from a practical standpoint. Based on the fact that plate theories overestimate the buckling loads for thick plates, 3D elasticity is implemented here to account for the thickness stretching effects, for the sake of accuracy, together with a more efficient mixed FE-GDQ method rather than conventional FEs. Three different porosity distributions are selected here in the thickness direction, namely, a nonlinear symmetric, a nonlinear asymmetric and uniform distribution. The objective of the work is to check the effects of different porosity distributions, as well as the porosity and Skempton coefficients on the critical buckling load of undrained rectangular plates with different geometrical dimensions and BCs, as useful for many engineering applications.

The remainder of the paper is structured as follows. In Section 2, the geometrical and mechanical properties of porous rectangular plates are briefly described, together with the governing equations of the problem, as determined by means of the virtual work principle and Biot's constitutive poroelastic law. Section 3 presents the main basics of the mixed FE-GDQ numerical formulation, as proposed here to solve the problem, whose numerical examples are tested and discussed in Section 4 among a large systematic investigation. Conclusions are finally drawn in Section 5.

2. Theoretical Definition of the Problem

2.1. Poroelastic Modeling of Plates

Let us consider a rectangular FG porous plate, with in-plane dimensions a and b, and thickness h, as depicted in Figure 1, along with three different porosity patterns throughout the thickness direction ($0 \leq z \leq h$), namely, a non-symmetric nonlinear porous distribution (PNND), a symmetric nonlinear porous distribution (PNSD) and a uniform porous distribution (PUD). Except for uniform porosities, the mechanical properties of the material in terms of shear modulus, Young's modulus and mass density, for a PNND and PNSD, are defined as in Equation (1) [36–40].

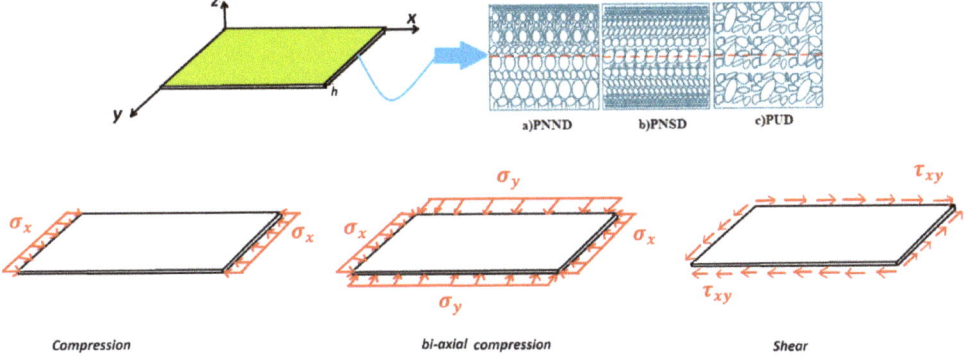

Figure 1. Geometrical scheme and loading conditions for a FG-saturated porous plate.

$$E = E_1[1 - e_0 Q]$$
$$G = G_1[1 - e_0 Q] \tag{1}$$
$$\rho = \rho_1[1 - e_m Q]$$

In which

$$Q(z) = \begin{cases} (a) \text{PNND} & \cos\left(\dfrac{\pi z}{2h}\right) \\ (b) - \text{PNSD} & \cos\left(\dfrac{\pi}{2} - \dfrac{\pi z}{h}\right) \end{cases} \tag{2}$$

and $0 \leq e_0 \leq 1$ is the porosity coefficient. Moreover, E_1, G_1 and ρ_1 denote the Young's modulus, the shear modulus and the mass density at $z = h$ (for a PNND) and at $z = 0$ (for a PNSD), whereby $E_j = 2G_j(1+\nu)$, $j = 0, 1$ and the Poisson's ratio, ν, is assumed to be constant in the z-direction. The constitutive equations of FG-saturated porous rectangular plates are derived from Biot's theory, which accounts for the displacements field of the solid, the pore fluid movement as well as their interactions owing to the applied loads [41]. Based on Biot's theory, the constitutive law is thus written as [42]:

$$\sigma_{ij} = 2G\varepsilon_{ij} + \lambda \varepsilon_{kk} \delta_{ij} - p\alpha \delta_{ij} \tag{3}$$

where

$$p = \overline{M}(\Psi - \alpha \varepsilon_{kk})$$

$$\overline{M} = \dfrac{2G(\nu_u - \nu)}{\alpha^2(1 - 2\nu_u)(1 - 2\nu)}$$

$$\nu_u = \dfrac{\nu + \alpha \beta(1 - 2\nu)/3}{1 - \alpha \beta(1 - 2\nu)/3}$$

$$\nu = \dfrac{\varepsilon_{jj}}{\varepsilon_{ii}}\bigg|_{\sigma_{ii}=0}, p = 0, i \neq j$$

$$\nu_u = \dfrac{\varepsilon_{jj}}{\varepsilon_{ii}}\bigg|_{\sigma_{ii}=0}, \Psi = 0, i \neq j$$

$$\alpha = 1 - \dfrac{K}{K_S}$$

$$K = \dfrac{2(1+\nu)}{3(1-2\nu)}G$$

$$K_u = \dfrac{2(1+\nu_u)}{3(1-2\nu_u)}G$$

Note that p is the pore fluid pressure, such that, for $p = 0$, Biot's law reverts to the classical Hooke's law (or drained condition). In addition, λ denotes the Lamè constant, δ_{ij} is the Kronecker delta and α is the Biot's effective stress coefficient (with $0 < \alpha < 1$). This parameter accounts for the effect of porosity on the structural behavior and resistance of porous materials in the absence of an internal fluid. At the same time, $\overline{M}$, G, ν_u, ε_{kk}, Ψ, K_s and β stand for the Biot's modulus, shear modulus, undrained Poisson's ratio ($\nu < \nu_u < 0.5$), volumetric strain, variation of fluid volume content, bulk modulus of a homogeneous material and the Skempton coefficient, which introduces the pore fluid property, respectively. This last coefficient, β, in particular, denotes a dimensionless parameter to include the impact of a fluid within cavities on the overall response of a porous material in undrained condition ($\Psi = 0$), and it is described as the ratio of the cavity pressure to the total body stress, namely,

$$\beta = \dfrac{dp}{d\sigma}\bigg|_{\Psi=0} = \dfrac{1}{1 + e_0 \dfrac{C_P}{C_S}} = \dfrac{K_u - K}{\alpha K_u} \tag{4}$$

In which K_u, K refer to the bulk modulus in undrained and drained conditions respectively, and C_P and C_S stand for the fluid and solid compressibility in pores. Thus, the Skempton coefficient defines the effect of fluid compressibility on the elastic modulus and compressibility of the whole porous material [42].

2.2. Governing Equations

The governing equations of the problem are derived here from the principle of virtual work, as follows:

$$\delta U - \delta V_g = 0 \tag{5}$$

where U is the total strain potential energy of the plate defined on the domain Ω as:

$$U = \frac{1}{2} \int_\Omega \sigma_{ij} \varepsilon_{ij} d\Omega \tag{6}$$

and V_g is the potential energy related to geometry, which takes the following form:

$$V_g = \frac{1}{2} \int_\Omega \left\{ P_x \left[\left(\frac{\partial u}{\partial x}\right)^2 + \left(\frac{\partial w}{\partial x}\right)^2 + \left(\frac{\partial v}{\partial x}\right)^2 \right] + P_y \left[\left(\frac{\partial u}{\partial y}\right)^2 + \left(\frac{\partial w}{\partial y}\right)^2 + \left(\frac{\partial v}{\partial y}\right)^2 \right] + P_{xy} \left(\frac{\partial w}{\partial x} \frac{\partial w}{\partial y} + \frac{\partial v}{\partial x} \frac{\partial v}{\partial y} - \frac{\partial u}{\partial x} \frac{\partial u}{\partial y} \right) \right\} d\Omega \tag{7}$$

By substitution of Equations (6) and (7) into Equation (5), the following relation is obtained:

$$\int_0^h \int_0^a \int_0^b \left[\begin{array}{l} (\sigma_{xx}\delta\varepsilon_{xx} + \sigma_{yy}\delta\varepsilon_{yy} + \sigma_{zz}\delta\varepsilon_{zz} + \sigma_{yz}\delta\gamma_{yz} + \sigma_{xz}\delta\gamma_{xz} + \sigma_{xy}\delta\gamma_{xy}) \\ -P_x \left(\frac{\partial u}{\partial x} \frac{\partial \delta u}{\partial x} + \frac{\partial v}{\partial x} \frac{\partial \delta v}{\partial x} + \frac{\partial w}{\partial x} \frac{\partial \delta w}{\partial x} \right) - P_y \left(\frac{\partial u}{\partial y} \frac{\partial \delta u}{\partial y} + \frac{\partial v}{\partial y} \frac{\partial \delta v}{\partial y} + \frac{\partial w}{\partial y} \frac{\partial \delta w}{\partial y} \right) \\ -P_{xy} \left(\frac{\partial \delta w}{\partial x} \frac{\partial w}{\partial y} + \frac{\partial \delta w}{\partial y} \frac{\partial w}{\partial x} + \frac{\partial \delta v}{\partial x} \frac{\partial v}{\partial y} + \frac{\partial \delta v}{\partial y} \frac{\partial v}{\partial x} - \frac{\partial \delta u}{\partial x} \frac{\partial u}{\partial y} - \frac{\partial \delta u}{\partial y} \frac{\partial u}{\partial x} \right) \end{array} \right] dv = 0 \tag{8}$$

where the constitutive relations for FG-saturated porous plates in 3D poroelasticity can be defined according to Biot's constitutive law, as $[\sigma_{ij}] = [C][\varepsilon_{ij}]$. For FG-saturated porous rectangular plates, the elasticity matrix reads as follows:

$$C = \frac{E(1-\nu)}{(1+\nu)(1-2\nu)} \begin{pmatrix} 1 & \frac{\nu}{1-\nu} & \frac{\nu}{1-\nu} & 0 & 0 & 0 \\ \frac{\nu}{1-\nu} & 1 & \frac{\nu}{1-\nu} & 0 & 0 & 0 \\ \frac{\nu}{1-\nu} & \frac{\nu}{1-\nu} & 1 & 0 & 0 & 0 \\ 0 & 0 & 0 & \frac{1-2\nu}{2(1-\nu)} & 0 & 0 \\ 0 & 0 & 0 & 0 & \frac{1-2\nu}{2(1-\nu)} & 0 \\ 0 & 0 & 0 & 0 & 0 & \frac{1-2\nu}{2(1-\nu)} \end{pmatrix} + \begin{pmatrix} \overline{M}\alpha^2 & 0 & 0 & 0 & 0 & 0 \\ 0 & \overline{M}\alpha^2 & 0 & 0 & 0 & 0 \\ 0 & 0 & \overline{M}\alpha^2 & 0 & 0 & 0 \\ 0 & 0 & 0 & 0 & 0 & 0 \\ 0 & 0 & 0 & 0 & 0 & 0 \\ 0 & 0 & 0 & 0 & 0 & 0 \end{pmatrix} =$$

$$= E \Lambda + \Upsilon = \begin{bmatrix} C_{11} & C_{12} & C_{12} & 0 & 0 & 0 \\ C_{12} & C_{11} & C_{12} & 0 & 0 & 0 \\ C_{12} & C_{12} & C_{11} & 0 & 0 & 0 \\ 0 & 0 & 0 & C_{44} & 0 & 0 \\ 0 & 0 & 0 & 0 & C_{55} & 0 \\ 0 & 0 & 0 & 0 & 0 & C_{66} \end{bmatrix} \tag{9}$$

The elasticity modulus, E, is assumed to vary along the z-direction, whereas the Poisson's ratio, ν, remains constant.

3. Mixed FE-GDQ Numerical Formulation

A set of rectangular-quadratic elements, N_e, is considered to discretize the x–y plane of the domain. Each element should be differentiable at both in-plane and transverse displacements for solving the governing equations. Hence, the displacement components are approximated as:

$$u(x,y,z,t) = \sum_{j=1}^{N} \varphi_j(x,y) U_j(z,t),$$
$$v(x,y,z,t) = \sum_{j=1}^{N} \varphi_j(x,y) V_j(z,t), \quad j = 1, 2, \ldots, N \qquad (10)$$
$$w(x,y,z,t) = \sum_{j=1}^{N} \varphi_j(x,y) W_j(z,t)$$

where N stands for the total number of nodes in the discretized x–y plane, and $\varphi_j(x,y)$ denotes the global Lagrange interpolation functions. By combination of Equations (3) and (8), and integrating by parts in the z-direction, the following governing equations per each node i ($i = 1, 2, \ldots, N$) are obtained:

$$\delta U_i : \sum_{j=1}^{N} A^{ij} C_{55}(z_j) \frac{\partial^2 U_j}{\partial z^2} + \sum_{j=1}^{N} \frac{\partial C_{55}(z_j)}{\partial z} \frac{\partial U_j}{\partial z} + \sum_{j=1}^{N} \left(D^{ij} C_{55}(z_j) - D^{ji} C_{12}(z_j) \right) \frac{\partial W_j}{\partial z}$$
$$+ \sum_{j=1}^{N} \frac{\partial C_{44}}{\partial z}(z_j) W_j - \sum_{j=1}^{N} \left(E^{ji} C_{12}(z_j) + E^{ij} C_{66}(z_j) \right) U_j - \sum_{j=1}^{N} \left(F^{ij} C_{11}(z_j) + H^{ji} C_{66}(z_j) \right) V_j \qquad (11)$$
$$- P_x \sum_{j=1}^{N} D_b^{ij} V_j - P_y \sum_{j=1}^{N} B_b^{ij} V_j - P_{xy} \sum_{j=1}^{N} S_b^{ij} V_j - P_{xy} \sum_{j=1}^{N} K_b^{ij} V_j = 0$$

$$\delta V_i : \sum_{j=1}^{N} A^{ij} C_{44}(z_j) \frac{\partial^2 V_j}{\partial z^2} + \sum_{j=1}^{N} \frac{\partial C_{44}(z_j)}{\partial z} \frac{\partial V_j}{\partial z} + \sum_{j=1}^{N} \left(D^{ij} C_{44}(z_j) - D^{ji} C_{12}(z_j) \right) \frac{\partial W_j}{\partial z}$$
$$+ \sum_{j=1}^{N} \frac{\partial C_{44}}{\partial z}(z_j) W_j - \sum_{j=1}^{N} \left(E^{ji} C_{12}(z_j) + E^{ij} C_{66}(z_j) \right) U_j - \sum_{j=1}^{N} \left(F^{ij} C_{11}(z_j) + H^{ji} C_{66}(z_j) \right) V_j \qquad (12)$$
$$- P_x \sum_{j=1}^{N} D_b^{ij} V_j - P_y \sum_{j=1}^{N} B_b^{ij} V_j - P_{xy} \sum_{j=1}^{N} S_b^{ij} V_j - P_{xy} \sum_{j=1}^{N} K_b^{ij} V_j = 0$$

$$\delta W_i : \sum_{j=1}^{N} A^{ij} C_{11}(z_j) \frac{\partial^2 W_j}{\partial z^2} + \sum_{j=1}^{N} \left(B^{ij} C_{12}(z_j) - B^{ij} C_{55}(z_j) \right) \frac{\partial U_j}{\partial z}$$
$$+ \sum_{j=1}^{N} \left(D^{ij} C_{12}(z_j) - D^{ji} C_{44}(z_j) \right) \frac{\partial V_j}{\partial z} - \sum_{j=1}^{N} B^{ij} \frac{\partial C_{12}(z)}{\partial z} U_j$$
$$+ \sum_{j=1}^{N} D^{ij} \frac{\partial C_{12}(z)}{\partial z} V_j + \sum_{j=1}^{N} \left(F^{ij} C_{44}(z_j) + H^{ij} C_{55}(z_j) + A^{ij} \frac{\partial C_{11}(z)}{\partial z} \right) W_j \qquad (13)$$
$$- P_x \sum_{j=1}^{N} D_b^{ij} W_j - P_y \sum_{j=1}^{N} B_b^{ij} W_j - P_{xy} \sum_{j=1}^{N} S_b^{ij} W_j - P_{xy} \sum_{j=1}^{N} K_b^{ij} W_j = 0$$

where,

$$H^{ij} = \int_0^a \int_0^b \frac{\partial \varphi_i}{\partial x} \frac{\partial \varphi_j}{\partial x} dxdy, \quad F^{ij} = \int_0^a \int_0^b \frac{\partial \varphi_i}{\partial y} \frac{\partial \varphi_j}{\partial y} dxdy, \quad E^{ij} = \int_0^a \int_0^b \frac{\partial \varphi_i}{\partial x} \frac{\partial \varphi_j}{\partial y} dxdy$$
$$H^{ij} = \int_0^a \int_0^b \frac{\partial \varphi_i}{\partial x} \frac{\partial \varphi_j}{\partial x} dxdy, \quad F^{ij} = \int_0^a \int_0^b \frac{\partial \varphi_i}{\partial y} \frac{\partial \varphi_j}{\partial y} dxdy, \quad E^{ij} = \int_0^a \int_0^b \frac{\partial \varphi_i}{\partial x} \frac{\partial \varphi_j}{\partial y} dxdy \qquad (14)$$
$$D_b^{ij} = \int_0^a \int_0^b \frac{\partial \varphi_i}{\partial x} \frac{\partial \varphi_j}{\partial x} dydx, \quad B_b^{ij} = \int_0^a \int_0^b \frac{\partial \varphi_i}{\partial y} \frac{\partial \varphi_j}{\partial y} dydx, \quad S_b^{ij} = \int_0^a \int_0^b \frac{\partial \varphi_i}{\partial x} \frac{\partial \varphi_j}{\partial y} dydx, \quad K_b^{ij} = \int_0^a \int_0^b \frac{\partial \varphi_i}{\partial y} \frac{\partial \varphi_j}{\partial x} dydx$$

The BCs at the lower and upper surfaces ($z = 0$ and $z = h$) associated with Equations (11)–(13) are defined as:

$$\text{Either } \delta U_i = 0, \text{ or} \begin{cases} \sum_{j=1}^{N} A^{ij} C_{55}(z_j) \frac{\partial U_j}{\partial z} + \sum_{j=1}^{N} B^{ij} C_{55}(z_{ij}) W_j = 0 & \text{at } z = 0, \\ \sum_{j=1}^{N} A^{ij} C_{55}(z_j) \frac{\partial U_j}{\partial z} + \sum_{j=1}^{N} B^{ij} C_{55}(z_{ij}) W_j = 0 & \text{at } z = h, \end{cases} \quad (15)$$

$$\text{Either } \delta V_i = 0, \text{ or} \begin{cases} \sum_{j=1}^{N} A^{ij} C_{44}(z_j) \frac{\partial V_j}{\partial z} + \sum_{j=1}^{N} D^{ij} C_{44}(z_j) W_j = 0 & \text{at } z = 0, \\ \sum_{j=1}^{N} A^{ij} C_{44}(z_j) \frac{\partial V_j}{\partial z} + \sum_{j=1}^{N} D^{ij} C_{44}(z_j) W_j = 0 & \text{at } z = h, \end{cases} \quad (16)$$

$$\text{Either } \delta W_i = 0, \text{ or} \begin{cases} \sum_{j=1}^{N} A^{ij} C_{11}(z_j) \frac{\partial W_j}{\partial z} + \sum_{j=1}^{N} D^{ij} C_{12}(z_j) V_j + \sum_{j=1}^{N} B^{ij} C_{12}(z_j) U_j = 0 & \text{at } z = 0 \\ \sum_{j=1}^{N} A^{ij} C_{11}(z_j) \frac{\partial W_j}{\partial z} + \sum_{j=1}^{N} D^{ij} C_{12}(z_j) V_j + \sum_{j=1}^{N} B^{ij} C_{12}(z_j) U_j = 0 & \text{at } z = h \end{cases} \quad (17)$$

As far as the GDQ method is concerned, this approach discretizes the spatial derivatives of a function $f(z,t)$ as a weighted linear sum of the functional values at all nodes in the solution domain, by means of some fixed weighting coefficients. Thus, the first- and second-order derivatives of a one-dimensional function read as follows:

$$\left. \frac{\partial f(z,t)}{\partial z} \right|_{z=z_i} = \sum_{j=1}^{N_z} A^z_{ij} f(z_j,t) = \sum_{j=1}^{N_z} A^z_{ij} f_j(t)$$

$$\left. \frac{\partial^2 f(z,t)}{\partial z^2} \right|_{z=z_i} = \sum_{j=1}^{N_z} B^z_{ij} f(z_j,t) = \sum_{j=1}^{N_z} B^z_{ij} f_j(t) \quad (18)$$

where A^z_{ij} and B^z_{ij} are the weighted coefficients at the grid nodes of the solution domain. To derive the weighting coefficients, the following relations are employed:

$$A^z_{ij} = \begin{cases} \dfrac{M(z_i)}{(z_i - z_j) M(z_i)} & \text{for } i \neq j \\ -\sum_{k=1, k \neq i}^{N_z} A^z_{ik} & \text{for } i = j \end{cases} \quad i,j = 1,2,\ldots,N_z, \quad (19)$$

$$B^z_{ij} = \begin{cases} 2\left[A^z_{ii} A^z_{ij} - \dfrac{A^z_{ij}}{z_i - z_j} \right] & \text{for } i \neq j, \\ -\sum_{k=1, k \neq i}^{N_z} B^z_{ik} & \text{for } i = j \end{cases} \quad i,j = 1,2,\ldots,N_z, \quad (20)$$

being

$$M^{(1)}(z_i) = \prod_{j=1, j \neq i}^{N} (z_i - z_j) \text{ for } i = 1,2,\ldots,N$$

To obtain more accurate results, a Chebyshev–Gauss–Lobatto quadrature-mesh size is assumed here, in line with findings by Malik and Bert [43]. At the current stage, the GDQ method is employed to discretize the system of equations through the thickness direction (i.e., along the z-axis). A set of N_z grid points is assumed to discretize the domain along the thickness direction for each quadratic grid point. This means that Equations (11)–(13) can be rewritten in the domain (i.e., for each node $k = 2, 3, \ldots, N_z - 1$), as follows:

$$
\begin{aligned}
\delta U_i : & \sum_{j=1}^{N}\sum_{m=1}^{N_z} A^{ij} C_{55}(z_j) B_{km} U_{jm} + \sum_{j=1}^{N} \frac{\partial C_{55}(z_j)}{\partial z} A^z_{km} U_{jm} + \sum_{j=1}^{N}\sum_{m=1}^{N_z} \left(B^{ij} C_{55}(z_j) - B^{ij} C_{12}(z_j)\right) A^z_{km} W_{jm} \\
& + \sum_{j=1}^{N} \frac{\partial C_{44}}{\partial z}(z_j) W_j - \sum_{j=1}^{N} \left(E^{ji} C_{12}(z_j) + E^{ij} C_{66}(z_j)\right) U_j - \sum_{j=1}^{N} \left(F^{ij} C_{11}(z_j) + H^{ji} C_{66}(z_j)\right) V_j \\
& - P_x \sum_{j=1}^{N} D_b^{ij} V_j - P_y \sum_{j=1}^{N} B_b^{ij} V_j - P_{xy} \sum_{j=1}^{N} S_b^{ij} V_j - P_{xy} \sum_{j=1}^{N} K_b^{ij} V_j = 0
\end{aligned}
\quad (21)
$$

$$
\begin{aligned}
\delta V_i : & \sum_{j=1}^{N}\sum_{m=1}^{N_z} A^{ij} C_{44}(z_j) B_{km} V_{jm} + \sum_{j=1}^{N}\sum_{m=1}^{N_z} \frac{\partial C_{44}(z_j)}{\partial z} A^z_{km} V_{km} + \sum_{j=1}^{N}\sum_{m=1}^{N_z} \left(D^{ij} C_{44}(z_j) - D^{ji} C_{12}(z_j)\right) A^z_{km} W_{km} \\
& + \sum_{j=1}^{N} \frac{\partial C_{44}}{\partial z}(z_j) W_j - \sum_{j=1}^{N} \left(E^{ji} C_{12}(z_j) + E^{ij} C_{66}(z_j)\right) U_j - \sum_{j=1}^{N} \left(F^{ij} C_{11}(z_j) + H^{ji} C_{66}(z_j)\right) V_j \\
& - P_x \sum_{j=1}^{N} D_b^{ij} V_j - P_y \sum_{j=1}^{N} B_b^{ij} V_j - P_{xy} \sum_{j=1}^{N} S_b^{ij} V_j - P_{xy} \sum_{j=1}^{N} K_b^{ij} V_j = 0
\end{aligned}
\quad (22)
$$

$$
\begin{aligned}
\delta W_i : & \sum_{j=1}^{N}\sum_{m=1}^{N_z} A^{ij} C_{11}(z_j) B^z_{km} W_{jm} + \sum_{j=1}^{N}\sum_{m=1}^{N_z} \left(B^{ij} C_{12}(z_j) - B^{ij} C_{55}(z_j)\right) A^z_{km} U_{jm} \\
& + \sum_{j=1}^{N}\sum_{m=1}^{N_z} \left(D^{ij} C_{12}(z_j) - D^{ji} C_{44}(z_j)\right) A^z_{km} V_{jm} - \sum_{j=1}^{N} B^{ij} \frac{\partial C_{12}(z)}{\partial z} U_j + \sum_{j=1}^{N} D^{ij} \frac{\partial C_{12}(z)}{\partial z} V_j \\
& + \sum_{j=1}^{N} \left(F^{ij} C_{44}(z_j) + H^{ij} C_{55}(z_j) + A^{ij} \frac{\partial C_{11}(z)}{\partial z}\right) W_j - P_x \sum_{j=1}^{N} D_b^{ij} W_j - P_y \sum_{j=1}^{N} B_b^{ij} W_j \\
& - P_{xy} \sum_{j=1}^{N} S_b^{ij} W_j - P_{xy} \sum_{j=1}^{N} K_b^{ij} W_j = 0
\end{aligned}
\quad (23)
$$

Likewise, the BCs in Equations (15)–(17) at the top and bottom sides of the structures take the following form:

$$
\text{Either } U_{ik} = 0, \text{ or } \begin{cases} \sum_{j=1}^{N}\sum_{m=1}^{N_z} A^{ij} C_{55}(z_j) A^z_{km} U_{jm} + \sum_{j=1}^{N} B^{ij} C_{55}(z_j) W_{jk} = 0 & \text{for } k=1 \\ \sum_{j=1}^{N}\sum_{m=1}^{N_z} A^{ij} C_{55}(z_j) A^z_{km} U_{jm} + \sum_{j=1}^{N} B^{ij} C_{55}(z_j) W_{jk} = 0 & \text{for } k=N_Z \end{cases}
\quad (24)
$$

$$
\text{Either } V_{ik} = 0, \text{ or } \begin{cases} \sum_{j=1}^{N}\sum_{m=1}^{N_z} A^{ij} C_{44}(z_j) A^z_{km} V_{jm} + \sum_{j=1}^{N} D^{ij} C_{44}(z_j) W_{jk} = 0 & \text{for } k=1 \\ \sum_{j=1}^{N}\sum_{m=1}^{N_z} A^{ij} C_{44}(z_j) A^z_{km} V_{jm} + \sum_{j=1}^{N} D^{ij} C_{44}(z_j) W_{jk} = 0 & \text{for } k=N_Z \end{cases}
\quad (25)
$$

$$
\text{Either } W_{ik} = 0, \text{ or } \begin{cases} \sum_{j=1}^{N}\sum_{m=1}^{N_z} A^{ij} C_{11}(z_j) A^z_{km} W_{jm} + \sum_{j=1}^{N} D^{ij} C_{12}(z_j) V_{jk} + \sum_{j=1}^{N} B^{ij} C_{12}(z_j) U_{jk} = 0, & \text{for } k=1 \\ \sum_{j=1}^{N}\sum_{m=1}^{N_z} A^{ij} C_{11}(z_j) A^z_{km} W_{jm} + \sum_{j=1}^{N} D^{ij} C_{12}(z_j) V_{jk} + \sum_{j=1}^{N} B^{ij} C_{12}(z_j) U_{jk} = 0, & \text{for } k=N_Z \end{cases}
\quad (26)
$$

For a unified treatment of the problem, the degrees of freedom (DOFs) can be divided into the domain- and boundary-type DOFs, as follows:

$$\mathbf{U}_d = \begin{Bmatrix} U_{12} \\ U_{13} \\ \vdots \\ U_{N(N_z-1)} \end{Bmatrix}, \mathbf{V}_d = \begin{Bmatrix} V_{12} \\ V_{13} \\ \vdots \\ V_{N(N_z-1)} \end{Bmatrix}, \mathbf{W}_d = \begin{Bmatrix} W_{12} \\ W_{13} \\ \vdots \\ W_{N(N_z-1)} \end{Bmatrix},$$

$$\mathbf{U}_b = \begin{Bmatrix} U_{11} \\ U_{21} \\ \vdots \\ U_{NN_z} \end{Bmatrix}, \mathbf{V}_b = \begin{Bmatrix} V_{11} \\ V_{21} \\ \vdots \\ V_{NN_z} \end{Bmatrix}, \mathbf{W}_b = \begin{Bmatrix} W_{11} \\ W_{21} \\ \vdots \\ W_{NN_z} \end{Bmatrix}, \quad (27)$$

In which $W_{mn} = W_m(z_n, t)$, $V_{mn} = V_m(z_n, t)$, $U_{mn} = U_m(z_n, t)$.

This means that Equations (21)–(23) can be rearranged in matrix form as:

$$\begin{bmatrix} \mathbf{k}_{bb} & \mathbf{k}_{bd} \\ \mathbf{k}_{db} & \mathbf{k}_{dd} \end{bmatrix} \begin{bmatrix} \mathbf{d}_b \\ \mathbf{d}_d \end{bmatrix} = P \begin{bmatrix} 0 & 0 \\ 0 & \mathbf{G} \end{bmatrix} \begin{bmatrix} \mathbf{d}_b \\ \mathbf{d}_d \end{bmatrix} \quad (28)$$

where the stiffness quantities $\mathbf{k}_{bb}$, $\mathbf{k}_{db}$, $\mathbf{k}_{bd}$, $\mathbf{k}_{dd}$ refer to the boundary, b, and domain, d, weighting coefficients of the plate respectively, $[\mathbf{d}_b \ \mathbf{d}_d]^T$ is the displacement vector, P refers to the buckling load and $\mathbf{G}$ is the stability matrix due to the in-plane stresses.

At the same time, from Equations (23)–(26), the boundary weight coefficients can be replaced by the domain weight coefficients as follows:

$$\mathbf{d}_b = \mathbf{k}_{bb}^{-1} \mathbf{k}_{bd} \mathbf{d}_d \quad (29)$$

By substitution of $\mathbf{d}_b$ from Equation (29) into Equation (28), and considering the harmonic solution $\begin{bmatrix} \mathbf{U}_d \\ \mathbf{V}_d \\ \mathbf{W}_d \end{bmatrix} = \begin{bmatrix} \overline{\mathbf{U}}_d \\ \overline{\mathbf{V}}_d \\ \overline{\mathbf{W}}_d \end{bmatrix} e^{iwt}$, the governing equations of the problem can be rewritten in terms of the domain unknowns, as follows:

$$\mathbf{K} \begin{bmatrix} \overline{\mathbf{U}}_d \\ \overline{\mathbf{V}}_d \\ \overline{\mathbf{W}}_d \end{bmatrix} - P\mathbf{G} \begin{bmatrix} \overline{\mathbf{U}}_d \\ \overline{\mathbf{V}}_d \\ \overline{\mathbf{W}}_d \end{bmatrix} = 0 \quad (30)$$

where $\mathbf{K}$ is the stiffness matrix, defined as:

$$\mathbf{K} = \mathbf{K}_{dd} - \mathbf{K}_{db} \mathbf{K}_{bb}^{-1} \mathbf{K}_{bd} \quad (31)$$

The solution of Equation (30) corresponds to the critical buckling load of the structure under in-plane conditions, labeled hereafter as λ.

In what follows, three different BCs for the buckling analysis of the plate structure are considered:

(i) Simply supported BCs at all edges (SSSS):

$$\begin{aligned} w(0,y,z) &= w(a,y,z) = w(x,0,z) = w(x,b,z) = 0 \\ u(0,y,z) &= u(a,y,z) = v(x,0,z) = v(x,b,z) = 0 \end{aligned} \quad (32)$$

(ii) Clamped BCs at edges parallel to the y-axis (i.e., at $x = 0$, a) and free BCs at edges parallel to the x-axis (i.e., at $y = 0$, b) (CFCF):

$$u, v, w(0, y, z) = u, v, w(a, y, z) = 0 \quad (33)$$

(iii) Clamped BCs at edges parallel to the x-axis (i.e., at $y = 0$, b) and free BCs at edges parallel to the y-axis (i.e., at $x = 0$, a) (FCFC):

$$u, v, w(x, 0, z) = u, v, w(x, b, z) = 0 \quad (34)$$

Note that the (i)-type BC is considered for both uniaxial and biaxial loading conditions.

4. Numerical Investigation

The numerical study starts with a preliminary validation of the proposed formulation against the classical FE results (Ansys Workbench). At the present stage, the porosity coefficient is assumed to be zero ($e_0 = 0$), together with a null Skempton coefficient, a null Biot's modulus $\overline{M} = 0$, a null pore fluid pressure $p = 0$ and $\nu_u = \nu = 1/3$, $E_1 = 210$ GPa. The rectangular plate selected here for the analysis has $b = 1$ m, $a/b = 2$ and $h = 0.1$ m. Table 1 summarizes the results for the first three buckling loads (λ_1, λ_2, λ_3), as provided by our proposed FE-GDQ formulation and FEs, while considering the three different BCs (32)–(34) alternatively, as well as a uniaxial or biaxial loading condition. To model the problem in Ansys Workbench, the most accurate 20-node hexahedral quadratic elements were chosen to mesh the plate. First, a linear static analysis for edge loads equal to 1 Pa was performed in a static structural environment, and then the solutions were transferred to an eigenvalue buckling environment. Based on the results from this table, the good correspondence among predictions from the two alternative computational strategies proves the reliability and accuracy of the proposed FE-GDQ method to handle the problem. This is also confirmed in terms of mode shapes, as visible in the contour plots of Figures 2 and 3, at least for the SSSS rectangular plate under a shear and biaxial loading, respectively.

Table 1. Comparative evaluation of the first three buckling loads (10 GPa), as provided by our formulation and from FEM for different boundary conditions.

	BC	FE-GDQ			FE			Difference (%)		
		λ_1	λ_2	λ_3	λ_1	λ_2	λ_3	λ_1	λ_2	λ_3
Shear load	CFCF	0.314	0.529	0.624	0.317	0.574	0.678	0.94	7.83	7.96
Shear load	FCFC	1.405	1.406	2.168	1.430	1.430	2.117	1.74	1.67	2.35
Shear load	SSSS	2.534	2.642	3.062	2.575	2.691	3.0628	1.59	1.82	0
Axial load (x-direction)	SSSS	1.197	1.231	2.381	1.206	1.240	2.401	0.74	0.72	6.7
Biaxial load	SSSS	0.259	0.489	0.790	0.264	0.494	0.794	1.89	1.01	0.506

After this validation step, the numerical study aimed at computing the buckling load of FG-saturated porous plates in undrained conditions, accounting for the effects of different BCs, aspect ratios, Skempton coefficients, porosity distributions and porosity coefficients, while keeping the geometry and material properties of the structure fixed. In Tables 2–4, the first four shear buckling loads are summarized for a FG-saturated plate under the SSSS, FCFC and CFCF BCs respectively, while keeping the Skempton coefficient constant. The systematic study starts by considering a square plate with aspect ratio $a/b = 1$, thus extending the analysis to a rectangular plate with $a/b = 2$. In each case, the porosity coefficient, e_0, is gradually increased from 0.3 up to 0.6, by steps of 0.3, to check for the sensitivity of the buckling response to porosity. As expected, when the porosity coefficient increases, the stiffness of the structure decreases, and the buckling load decreases as well. The results demonstrate that the maximum and minimum values of the buckling load are associated with the symmetric (PNSD) and uniform (PUD) porosity distributions respectively, due to the highest and lowest stiffness reached in the structure. An intermediate buckling load level, instead, is always obtained for a PNND porosity distribution within the material. It also seems that the effect of the porosity coefficient, e_0, on the buckling load becomes more pronounced for a uniform porosity distribution than the other ones. Moreover, by increasing the aspect ratio a/b, the buckling load can decrease or increase, under the same assumptions of porosity coefficient and distribution, depending on the selected BC. Differently from the FCFC case, a clear reduction of the buckling load is noticed for a SSSS and CFCF porous plate with an increased aspect ratio. This confirms the strict dependence of the stiffness on the geometrical dimensions and BCs of the structural member.

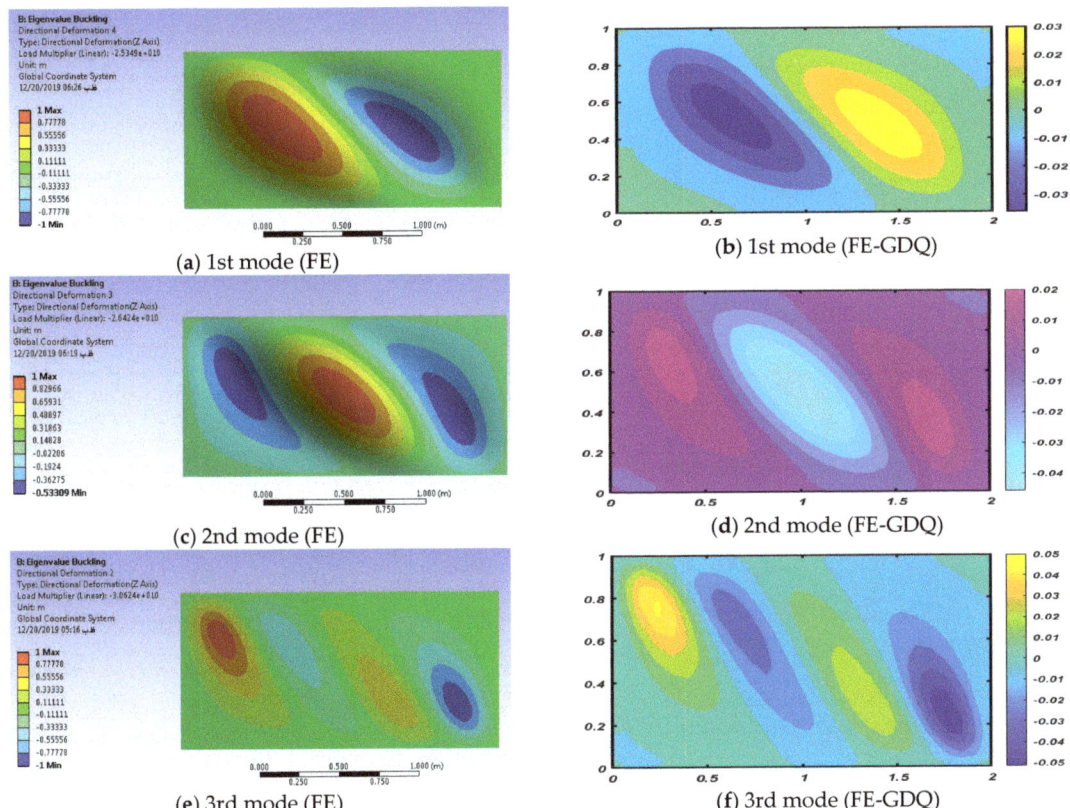

(a) 1st mode (FE) (b) 1st mode (FE-GDQ)
(c) 2nd mode (FE) (d) 2nd mode (FE-GDQ)
(e) 3rd mode (FE) (f) 3rd mode (FE-GDQ)

Figure 2. Comparative evaluation of the first three buckling mode shapes for a SSSS homogenous plate under a shear loading condition.

Table 2. First four shear buckling loads for a SSSS FG-saturated porous plate with different aspect ratios, a/b, porosity distributions and porosity coefficients, e_0, keeping $\beta = 0.6$ fixed.

Aspect Ratio	Buckling Load (10 GPa)	PNND			PNSD			PUD		
		$e_0 = 0.3$	$e_0 = 0.6$	$e_0 = 0.9$	$e_0 = 0.3$	$e_0 = 0.6$	$e_0 = 0.9$	$e_0 = 0.3$	$e_0 = 0.6$	$e_0 = 0.9$
$a/b = 1$	λ_1	2.666	1.935	1.226	2.674	1.961	1.263	2.269	1.212	0.275
	λ_2	2.666	1.936	1.226	2.674	1.961	1.263	2.269	1.212	0.275
	λ_3	2.958	2.255	1.442	2.998	2.265	1.390	2.595	1.482	0.354
	λ_4	2.984	2.261	1.447	3.016	2.266	1.390	2.607	1.483	0.354
$a/b = 2$	λ_1	2.281	1.863	1.162	2.392	2.005	1.269	2.083	1.281	0.2952
	λ_2	2.339	1.865	1.162	2.438	2.029	1.271	2.105	1.282	0.2954
	λ_3	2.558	1.968	1.237	2.627	2.068	1.379	2.248	1.334	0.3464
	λ_4	2.719	2.044	1.307	2.761	2.094	1.388	2.350	1.356	0.3467

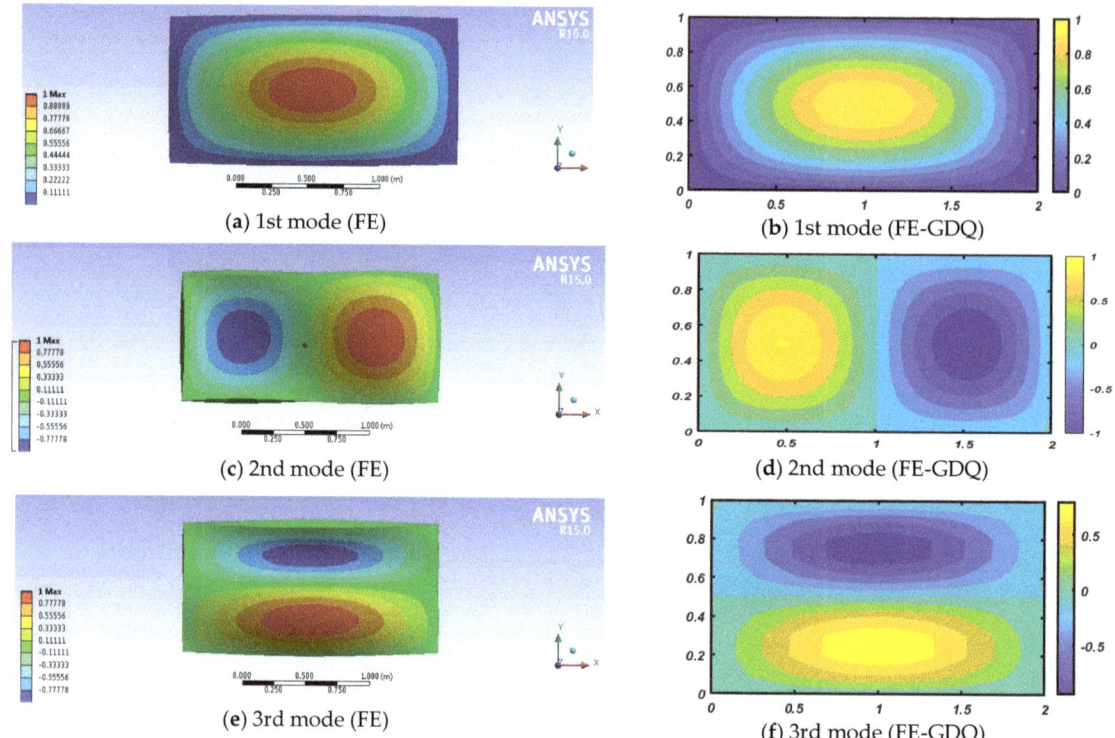

Figure 3. Comparative evaluation of the first three buckling mode shapes for a SSSS homogenous plate under a biaxial compression.

Table 3. First four shear buckling loads for a FCFC FG-saturated porous plate with different aspect ratios, a/b, porosity distributions and porosity coefficients, e_0, keeping $\beta = 0.6$ fixed.

Aspect Ratio	Buckling Load (10 GPa)	PNND			PNSD			PUD		
		$e_0 = 0.3$	$e_0 = 0.6$	$e_0 = 0.9$	$e_0 = 0.3$	$e_0 = 0.6$	$e_0 = 0.9$	$e_0 = 0.3$	$e_0 = 0.6$	$e_0 = 0.9$
$a/b = 1$	λ_1	1.143	0.897	0.536	1.199	0.978	0.692	1.020	0.636	0.177
	λ_2	1.150	0.900	0.543	1.208	0.983	0.693	1.020	0.645	0.182
	λ_3	1.653	1.267	0.773	1.705	1.337	0.845	1.455	0.865	0.219
	λ_4	1.664	1.275	0.778	1.717	1.343	0.849	1.464	0.868	0.219
$a/b = 2$	λ_1	1.216	0.954	0.576	1.276	1.039	0.737	1.082	0.680	0.191
	λ_2	1.217	0.955	0.576	1.277	1.040	0.738	1.083	0.680	0.191
	λ_3	1.768	1.360	0.834	1.824	1.401	0.912	1.558	0.931	0.238
	λ_4	1.768	1.360	0.834	1.824	1.402	0.912	1.558	0.931	0.238

Table 4. First four shear buckling loads for a CFCF FG-saturated porous plate with different aspect ratios, a/b, porosity distributions and porosity coefficients, e_0, keeping $\beta = 0.6$ fixed.

Aspect Ratio	Buckling Load (10 GPa)	PNND			PNSD			PUD		
		$e_0 = 0.3$	$e_0 = 0.6$	$e_0 = 0.9$	$e_0 = 0.3$	$e_0 = 0.6$	$e_0 = 0.9$	$e_0 = 0.3$	$e_0 = 0.6$	$e_0 = 0.9$
$a/b = 1$	λ_1	1.143	0.897	0.536	1.199	1.006	0.692	1.020	0.636	0.177
	λ_2	1.150	0.900	0.543	1.208	1.007	0.693	1.020	0.645	0.182
	λ_3	1.653	1.267	0.773	1.705	1.434	0.845	1.455	0.865	0.219
	λ_4	1.664	1.275	0.778	1.717	1.350	0.849	1.464	0.868	0.219
$a/b = 2$	λ_1	0.310	0.255	0.153	0.330	0.292	0.241	0.287	0.186	0.055
	λ_2	0.497	0.393	0.229	0.538	0.470	0.375	0.449	0.281	0.078
	λ_3	0.598	0.466	0.281	0.610	0.520	0.405	0.528	0.325	0.096
	λ_4	0.805	0.622	0.374	0.817	0.682	0.504	0.708	0.430	0.116

The same parametric investigation was then repeated for a biaxial loading (Table 5) and a uniaxial loading acting in the x-direction of a SSSS structure (Table 6). Based on a comparative evaluation of results in Tables 5 and 6 with Table 2, a uniaxial or biaxial loading condition clearly reduces the buckling load of the structure under the same values of e_0, a/b and porosity distribution.

Table 5. First four biaxial buckling loads for a SSSS FG-saturated porous plate with different aspect ratios, a/b, porosity distributions and porosity coefficients, e_0, keeping $\beta = 0.6$ fixed.

Aspect Ratio	Buckling Load (10 GPa)	PNND			PNSD			PUD		
		$e_0 = 0.3$	$e_0 = 0.6$	$e_0 = 0.9$	$e_0 = 0.3$	$e_0 = 0.6$	$e_0 = 0.9$	$e_0 = 0.3$	$e_0 = 0.6$	$e_0 = 0.9$
a/b = 1	λ_1	0.392	0.303	0.166	0.406	0.362	0.309	0.346	0.216	0.069
	λ_2	0.929	0.723	0.408	0.949	0.830	0.658	0.827	0.523	0.151
	λ_3	0.929	0.723	0.408	0.949	0.830	0.658	0.827	0.523	0.151
	λ_4	0.141	0.108	0.604	0.144	0.127	0.926	1.245	0.763	0.210
a/b = 2	λ_1	0.233	0.186	0.106	0.240	0.221	0.193	0.209	0.138	0.0420
	λ_2	0.434	0.336	0.188	0.405	0.400	0.329	0.384	0.239	0.0680
	λ_3	0.698	0.564	0.333	0.703	0.634	0.519	0.634	0.426	0.120
	λ_4	0.778	0.603	0.341	0.804	0.702	0.570	0.688	0.427	0.131

Table 6. First four uniaxial buckling loads (in the x-direction) for a SSSS FG-saturated porous plate with different aspect ratios, a/b, porosity distributions and porosity coefficients, e_0, keeping $\beta = 0.6$ fixed.

Aspect Ratio	Buckling Load (10 GPa)	PNND			PNSD			PUD		
		$e_0 = 0.3$	$e_0 = 0.6$	$e_0 = 0.9$	$e_0 = 0.3$	$e_0 = 0.6$	$e_0 = 0.9$	$e_0 = 0.3$	$e_0 = 0.6$	$e_0 = 0.9$
a/b = 1	λ_1	0.781	0.603	0.330	0.808	0.721	0.614	0.690	0.430	0.122
	λ_2	0.981	0.774	0.446	0.983	0.875	0.703	0.879	0.568	0.168
	λ_3	1.554	1.225	0.724	1.526	1.320	0.981	1.394	0.895	0.261
	λ_4	2.147	1.688	1.017	2.081	1.856	1.218	1.925	1.225	0.350
a/b = 2	λ_1	1.053	0.816	0.462	1.073	0.943	0.771	0.930	0.574	0.159
	λ_2	1.101	0.860	0.495	1.117	0.982	0.795	0.977	0.613	0.173
	λ_3	2.097	1.553	0.880	2.082	1.725	1.314	1.801	1.020	0.260
	λ_4	2.331	1.771	1.042	2.293	1.907	1.405	2.035	1.204	0.319

As expected, this reduction is much more pronounced for rectangular plates subjected to a biaxial loading condition, due to the overall decay of the structural stiffness. A further investigation considered the effect of the porosity distribution and Skempton coefficient on the first four buckling loads of FG-saturated plates, as listed in Tables 7–9 (for a shear loading condition), in Table 10 (for an axial loading condition) and in Table 11 (for a biaxial loading condition), while keeping the porosity coefficient fixed at $e_0 = 0.6$. The same BCs from Table 1 are accounted here for the analyses. Based on a comparative estimation of results from these tables, it is confirmed that the maximum and minimum buckling loads are always associated with a symmetric (PNSD) and uniform (PUD) porosity distribution, respectively.

Table 7. First four shear buckling loads for a SSSS FG-saturated porous plate with different aspect ratios, a/b, porosity distributions and Skempton coefficients, β, while keeping $e_0 = 0.6$ fixed.

Aspect Ratio	Buckling Load (10 GPa)	PNND			PNSD			PUD		
		$\beta = 0.0$	$\beta = 0.6$	$\beta = 0.9$	$\beta = 0.0$	$\beta = 0.6$	$\beta = 0.9$	$\beta = 0.0$	$\beta = 0.6$	$\beta = 0.9$
a/b = 1	λ_1	1.839	1.935	1.941	1.954	1.961	1.982	1.196	1.212	1.303
	λ_2	1.913	1.944	1.951	1.957	1.961	1.986	1.196	1.212	1.303
	λ_3	2.216	2.255	2.264	2.236	2.265	2.274	1.478	1.480	1.486
	λ_4	2.240	2.261	2.273	2.250	2.266	2.275	1.478	1.483	1.486
a/b = 2	λ_1	1.689	1.863	1.963	1.942	2.005	2.012	1.190	1.281	1.326
	λ_2	1.738	1.865	1.964	1.953	2.029	2.043	1.203	1.283	1.326
	λ_3	1.913	1.968	2.018	2.050	2.068	2.094	1.284	1.334	1.341
	λ_4	2.038	2.044	2.048	2.130	2.139	2.147	1.343	1.356	1.386

Table 8. First four shear buckling loads for a FCFC FG-saturated porous plate with different aspect ratios, a/b, porosity distributions and Skempton coefficients, β, while keeping $e_0 = 0.6$ fixed.

Aspect Ratio	Buckling Load (10 GPa)	PNND			PNSD			PUD		
		$\beta = 0.0$	$\beta = 0.6$	$\beta = 0.9$	$\beta = 0.0$	$\beta = 0.6$	$\beta = 0.9$	$\beta = 0.0$	$\beta = 0.6$	$\beta = 0.9$
$a/b = 1$	λ_1	0.848	0.897	0.955	0.969	0.978	0.989	0.583	0.636	0.709
	λ_2	0.853	0.900	0.967	0.978	0.983	0.993	0.583	0.645	0.728
	λ_3	1.231	1.267	1.299	1.329	1.337	1.346	0.831	0.865	0.878
	λ_4	1.240	1.275	1.306	1.338	1.343	1.371	0.836	0.868	0.879
$a/b = 2$	λ_1	0.903	0.954	1.021	1.031	1.039	1.042	0.618	0.680	0.765
	λ_2	0.904	0.955	1.026	1.032	1.040	1.068	0.619	0.680	0.765
	λ_3	1.318	1.360	1.399	1.423	1.401	1.414	0.890	0.931	0.955
	λ_4	1.318	1.360	1.399	1.423	1.402	1.416	0.890	0.931	0.955

Table 9. First four shear buckling loads for a CFCF FG-saturated porous plate with different aspect ratios, a/b, porosity distributions and Skempton coefficients, β, while keeping $e_0 = 0.6$ fixed.

Aspect Ratio	Buckling Load (10 GPa)	PNND			PNSD			PUD		
		$\beta = 0.0$	$\beta = 0.6$	$\beta = 0.9$	$\beta = 0.0$	$\beta = 0.6$	$\beta = 0.9$	$\beta = 0.0$	$\beta = 0.6$	$\beta = 0.9$
$a/b = 1$	λ_1	0.848	0.897	0.955	0.969	0.978	0.989	0.583	0.636	0.709
	λ_2	0.853	0.900	0.967	0.978	0.983	0.993	0.583	0.645	0.728
	λ_3	1.231	1.267	1.299	1.329	1.337	1.346	0.831	0.865	0.878
	λ_4	1.240	1.275	1.306	1.338	1.343	1.371	0.836	0.868	0.879
$a/b = 2$	λ_1	0.237	0.255	0.280	0.284	0.292	0.296	0.164	0.186	0.222
	λ_2	0.367	0.393	0.425	0.494	0.470	0.482	0.254	0.281	0.314
	λ_3	0.458	0.466	0.500	0.520	0.524	0.529	0.302	0.325	0.387
	λ_4	0.600	0.622	0.615	0.682	0.689	0.798	0.404	0.430	0.465

Table 10. First four biaxial buckling loads for a SSSS FG-saturated porous plate with different aspect ratios, a/b, porosity distributions and Skempton coefficients, β, while keeping $e_0 = 0.6$ fixed.

Aspect Ratio	Buckling Load (10 GPa)	PNND			PNSD			PUD		
		$\beta = 0.0$	$\beta = 0.6$	$\beta = 0.9$	$\beta = 0.0$	$\beta = 0.6$	$\beta = 0.9$	$\beta = 0.0$	$\beta = 0.6$	$\beta = 0.9$
$a/b = 1$	λ_1	0.288	0.303	0.325	0.361	0.362	0.389	0.198	0.216	0.246
	λ_2	0.683	0.723	0.781	0.825	0.830	0.846	0.472	0.523	0.603
	λ_3	0.683	0.723	0.781	0.825	0.830	0.846	0.472	0.523	0.603
	λ_4	0.103	0.108	0.146	0.123	0.127	0.131	0.711	0.763	0.843
$a/b = 2$	λ_1	0.171	0.186	0.207	0.213	0.221	0.229	0.119	0.138	0.168
	λ_2	0.319	0.369	0.362	0.397	0.400	0.405	0.219	0.239	0.273
	λ_3	0.513	0.564	0.635	0.616	0.634	0.638	0.362	0.426	0.480
	λ_4	0.573	0.603	0.644	0.697	0.702	0.713	0.393	0.427	0.526

Results denote that in drained conditions (i.e., $\beta = 0$), the plate always features the smallest buckling load, under a fixed aspect ratio, porosity coefficient and distribution. An increasing value of the Skempton coefficient, instead, enables an increased buckling load, because of a decreased compressibility of the fluid within pores. In other words, if the compressibility of the pore fluid becomes high ($\beta \to 0$), the mechanical response of the plate resembles that of a porous plate in drained conditions (i.e., in the absence of fluid). In this condition, the structural stiffness reaches its minimum value along with the lowest buckling load. Differently, when the compressibility of a pore fluid becomes small ($\beta \to 1$), the plate behaves as a rigid solid, thus reaching its highest load magnitude. Furthermore, the effect of the Skempton coefficient on the buckling load, for a uniform distribution, seems to be more pronounced than other porosity distributions. By comparing Tables 2–6 with Tables 7–11, it is worth observing the higher sensitivity of the buckling response to the porosity coefficient than the Skempton coefficient. The first four buckling mode shapes are finally plotted in Figures 4–8, for a rectangular plate with $a = 2$ m, $b = 1$ m, under different loading and boundary conditions, and a fixed value of $e_0 = \beta = 0.6$. More specifically, in Figures 4–6, the rectangular plate is subjected to a shear loading condition, with a clear compatibility among the displacement field and the selected BCs (i.e., CFCF, FCFC and

SSSS, respectively). Figures 7 and 8 plot the four mode shapes for the same plate under a uniaxial (Figure 7) and biaxial (Figure 8) loading, where the kinematic response clearly changes depending on the selected loading condition.

Table 11. First four uniaxial buckling loads (x-direction) for a SSSS FG-saturated porous plate with different aspect ratios, a/b, porosity distributions and Skempton coefficients, β, while keeping $e_0 = 0.6$ fixed.

Aspect Ratio	Buckling Load (10 GPa)	PNND			PNSD			PUD		
		$\beta = 0.0$	$\beta = 0.6$	$\beta = 0.9$	$\beta = 0.0$	$\beta = 0.6$	$\beta = 0.9$	$\beta = 0.0$	$\beta = 0.6$	$\beta = 0.9$
$a/b = 1$	λ_1	0.572	0.603	0.648	0.719	0.721	0.737	0.394	0.430	0.491
	λ_2	0.722	0.774	0.848	0.867	0.875	0.884	0.502	0.568	0.672
	λ_3	1.147	1.225	1.331	1.318	1.320	1.384	0.796	0.895	1.046
	λ_4	1.589	1.688	1.817	1.756	1.856	1.878	1.100	1.225	1.403
$a/b = 2$	λ_1	0.777	0.816	0.868	0.938	0.943	0.941	0.531	0.574	0.636
	λ_2	0.813	0.860	0.923	0.977	0.982	0.993	0.558	0.613	0.693
	λ_3	1.560	1.553	1.550	1.720	1.725	1.749	1.009	1.020	1.042
	λ_4	1.735	1.771	1.817	1.901	1.907	1.923	1.163	1.204	1.276

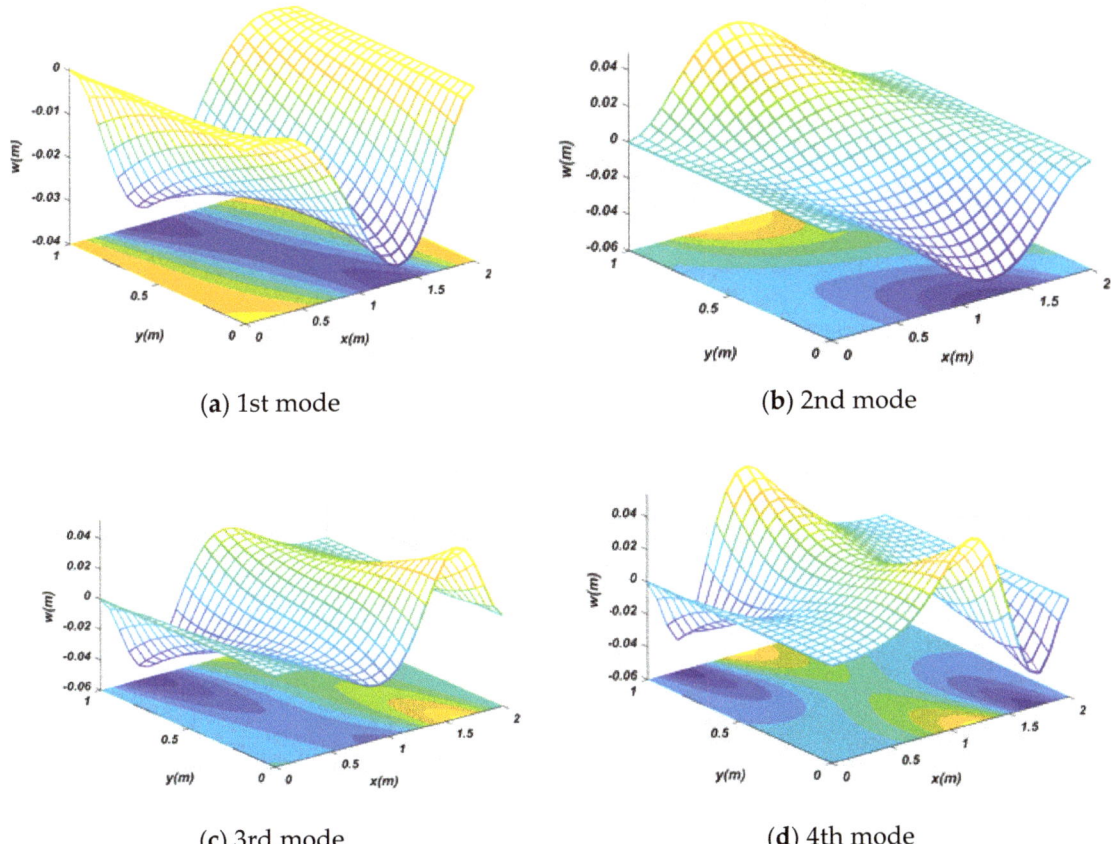

(a) 1st mode (b) 2nd mode

(c) 3rd mode (d) 4th mode

Figure 4. First four buckling mode shapes of a FG-saturated porous plate subjected to a shear load (CFCF, $a = 2$ m, $b = 1$ m).

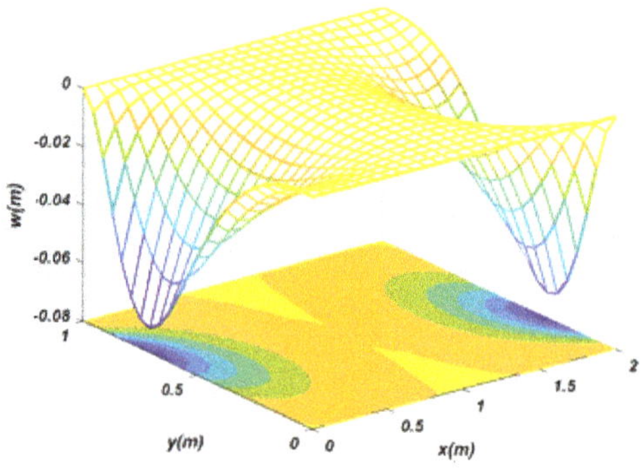

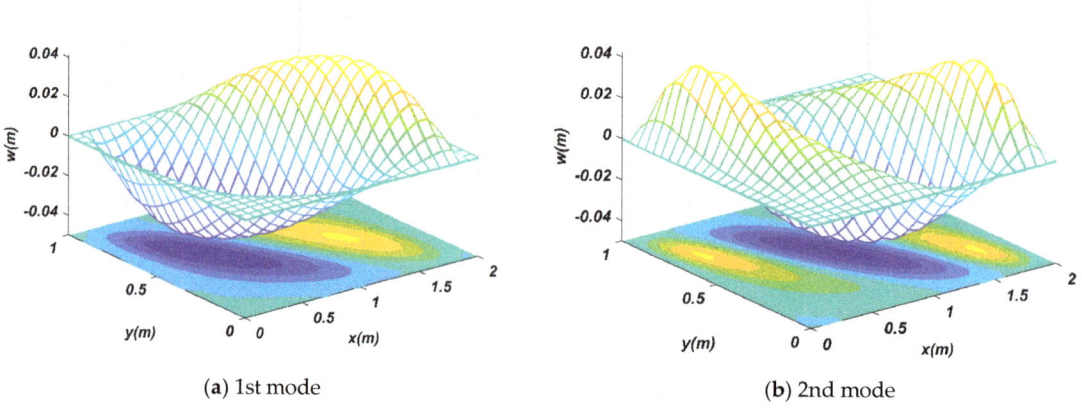

(a) 1st mode 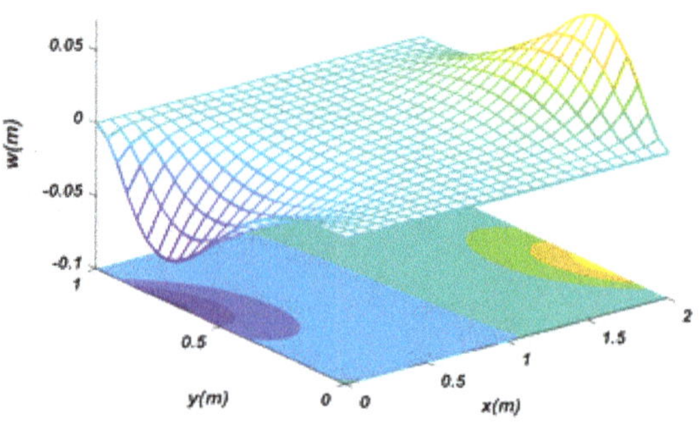 (b) 2nd mode 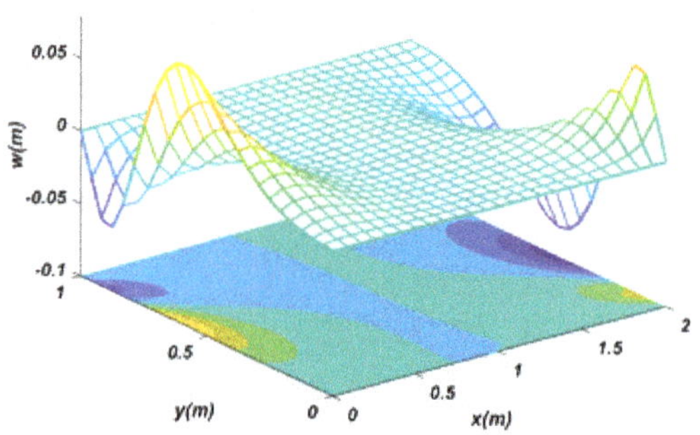 (c) 3rd mode 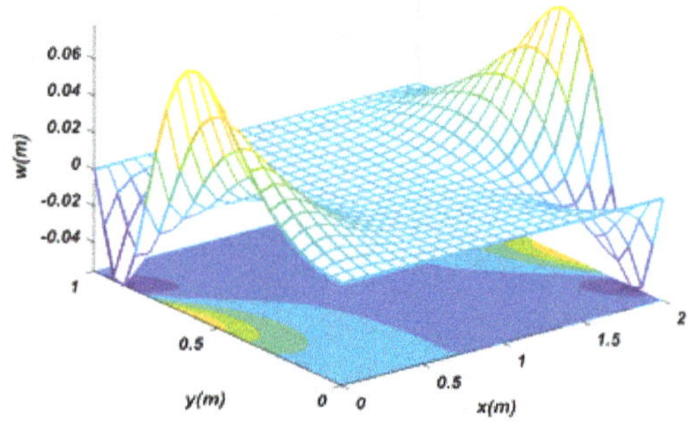 (d) 4th mode

Figure 5. First four buckling mode shapes of a FG-saturated porous plate subjected to a shear load (FCFC, $a = 2$ m, $b = 1$ m).

(a) 1st mode (b) 2nd mode

Figure 6. *Cont.*

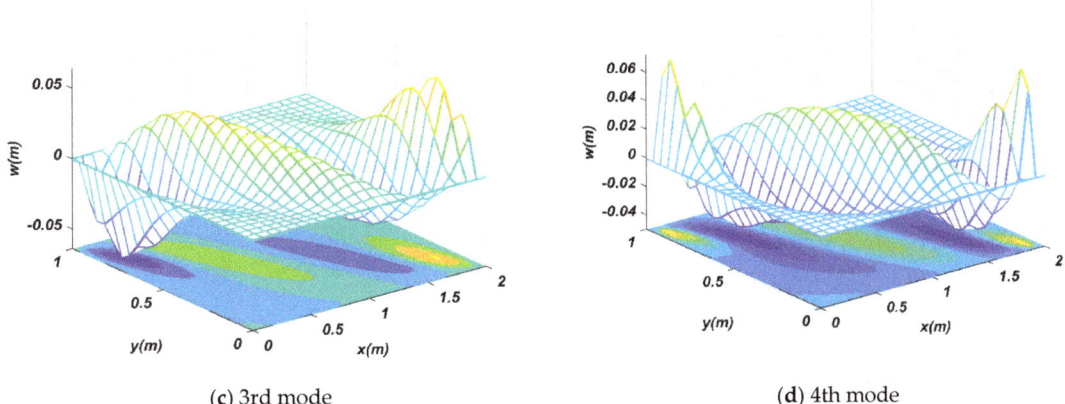

(c) 3rd mode   (d) 4th mode

Figure 6. First four buckling mode shapes of a FG-saturated porous plate subjected to a shear load (SSSS, $a = 2$ m, $b = 1$ m).

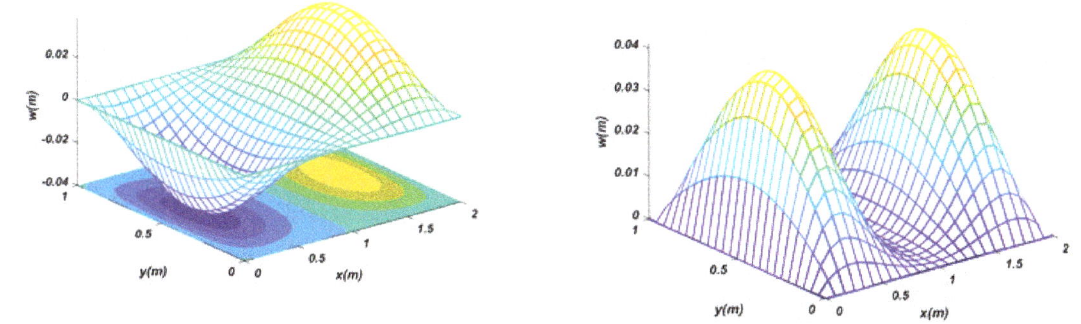

(a) 1st mode   (b) 2nd mode

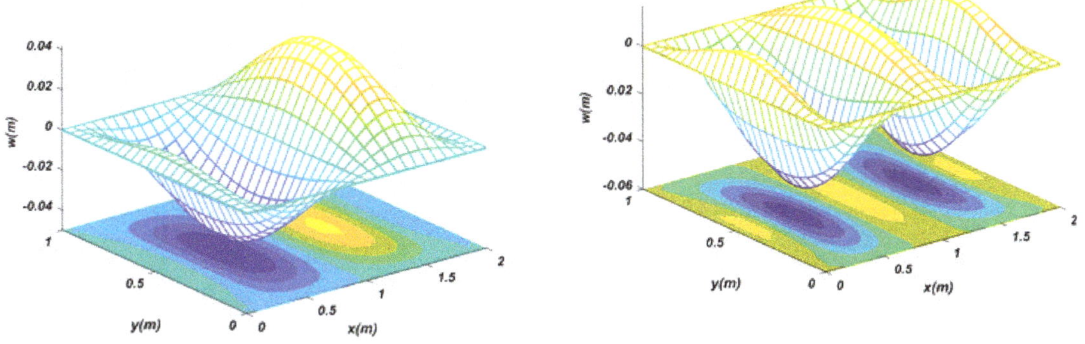

(c) 3rd mode   (d) 4th mode

Figure 7. First four buckling mode shapes of a FG-saturated porous plate subjected to uniaxial load (SSSS, $a = 2$ m, $b = 1$ m).

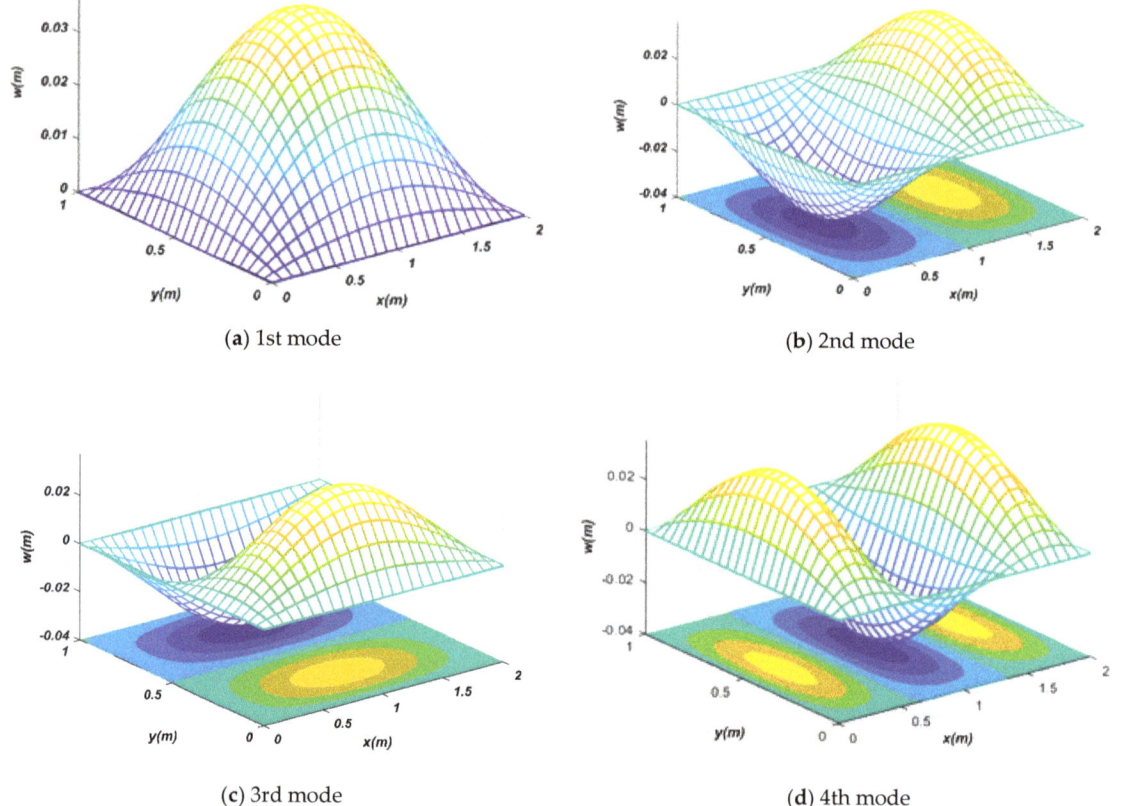

Figure 8. First four buckling mode shapes of a FG-saturated porous plate subjected to a biaxial load (SSSS, $a = 2$ m, $b = 1$ m).

5. Conclusions

The present work focused on the buckling behavior of FG-saturated porous rectangular plates subjected to normal and shear loads, while adopting Biot's constitutive law and proposing a combined FE-GDQ as an efficient computational tool to solve the problem. This means that the in-plane problem has been discretized in the x–y-directions by means of classical FEs, and follows a weak formulation. Along the thickness direction (the z-direction), instead, the problem is defined in a strong form based on a GDQ approximation. This mixed method deals with a three-dimensional theory of elasticity without any additional kinematic assumption for the plate deformability. Various numerical examples have been considered and solved systematically to check for the reliability of the proposed method against a pure FE response, as well as to study the sensitivity of the response to some input parameters, i.e., the geometrical aspect ratio, the Skempton coefficients, the porosity distribution and coefficient and the BCs. Based on the parametric analysis, the main conclusions can be summarized as follows:

- The porosity coefficient more significantly affects the buckling load than the Skempton coefficient. In detail, an increased porosity coefficient and a decreased Skempton coefficient yield an overall decrease of the buckling load.
- Among different boundary and loading conditions, the maximum and minimum values of the buckling load are reached for a FCFC plate under a shear loading and a SSSS plate under a biaxial loading condition, respectively.

- The influence of the porosity coefficient on the buckling load for a uniform distribution is larger than other types of non-uniform porosity distributions.
- The effect of the Skempton coefficient on the buckling load, for a uniform distribution, is larger than other types of porosity distributions.
- By increasing the ratio, the buckling load generally decreases, except for a FCFC plate under a shear load and a SSSS plate under a normal uniaxial load, because of the variability in stiffness of the overall structure.
- The proposed method is verified to be a reliable tool for the computational study of saturated porous materials and structures, even from a design standpoint.

Author Contributions: Conceptualization, F.T.; Data curation, F.T.; Formal analysis, F.K., M.B., K.A., R.D. and F.T.; Investigation, F.K., M.B., K.A., R.D. and F.T.; Methodology, R.D. and F.T.; Supervision, R.D. and F.T.; Validation, R.D. and F.T.; Writing—original draft, F.K., M.B. and K.A.; Writing—review & editing, R.D. and F.T. All authors have read and agreed to the published version of the manuscript.

Funding: This research received no external funding.

Conflicts of Interest: The authors declare no conflict of interest.

Nomenclature

Fluid compressibility in pores	C_P
Solid compressibility in pores	C_s
Shear modulus	G
Porosity coefficient	e_0
Biot's effective stress coefficient	α
Skempton coefficient	β
Variation of fluid volume content	Ψ
Volumetric strain	ε_{kk}
Poisson's ratio	ν
Undrained Poisson's ratio	ν_u
Lamè constant	λ
Pore fluid pressure	P
Bulk modules	K
Undrained bulk modules	K_u
Biot's modulus	M
Total strain potential energy	U
Potential energy related to geometry	V_g
Elasticity modulus	E
Stress tensor	$[\sigma_{ij}]$
Strain tensor	$[\varepsilon_{ij}]$
Elasticity matrix	$[C]$
Displacement components along x, y and z directions	u, v, w
Global Lagrange interpolation functions	$\varphi_j(x,y)$
Weighted coefficients at the grid nodes of the solution domain	A_{ij}^z, B_{ij}^z
Buckling load	λ
Stability matrix due to the in-plane stresses	**G**
Stiffness matrix	**K**

References

1. Biot, M.A. Theory of buckling of a porous slab and its thermoelastic analogy. *J. Appl. Mech. ASME* **1964**, *31*, 94–198. [CrossRef]
2. Magnucki, K.; Stasiewicz, P. Elastic buckling of a porous beam. *J. Theor. Appl. Mech.* **2004**, *42*, 859–868.
3. Magnucka-Blandzi, E. Axi-symmetrical deflection and buckling of circular porous-cellular plate. *Thin-Walled Struct.* **2008**, *46*, 333–337. [CrossRef]
4. Magnucki, K.; Malinowski, M.; Kasprzak, J. Bending and buckling of a rectangular porous plate. *Steel Compos. Struct.* **2006**, *6*, 319–333. [CrossRef]
5. Chen, D.; Yang, J.; Kitipornchai, S. Elastic buckling and static bending of shear deformable functionally graded porous beam. *Compos. Struct.* **2015**, *133*, 54–61. [CrossRef]

6. Jasion, P.; Magnucka-Blandzi, E.; Szyc, W.; Magnucki, K. Global and local buckling of sandwich circular and beam-rectangular plates with metal foam core. *Thin-Walled Struct.* **2012**, *61*, 154–161. [CrossRef]
7. Gao, K.; Huang, Q.; Kitipornchai, S.; Yang, J. Nonlinear dynamic buckling of functionally graded porous beams. *Mech. Adv. Mater. Struct.* **2019**, *28*, 418–429. [CrossRef]
8. Tang, H.; Li, L.; Hu, Y. Buckling analysis of two-directionally porous beam. *Aerosp. Sci. Techn.* **2018**, *78*, 471–479. [CrossRef]
9. Ebrahimi, F.; Jafari, A. Buckling behavior of smart MEE-FG porous plate with various boundary conditions based on refined theory. *Adv. Mater. Res.* **2016**, *5*, 279–298. [CrossRef]
10. Kumar, R.; Lal, A.; Singh, B.N.; Singh, J. Meshfree approach on buckling and free vibration analysis of porous FGM plate with proposed IHHSDT resting on the foundation. *Curved Layer. Struct.* **2019**, *6*, 192–211. [CrossRef]
11. Cong, P.H.; Chien, T.M.; Khoa, N.D.; Duc, N.D. Nonlinear thermomechanical buckling and post-buckling response of porous FGM plates using Reddy's HSDT. *Aerosp. Sci. Technol.* **2018**, *77*, 419–428. [CrossRef]
12. Bourada, M.; Bouadi, A.; Bousahla, A.A.; Senouci, A.; Bourada, F.; Tounsi, A.; Mahmoud, S.R. Buckling behavior of rectangular plates under uniaxial and biaxial compression. *Struct. Eng. Mech.* **2019**, *70*, 113–123.
13. Thang, P.T.; Nguyen-Thoi, T.; Lee, D.; Kang, J.; Lee, J. Elastic buckling and free vibration analyses of porous-cellular plates with uniform and non-uniform porosity distributions. *Aerosp. Sci. Techn.* **2018**, *79*, 278–287. [CrossRef]
14. Chen, D.; Yang, J.; Kitipornchai, S. Buckling and bending analyses of a novel functionally graded porous plate using Chebyshev-Ritz method. *Arch. Civ. Mech. Eng.* **2019**, *19*, 157–170. [CrossRef]
15. Yang, J.; Chen, D.; Kitipornchai, S. Buckling and free vibration analyses of functionally graded graphene reinforced porous nanocomposite plates based on Chebyshev-Ritz method. *Compos. Struct.* **2018**, *193*, 281–294. [CrossRef]
16. Tu, T.M.; Hoa, L.K.; Hung, D.X.; Hai, L.T. Nonlinear buckling and post-buckling analysis of imperfect porous plates under mechanical loads. *J. Sandw. Struct. Mater.* **2018**, *22*, 1910–1930. [CrossRef]
17. Sekkal, M.; Fahsi, B.; Tounsi, A.; Mahmoud, S.R. A new quasi-3D HSDT for buckling and vibration of FG plate. *Struct. Eng. Mech.* **2017**, *64*, 737–749.
18. Shahsavari, D.; Karami, B.; Fahham, H.R.; Li, L. On the shear buckling of porous nanoplates using a new size-dependent quasi-3D shear deformation theory. *Acta Mech.* **2018**, *229*, 4549–4573. [CrossRef]
19. Safaei, B.; Moradi-Dastjerdi, R.; Behdinan, K.; Chu, F. Critical buckling temperature and force in porous sandwich plates with CNT-reinforced nanocomposite layers. *Aerosp. Sci. Techn.* **2019**, *91*, 175–185. [CrossRef]
20. Li, Q.; Wu, D.; Chen, X.; Liu, L.; Yu, Y.; Gao, W. Nonlinear vibration and dynamic buckling analyses of sandwich functionally graded porous plate with graphene platelet reinforcement resting on Winkler–Pasternak elastic foundation. *Int. J. Mech. Sci.* **2018**, *148*, 596–610. [CrossRef]
21. Shahgholian, D.; Safarpour, M.; Rahimi, A.R.; Aligeigloo, A. Buckling analyses of functionally graded graphene-reinforced porous cylindrical shell using the Rayleigh–Ritz method. *Acta Mech.* **2020**, *231*, 1887–1902. [CrossRef]
22. Zhao, S.; Yang, Z.; Kitipornchai, S.; Yang, J. Dynamic instability of functionally graded porous arches reinforced by graphene platelets. *Thin-Walled Struct.* **2020**, *147*, 106491. [CrossRef]
23. Jabbari, M.; Rezaei, M.; Mojahedin, A.; Eslami, M. Mechanical buckling of FG saturated porous rectangular plate under temperature field. *Iranian J. Mech. Eng. Tran. ISME* **2016**, *17*, 61–78.
24. Jabbari, M.; Rezaei, M.; Mojahedin, A. Mechanical buckling of FG saturated porous rectangular plate with piezoelectric actuators. *Iranian J. Mech. Eng. Tran. ISME* **2016**, *17*, 46–66.
25. Mojahedin, A.; Jabbari, M.; Salavati, M. Axisymmetric buckling of saturated circular porous-cellular plate based on first-order shear deformation theory. *Int. J. Hydromech.* **2019**, *2*, 144–158. [CrossRef]
26. Jabbari, M.; Mojahedin, A.; Khorshidvand, A.R.; Eslami, M.R. Buckling analysis of functionally graded thin circular plate made of saturated porous materials. *ASCE J. Eng. Mech.* **2014**, *34*, 287–295. [CrossRef]
27. Jabbari, M.; Hashemitaheri, M.; Mojahedin, A.; Eslami, M.R. Thermal buckling analysis of functionally graded thin circular plate made of saturated porous materials. *J. Therm. Stresses* **2014**, *37*, 202–220. [CrossRef]
28. Jabbari, M.; Mojahedin, A.; Haghi, M. Buckling analysis of thin circular FG plates made of saturated porous-soft ferromagnetic materials in transverse magnetic field. *Thin-Walled Struct.* **2014**, *85*, 50–56. [CrossRef]
29. Khorshidvand, A.R.; Joubaneh, E.F.; Jabbari, M.; Eslami, M.R. Buckling analysis of a porous circular plate with piezoelectric sensor–actuator layers under uniform radial compression. *Acta Mech.* **2014**, *225*, 179–193. [CrossRef]
30. Feyzi, M.R.; Khorshidvand, A.R. Axisymmetric post-buckling behavior of saturated porous circular plates. *Thin-Walled Struct.* **2017**, *1*, 149–158. [CrossRef]
31. Rezaei, A.S.; Saidi, A.R. Buckling response of moderately thick fluid-infiltrated porous annular sector plates. *Acta Mech.* **2017**, *228*, 3929–3945. [CrossRef]
32. Anyfantis, K.N. Evaluating the influence of geometric distortions to the buckling capacity of stiffened panels. *Thin-Walled Struct.* **2014**, *140*, 450–465. [CrossRef]
33. Shahani, A.R.; Kiarasi, F. Numerical and experimental investigation on post-buckling behavior of stiffened cylindrical shells with cutout subject to uniform axial compression. *J. Appl. Comput. Mech.* **2021**, in press. [CrossRef]
34. Zhang, Y.; Huang, Y.; Meng, F. Ultimate strength of hull structural stiffened plate with pitting corrosion damage under uniaxial compression. *Mar. Struct.* **2017**, *56*, 117–136. [CrossRef]

35. Kiarasi, F.; Babaei, M.; Dimitri, R.; Tornabene, F. Hygrothermal modeling of the buckling behavior of sandwich plates with nanocomposite face sheets resting on a Pasternak foundation. *Contin. Mech. Thermodyn.* **2021**, *33*, 911–932. [CrossRef]
36. Babaei, M.; Hajmohammad, M.H.; Asemi, K. Natural frequency and dynamic analyses of functionally graded saturated porous annular sector plate and cylindrical panel based on 3D elasticity. *Aerosp. Sci. Technol.* **2020**, *96*, 105524. [CrossRef]
37. Babaei, M.; Asemi, K.; Safarpour, P. Natural frequency and dynamic analyses of functionally graded saturated porous beam resting on viscoelastic foundation based on higher order beam theory. *J. Solid Mech.* **2019**, *11*, 615–634.
38. Babaei, M.; Asemi, K.; Safarpour, P. Buckling and static analyses of functionally graded saturated porous thick beam resting on elastic foundation based on higher order beam theory. *Iranian J. Mech. Eng. Tran. ISME* **2019**, *20*, 94–112.
39. Babaei, M.; Asemi, K. Stress analysis of functionally graded saturated porous rotating thick truncated cone. *Mech. Based Des. Struct. Mach.* **2020**, 1–28. [CrossRef]
40. Babaei, M.; Asemi, K.; Kiarasi, F. Static response and free-vibration analysis of a functionally graded annular elliptical sector plate made of saturated porous material based on 3D finite element method. *Mech. Based Des. Struct. Mach.* **2020**, 1–25. [CrossRef]
41. Detournay, E.; Cheng, A.H.D. Fundamentals of poroelasticity. In *Analysis and Design Methods*; Pergamon: Oxford, UK, 1993; pp. 113–171.
42. Babaei, M.; Asemi, K.; Kiarasi, F. Dynamic analysis of functionally graded rotating thick truncated cone made of saturated porous materials. *Thin-Walled Struct.* **2021**, *164*, 107852. [CrossRef]
43. Arshid, E.; Khorshidvand, A.R. Free vibration analysis of saturated porous FG circular plates integrated with piezoelectric actuators via differential quadrature method. *Thin-Walled Struct.* **2018**, *125*, 220–233. [CrossRef]

Article

3D Stress Analysis of Multilayered Functionally Graded Plates and Shells under Moisture Conditions

Salvatore Brischetto * and Roberto Torre

Department of Mechanical and Aerospace Engineering, Politecnico di Torino, Corso Duca degli Abruzzi 24, 10129 Torino, Italy; roberto.torre@polito.it
* Correspondence: salvatore.brischetto@polito.it; Tel.: +39-011-090-6813; Fax: +39-011-090-6899

Abstract: This paper presents the steady-state stress analysis of single-layered and multilayered plates and shells embedding Functionally Graded Material (FGM) layers under moisture conditions. This solution relies on an exact layer-wise approach; the formulation is unique despite the geometry. It studies spherical and cylindrical shells, cylinders, and plates in an orthogonal mixed curvilinear coordinate system (α, β, z). The moisture conditions are defined at the external surfaces and evaluated in the thickness direction under steady-state conditions following three procedures. This solution handles the 3D Fick diffusion equation, the 1D Fick diffusion equation, and the a priori assumed linear profile. The paper discusses their assumptions and the different results they deliver. Once defined, the moisture content acts as an external load; this leads to a system of three non-homogeneous second-order differential equilibrium equations. The 3D problem is reduced to a system of partial differential equations in the thickness coordinate, solved via the exponential matrix method. It returns the displacements and their z-derivatives as a direct result. The paper validates the model by comparing the results with 3D analytical models proposed in the literature and numerical models. Then, new results are presented for one-layered and multilayered FGM plates, cylinders, and cylindrical and spherical shells, considering different moisture contents, thickness ratios, and material laws.

Keywords: functionally graded materials; 3D shell model; steady-state hygro-elastic analysis; Fick moisture diffusion equation; moisture content profile; layer-wise approach

1. Introduction

The environmental conditions characterizing the service life of structural components can be adverse in many applications. The aerospace field gives several examples of changing environments, which results in temperature gradients and moisture concentration variability. It is crucial to consider such two factors and to include them in a proper structural analysis: the thermal and hygrometric fields induce an internal stress distribution, which changes as soon as the environmental conditions change. As with any stress field, it might induce failures on the structure [1], either on its own or because it sums up to that caused by classical mechanical loads. Composite, multilayered, and FGM-embedding structures require a special focus. Composite materials can also be degraded by moisture absorption and the following diffusion through the matrix [2]; multilayered structures highlight a strong heterogeneity in the hygro/thermal/mechanical properties; FGMs induce a further complication due to non-constant terms in governing equations. However, their boosted implementation in critical structural applications recently increased the attention of the researchers on these effects.

Laminated structures and composite materials suffer from a clear variation of the properties at the interfaces, which is the critical source of the delamination process [3]. Eliminating this discontinuity and substituting it with a smooth trend is the key achievement of Functionally Graded Materials (FGMs). They are advanced composite materials made by two or more different phases mixed with a continuous graded distribution. As a result, they are heterogeneous materials, delivering optimized responses for each application: the advancements in processing technologies made it possible to control a unidirectional and even multidirectional variation. Combinations of a metallic and a ceramic phase are classic examples of FGMs finding application in severe thermal environments: they overcome the differences in thermal properties of the two constituents and, at the same time, deliver a reduced thermal stress distribution. Stiffness coefficients, hardness, thermal conductivity, moisture diffusivity, and corrosion resistance are just some examples of the performance characteristics that can be combined and enhanced [4].

Several recent and relevant works focused on structures embedding FG layers underline their importance in practical applications. From an analytical perspective, Sobhani et al. recently proposed several analytical results in the frame of the vibration analysis of FG shells. First, the governing equations followed the First-order Shear Deformation Theory (FSDT); then, they used the Generalized Differential Quadrature Method (GDQM), a semi-analytical solution method, to solve the system of partial differential equations. They studied conical shells embedding hybrid matrix/fiber nanocomposites in [5]; the approach was then extended to paraboloidal and hyperboloidal shells embedding polymer matrix, carbon fiber, and graphene nanoplatelets in [6]. Using the same approach, they studied sandwich conical-cylindrical-conical shells [7]. The layers are reinforced with functionally graded carbon nanotubes and graphene nanoplatelets; they described the elastic coefficients following an Equivalent Single Layer approach. They also studied five different patterns of CNT fibers distribution inside of the matrix while defining the vibrational behavior of coupled conical-conical shells [8]. They used the five-parameter shell theory and solved the differential equations through GDQM. Three-phase nanocomposites were also studied in hemispherical-cylindrical shells exploiting a similar approach [9] and defining the governing equations through the first-order shear deformation theory. From a numerical perspective, Rezaiee-Pajand et al. studied sandwich beams embedding FGM in two ways: they applied the Ritz method and the principle of minimum total potential energy within the framework of Timoshenko and Reddy beam theories in [10] to assess the bending of beams with different cross sections; they also developed a four-node isoparametric beam element to study porous beams with FGM [11]. Concerning two-dimensional geometries, the same authors proposed the nonlinear analysis of FG shells in [12,13] by improving an isoparametric six-node TRIA element with strain interpolation functions; the mechanical properties grading followed a power low. They also proposed a three-node TRIA element [14] using a mixed strain finite element approach and demonstrated that it is possible to get the exact response of the beams with a low number of elements under large deformations.

The solutions available in the literature seem to be promising. However, the results of a hygrometric stress analysis depend not only on the capabilities of the elastic model implemented but also on how the moisture content field has been evaluated, given the external boundary conditions. The hygrometric field acts as a field load; its quantification necessarily influences the stress analysis results. Aerospace applications usually involve thin components. Consequently, evaluating the moisture content field translates into determining its profile along with the thickness direction, generally coinciding with the grading direction of the mechanical/hygrometric properties. Developing a mechanical/elastic model for a thin component can be done at different levels of detail: 3D or 2D approaches, coupled with an analytical or numerical method. However, this might not be sufficient in defining the refinement of a model when hygrometric stresses are concerned. Even an analytical, exact 3D model would give wrong results when fed from an inaccurate moisture content field. The molecular diffusion depends on the gradient of the concentration and is described by Fick's law. A solid mathematical analogy between Fick's law and the Fourier heat conduction equation strongly simplifies the analysis and the dissertation. An exact solution comes from resolving three-dimensionally the diffusion/conduction equations; however, simplified solutions might benefit from their unidirectional simplification or an assumed-linear profile. Those three situations require an external evaluation of the moisture content field before the mechanical analysis. An alternative consists in defining a coupled hygro-mechanical model, in which the moisture content field is a primary variable of the problem in analogy with the thermal field [15–20].

This paper discusses an hygro-elastic shell model, which relies on the exponential matrix method. This model does not limit to a specific geometry or lamination scheme; on the contrary, it handles plates, cylindrical, and spherical shells. Furthermore, it accepts one or more FGM layers on their own or coupled with homogeneous layers (sandwiches give an example). It extends the authors' 3D exact thermo-elastic shell model discussed in [21] to hygro-elastic stress analysis. The problem is defined under steady-state conditions only; the solution requires the moisture content amplitudes at the top and bottom external surfaces to be specified. As a first step, the solution evaluates the moisture content profile through the thickness direction. The authors considered three possible options: the 3D Fick diffusion law, its 1D simplified version, and an a priori assumed linear trend. Despite this, the model handles the solution via a layer-wise approach and through the exponential matrix method. The present authors are not the only ones using the exponential matrix method for solving the differential equilibrium equations. Soldatos and Ye [22] already used it in analyzing the free vibrations of cylinders; Messina [23] applied it to study multilayered plates. The orthogonal mixed curvilinear coordinate system helped study spherical shells in [24] using three transverse stress and three displacement components as primary variables of the problem. The analytical procedure is in analogy with the free vibration analysis and the static analysis under mechanical load discussed by the first author in [25–30]. Those previous formulations already handled different materials and geometries but lacked loads other than the mechanical ones. In those simpler cases, a set of three homogeneous differential equations are at the basis of the problem. However, the hygrometric load adds an additional term that makes them not homogeneous. The exponential matrix method also handles this feature, as discussed in [21] for the thermoelastic stress analysis of shells with FGMs. The closed-form solution of the problem is possible given the harmonic forms imposed on the variables (displacements and hygrometric field) and the simply supported boundary conditions. Moreover, the whole formulation benefits an orthogonal mixed curvilinear coordinate system, following the suggestions in [31–34]. This strategy introduces a set of curvature terms, the elements through which the equations automatically adapt to the different geometries. Such terms are a function of the thickness coordinate, together with the elastic and hygrometric coefficients in FGM layers. Introducing a set of fictitious (mathematical) layers allows obtaining constant coefficients and using the method discussed in [35] to solve the problem.

The literature overview offers different analytical and numerical 2D models handling the hygrometric stress analysis; they specifically focus on multilayered structures. Far fewer discuss the problem of structures embedding FGMs. The literature overview will demonstrate that, to the authors' knowledge, no analytical 3D shell model exists, in which the structures are different, provided that they have constant radii of curvature, and the moisture content evaluation follows three different approaches. Laoufi [36] studied rectangular plates embedding FGMs when subjected to different boundary conditions, including moisture content and temperature field. They developed an analytical method through the hyperbolic shear deformation plate model, which satisfied the stress boundary conditions and required no shear corrections. The volume fractions of the ceramic-metal constituents were used to define the materials grading in the thickness direction following a power law. The same power law was included in Inala's work [37], devoted to studying how the hygrothermal environment affects the vibration characteristics of plates embedding FGMs. To do this, the author developed a finite element model of the structures under investigation. Dai [38] and his colleagues focused on circular plates with a variable thickness along with the radial direction. They derived nonlinear governing equations for temperature, moisture content, and displacement fields and solved them through the differential quadrature method. Zenkour also studied how the thermal and hygrometric fields affect the bending and the buckling of plates embedding FGMs [39]. In this case, also, the authors considered a material grading in the thickness direction of the structure. Their formulation relies on an exponential shear deformation theory and applies to plates resting on elastic foundations. Hamilton's principle was used to derive the equilibrium equations, and Navier's method to obtain the results. Boukelf also studied plates resting on elastic foundations, embedding FGMs [40]. The author developed a novel higher-order shear deformation theory model, deducing the problem equations through the virtual work principle. Hygro, thermal, and mechanical loads are all properties that can be handled. Analogously, Zidi [41] developed a further model for the same issue. In this case, the author used a four-variable refined plate theory. Analogous research has been proposed in [42] concerning functionally graded beams. The author deepened the influence of moisture content field and temperature on the bost-buckling response of beams. A nonlinear finite element solution was considered, in which the authors handled the kinematics of the bost-buckling through the Lagrangian approach.

More substantial is the literature discussing the moisture content effects on isotropic, orthotropic, and laminated structures. Chiba and Sugano [43] studied multilayered plates and proposed a two-dimensional analytical solution for hygrothermal effects on multilayered plates. The Classical Lamination Plate Theory and an analytical 3D plate model were compared in [44,45] while performing the hygro-thermal stress analysis. The CLT also received the attention of Kalil [46] when he investigated composite plates. Kollar [47] made an analytical investigation of composite cylindrical shells, while Shen [48] focused on buckling and post-buckling behaviors. The CLT was found to be inadequate in hygrothermal mechanical analysis by Lee and his colleagues [49]. There is no shortage of numerical models in this field; an example is given in [50], where plates with a central hole were studied. Multilayered structures under hygrometric fields have been the target of the finite element model by Khoshbakht et al. [51]; Kundu and Han studied buckling and vibration in multilayered shells with double curvature [52] through a finite element model grasping the hygrothermal effects. They extended this research to dynamic instability of doubly-curved shells embedding composite materials through a nonlinear finite element under orthogonal curvilinear coordinates [53]. Marques and Creus included the Fick diffusion law [54] into a FE shell model devoted to isotropic and multilayered structures under a hygrothermal environment. With the same aim, Naidu and Sinha [55] investigated cylindrical shell panels under large deflections through a higher-order shear deformation theory. Patel also proposed a higher-order FE for laminated parts [56]. Sereir et al. [57–59] considered the elastic properties variation in composite plates with the temperature and the moisture content and discussed a transient hygroscopic stress analysis. In [60–63], simplified solutions for hygro-thermal stresses analysis of composite plates considering the variation in elastic properties due to the hygrothermal environment are also considered. An analogous evaluation devoted to dynamic behavior has been proposed in [64,65]. Ghosh [66] used a FE model to investigate how a severe hygrothermal environment can affect the initiation and progress of damages in composites.

The literature survey highlighted that the hygromechanical stress analysis of structures embedding FGM layers still misses general and exact solutions as benchmarks in new solutions. Instead, the only analytical models work on defined and specific boundary conditions, laminations, and geometry. This paper intends to fill this gap by extending the authors' previous work on multilayered structures to FGMs. The manuscript is organized as follows. Section 3 describes the hygro-elastic shell problem and its solution with the exponential matrix method. Section 2 explores the moisture diffusion problem with the three approaches introduced before. The Fick law of diffusion has the same mathematical formulation of the Fourier heat conduction equation; the moisture diffusion problem and the heat conduction problem are in analogy, indeed, as demonstrated in [67], and further confirmed by Tay and Goh [68,69]. Therefore, Section 2 quantifies the moisture content profiles. Solving the 3D Fick diffusion law is possible by exploiting the analogy with the heat conduction problem. Tungikar and Rao [35] give a methodology that can also be applied in this context; the unidimensional Fick diffusion law disregards the diffusion fluxes in directions other than the thickness coordinate and can be calculated in analogy. Section 4 gives two sets of results. The first set, labeled as Assessments, is introduced to validate the problem. In the absence of 3D exact solutions for hygro-elastic problems in shells with FGM layers, the section exploits existing results for thermal stress analysis. After validating the present model and an additional FE model, the section uses this last FE model to verify the results when the hygrometric load replaces the thermal one. The second set, labeled as Benchmarks, discusses new results, which introduce further comments on the effects of moisture content profiles, thickness ratio, material, and lamination scheme, together with the impact of the geometry. The main conclusions and the further development are then summarized in Section 5.

2. Fick Moisture Diffusion Equation

This model relies on a decoupled solution, which means that the moisture content is evaluated separately and enters the elastic part of the problem as a known term. The essential hypothesis to obtain exact closed-form solutions is that all the problem variables have a harmonic form. For what concerns the moisture content, this means

$$\mathcal{M}^k(\alpha,\beta,z) = M^k(z)\sin(\bar{\alpha}\alpha)\sin(\bar{\beta}\beta),\qquad(1)$$

$M^k(z)$ indicates the moisture content amplitude. Referring to Figures 1 and 2: m and n are the half-wave numbers in the two in-plane directions α and β, respectively. a and b are the shell dimensions referred to the mid-surface Ω_0; they allow calculating the terms $\bar{\alpha} = \frac{m\pi}{a}$ and $\bar{\beta} = \frac{n\pi}{b}$. Note that the harmonic form of the moisture content reduces its assessment to its profile along with the thickness direction z. What happens in α and β directions is already defined through the sinusoidal functions.

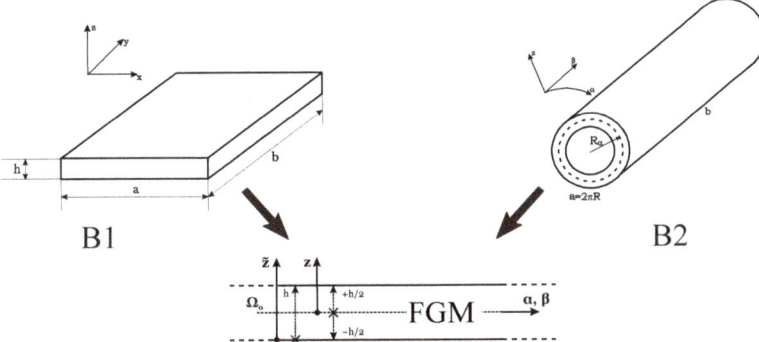

Figure 1. Geometrical data and coordinate system for plates and cylinders. The figure also shows the stacking sequence used in Benchmarks 1 and 2.

The hygrometric boundary conditions are the moisture content amplitudes at the bottom and the top of the shell, M_b, and M_t, respectively. Three approaches exist to evaluate the moisture content profile; this solution implements all three, allowing a homogeneous comparison among them and with other methods. The solution is accomplished in analogy with what the authors achieved in [21] concerning the temperature field. Indeed, the Fourier heat conduction equation, regulating the thermal phenomenon, features the same mathematical expression of the Fick diffusion law, which applies to the moisture diffusion problem. The three approaches are here listed:

- the moisture content profile is defined by solving the 3D Fick diffusion law; the hygro-elastic model considering it takes the name of 3D($\mathcal{M}_c$,3D);
- the moisture content profile is defined by solving the 1D version of the Fick diffusion law; the hygro-elastic model considering it takes the name of 3D($\mathcal{M}_c$,1D);
- the moisture content profile is "a-priori" assumed as linear along with the thickness direction; the hygro-elastic model considering it takes the name of 3D($\mathcal{M}_a$).

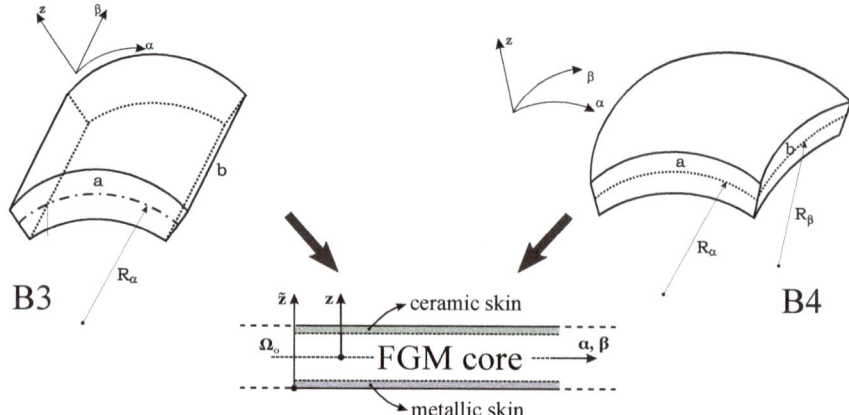

Figure 2. Geometrical data and coordinate system for cylindrical and spherical shell panels. The figure also shows the stacking sequence used in Benchmarks 3 and 4.

All the acronyms report the term 3D: it states that the elastic part of the shell model relies on a three-dimensional solution. Instead, the part of the acronym inside the parentheses defines how the moisture content profile has been quantified. "c" means calculated, either via the 3D or the 1D Fick diffusion laws; "a" means linear assumed.

2.1. 3D Fick Equation

By analogy with the heat flux q, suppose to define a moisture flux g. Then, the differential equation of moisture diffusion into a homogeneous solid, in the absence of chemical reactions and under steady-state conditions ($\frac{\partial M}{\partial t} = 0$) reads:

$$\nabla \cdot g(u_1, u_2, u_3) = 0. \tag{2}$$

Equation (2) is written in a general orthogonal curvilinear coordinate system (u_1, u_2, u_3); however, in these conditions, the divergence of the moisture flux takes the following expression:

$$\nabla \cdot g = \frac{1}{a}\left[\frac{\partial}{\partial u_1}\left(\frac{a}{a_1}g_1\right) + \frac{\partial}{\partial u_2}\left(\frac{a}{a_2}g_2\right) + \frac{\partial}{\partial u_3}\left(\frac{a}{a_3}g_3\right)\right], \tag{3}$$

g_1, g_2, g_3 express the components of the 3D flux in directions 1, 2, and 3; their explicit form is

$$g_i = -\mathcal{D}_i \frac{1}{a_i}\frac{\partial M}{\partial u_i}, \tag{4}$$

$\mathcal{D}_i$ is the diffusion coefficient along with direction i; a_1, a_2, and a_3 are the so-called scale factors, with a the product of the three factors ($a = a_1 a_2 a_3$). The problem takes place in a mixed curvilinear orthogonal coordinate system (α, β, z); Povstenko [32] discussed that, in this context, Equation (3) might be rewritten as follows:

$$\frac{1}{H_\alpha H_\beta}\left[\frac{\partial}{\partial \alpha}\left(\frac{H_\alpha H_\beta}{H_\alpha}\mathcal{D}_1 \frac{1}{H_\alpha}\frac{\partial M}{\partial \alpha}\right) + \frac{\partial}{\partial \beta}\left(\frac{H_\alpha H_\beta}{H_\beta}\mathcal{D}_2 \frac{1}{H_\beta}\frac{\partial M}{\partial \beta}\right)\right] + \mathcal{D}_3 \frac{\partial^2 M}{\partial z^2} = 0, \tag{5}$$

FGM layers distinguish from classical isotropic and orthotropic layers as the diffusion coefficients $\mathcal{D}_1(z)$, $\mathcal{D}_2(z)$, and $\mathcal{D}_3(z)$ are a function of the thickness coordinate z. Despite the lamination scheme, $H_\alpha(z)$ and $H_\beta(z)$ are two parametric coefficients, defined as follows:

$$H_\alpha(z) = \left(1 + \frac{z}{R_\alpha}\right) = H_\alpha(\tilde{z}) = \left(1 + \frac{\tilde{z} - h/2}{R_\alpha}\right), \tag{6}$$

$$H_\beta(z) = \left(1 + \frac{z}{R_\beta}\right) = H_\beta(\tilde{z}) = \left(1 + \frac{\tilde{z} - h/2}{R_\beta}\right). \tag{7}$$

When considering shells with constant radii of curvature, R_α and R_β, they are a linear function of the thickness coordinate z, which varies from $-h/2$ to $+h/2$ or $\tilde{z}$, which varies from 0 to h. h is the global thickness. The thickness coordinate z (or $\tilde{z}$) is rectilinear; for consistency, a further coefficient might be defined: $H_z = 1$. Equation (5) points out two main blocks; the second block has a simpler formulation as the third coordinate z is rectilinear. Actually, this point is further confirmed by the work of Leissa [70], which showed that

$$a_1 = H_\alpha, \; a_2 = H_\beta, \; a_3 = H_z = 1. \tag{8}$$

The coefficients of Equation (5) are not constant even inside a k-th physical layer. This issue is due to the parametric coefficients H_α and H_β, a function of the thickness coordinate z, and to $\mathcal{D}_1$, $\mathcal{D}_2$, and $\mathcal{D}_3$ which are not constant in FGM layers. A possible solution consists in dividing each physical layer into sufficiently thin mathematical (fictitious) layers. There, calculating the parametric and the diffusion coefficients at its middle allows moving the differential operators to the moisture content only. Inside each mathematical layers it holds that

$$\mathcal{D}_1^{*j}\frac{\partial^2 \mathcal{M}}{\partial \alpha^2} + \mathcal{D}_2^{*j}\frac{\partial^2 \mathcal{M}}{\partial \beta^2} + \mathcal{D}_3^{*j}\frac{\partial^2 \mathcal{M}}{\partial z^2} = 0. \tag{9}$$

where

$$\mathcal{D}_1^{*j} = \frac{\mathcal{D}_1^j}{H_\alpha^{2j}}, \quad \mathcal{D}_2^{*j} = \frac{\mathcal{D}_2^j}{H_\beta^{2j}}, \quad \mathcal{D}_3^{*j} = \mathcal{D}_3^j. \tag{10}$$

Equation (9) is automatically satisfied by the harmonic expression of the moisture content $\mathcal{M}(\alpha, \beta, z)$ already discussed in Equation (1). Still, the form of the moisture content amplitude in the thickness direction $M(z)$ needs to be clarified. A tentative function is

$$M^j(z) = S_1^j \cosh(s_1^j z) + S_2^j \sinh(s_1^j z) \tag{11}$$

Each mathematical layer feature a coefficients s^j and a pair of coefficients M_0^j. The first can be computed, introducing the harmonic form of moisture content, provided with the assumption of Equation (11) on its amplitude, in Equation (9):

$$s_{1,2}^j = \pm\sqrt{\frac{\mathcal{D}_1^{*j}\bar{\alpha}^2 + \mathcal{D}_2^{*j}\bar{\beta}^2}{\mathcal{D}_3^{*j}}}, \tag{12}$$

and choosing the positive solution, referred to as s_1^j. Equation (11) shows that a pair of coefficients per mathematical layer j is needed, in addition to s_1^j. Each layer features its own set of coefficients, which means that still $2 \times G$ unknowns are involved. However, at each interface between the fictitious layers, the following continuity conditions must hold:

$$M_b^{(j+1)} = M_t^j, \tag{13}$$

$$\mathcal{D}_3^{*j+1} M_{z_b}^{(j+1)} = \mathcal{D}_3^{*j} M_{z_t}^{j}. \tag{14}$$

The physical meaning of both the equations is obvious; Equation (13) states that the moisture content at the bottom of any $(j+1)$-th layer must equal that at the top of the j-th layer. Likewise, Equation (14) states that the moisture flux component in the thickness direction at the bottom of any $(j+1)$-th layer must equal that at the top of the j-th layer. The matrix form of Equations (13) and (14) allows compacting the analysis:

$$\begin{bmatrix} S_1 \\ S_2 \end{bmatrix}^{j+1} = \begin{bmatrix} V_{M_1}^{j+1,j} & V_{M_2}^{j+1,j} \\ V_{M_3}^{j+1,j} & V_{M_4}^{j+1,j} \end{bmatrix} \begin{bmatrix} S_1 \\ S_2 \end{bmatrix}^{j}. \tag{15}$$

Both the continuity conditions hold at the interfaces between layers; consequently, $2 \times (G-1)$ conditions can be imposed. The recursive use of Equation (15) allows linking the coefficients of the bottom layer ($j = 1$) with those of the top ($j = G$); such an idea can be compacted by identifying the transfer matrices of Equation (15) with the name $V_M^{(j+1,j)}$.

$$\begin{bmatrix} S_1 \\ S_2 \end{bmatrix}^{G} = V_M^{(G,G-1)} V_M^{(G-1,G-2)} \ldots\ldots V_M^{(3,2)} V_M^{(2,1)} \begin{bmatrix} S_1 \\ S_2 \end{bmatrix}^{1} = V_M^{(G,1)} \begin{bmatrix} S_1 \\ S_2 \end{bmatrix}^{1}. \tag{16}$$

Two conditions are still missing to quantify all the $2 \times G$ coefficients. However, simply imposing the moisture content at the top and the bottom of the structure gives the missing information. The coefficients for the external layers derive from this input and Equation (16). Once they are known, Equation (15) allows calculating all the remaining ones. Then, the moisture content profile is determined along with the thickness direction. The 3D hygro-elastic model including this hygrometric profile is referred to as 3D($\mathcal{M}_c$,3D).

2.2. 1D Fick Equation

The three-dimensional problem of defining the moisture content field can be simplified when the structure under investigation is thin enough; such condition is expressed through sufficiently high thickness ratios. By recalling Equation (4) and the harmonic form for the moisture content field, the three components of the moisture content flux inside a k-th layer are

$$g_1^k = \mathcal{D}_1^{*k}(z)\tilde{\alpha} M^k(z) cos(\tilde{\alpha}\alpha) sin(\tilde{\beta}\beta), \tag{17}$$

$$g_2^k = \mathcal{D}_2^{*k}(z)\tilde{\beta} M^k(z) sin(\tilde{\alpha}\alpha) cos(\tilde{\beta}\beta), \tag{18}$$

$$g_3^k = \mathcal{D}_3^{*k}(z) M_{,z}^k(z) sin(\tilde{\alpha}\alpha) sin(\tilde{\beta}\beta). \tag{19}$$

As already seen, the diffusion coefficients are, in general, a function of the thickness coordinate as they are not constant inside FGM layers. The relative weight of the moisture content fluxes g_1 and g_2 decreases if compared to g_3 the thinner is the shell. When this happens, the first two components can be disregarded; Equation (19) now becomes

$$g_3(z) = \left(\mathcal{D}_3^* \frac{\partial M}{\partial z} \right) = const. \tag{20}$$

As already seen in the previous section, there is no practical difference between $\mathcal{D}_3^*$ and $\mathcal{D}_3$ because the third coordinate $3 \equiv z$ is rectilinear, and its parametric coefficient H_z equals 1. Equation (20) implies that $g_3(z)$ is actually constant along with z; it can be specialized for a generic j-th mathematical layer. The coefficient $\mathcal{D}_3^*$ does not change inside it; this implies that the derivative of the moisture content to z is constant: the moisture content is linear inside each mathematical layer. Consequently Equation (20) can be simplified and rewritten in an algebraic form:

$$g_3^j = -\mathcal{D}_3^{*j} \frac{\partial M^j}{\partial z} = -\frac{\mathcal{D}_3^{*j}}{h^j} (M_t^j - M_b^j). \tag{21}$$

For the term $\mathcal{D}_3^{*j}/h^j$, it is possible to exploit the analogy with the electrical conductance, as already done in [21] where a sort of thermal conductance was defined: the layer diffuses more, becoming thinner, increasing $\mathcal{D}_3^*$. This analogy is beneficial, as it enables defining an equivalent moisture content diffusion resistance for the overall structure:

$$R_{z_{eq}} = \sum_{j=1}^{G} \frac{h^j}{\mathcal{D}_3^{*j}}. \qquad (22)$$

The moisture flux in the thickness direction can now be easily quantified, as the dissertation demonstrated it is constant and quantified the equivalent diffusion resistance. Defined M_t the moisture content amplitude on the top external surface and M_b that on the bottom, g_3 is

$$g_3 = -\frac{1}{R_{z_{eq}}}(M_t - M_b) = const. \qquad (23)$$

As g_3 is constant across all the layers, the moisture content at any interface (and even at any z coordinate) is easily obtained:

$$g_3^j = -\mathcal{D}_3^{*j}\frac{(M_t^j - M_b^j)}{h^j} = g_3^{j+1} = -\mathcal{D}_3^{*j+1}\frac{(M_t^{j+1} - M_b^{j+1})}{h^{j+1}} = g_3 = const., \qquad (24)$$

The coefficient $\mathcal{D}_3^{*j}$ changes in the thickness direction when FGM layers are considered in the stacking sequence. As a consequence, the moisture flux can keep constant only if the slope of the moisture content profile modifies accordingly. In this perspective, the effect of the material is considered (which means the stacking sequence and the FGM law), but the impact of the thickness is disregarded. The 3D hygro-elastic model including this hygrometric profile is referred to as 3D($\mathcal{M}_c$,1D).

2.3. Assumed Linear Moisture Content through the Thickness Direction

The analysis is further simplified if also the effect of the material is disregarded, in addition to that of the thickness. A common assumption in the literature is that the moisture content profile is linear throughout the thickness direction of the shell: it does not take into account any change in the hygroscopic properties of the layers, as well as the fluxes in α and β directions. It is close to reality only when the shell is really thin and embeds a single layer or even different, but with homogeneous hygroscopic properties. The profile is immediately determined once the top and bottom external sovra-temperatures are set. The 3D hygroelastic model including this assumed hygrometric profile is referred to as 3D($\mathcal{M}_a$).

3. 3D Exact Shell Model for Hygro-Elastic Stress Analysis

The 3D equilibrium equations for shells are written in the orthogonal mixed curvilinear coordinate system (α, β, z) shown in Figures 1 and 2. These equations are modified using 3D constitutive equations for Functionally Graded Materials (FGMs) and general geometrical relations for shells in (α, β, z) coordinates. Therefore, the system includes three differential equations of second order in z, and the related coefficients are not constant because of the radii of curvature and elastic coefficients for FGMs. A reasonable number of mathematical layers is introduced to obtain constant-coefficient equations; redoubling the number of variables allows reducing the order of the differential equations. Simply-supported sides and harmonic variables allow the analytical calculation of the partial derivatives in α and β directions. The final system, including the moisture content profile, has only first-order partial derivatives in z, and the exponential matrix method allows determining both general and particular solutions.

The investigated multilayered shells and plates subjected to a moisture content $\mathcal{M}(\alpha, \beta, z)$ at the external surfaces have k classical/composite layers and/or functionally graded material layers. Stains defined in an orthogonal mixed curvilinear reference system (α,β,z) are the algebraic summation of mechanical elastic parts (subscript m) and hygroscopic parts (subscript $\mathcal{M}$). The 6×1 vector ($\epsilon_{\alpha\alpha}^k$, $\epsilon_{\beta\beta}^k$, ϵ_{zz}^k, $\gamma_{\beta z}^k$, $\gamma_{\alpha z}^k$, $\gamma_{\alpha\beta}^k$) for each k layer is defined as

$$\epsilon_{\alpha\alpha}^k = \epsilon_{\alpha\alpha m}^k - \epsilon_{\alpha\alpha\mathcal{M}}^k = \frac{1}{H_\alpha(z)}\frac{\partial u^k}{\partial \alpha} + \frac{w^k}{H_\alpha(z)R_\alpha} - \eta_\alpha^k(z)\mathcal{M}^k, \tag{25}$$

$$\epsilon_{\beta\beta}^k = \epsilon_{\beta\beta m}^k - \epsilon_{\beta\beta\mathcal{M}}^k = \frac{1}{H_\beta(z)}\frac{\partial v^k}{\partial \beta} + \frac{w^k}{H_\beta(z)R_\beta} - \eta_\beta^k(z)\mathcal{M}^k, \tag{26}$$

$$\epsilon_{zz}^k = \epsilon_{zzm}^k - \epsilon_{zz\mathcal{M}}^k = \frac{\partial w^k}{\partial z} - \eta_z^k(z)\mathcal{M}^k, \tag{27}$$

$$\gamma_{\beta z}^k = \gamma_{\beta z m}^k = \frac{1}{H_\beta(z)}\frac{\partial w^k}{\partial \beta} + \frac{\partial v^k}{\partial z} - \frac{v^k}{H_\beta(z)R_\beta}, \tag{28}$$

$$\gamma_{\alpha z}^k = \gamma_{\alpha z m}^k = \frac{1}{H_\alpha(z)}\frac{\partial w^k}{\partial \alpha} + \frac{\partial u^k}{\partial z} - \frac{u^k}{H_\alpha(z)R_\alpha}, \tag{29}$$

$$\gamma_{\alpha\beta}^k = \gamma_{\alpha\beta m}^k = \frac{1}{H_\alpha(z)}\frac{\partial v^k}{\partial \alpha} + \frac{1}{H_\beta(z)}\frac{\partial u^k}{\partial \beta}, \tag{30}$$

In the hygro-elastic strains, the three displacement components u^k, v^k, and w^k and the scalar moisture content $\mathcal{M}^k$ are defined in the reference system (α, β, z). Moisture expansion coefficients $\eta_\alpha^k(z)$, $\eta_\beta^k(z)$ and $\eta_z^k(z)$ in the k physical layer could depend on the z coordinate in the case of Functionally Graded Material (FGM) layers; they are defined in the structural reference system (α, β, z) starting from the moisture expansion coefficients $\eta_1^k(z)$, $\eta_2^k(z)$ and $\eta_3^k(z)$ in the material reference system (1, 2, 3). The partial derivatives are defined via the symbol ∂.

As anticipated, an essential hypothesis for exact closed-form solutions is the harmonic forms for all the variables: displacement components and moisture content. While the latter has already been discussed, for the displacement components it holds that

$$u^k(\alpha,\beta,z) = U^k(z)cos(\bar{\alpha}\alpha)sin(\bar{\beta}\beta), \tag{31}$$

$$v^k(\alpha,\beta,z) = V^k(z)sin(\bar{\alpha}\alpha)cos(\bar{\beta}\beta), \tag{32}$$

$$w^k(\alpha,\beta,z) = W^k(z)sin(\bar{\alpha}\alpha)sin(\bar{\beta}\beta), \tag{33}$$

displacement amplitudes are indicated as ($U^k(z), V^k(z), W^k(z)$). As already described, m and n are the half-wave numbers, a and b the mid-surface dimensions, and $\bar{\alpha} = \frac{m\pi}{a}$ and $\bar{\beta} = \frac{n\pi}{b}$.

The three-dimensional equilibrium equations for spherical shells having constant radii of curvature $R_\alpha = R_\beta$ and a total number N_L of physical k (either classical or FGM) layers are

$$H_\beta(z)\frac{\partial \sigma_{\alpha\alpha}^k}{\partial \alpha} + H_\alpha(z)\frac{\partial \sigma_{\alpha\beta}^k}{\partial \beta} + H_\alpha(z)H_\beta(z)\frac{\partial \sigma_{\alpha z}^k}{\partial z} + \left(\frac{2H_\beta(z)}{R_\alpha} + \frac{H_\alpha(z)}{R_\beta}\right)\sigma_{\alpha z}^k = 0, \tag{34}$$

$$H_\beta(z)\frac{\partial \sigma_{\alpha\beta}^k}{\partial \alpha} + H_\alpha(z)\frac{\partial \sigma_{\beta\beta}^k}{\partial \beta} + H_\alpha(z)H_\beta(z)\frac{\partial \sigma_{\beta z}^k}{\partial z} + \left(\frac{2H_\alpha(z)}{R_\beta} + \frac{H_\beta(z)}{R_\alpha}\right)\sigma_{\beta z}^k = 0, \tag{35}$$

$$H_\beta(z)\frac{\partial \sigma_{\alpha z}^k}{\partial \alpha} + H_\alpha(z)\frac{\partial \sigma_{\beta z}^k}{\partial \beta} + H_\alpha(z)H_\beta(z)\frac{\partial \sigma_{zz}^k}{\partial z} - \frac{H_\beta(z)}{R_\alpha}\sigma_{\alpha\alpha}^k - \frac{H_\alpha(z)}{R_\beta}\sigma_{\beta\beta}^k + \left(\frac{H_\beta(z)}{R_\alpha} + \frac{H_\alpha(z)}{R_\beta}\right)\sigma_{zz}^k = 0. \tag{36}$$

Three-dimensional equilibrium equations for cylinders/cylindrical panels and plates are obtained from Equations (34)–(36) when one of the two radii of curvature is infinite and when both the radii of curvature are infinite, respectively.

The constitutive equations are developed by considering the algebraic summation of mechanical elastic strains and hygroscopic strains:

$$\sigma^k = C^k(z)\epsilon^k = C^k(z)(\epsilon_m^k - \epsilon_\mathcal{M}^k), \quad (37)$$

σ^k is the stress vector having a 6×1 dimension, $C^k(z)$ is the 6×6 elastic coefficient matrix and it could depend on the z coordinate in the case of a kth FGM layer. The strains are included in the constitutive equations by using the form shown in Equations (25)–(30). Closed-form solutions are possible when orthotropic angles are $0°$ or $90°$ to obtain $C_{16}^k(z) = C_{26}^k(z) = C_{36}^k(z) = C_{45}^k(z) = 0$ in the structural reference system (α, β, z). Therefore,

$$C^k(z) = \begin{bmatrix} C_{11}^k(z) & C_{12}^k(z) & C_{13}^k(z) & 0 & 0 & 0 \\ C_{12}^k(z) & C_{22}^k(z) & C_{23}^k(z) & 0 & 0 & 0 \\ C_{13}^k(z) & C_{23}^k(z) & C_{33}^k(z) & 0 & 0 & 0 \\ 0 & 0 & 0 & C_{44}^k(z) & 0 & 0 \\ 0 & 0 & 0 & 0 & C_{55}^k(z) & 0 \\ 0 & 0 & 0 & 0 & 0 & C_{66}^k(z) \end{bmatrix}. \quad (38)$$

The explicit form of the constitutive equations develops thanks to the introduction of the geometrical Equations (25)–(30) into the constitutive Equation (37):

$$\sigma_{\alpha\alpha}^k = \frac{C_{11}^k(z)}{H_\alpha(z)} u_{,\alpha}^k + \frac{C_{11}^k(z)}{H_\alpha(z)R_\alpha} w^k + \frac{C_{12}^k(z)}{H_\beta(z)} v_{,\beta}^k + \frac{C_{12}^k(z)}{H_\beta(z)R_\beta} w^k + C_{13}^k(z)w_{,z}^k - \xi_\alpha^k(z)\mathcal{M}^k, \quad (39)$$

$$\sigma_{\beta\beta}^k = \frac{C_{12}^k(z)}{H_\alpha(z)} u_{,\alpha}^k + \frac{C_{12}^k(z)}{H_\alpha(z)R_\alpha} w^k + \frac{C_{22}^k(z)}{H_\beta(z)} v_{,\beta}^k + \frac{C_{22}^k(z)}{H_\beta(z)R_\beta} w^k + C_{23}^k(z)w_{,z}^k - \xi_\beta^k(z)\mathcal{M}^k, \quad (40)$$

$$\sigma_{zz}^k = \frac{C_{13}^k(z)}{H_\alpha(z)} u_{,\alpha}^k + \frac{C_{13}^k(z)}{H_\alpha(z)R_\alpha} w^k + \frac{C_{23}^k(z)}{H_\beta(z)} v_{,\beta}^k + \frac{C_{23}^k(z)}{H_\beta(z)R_\beta} w^k + C_{33}^k(z)w_{,z}^k - \xi_z^k(z)\mathcal{M}^k, \quad (41)$$

$$\sigma_{\beta z}^k = \frac{C_{44}^k(z)}{H_\beta(z)} w_{,\beta}^k + C_{44}^k(z)v_{,z}^k - \frac{C_{44}^k(z)}{H_\beta(z)R_\beta} v^k, \quad (42)$$

$$\sigma_{\alpha z}^k = \frac{C_{55}^k(z)}{H_\alpha(z)} w_{,\alpha}^k + C_{55}^k(z)u_{,z}^k - \frac{C_{55}^k(z)}{H_\alpha(z)R_\alpha} u^k, \quad (43)$$

$$\sigma_{\alpha\beta}^k = \frac{C_{66}^k(z)}{H_\alpha(z)} v_{,\alpha}^k + \frac{C_{66}^k(z)}{H_\beta(z)} u_{,\beta}^k, \quad (44)$$

partial derivatives ($\frac{\partial}{\partial \alpha}$), ($\frac{\partial}{\partial \beta}$), and ($\frac{\partial}{\partial z}$) are indicated in Equations (39)–(44) through subscripts $(,\alpha)$, $(,\beta)$, and $(,z)$, respectively. Terms $\xi_\alpha^k(z)$, $\xi_\beta^k(z)$, and $\xi_z^k(z)$ designate the hygromechanical coupling coefficients in the structural reference system (α, β, z), and they could depend on z in the case of FGM layers:

$$\xi_\alpha^k(z) = C_{11}^k(z)\eta_\alpha^k(z) + C_{12}^k(z)\eta_\beta^k(z) + C_{13}^k(z)\eta_z^k(z), \quad (45)$$

$$\xi_\beta^k(z) = C_{12}^k(z)\eta_\alpha^k(z) + C_{22}^k(z)\eta_\beta^k(z) + C_{23}^k(z)\eta_z^k(z), \quad (46)$$

$$\xi_z^k(z) = C_{13}^k(z)\eta_\alpha^k(z) + C_{23}^k(z)\eta_\beta^k(z) + C_{33}^k(z)\eta_z^k(z). \quad (47)$$

The final system is obtained thanks to the substitution of the harmonic form equations for displacements and moisture content Equations (1)–(31) and the modified constitutive relations Equations (39)–(44) into the three-dimensional equilibrium Equations (34)–(36) developed for spherical shells:

$$A_1^j U^j + A_2^j V^j + A_3^j W^j + A_4^j U_{,z}^j + A_5^j W_{,z}^j + A_6^j U_{,zz}^j + L_1^j M^j = 0, \tag{48}$$

$$A_7^j U^j + A_8^j V^j + A_9^j W^j + A_{10}^j V_{,z}^j + A_{11}^j W_{,z}^j + A_{12}^j V_{,zz}^j + L_2^j M^j = 0, \tag{49}$$

$$A_{13}^j U^j + A_{14}^j V^j + A_{15}^j W^j + A_{16}^j U_{,z}^j + A_{17}^j V_{,z}^j + A_{18}^j W_{,z}^j + A_{19}^j W_{,zz}^j + L_3^j M_{,z}^j + L_4^j M^j = 0. \tag{50}$$

The above equations have no constant coefficients because the parametric coefficients H_α and H_β depend on z in the case of shell geometries and/or elastic and hygrometric coefficients depend on z in FGM layers. For these reasons, each k physical layer is divided into a suitable number of mathematical layers indicated with a new index j which changes from 1 to the global number of mathematical layers G. In each j mathematical layer, the parametric coefficients H_α and H_β and the variable elastic and hygrometric coefficients for FGM layers can be exactly defined by using the z coordinate in the middle of each j layer. Therefore, coefficients A_s^j (s from 1 to 19) and L_r^j (r from 1 to 4) become constant terms in the compact form of the system of differential equations in z defined in Equations (48)–(50).

Equations (48)–(50) indicate a system of three differential equations of second order in z. The unknowns are the displacement and moisture content amplitudes and the associated derivatives calculated to z. The derivatives in α and β have already been calculated via the harmonic forms for displacements and moisture content previously introduced. In this system, decoupling the variables is possible: this means a separate quantification of the moisture content profile through the thickness direction z, addressed in an appropriate section. Therefore, it becomes a known term.

The consequence of these choices is that the system contains second-order differential equations only, in the displacement amplitudes U^j, V^j, W^j and their derivatives in z. Redoubling the variables as proposed in [71,72] allows reducing the system into a first-order one in z. In each j layer, the 3×1 vector of unknowns (U^j, V^j, W^j) becomes a 6×1 unknown vector (U^j, V^j, W^j, $U^{j'}$, $V^{j'}$, $W^{j'}$); superscript $'$ indicates derivatives performed to z (also written as $\frac{\partial}{\partial z}$). Terms M^j and $M^{j'}$ can be considered as known terms because they can be opportunely calculated, as will be demonstrated in the next section:

$$\begin{bmatrix} A_6^j & 0 & 0 & 0 & 0 & 0 \\ 0 & A_{12}^j & 0 & 0 & 0 & 0 \\ 0 & 0 & A_{19}^j & 0 & 0 & 0 \\ 0 & 0 & 0 & A_6^j & 0 & 0 \\ 0 & 0 & 0 & 0 & A_{12}^j & 0 \\ 0 & 0 & 0 & 0 & 0 & A_{19}^j \end{bmatrix} \begin{bmatrix} U^j \\ V^j \\ W^j \\ U^{j'} \\ V^{j'} \\ W^{j'} \end{bmatrix}' =$$

$$\begin{bmatrix} 0 & 0 & 0 & A_6^j & 0 & 0 \\ 0 & 0 & 0 & 0 & A_{12}^j & 0 \\ 0 & 0 & 0 & 0 & 0 & A_{19}^j \\ -A_1^j & -A_2^j & -A_3^j & -A_4^j & 0 & -A_5^j \\ -A_7^j & -A_8^j & -A_9^j & 0 & -A_{10}^j & -A_{11}^j \\ -A_{13}^j & -A_{14}^j & -A_{15}^j & -A_{16}^j & -A_{17}^j & -A_{18}^j \end{bmatrix} \begin{bmatrix} U^j \\ V^j \\ W^j \\ U^{j'} \\ V^{j'} \\ W^{j'} \end{bmatrix} + \begin{bmatrix} 0 & 0 & 0 & 0 & 0 \\ 0 & 0 & 0 & 0 & 0 \\ 0 & 0 & 0 & 0 & 0 \\ -L_1^j & 0 & 0 & 0 & 0 \\ -L_2^j & 0 & 0 & 0 & 0 \\ -L_4^j & -L_3^j & 0 & 0 & 0 \end{bmatrix} \begin{bmatrix} M^j \\ M^{j'} \\ 0 \\ 0 \\ 0 \end{bmatrix}, \tag{51}$$

By defining vectors $\boldsymbol{U}^j = [U^j\ V^j\ W^j\ U^{j'}\ V^{j'}\ W^{j'}]^T$, $\boldsymbol{U}^{j'} = \frac{\partial \boldsymbol{U}^j}{\partial z}$ and $\boldsymbol{M}^j = [M^j\ M^{j'}\ 0\ 0\ 0\ 0]^T$ (where T means the transpose of a vector), the following compact form is possible:

$$\boldsymbol{D}^j \boldsymbol{U}^{j'} = \boldsymbol{A}^j \boldsymbol{U}^j + \boldsymbol{L}^j \boldsymbol{M}^j, \tag{52}$$

the above equation, thanks to the definitions $A^{*j} = D^{j^{-1}} A^j$ and $L^{*j} = D^{j^{-1}} L^j$, can be rewritten as

$$u^{j'} = D^{j^{-1}} A^j u^j + D^{j^{-1}} L^j M^j, \tag{53}$$

$$u^{j'} = A^{*j} u^j + L^{*j} M^j. \tag{54}$$

The moisture content profile through the thickness direction z can be calculated using one of the three methods proposed in the previous section. This profile can be reconstructed using a linear approximation of the moisture content in each j mathematical layer. This reconstruction can be defined as

$$M^j(\tilde{z}^j) = a_M^j \tilde{z}^j + b_M^j, \tag{55}$$

in a jth mathematical layer, a_M^j and b_M^j are two constant coefficients. The first represents the slope of the moisture content profile inside a mathematical layer; the second the moisture content at the bottom. $\tilde{z}^j$ is a local thickness coordinate for each j mathematical layer, and it changes from 0 at the bottom of the considered j mathematical layer to the top value h^j of the same mathematical layer.

The system of first-order differential equations in $\tilde{z}$ or z shown in Equation (54) is not homogeneous because the hygroscopic term $L^{*j} M^j$ depends on $\tilde{z}^j$ or z^j. In the case of a generic system of non-homogeneous first-order differential equations having an unknown $G \times 1$ vector x, a $G \times G$ $|A$ matrix containing constant coefficients, and a known function vector $f(t) = [f_1(t) ... f_G(t)]^T$, we can write

$$\frac{dx}{dt} = Ax + f(t), \tag{56}$$

a possible solution of the Equation (56) can be obtained via the exponential matrix method:

$$x(t) = e^{A(t-t_0)} x_0 + \int_{t_0}^{t} e^{A(t-s)} f(s) ds. \tag{57}$$

The known term in Equation (54) can be given in the following complete form:

$$M^{*j} = L^{*j} M^j = \begin{bmatrix} 0 & 0 & 0 & 0 & 0 & 0 \\ 0 & 0 & 0 & 0 & 0 & 0 \\ 0 & 0 & 0 & 0 & 0 & 0 \\ -L_1^{*j} & 0 & 0 & 0 & 0 & 0 \\ -L_2^{*j} & 0 & 0 & 0 & 0 & 0 \\ -L_4^{*j} & -L_3^{*j} & 0 & 0 & 0 & 0 \end{bmatrix} \begin{bmatrix} a_M^j \tilde{z}^j + b_M^j \\ a_M^j \\ 0 \\ 0 \\ 0 \\ 0 \end{bmatrix} = \begin{bmatrix} 0 \\ 0 \\ 0 \\ -L_1^{*j}(a_M^j \tilde{z}^j + b_M^j) \\ -L_2^{*j}(a_M^j \tilde{z}^j + b_M^j) \\ -L_4^{*j}(a_M^j \tilde{z}^j + b_M^j) - L_3^{*j} a_M^j \end{bmatrix}. \tag{58}$$

Therefore, Equation (54) can be rewritten as

$$u^{j'} = A^{*j} u^j + M^{*j}, \tag{59}$$

M^{*j} includes only linear and known functions in $\tilde{z}^j$ coordinate. The Equation (59) can be solved through the exponential matrix method:

$$u^j(\tilde{z}^j) = e^{(A^{*j} \tilde{z}^j)} u^j(0) + \int_0^{\tilde{z}^j} e^{(A^{*j}(\tilde{z}^j - s))} M^{*j}(s) ds. \tag{60}$$

$A^{**j} = e^{(A^{*j} h^j)}$ and $L^{**j} = \int_0^{h^j} e^{(A^{*j}(h^j - s))} M^{*j}(s) ds$ must be opportunely defined for each j layer having thickness h^j. In this way, the displacement vector at the top of each j mathematical layer is calculated. Both terms are defined via the exponential matrix opportunely expanded and evaluated in each j mathematical layer with thickness h^j:

$$A^{**j} = e^{(A^{*j}h^j)} = I + A^{*j}h^j + \frac{A^{*j^2}}{2!}h^{j^2} + \frac{A^{*j^3}}{3!}h^{j^3} + \cdots + \frac{A^{*j^N}}{N!}h^{j^N}, \quad (61)$$

$$L^{**j} = \int_0^{h^j} e^{(A^{*j}(h^j-s))} M^{*j}(s)ds = \int_0^{h^j} \left(I + A^{*j}(h^j - s) + \frac{A^{*j^2}}{2!}(h^j - s)^2 + \frac{A^{*j^3}}{3!}(h^j - s)^3 + \right.$$
$$\left. \cdots + \frac{A^{*j^N}}{N!}(h^j - s)^N\right) M^{*j}(s)ds, \quad (62)$$

where I is the 6 × 6 identity matrix, and the integral given in Equation (60) can be calculated in each j layer having thickness h^j by using the exponential matrix and the same order N already shown in Equation (61). By using Equations (61) and (62), Equation (60) is modified as

$$U^j(h^j) = A^{**j} U^j(0) + L^{**j}, \quad (63)$$

where U_t^j indicates $U^j(h^j)$ and it is defined at the top t of each j layer, and U_b^j indicates $U^j(0)$ and it is defined at the bottom of each j layer. In this way, Equation (63) is defined as

$$U_t^j = A^{**j} U_b^j + L^{**j}. \quad (64)$$

Using Equation (64) allows to connect displacements and their derivatives in z defined at the top of the j mathematical layer with the same variables defined at the bottom of the same j layer.

The general three-dimensional shell model is developed using a layer-wise approach. Inter-laminar continuity conditions in displacements and transverse stresses must be defined at each interface between the two adjacent mathematical layers. The inter-laminar continuity conditions for displacements come through congruence hypotheses:

$$u_b^j = u_t^{j-1}, \quad v_b^j = v_t^{j-1}, \quad w_b^j = w_t^{j-1}. \quad (65)$$

The inter-laminar continuity conditions for transverse shear and normal transverse stresses are defined employing equilibrium hypotheses:

$$\sigma_{zz_b}^j = \sigma_{zz_t}^{j-1}, \quad \sigma_{\alpha z_b}^j = \sigma_{\alpha z_t}^{j-1}, \quad \sigma_{\beta z_b}^j = \sigma_{\beta z_t}^{j-1}. \quad (66)$$

The formulation of this solution requires rewriting the inter-laminar continuity conditions, Equations (65) and (66), in a displacement form. Achieving this task is possible by recalling the constitutive Equations (39)–(44) and the harmonic form of the variables of the problem, Equations (1) and (31)–(33). This procedure allows writing the inter-laminar continuity condition in the amplitude displacements and their derivatives to the thickness coordinate. Then, compacting the notation is possible, recalling the vectors U^j and M^j, and introducing a pair of transfer matrices. The procedure is similar to that employed in [25]; exception made for an additional coefficient multiplying the moisture content amplitude:

$$\begin{bmatrix} U \\ V \\ W \\ U' \\ V' \\ W' \end{bmatrix}_b^j = \begin{bmatrix} 1 & 0 & 0 & 0 & 0 & 0 \\ 0 & 1 & 0 & 0 & 0 & 0 \\ 0 & 0 & 1 & 0 & 0 & 0 \\ T_1 & 0 & T_2 & T_3 & 0 & 0 \\ 0 & T_4 & T_5 & 0 & T_6 & 0 \\ T_7 & T_8 & T_9 & 0 & 0 & T_{10} \end{bmatrix}^{j-1,j} \begin{bmatrix} U \\ V \\ W \\ U' \\ V' \\ W' \end{bmatrix}_t^{j-1} + \begin{bmatrix} 0 & 0 & 0 & 0 & 0 \\ 0 & 0 & 0 & 0 & 0 \\ 0 & 0 & 0 & 0 & 0 \\ 0 & 0 & 0 & 0 & 0 \\ 0 & 0 & 0 & 0 & 0 \\ T_{11} & 0 & 0 & 0 & 0 \end{bmatrix}^{j-1,j} \begin{bmatrix} M \\ M' \\ 0 \\ 0 \\ 0 \\ 0 \end{bmatrix}_t^{j-1}. \quad (67)$$

The first three rows of Equation (67) denote the displacement continuity equations; the last three are the continuity conditions for stresses. The compact form of Equation (67) follows:

$$u_b^j = T_U^{j-1,j} u_t^{j-1} + T_M^{j-1,j} M_t^{j-1}.\tag{68}$$

Equation (68) expresses the link between the displacements and their z derivatives, calculated at the bottom of the *j*th layer, with their corresponding values plus the moisture content (and its z derivative), at the top of the $(j-1)$th layer.

The harmonic form implemented in all the variables of the problem automatically satisfies the simply supported boundary conditions, here reported for exhaustiveness:

$$w = v = 0,\ \sigma_{\alpha\alpha} = 0 \quad \text{for} \quad \alpha = 0, a,\tag{69}$$

$$w = u = 0,\ \sigma_{\beta\beta} = 0 \quad \text{for} \quad \beta = 0, b\tag{70}$$

This solution extends the model already seen in [25] for the static analysis of shells subjected to static loads. In that context, mechanical loads can act on the external top and bottom surfaces, with components defined in the three directions α, β, and z. They are grouped into vectors $\boldsymbol{P} = (P_\alpha\ P_\beta\ P_z)^T$; the superscript G means the one acting on the top surface; the one with superscript 1 identifies the one acting on the lower one. The effect of the external loads affects the displacements:

$$B_t^G U_t^G = P_t^G,\tag{71}$$

$$B_b^1 U_b^1 = P_b^1,\tag{72}$$

more details are available in [25]. The hygro-elastic analysis does not involve anything different; once applied, the moisture content induces an equivalent load, which sums up the (possible) mechanical load. The matrices $\boldsymbol{B}$ convert the mechanical and/or hygrometric loads into displacements. B_t^G, in Equation (71) handles the top (t) of the last layer (G); B_b^1, in Equation (72), the bottom (b) of the first layer (1).

The algebraic system of Equations (71) and (72) can be reformulated into a matrix form, rewriting the displacements $\boldsymbol{U}_t^G = \boldsymbol{U}^G(h^G)$ as a function of $\boldsymbol{U}_b^1 = \boldsymbol{U}^1(0)$. This rewriting is possible through a recursive substitution of Equation (68) into Equation (64), linking the displacements at the top of the last layer (and their z derivatives) to those at the bottom of the first layer:

$$\begin{aligned}U_t^G =& \left(A^{**G} T_U^{G-1,G} A^{**G-1} T_U^{G-2,G-1} \ldots\ldots A^{**2} T_U^{1,2} A^{**1}\right) U_b^1 + \\ & \left(A^{**G} T_U^{G-1,G} A^{**G-1} \ldots\ldots A^{**2} T_U^{1,2} L^{**1} + \right.\\ & A^{**G} T_U^{G-1,G} A^{**G-1} \ldots\ldots A^{**3} T_U^{2,3} L^{**2} + \\ & \vdots \\ & A^{**G} T_U^{G-1,G} L^{**G-1} + \\ & L^{**G} + \\ & A^{**G} T_U^{G-1,G} A^{**G} \ldots\ldots A^{**2} T_M^{1,2} M_t^1 + \\ & A^{**G} T_U^{G-1,G} A^{**G} \ldots\ldots A^{**3} T_M^{2,3} M_t^2 + \\ & \vdots \\ & A^{**G} T_U^{G-1,G} A^{**G-1} T_M^{G-2,G-1} M_t^{G-2} + \\ & \left. A^{**G} T_M^{G-1,G} M_t^{G-1}\right).\end{aligned}\tag{73}$$

Equation (73) consists of two main blocks. The first has as a common multiplication factor the bottom displacements of the first layer U_b^1; the content in the brackets is a 6×6 matrix, which is identical to that defined H_m for classical elastic analysis in [25]. The hygrometric field brings with it additional terms included in the second block. The terms L^{**j} explicitly include the hygrometric profile; they are G as each mathematical layer features its own profile. The terms M_t^j set the moisture content at each interface; consequently, they are $G - 1$. This block is a known and constant term; it takes the form of a 6×1 vector, from now on referred to as H_M. Therefore, the compact expression of Equation (73) takes the following form:

$$U_t^G = H_m U_b^1 + H_M . \tag{74}$$

Given this result, Equation (71) can be rewritten in terms of U_b^1:

$$B_t^G H_m U_b^1 = -B_t^G H_M . \tag{75}$$

Equations (72) and (75) now share the same unknown; they can be put to system as follows:

$$E U_b^1 = P_M , \tag{76}$$

where

$$E = \begin{bmatrix} B_t^G H_m \\ B_b^1 \end{bmatrix} \tag{77}$$

and

$$P_M = \begin{bmatrix} -B_t^G H_M \\ 0 \end{bmatrix} . \tag{78}$$

One of the main advantages of this solution is that the dimensions of the matrix E keep low, 6×6, despite the number G of mathematical layers and the layer-wise approach to the problem. Furthermore, E is the same as that needed for the classical elastic analysis (see in [25]). The vector P_M adds the hygrometric load as an equivalent mechanical action and sums up the (possible) mechanical load P. Solving the system allows getting the bottom displacement components (and their z derivatives); Equations (64) and (68) allow then calculating their values at any coordinate in the thickness direction.

4. Results

This section is of fundamental importance. First of all, it defines the properties of the Functionally Graded (FM) layers, presenting the mechanical and hygrometric properties of their constituents and the law defining their variation in composition. Then, it features two subsections. The first one validates this exact 3D solution for shells embedding layers made of FGM. Validations of new solutions are often conducted comparing the new outputs with those already available in the literature. Besides verifying the accuracy of this solution, this phase helps define how many mathematical layers should be used to consider with confidence the effects of the curvature and those of the FGM law and the order of expansion to use in the exponential matrix calculation. Strengthened by those results, the second subsection presents a set of new results. The effect of the moisture content field is studied on different geometries, featuring different stacking sequences, thickness ratios, and moisture content boundary conditions.

In all the assessments and benchmarks, the FGM layers rely on two constituents: a metal and ceramic. The metal is Monel, 70Ni30Cu, a nickel-based alloy; the ceramic is Zirconia. An estimate of the mechanical properties is given in terms of the bulk modulus K

and the shear modulus μ of the two materials; the moisture expansion coefficients η and the moisture diffusion coefficients $\mathcal{D}$ are given explicitly:

$$K_m = 227.24\,\text{GPa}, \quad \mu_m = 65.55\,\text{GPa}, \quad \eta_m = 2 \times 10^{-3}\frac{1}{\%}, \quad \mathcal{D}_m = 10^{-9}\frac{\text{kg}}{\text{ms}}, \tag{79}$$

$$K_c = 125.83\,\text{GPa}, \quad \mu_c = 58.077\,\text{GPa}, \quad \eta_c = 1 \times 10^{-3}\frac{1}{\%}, \quad \mathcal{D}_c = 10^{-10}\frac{\text{kg}}{\text{ms}}, \tag{80}$$

The metal and the ceramic phases are denoted by the subscripts m and c. The estimate of the thermal properties is also given, as they will be necessary in the assessment phase:

$$\gamma_m = 15 \times 10^{-6}\frac{1}{K}, \quad \kappa_m = 25\frac{\text{W}}{\text{mK}} \tag{81}$$

$$\gamma_c = 10 \times 10^{-6}\frac{1}{K}, \quad \kappa_c = 2.09\frac{\text{W}}{\text{mK}} \tag{82}$$

γ denotes the thermal expansion coefficient, while κ the conductivity coefficient. This work assumes that the volume fraction of the ceramic phase follows a power law of order p. Introducing the thickness of the FG layer h_{FG} and a local thickness coordinate $\tilde{z}_{FG}$ inside it (0 at its bottom, h at its top), the FG law takes the following expression:

$$V_c = (\tilde{z}_{FG}/h_{FG})^p \tag{83}$$

At the bottom of the FG layer, where $\tilde{z}_{FG} = 0$, Equation (83) implies $V_c = 0$, meaning that it is made of metal only. At the top of the FG layer, where $\tilde{z}_{FG} = h_{FG}$, Equation (83) implies $V_c = 1$, meaning that it is made of ceramic only. The bulk and the shear moduli of the FG layer evolve along with the thickness direction following the Mori–Tanaka estimates:

$$\frac{K - K_m}{K_c - K_m} = \frac{V_c}{1 + (1 - V_c)\frac{K_c - K_m}{K_m + \frac{4}{3}\mu_m}}, \quad \frac{\mu - \mu_m}{\mu_c - \mu_m} = \frac{V_c}{1 + (1 - V_c)\frac{\mu_c - \mu_m}{\mu_m + f_m}}, \quad f_m = \frac{\mu_m(9K_m + 8\mu_m)}{6(K_m + 2\mu_m)} \tag{84}$$

The same applies to the effective moisture expansion and moisture diffusion coefficients, by analogy with the estimates of Hatta and Taya for the corresponding thermal properties:

$$\frac{\mathcal{D} - \mathcal{D}_m}{\mathcal{D}_c - \mathcal{D}_m} = \frac{V_c}{1 + (1 - V_c)\frac{\mathcal{D}_c - \mathcal{D}_m}{3\mathcal{D}_m}}, \quad \frac{\eta - \eta_m}{\eta_c - \eta_m} = \frac{\frac{1}{K} - \frac{1}{K_m}}{\frac{1}{K_c} - \frac{1}{K_m}} \tag{85}$$

The estimates reported in Equation (85) are also valid for the thermal properties, following the parallel of the moisture diffusion coefficient $\mathcal{D}$ with the thermal conductivity coefficient κ, and that of the moisture expansion coefficient η with the the thermal expansion coefficient γ.

4.1. Assessments

The present solution handles several geometries and different load cases. As discussed, it can study plates, cylinders, cylindrical and spherical shells under mechanical, thermal, and hygrometric load. Furthermore, it is not limited to isotropic monolayer structures, but it also handles orthotropic and multi-layered lamination schemes, and it can go up to layers embedding FGM. However, confident use of the model is possible only after validated against established solutions already offered in the literature. The authors did not find hygro-elastic results from exact 3D solutions in the literature which were applied to FGMs. For this reason, the validation process of the model is built by separately validating its different sections against the results provided by other researchers, exploiting the parallel of the moisture content field and the thermal field, and complementing this process with the assistance of 3D FE (Finite Element) models. A static 3D FE model is solved through the Nastran solver SOL101. IsoMesh meshed the geometry of each plate with 3D HEX8 elements; the mesh did not change with the thickness ratio and included 25 elements in the thickness direction and 30 in both the in-plane ones. Solid elements were necessary to define the mechanical properties evolution in the thickness direction and build significant thermal, hygrometric, and mechanical variables profiles. Depending on the geometry, the coordinate system of the model has one, two, or none curvilinear coordinate. For consistency with the analytical model, the displacement-related boundary conditions are defined on the lateral surfaces of the structure following the harmonic form hypotheses: a pair of displacement coordinates is set 0 on each surface. First, the harmonic thermal/hygrometric field is introduced in an equation, following Equation (1). It is applied to the top and bottom surfaces of the structure; the preprocessor automatically calculates the field values at each node of the external surfaces through their coordinates. The first run solves the thermal part of the problem and returns the temperature (or moisture content) at each node. This field is then applied to a further (and identical) FE model, which solves the elastic part. The thermal, hygrometric, and mechanical properties of the FGM layer are given imagining to split the structures into a number of fictitious layers coinciding with the number of elements in the thickness direction. The properties of each fictitious layer are calculated at its midpoint, following Equations (83)–(85). Needless to say, they are not exact solutions, but they can guide in benchmarking the proposed model.

The first assessment considers a simply-supported one-layered FGM square plate. It investigates different thickness ratios ($a/h = 4, 10, 50$) in plates with in-plane dimensions of $a = b = 100$ m. The FGM layer relies on a metallic constituent, and a ceramic one, whose mechanical, thermal, and hygrometric properties are defined at the beginning of Section 4. The volume fraction power law considers $p = 2$ as the exponent. An external sovra-temperature field acts on the top ($\theta_t = +1$ K) and bottom ($\theta_b = 0$ K) surfaces. The thermal field has a harmonic form, with half-wave numbers $m = n = 1$. The reference solution is the asymptotic method of Reddy and Cheng [73], which considers a 3D temperature profile along with thickness direction. Table 1 proposes a pair of results for each thickness ratio at different coordinates along with direction z in terms of a displacement, w or u, and an in-plane shear component, $\sigma_{\alpha\alpha}$. The results show that the 3D shell model always coincides with Reddy and Cheng's asymptotic method, despite the thickness ratio and the considered variable, when the number of mathematical layers is sufficiently high. $NL = 300$, coupled with an order of expansion $N = 3$ for the exponential matrix, always delivers the correct results. Therefore, this assessment verified that the 3D shell model correctly handles the thermomechanical analysis of FGM plates. Furthermore, it simultaneously assessed the 3D FE model, which will be helpful in the following assessment to validate the hygromechanical analysis.

The second assessment is meant to validate the hygroelastic part of the 3D solution for plates embedding an FGM layer. To this end, it considers the previous test case as a reference and removes the thermal field in favor of a hygrometric one. Consequently, it focuses on a simply-supported one-layered FGM square plate; the in-plane dimensions are $a = b = 100$ m, the thickness varies to obtain different thickness ratios ($a/h = 4, 10, 50$). The FGM layer relies on the same metallic and ceramic constituents, whose volume fraction follows the same power-law with $p = 2$. An external hygrometric field acts on the top ($M_t = 1.0\%$) and bottom ($M_b = 0.5\%$) surfaces; it has a harmonic form, with half-wave numbers $m = n = 1$. The reference results are obtained through the same 3D FE model of the previous assessment, in which the hygrometric field replaces the thermal one. Its previous validation allows considering it as a reliable source for reference results. Table 2 proposes a pair of results for each thickness ratio at different coordinates along with direction z in terms of a displacement, w or u, and an in-plane shear component, $\sigma_{\alpha\alpha}$. Consistent with the previous test case, the results show that the 3D shell model always gives comparable results with the 3D FE model, despite the thickness ratio and the considered variable, when the number of mathematical layers is sufficiently high. $NL = 300$, coupled with an order of expansion $N = 3$ for the exponential matrix, always delivers the correct results. Therefore, this assessment confirmed the capabilities of the 3D shell model in handling the hygromechanical analysis of FGM plates.

Table 1. First assessment. One-layered FGM square plate ($a/b = 1$), featuring different thickness ratios. The volume fraction power law considers $p = 2$ as the exponent. An external sovra-temperature field acts on the top ($\theta_t = +1$ K) and bottom ($\theta_b = 0$ K) surfaces; $m = n = 1$. The reference solution is the asymptotic method of Reddy and Cheng [73], considering a 3D temperature profile along with thickness direction. A 3D FE model is also assessed. The results of the present solution are obtained with $N = 3$ and for a varying number of mathematical layers NL.

		Present Solution					3D FEM	Ref. 3D [73]
	NL →	10	50	100	200	300		
		$a/h = 4$						
$\tilde{w}$ at $z = h$	3D(θ_a)	4.2943	4.2842	4.2838	4.2837	4.2836		
	3D(θ_c, 1D)	3.3331	3.2130	3.2083	3.2071	3.2068	3.067	3.043
	3D(θ_c, 3D)	3.1791	3.0482	3.0431	3.0418	3.0415		
$\tilde{\sigma}_{\alpha\alpha}$ at $z = 0$	3D(θ_a)	288.99	290.12	290.21	290.23	290.24		
	3D(θ_c, 1D)	−27.47	−22.16	−21.92	−21.85	−21.84	−75.57	−73.53
	3D(θ_c, 3D)	−81.37	−73.99	−73.67	−73.58	−73.57		
		$a/h = 10$						
$\tilde{u}$ at $z = h/2$	3D(θ_a)	−1.3433	−1.3417	−1.3417	−1.3417	−1.3417		
	3D(θ_c, 1D)	−0.8287	−0.8026	−0.8016	−0.8014	−0.8014	−0.7910	−0.7862
	3D(θ_c, 3D)	−0.8134	−0.7871	−0.7861	−0.7859	−0.7858		
$\tilde{\sigma}_{\alpha\alpha}$ at $z = h$	3D(θ_a)	−599.6	−490.7	−476.4	−469.2	−466.8		
	3D(θ_c, 1D)	−1102	−1021	−1006	−999.3	−996.7	−1058	−1006
	3D(θ_c, 3D)	−1117	−1036	−1021	−1014	−1011		
		$a/h = 50$						
$\tilde{w}$ at $z = h/2$	3D(θ_a)	35.97	35.81	35.80	35.80	35.80		
	3D(θ_c, 1D)	29.64	28.50	28.46	28.45	28.45	28.57	28.45
	3D(θ_c, 3D)	29.64	28.50	28.45	28.44	28.44		
$\tilde{\sigma}_{\alpha\alpha}$ at $z = h/2$	3D(θ_a)	−759.3	−730.7	−726.8	−724.9	−724.2		
	3D(θ_c, 1D)	−269.9	−255.3	−253.5	−252.5	−252.2	−250.4	−251.2
	3D(θ_c, 3D)	−269.3	−254.7	−252.9	−252.0	−251.7		

The third assessment considers a simply-supported one-layered FGM cylindrical shell. Again, the dimensions of the reference surface are fixed, as in the previous cases; they are $a = 1$ m and $b = \frac{\pi}{3} R_\beta$, with $R_\beta = 10$ m. This test case also considers different thickness ratios $R_\beta = 50, 1000$) to evaluate their influence on the performance of the solution. The constituents of the FGM layer are the same as those defined at the beginning of Section 4 and considered in the previous assessments. The power law is also the same, with $p = 2$. The top and bottom external surfaces are subjected to an external sovra-temperature field with amplitudes $\theta_t = +1$ K and $\theta_b = 0$ K. The half-wave numbers of the thermal field are $m = n = 1$. The reference solution is a refined 2D layer-wise solution based on the Unified Formulation [74], which considers a 3D temperature profile along with thickness direction. Table 3 proposes six results for each thickness ratio: the transverse displacement w and an in-plane displacement, evaluated at three different coordinates along with direction z. The table also assesses a 3D FE model, solved through the Nastran solver, which helps evaluate the hygroelastic analysis of the following assessment. IsoMesh meshed the geometry of each shell with 3D HEX8 elements; the mesh did not change with the thickness ratio and included 25 elements in the thickness direction and 30 in both the in-plane ones. Solid elements were necessary to define the mechanical properties evolution in the thickness direction and build significant thermal and mechanical variables profiles. The results show that the 3D shell model always coincides with the quasi-3D method [74], despite the thickness ratio and the considered variable, when the number of mathematical layers is sufficiently high. $NL = 300$, coupled with an order of expansion $N = 3$ for the exponential matrix, always delivers the correct results. Therefore, this assessment verified that the 3D shell model correctly handles the thermomechanical analysis of FGM shells. Furthermore, it simultaneously assessed the 3D FE model, which will be helpful in the following assessment to validate the hygromechanical analysis.

Table 2. Second assessment. One-layered FGM square plate ($a/b = 1$), featuring different thickness ratios. The geometry, materials, and FGM power law are the same as the first assessment. The thermal load is substituted by an external moisture content acting on the top ($M_t = 1.0\%$) and bottom ($M_b = 0.5\%$) surfaces; $m = n = 1$. The reference solution is the 3D FE model, already validated in the previous assessment. The results of the present solution are obtained with $N = 3$ and for a varying number of mathematical layers NL.

			Present Solution				3D FEM
	NL →	10	50	100	200	300	
				$a/h = 4$			
$\tilde{w}$ at $z = h$	3D($\mathcal{M}_a$)	227.5	225.7	225.6	225.6	225.6	
	3D($\mathcal{M}_c, 1D$)	177.9	171.1	170.8	170.7	170.7	143.3
	3D($\mathcal{M}_c, 3D$)	151.0	141.9	141.6	141.5	141.5	
$\tilde{\sigma}_{\alpha\alpha}$ at $z = 0$	3D($\mathcal{M}_a$)	−32,309	−32,391	−32,383	−32,380	−32,380	
	3D($\mathcal{M}_c, 1D$)	−53,262	−53,245	−53,234	−53,230	−53,230	−751,662
	3D($\mathcal{M}_c, 3D$)	−75,532	−75,561	−75,551	−75,548	−75,548	
				$a/h = 10$			
$\tilde{u}$ at $z = h/2$	3D($\mathcal{M}_a$)	−262.4	−262.1	−262.1	−262.1	−262.1	
	3D($\mathcal{M}_c, 1D$)	−231.9	−230.5	−230.4	−230.4	−230.4	−226.0
	3D($\mathcal{M}_c, 3D$)	−227.3	−225.8	−225.7	−225.7	−225.7	
$\tilde{\sigma}_{\alpha\alpha}$ at $z = h$	3D($\mathcal{M}_a$)	−55,689	−35,256	−32,575	−31,224	−30,772	
	3D($\mathcal{M}_c, 1D$)	−83,669	−64,410	−61,712	−60,336	−59,873	−63,825
	3D($\mathcal{M}_c, 3D$)	−87,250	−68,062	−65,359	−63,979	−63,515	
				$a/h = 50$			
$\tilde{u}$ at $z = 0$	3D($\mathcal{M}_a$)	−257.3	−258.1	−258.1	−258.1	−258.1	
	3D($\mathcal{M}_c, 1D$)	−236.0	−237.0	−237.0	−237.0	−237.0	−236.8
	3D($\mathcal{M}_c, 3D$)	−235.9	−236.8	−236.8	−236.9	−236.9	
$\tilde{\sigma}_{\alpha\alpha}$ at $z = h/2$	3D($\mathcal{M}_a$)	−145,561	−137,645	−136,589	−136,056	−135,878	
	3D($\mathcal{M}_c, 1D$)	−114,235	−107,530	−106,648	−106,203	−106,054	−105,149
	3D($\mathcal{M}_c, 3D$)	−114,035	−107,338	−106,457	−106,013	−105,865	

The fourth assessment is meant to validate the hygro-elastic part of the 3D solution for shells embedding an FGM layer. To this end, it considers the previous test case as a reference and removes the thermal field in favor of a hygrometric one. Consequently, it focuses on a simply-supported one-layered FGM square shell. The dimensions of the reference surface are $a = 1$ m and $b = \frac{\pi}{3}R_\beta$, with $R_\beta = 10$ m, the thickness varies to obtain different thickness ratios ($R_\beta = 50, 100$). The FGM layer relies on the same metallic and ceramic constituents of the previous test cases, whose volume fraction follows the same power-law with $p = 2$. An external hygrometric field acts on the top ($M_t = 1.0\%$) and bottom ($M_b = 0.5\%$) surfaces; it has a harmonic form, with half-wave numbers $m = n = 1$. The reference results are obtained through the same 3D FE model of the previous assessment, in which the hygrometric field replaces the thermal one. Its previous validation allows considering it as a reliable source for reference results. Table 4 proposes six results for each thickness ratio: the transverse displacement w and an in-plane displacement, evaluated at three different coordinates along with direction z. Consistent with the previous test case, the results show that the 3D shell model always gives comparable results with the 3D FE model, despite the thickness ratio and the considered variable, when the number of mathematical layers is sufficiently high. $NL = 300$, coupled with an order of expansion $N = 3$ for the exponential matrix, always delivers the correct results. Therefore, this assessment confirmed the capabilities of the 3D shell model in handling the hygromechanical analysis of FGM shells.

Table 3. Third assessment. One-layered FGM cylindrical shell ($R_\beta = 10$ m), featuring different thickness ratios. The volume fraction power law considers $p = 2$ as the exponent. An external sovra-temperature field acts on the top ($\theta_t = +1$ K) and bottom ($\theta_b = 0$ K) surfaces; $m = n = 1$. The reference solution is a refined 2D layer-wise solution based on the Unified Formulation [74], considering a 3D temperature profile along with thickness direction. A 3D FE model is also assessed. The results of the present solution are obtained with $N = 3$ and for a varying number of mathematical layers NL.

		Present Solution					3D FEM	Ref. [74]
	NL →	10	50	100	200	300		
		$R_\beta/h = 50$						
$\tilde{w}$ at $z = h$	3D(θ_a)	9.8080	9.7773	9.7762	9.7759	9.7759		
	3D(θ_c, 1D)	7.5570	7.2809	7.2703	7.2676	7.2671	7.2722	7.1337
	3D(θ_c, 3D)	7.4312	7.1467	7.1358	7.1330	7.1325		
$\tilde{w}$ at $z = h/2$	3D(θ_a)	8.6672	8.6375	8.6365	8.6362	8.6362		
	3D(θ_c, 1D)	6.7886	6.5371	6.5274	6.5250	6.5245	6.5284	6.4131
	3D(θ_c, 3D)	6.6843	6.4250	6.4150	6.4125	6.4120		
$\tilde{w}$ at $z = 0$	3D(θ_a)	8.1689	8.1401	8.1391	8.1388	8.1388		
	3D(θ_c, 1D)	6.5471	6.3026	6.2932	6.2908	6.2903	6.2996	6.1942
	3D(θ_c, 3D)	6.4582	6.2058	6.1960	6.1936	6.1931		
$\tilde{u}$ at $z = h$	3D(θ_a)	−5.3944	−5.3825	−5.3821	−5.3820	−5.3820		
	3D(θ_c, 1D)	−3.7791	−3.6488	−3.6438	−3.6426	−3.6423	−3.6252	−3.5466
	3D(θ_c, 3D)	−3.6863	−3.5527	−3.5476	−3.5463	−3.5461		
$\tilde{u}$ at $z = h/2$	3D(θ_a)	−2.5669	−2.5643	−2.5642	−2.5642	−2.5642		
	3D(θ_c, 1D)	−1.5635	−1.5150	−1.5131	−1.5127	−1.5126	−1.4900	−1.4532
	3D(θ_c, 3D)	−1.5046	−1.4554	−1.4536	−1.4531	−1.4530		
$\tilde{u}$ at $z = 0$	3D(θ_a)	0.0300	0.0239	0.0237	0.0236	0.0236		
	3D(θ_c, 1D)	0.4850	0.4578	0.4567	0.4564	0.4564	0.4816	0.4833
	3D(θ_c, 3D)	0.5136	0.4846	0.4835	0.4832	0.4831		
		$R_\beta/h = 1000$						
$\tilde{w}$ at $z = h$	3D(θ_a)	69.477	69.348	69.344	69.343	69.342		
	3D(θ_c, 1D)	45.154	43.661	43.604	43.589	43.587	44.402	43.590
	3D(θ_c, 3D)	45.150	43.657	43.600	43.586	43.583		
$\tilde{w}$ at $z = h/2$	3D(θ_a)	69.424	69.294	69.290	69.289	69.289		
	3D(θ_c, 1D)	45.116	43.624	43.567	43.553	43.550	44.364	43.553
	3D(θ_c, 3D)	45.113	43.620	43.563	43.549	43.546		
$\tilde{w}$ at $z = 0$	3D(θ_a)	69.417	69.287	69.283	69.282	69.282		
	3D(θ_c, 1D)	45.117	43.625	43.568	43.554	43.551	44.365	43.554
	3D(θ_c, 3D)	45.114	43.621	43.564	43.550	43.547		
$\tilde{u}$ at $z = h$	3D(θ_a)	−2.994	−2.991	−2.990	−1.990	−2.990		
	3D(θ_c, 1D)	−1.848	−1.789	−1.787	−1.786	−1.786	−1.8211	−1.7868
	3D(θ_c, 3D)	−1.847	−1.789	−1.787	−1.786	−1.786		
$\tilde{u}$ at $z = h/2$	3D(θ_a)	−1.903	−1.901	−1.901	−1.901	−1.901		
	3D(θ_c, 1D)	−1.138	−1.103	−1.102	−1.102	−1.102	−1.1238	−1.1021
	3D(θ_c, 3D)	−1.138	−1.103	−1.102	−1.102	−1.102		
$\tilde{u}$ at $z = 0$	3D(θ_a)	−0.8129	−0.8132	−0.8132	−0.8132	−0.8132		
	3D(θ_c, 1D)	−0.4298	−0.4183	−0.4179	−0.4177	−0.4177	−0.4269	−0.4178
	3D(θ_c, 3D)	−0.4297	−0.4182	−0.4178	−0.4177	−0.4177		

Table 4. Fourth assessment. One-layered FGM cylindrical shell ($R_\beta = 10$ m) featuring different thickness ratios. The geometry, materials, and FGM power law are the same as the third assessment. The thermal load is substituted by an external moisture content acting on the top ($M_t = 1.0\%$) and bottom ($M_b = 0.5\%$) surfaces; $m = n = 1$. The reference solution is the 3D FE model, already validated in the previous assessment. The results of the present solution are obtained with $N = 3$ and for a varying number of mathematical layers NL.

		Present Solution					3D FEM
	NL →	10	50	100	200	300	
		$R_\beta/h = 50$					
$\tilde{w}$ at $z = h$	3D($\mathcal{M}_a$)	557.35	551.43	551.22	551.17	551.16	
	3D($\mathcal{M}_c$, 1D)	440.34	423.22	422.60	422.44	422.41	411.97
	3D($\mathcal{M}_c$, 3D)	417.06	398.20	397.51	397.34	397.31	
$\tilde{w}$ at $z = h/2$	3D($\mathcal{M}_a$)	399.25	393.43	393.22	393.16	393.15	
	3D($\mathcal{M}_c$, 1D)	303.27	287.38	286.80	286.66	286.63	279.92
	3D($\mathcal{M}_c$, 3D)	285.49	268.05	267.42	267.26	267.23	
$\tilde{w}$ at $z = 0$	3D($\mathcal{M}_a$)	228.40	222.78	222.58	222.53	222.52	
	3D($\mathcal{M}_c$, 1D)	148.66	133.28	132.72	132.58	132.55	131.17
	3D($\mathcal{M}_c$, 3D)	136.53	119.68	119.07	118.91	118.88	
$\tilde{u}$ at $z = h$	3D($\mathcal{M}_a$)	−654.1	−651.9	−651.8	−651.8	−651.8	
	3D($\mathcal{M}_c$, 1D)	−563.3	−555.6	−555.3	−555.2	−555.2	−538.01
	3D($\mathcal{M}_c$, 3D)	−539.7	−531.1	−530.8	−530.7	−530.7	
$\tilde{u}$ at $z = h/2$	3D($\mathcal{M}_a$)	−515.9	−515.5	−515.5	−515.5	−515.5	
	3D($\mathcal{M}_c$, 1D)	−456.3	−453.7	−453.6	−453.6	−453.6	−435.01
	3D($\mathcal{M}_c$, 3D)	−438.5	−435.4	−435.3	−435.3	−435.3	
$\tilde{u}$ at $z = 0$	3D($\mathcal{M}_a$)	−424.7	−426.0	−426.0	−426.1	−426.1	
	3D($\mathcal{M}_c$, 1D)	−392.8	−394.8	−394.9	−394.9	−394.9	−380.50
	3D($\mathcal{M}_c$, 3D)	−379.6	−381.7	−381.8	−381.8	−381.8	
		$R_\beta/h = 1000$					
$\tilde{w}$ at $z = h$	3D($\mathcal{M}_a$)	11,564	11,545	11,544	11,544	11,544	
	3D($\mathcal{M}_c$, 1D)	10,150	10,065	10,062	10,061	10,061	10,128
	3D($\mathcal{M}_c$, 3D)	10,149	10,064	10,061	10,060	10,060	
$\tilde{w}$ at $z = h/2$	3D($\mathcal{M}_a$)	11,558	11,539	11,538	11,538	11,538	
	3D($\mathcal{M}_c$, 1D)	10,145	10,060	10,057	10,056	10,056	10,123
	3D($\mathcal{M}_c$, 3D)	10,144	10,059	10,056	10,055	10,055	
$\tilde{w}$ at $z = 0$	3D($\mathcal{M}_a$)	11,551	11,532	11,531	11,531	11,531	
	3D($\mathcal{M}_c$, 1D)	10,139	10,054	10,051	10,050	10,050	10,117
	3D($\mathcal{M}_c$, 3D)	10,138	10,053	10,050	10,049	10,049	
$\tilde{u}$ at $z = h$	3D($\mathcal{M}_a$)	−589.0	−588.5	−588.5	−588.5	−588.5	
	3D($\mathcal{M}_c$, 1D)	−521.1	−518.0	−517.9	−517.8	−517.8	−520.37
	3D($\mathcal{M}_c$, 3D)	−521.1	−517.9	−517.8	−517.8	−517.8	
$\tilde{u}$ at $z = h/2$	3D($\mathcal{M}_a$)	−407.4	−407.2	−407.2	−407.2	−407.2	
	3D($\mathcal{M}_c$, 1D)	−361.8	−359.9	−359.9	−359.9	−359.8	−361.32
	3D($\mathcal{M}_c$, 3D)	−361.7	−359.9	−359.8	−359.8	−359.8	
$\tilde{u}$ at $z = 0$	3D($\mathcal{M}_a$)	−225.9	−226.0	−226.0	−226.0	−226.0	
	3D($\mathcal{M}_c$, 1D)	−202.5	−202.0	−202.0	−202.0	−202.0	−202.42
	3D($\mathcal{M}_c$, 3D)	−202.5	−202.0	−201.9	−201.9	−201.9	

4.2. New Benchmarks

This section proposes a set of four benchmarks; those new results examine simply supported structures that undergo different moisture content profiles in steady-state conditions. They follow the harmonic form previously defined, precondition to get an exact solution to the problem. The assessments of the previous subsection validated the results of this new

3D shell model when applied to FGM layers: the results converge and are exact when the order of expansion $N = 3$ for the exponential matrix is coupled with a minimum number of $M = 300$ mathematical layer for the through the thickness mechanical properties and curvature approximation. The results of all the following benchmarks consider $N = 3$ and $M = 300$ as an a priori prerequisite for results accuracy.

The first benchmark studies a square plate with a single FGM layer and simply-supported sides. The plate has $a = b = 10$ m as in-plane dimensions but comes with several and different thicknesses, which allow the effect of this geometrical parameter to be evaluated. In fact, the thickness ratio goes from $a/h = 2$ to $a/h = 100$, thus ranging from very thick to very thin plates. In this benchmark, the volume fraction V_c of the ceramic phase evolves linearly: $p = 1$ is set in the material law defined through Equation (83). This relation also implies a fully ceramic top surface and a fully metallic bottom one. The moisture content is imposed on the top and the bottom external surfaces; it has a harmonic form on both with amplitudes $\mathcal{M}_t = 1.0\%$ and $\mathcal{M}_b = 0.0\%$, on top and bottom, respectively. The harmonic form of the moisture content has $m = n = 1$ as half-wave numbers in α and β directions, respectively. The elastic and hygroscopic properties of the metallic and ceramic phases are the same as those introduced at the beginning of this section. The FGM nature of the layer makes its mechanical properties evolve through the thickness direction; Figure 3a,b, respectively, shows how the volume fraction V_c and the bulk modulus K evolve with respect to non-dimensional thickness coordinate $\tilde{z}/h$. Note that K is not linear as V_c due to Equation (85). Table 5 and Figures 4 report an extract of the main results. The results in tabular form give the amplitude of some variables of the problem; they reflect the three different ways of evaluating the moisture content. 3D($\mathcal{M}_a$) implies the assumed linear moisture content profile; 3D($\mathcal{M}_c, 1D$) relies on the 1D version of the Fick moisture diffusion equation; finally, 3D($\mathcal{M}_c, 3D$) relies on a 3D solution of the moisture diffusion problem. The prefix 3D underlines that the elastic part of the solution is three-dimensional in all three models. Such an analysis allows grasping the differences between the three approaches. The 3D($\mathcal{M}_c, 3D$) shell model considers both the mechanical/hygrometric properties evolution through z and the three-dimensional nature of the problem. As discussed in the previous section, it always delivers an accurate result. Table 5 underlines that the 3D($\mathcal{M}_c, 1D$) model results get closer to those of the 3D($\mathcal{M}_c, 3D$) model as the thickness decreases. It considers how the mechanical/hygrometric properties evolve through z but disregards the moisture diffusion through alpha and beta direction, which have a negligible weight in thin structures. The results of the 3D($\mathcal{M}_a$) model are always unreliable, as they are built on a moisture content profile that is far from the actual scenario in a layer embedding an FGM. Figure 3c,d further facilitates understanding these concepts; it compares the three moisture content profiles for a thick and a thin plate. In thick structures, the difference between the three profiles is very pronounced: the three-dimensionality of the problem and the mechanical/hygrometric properties variability in the thickness direction make the 3D profile differ from the linear assumption. Even the 1D profile differs from the linearity: the hygrometric properties vary through z, reflecting on the moisture content at different thickness coordinates. These concepts also apply to thin structures; however, the three-dimensionality of the problem is insignificant, and the 3D and 1D profiles coincide. Figure 4 shows the complete profile of the tree displacement components, two stresses, and a strain. Note that all the quantities evolve with continuity, which is essential as it demonstrates both the graded elastic/hygrometric properties and the correct introduction of the continuity conditions. The transverse stress σ_{zz} and the transverse shear strain $\gamma_{\beta z}$ satisfy the external mechanical boundary conditions: it equals 0 at both the top and the bottom surfaces as no external load acts on them.

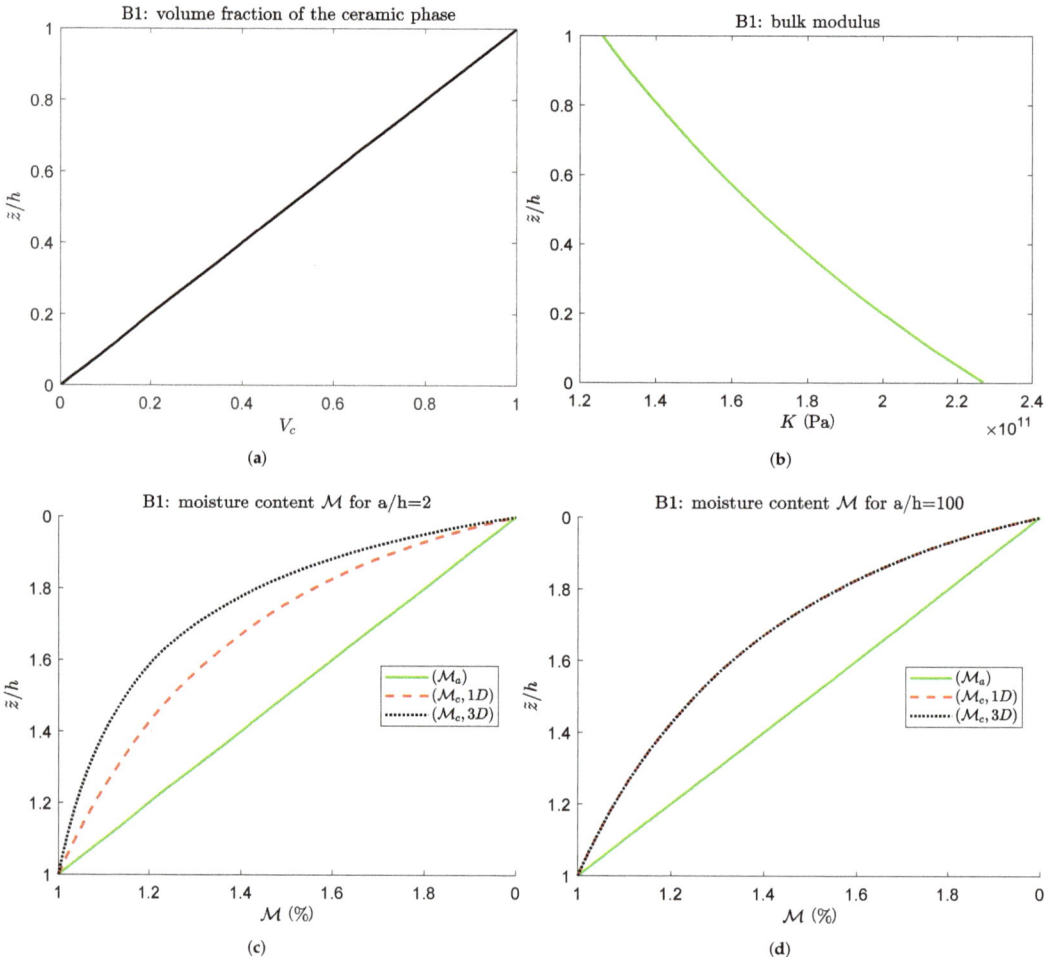

Figure 3. First benchmark: one-layered FGM ($p = 1$) square plate with an imposed moisture content on the top and bottom surfaces. The figures show the volume fraction of the ceramic phase, the bulk modulus, and the moisture content profiles for a thick and a thin structure through their thickness. (**a**) Volume fraction of the ceramic phase V_c. (**b**) Bulk modulus K. (**c**) Moisture content profile of the $a/h = 2$ plate. (**d**) Moisture content profile of the $a/h = 100$ plate.

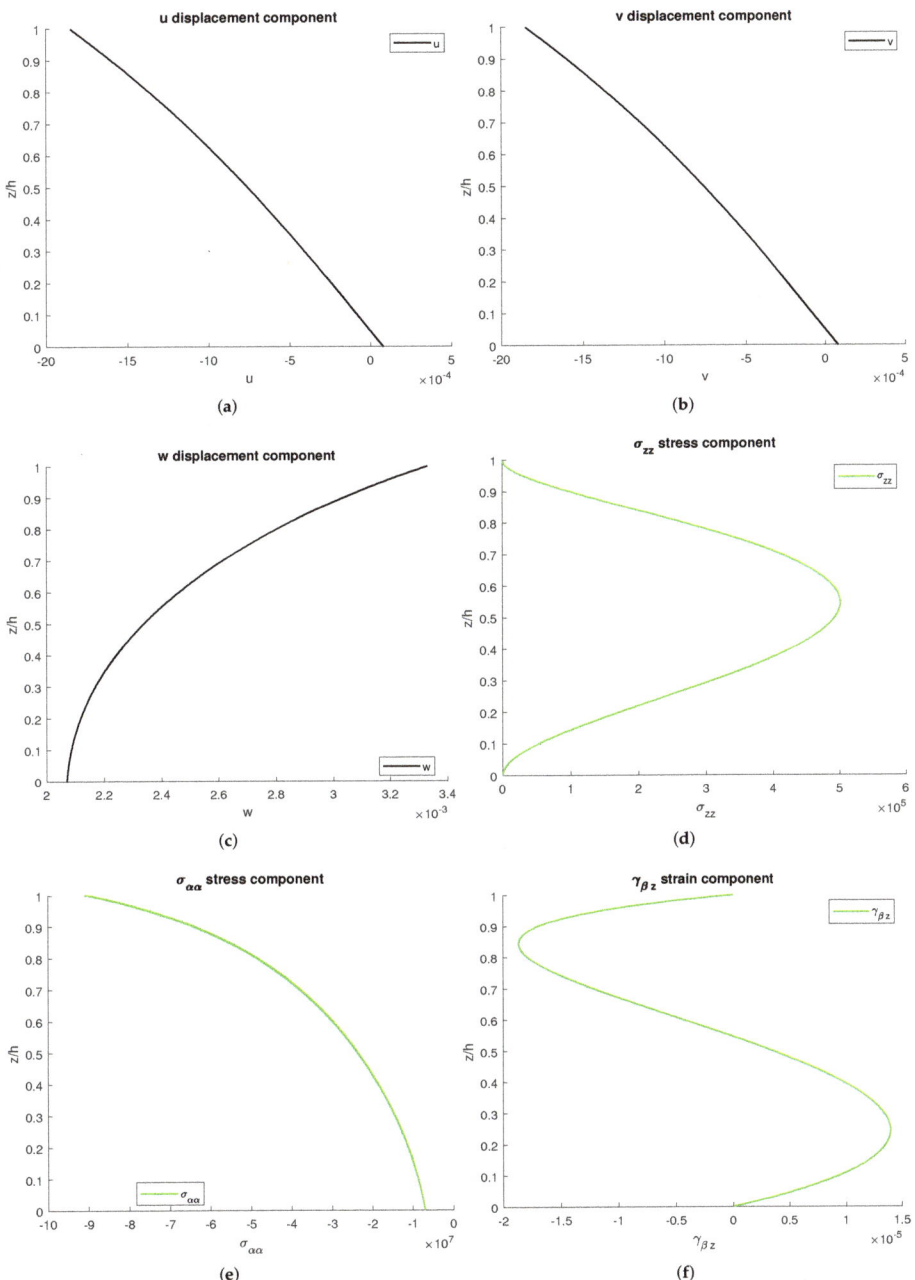

Figure 4. First benchmark: one-layered FGM square plate with an imposed moisture content on the top and bottom surfaces. The results are calculated for a moderately thick ($a/h = 4$) structure via the 3D($\mathcal{M}_c$, 3D) model. (**a**) Amplitude of u displacement component. (**b**) Amplitude of v displacement component. (**c**) Amplitude of w displacement component. (**d**) Amplitude of σ_{zz} stress component. (**e**) Amplitude of $\sigma_{\alpha\alpha}$ stress component. (**f**) Amplitude of $\gamma_{\beta z}$ strain component.

Table 5. First benchmark: square plate with a single FGM layer subject to an external moisture content applied to the top and bottom surfaces. The results of all the 3D models consider $N = 3$ and $NL = 300$.

a/h	2	4	10	50	100
$\mathcal{M}$ at $(\alpha = a/2, \beta = b/2; \tilde{z} = 4h/5)$ [−]					
3D($\mathcal{M}_a$)	0.8000	0.8000	0.8000	0.8000	0.8000
3D($\mathcal{M}_c$, 1D)	0.5595	0.5595	0.5595	0.5595	0.5595
3D($\mathcal{M}_c$, 3D)	0.4351	0.5197	0.5525	0.5592	0.5594
v at $(\alpha = a/2, \beta = 0; \tilde{z} = 4h/5)$ [mm]					
3D($\mathcal{M}_a$)	−2.094	−2.089	−2.087	−2.086	−2.086
3D($\mathcal{M}_c$, 1D)	−1.454	−1.461	−1.462	−1.462	−1.462
3D($\mathcal{M}_c$, 3D)	−1.148	−1.362	−1.445	−1.462	−1.462
w at $(\alpha = a/2, \beta = b/2; \tilde{z} = h/2)$ [mm]					
3D($\mathcal{M}_a$)	1.226	2.661	6.791	34.07	68.16
3D($\mathcal{M}_c$, 1D)	1.103	2.405	6.144	30.83	61.68
3D($\mathcal{M}_c$, 3D)	1.008	2.336	6.113	30.83	61.67
σ_{zz} at $(\alpha = a/2, \beta = b/2; \tilde{z} = h/5)$ [kPa]					
3D($\mathcal{M}_a$)	−5915	−982.6	−127.7	−4.879	−1.218
3D($\mathcal{M}_c$, 1D)	−1197	44.77	26.65	1.217	0.3054
3D($\mathcal{M}_c$, 3D)	515.5	173.0	30.09	1.222	0.3057
$\sigma_{\alpha\alpha}$ at $(\alpha = a/2, \beta = b/2; \tilde{z} = h)$ [MPa]					
3D($\mathcal{M}_a$)	−7.939	−34.86	−43.94	−45.68	−45.73
3D($\mathcal{M}_c$, 1D)	−63.24	−3.08	−89.55	−90.74	−90.78
3D($\mathcal{M}_c$, 3D)	−90.78	−90.95	−90.83	−90.79	−90.79
$\gamma_{\beta z}$ at $(\alpha = a/2, \beta = 0; \tilde{z} = h/3)$ [10^{-6}]					
3D($\mathcal{M}_a$)	−173.7	−57.29	−18.35	−3.489	−1.741
3D($\mathcal{M}_c$, 1D)	−39.06	3.440	4.709	1.072	0.5379
3D($\mathcal{M}_c$, 3D)	19.67	12.49	5.321	1.076	0.5384

The second benchmark focuses on a closed cylinder, featuring a single FGM layer and simply-supported sides. The dimensions of the reference mid-surface, $a = 2\pi R_\alpha$ and $b = 30$ m, are a function of the radii of curvature of the shell, one of which is infinite: $R_\alpha = 10$ m and $R_\beta = \infty$. Different thicknesses have been considered; the thickness ratio R_α/h is expressed with respect to R_α and ranges from 2 to 100 also in this second case study. The material volume fraction of the ceramic phase is a quadratic function of the thickness coordinate; the material law defined through Equation (83) consider $p = 2$. Given that a single layer is considered, the cylinder is metallic in the inner surface and ceramic in the outer. The moisture content is imposed on the outer external surface, $\mathcal{M}_t = 1.0\%$, and on the inner one, $\mathcal{M}_b = 0.0\%$. The half wave numbers of both the harmonic forms are the same, $m = 2$ and $n = 1$. The elastic and hygroscopic properties of both the phases introduced previously also apply here. Figure 5a,b, respectively, shows the volume fraction V_c and moisture diffusion coefficient $\mathcal{D}$ vs. the non-dimensional thickness coordinate $\tilde{z}/h$. V_c follows a power-law of order $p = 2$, $\mathcal{D}$ follows Equation (85). Table 6 and Figure 6 summarize an extract of the main results. This second benchmark also reports three different sets of results: the elastic model is the same (prefix 3D), but the moisture content profile follows the different approaches. This leads to models 3D($\mathcal{M}_a$), in which the moisture content is a priori assumed, and 3D($\mathcal{M}_c$, 1D)–3D($\mathcal{M}_c$, 3D), in which the profile is calculated following a monodimensional or three-dimensional approach. This analysis allows highlighting the distinctions between the three methods. The last one is the only model in which no assumptions are made concerning the three-dimensionality of the problem as the moisture content amplitude derives from Fick's law of diffusion. The results coming from 3D($\mathcal{M}_a$) are always wrong because the moisture content evaluation is inaccurate. The differences between 3D($\mathcal{M}_c$, 1D) and 3D($\mathcal{M}_c$, 3D) are less pronounced if compared with the previous benchmark and decrease with the thickness. The differences in the three moisture content profiles are shown in Figure 5c,d for two different cylinders: a thick and a thin one. The discrepancies between the calculated and assumed fields are really pronounced; the 1D and 3D computed profiles do not significantly differ, which is even more true as the thickness ratio increases. Figure 6 shows the profiles of the tree displacement components: two stresses and a strain. There is continuity in all the plots: this qualifies the correct introduction of the continuity conditions and elastic/hygrometric properties grading. No external mechanical loads are applied, and this is coherent with the transverse stress values at the bottom and top surfaces, 0.

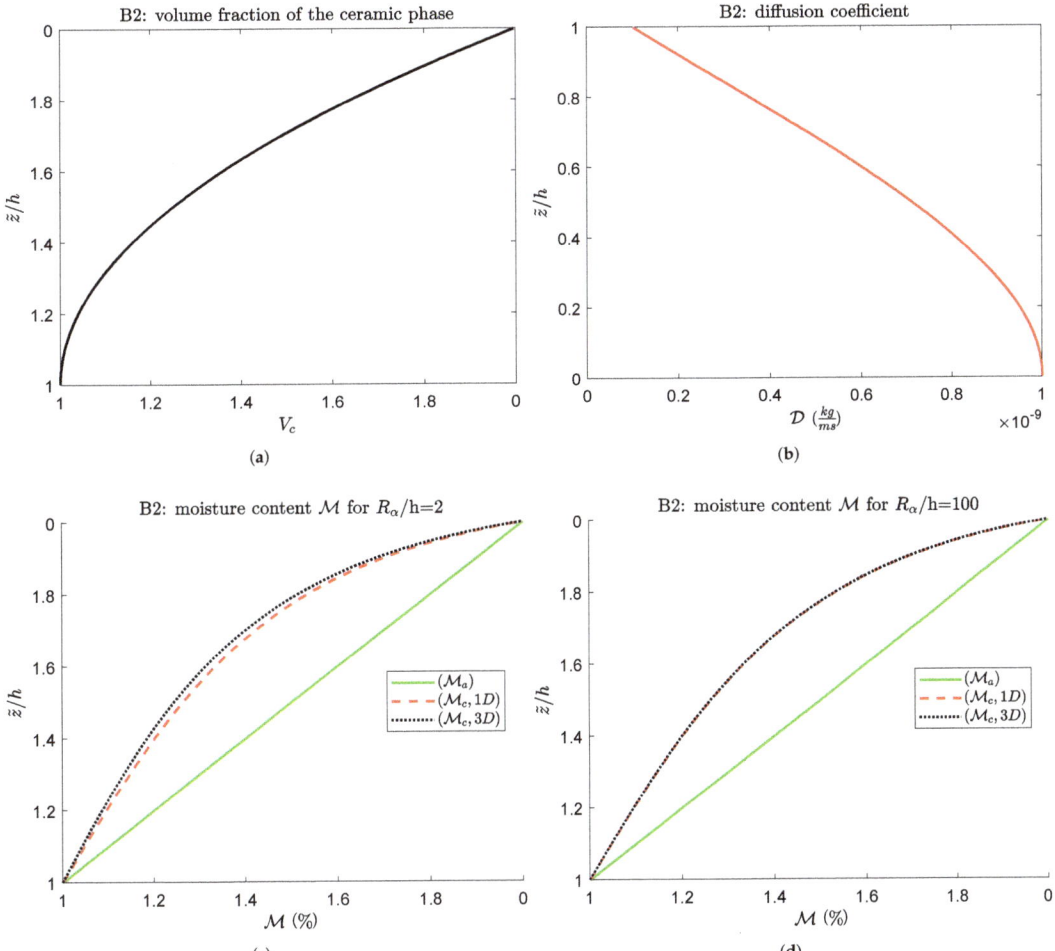

Figure 5. Second benchmark: one-layered FGM ($p = 2$) cylinder with an imposed moisture content on the top and bottom surfaces. The figures show the volume fraction of the ceramic phase, the diffusion coefficient, and the moisture content profiles for a thick and a thin structure through their thickness. (**a**) Volume fraction of the ceramic phase V_c. (**b**) Diffusion coefficient $\mathcal{D}$. (**c**) Moisture content profile of the $R_\alpha/h = 2$ cylinder. (**d**) Moisture content profile of the $R_\alpha/h = 100$ cylinder.

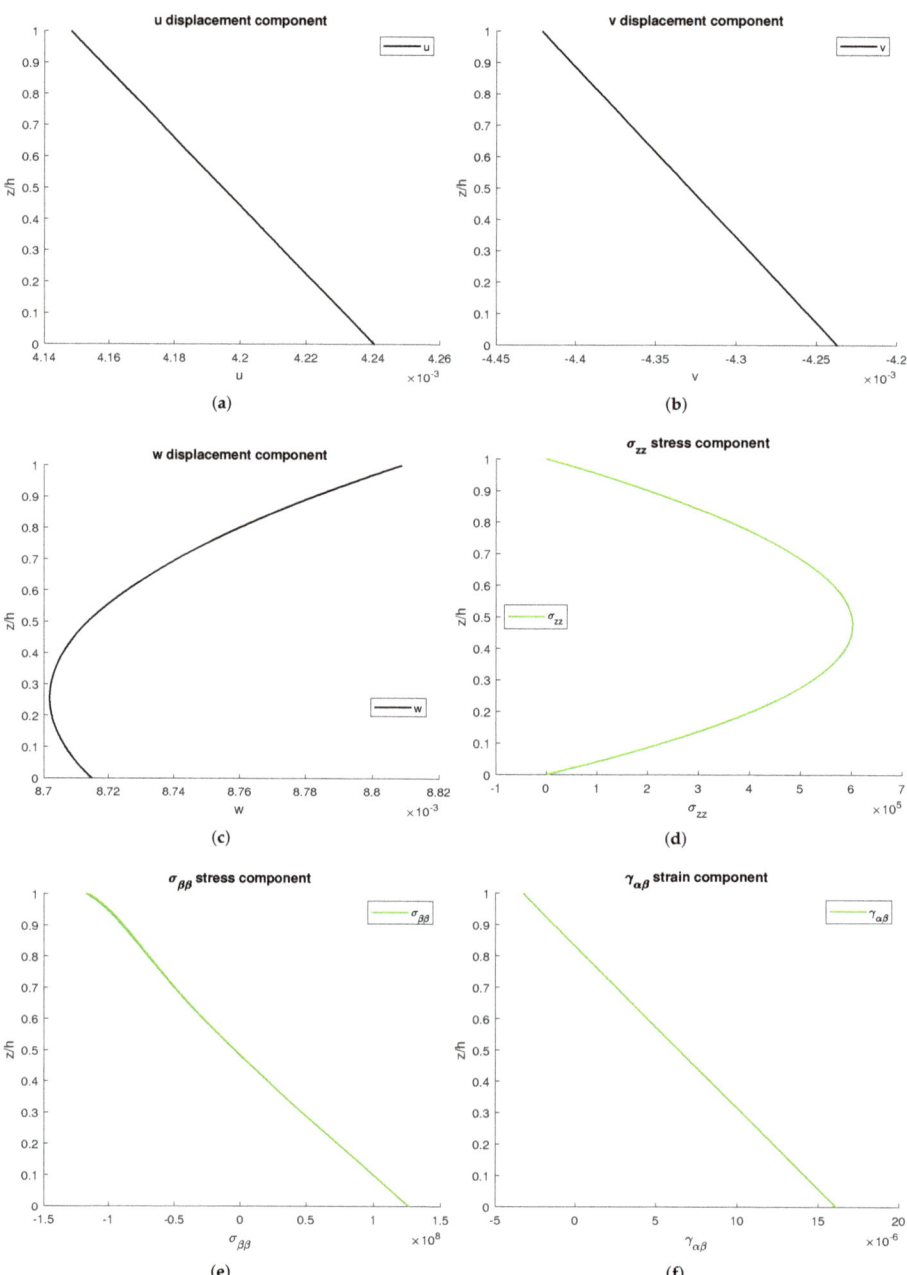

Figure 6. Second benchmark: one-layered FGM closed cylinder with an imposed moisture content on the top and bottom surfaces. The results are calculated for a moderately thin ($R_\alpha/h = 5$) structure via the 3D($\mathcal{M}_c$, 3D) model. (**a**) Amplitude of u displacement component. (**b**) Amplitude of v displacement component. (**c**) Amplitude of w displacement component. (**d**) Amplitude of σ_{zz} stress component. (**e**) Amplitude of $\sigma_{\beta\beta}$ stress component. (**f**) Amplitude of $\gamma_{\alpha\beta}$ strain component.

Table 6. Second benchmark: closed cylinder with a single FGM layer subject to an external moisture content applied to the top and bottom surfaces. The results of all the 3D models consider $N = 3$ and $NL = 300$.

R_α/h	2	4	10	50	100
$\mathcal{M}$ at $(\alpha = a/2, \beta = b/2; \tilde{z} = h/2)$ [−]					
3D($\mathcal{M}_a$)	0.5000	0.5000	0.5000	0.5000	0.5000
3D($\mathcal{M}_c$, 1D)	0.2627	0.2627	0.2627	0.2627	0.2627
3D($\mathcal{M}_c$, 3D)	0.2439	0.2577	0.2619	0.2626	0.2627
u at $(\alpha = 0, \beta = b/2; \tilde{z} = h/3)$ [mm]					
3D($\mathcal{M}_a$)	8.248	7.982	7.370	6.902	6.844
3D($\mathcal{M}_c$, 1D)	5.538	5.125	4.582	4.210	4.161
3D($\mathcal{M}_c$, 3D)	5.332	5.067	4.573	4.210	4.161
w at $(\alpha = a/2, \beta = b/2; \tilde{z} = h/2)$ [mm]					
3D($\mathcal{M}_a$)	14.95	15.14	14.72	14.32	14.27
3D($\mathcal{M}_c$, 1D)	9.600	9.488	9.056	8.714	8.667
3D($\mathcal{M}_c$, 3D)	9.187	9.371	9.037	8.713	8.666
σ_{zz} at $(\alpha = a/2, \beta = b/2; \tilde{z} = 2h/3)$ [MPa]					
3D($\mathcal{M}_a$)	4.442	4.370	2.346	0.5349	0.2716
3D($\mathcal{M}_c$, 1D)	7.308	5.035	2.383	0.5161	0.2605
3D($\mathcal{M}_c$, 3D)	7.578	5.055	2.384	0.5161	0.2605
$\sigma_{\beta\beta}$ at $(\alpha = a/2, \beta = b/2; \tilde{z} = h)$ [MPa]					
3D($\mathcal{M}_a$)	37.63	−6.103	−36.31	−52.37	−54.33
3D($\mathcal{M}_c$, 1D)	−53.79	−85.52	−106.3	−116.9	−118.2
3D($\mathcal{M}_c$, 3D)	−60.83	−87.15	−106.5	−116.9	−118.2
$\gamma_{\alpha\beta}$ at $(\alpha = 0, \beta = 0; \tilde{z} = 0)$ [10^{-6}]					
3D($\mathcal{M}_a$)	516.7	292.5	119.2	23.61	11.78
3D($\mathcal{M}_c$, 1D)	390.8	209.0	82.62	16.12	8.025
3D($\mathcal{M}_c$, 3D)	381.8	207.4	82.50	16.11	8.205

The third benchmark considers a cylindrical sandwich shell panel with an FGM core and simply-supported edges. The top and the bottom skin are in line with the FGM law: the top skin is ceramic as the top surface of the core is; at the same time, the bottom skin is metallic as the bottom surface of the core is. $p = 0.5$ is the coefficient for the volume fraction law across the FGM core, which defines how the elastic and hygrometric properties evolve in the thickness direction. The elastic and hygroscopic properties of both the phases already introduced in the assessments also apply here. The radii of curvature are coherent with those proposed in the previous benchmark, $R_\alpha = 10$ m and $R_\beta = \infty$. The dimension of the reference mid-surface in α direction is a function of the radius of curvature R_α and equals $a = \frac{\pi}{3} R_\alpha$; the dimension in the remaining direction β is fixed and equals $b = 30$. $m = 2$ and $n = 0$ have been chosen as half-wave numbers for the harmonic form of the moisture content imposed at the bottom and the top of the shell. The amplitude of the external fields discussed so far are as follows: the moisture content amplitude is $\mathcal{M}_t = 1.0\%$ on the top and $\mathcal{M}_b = 0.0\%$ on the bottom. This third case study also considers different thickness ratios to evaluate the effects of this parameter; as in the previous case, it is expressed with respect to R_α and ranges from 2 to 100. The volume fraction V_c of the ceramic phase runs from 0 to 1 inside the core; it equals 0 inside the bottom skin as it is fully metallic, 1 inside the top coat as it is fully ceramic. This is visible in Figure 7a,b, showing the volume fraction and the shear modulus along with the thickness coordinate z; the shear modulus of the top skin coincides with that of the ceramic; the shear modulus of the bottom skin coincides with that of the metal. The amplitudes of some variables are reported for all the thickness ratios and at different thickness coordinates in Table 7; Figure 8 explores six variables and shows their trend through z. The Figures rely on the 3D calculated moisture content profile; the table also reports the results obtained through $\mathcal{M}_a$ and $\mathcal{M}_c$, 1D. The differences between the three models can be already seen at the moisture content level and directly reflect the mechanical quantities. Figure 7c,d compares the three moisture content profiles for a thick and a thin cylindrical shell panel and confirms that the differences are sharp not only at a specific thickness coordinate, but throughout all the thickness. The 3D($\mathcal{M}_c$, 1D) model results get closer to those of the 3D($\mathcal{M}_c$, 3D) model as the thickness decreases, and this is clear from the results of Table 7. Figure 8 gives the profile of the tree displacement components, two stresses, and a strain. As in the previous cases, all the quantities evolve with continuity: the mechanical properties are introduced into the model with continuity. The transverse stress σ_{zz} satisfies the external mechanical boundary conditions: it equals 0 at both the top and the bottom surfaces as no external load acts on them.

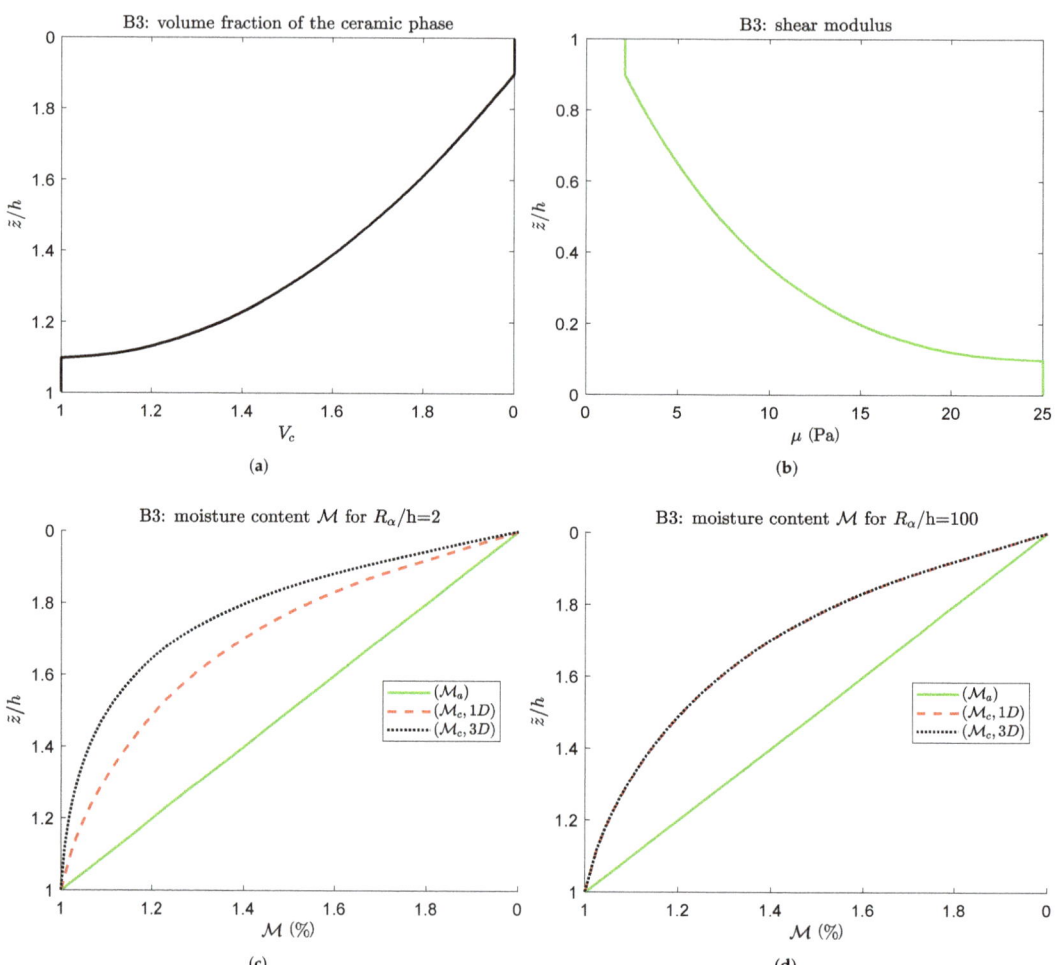

Figure 7. Third benchmark: cylindrical sandwich shell panel featuring an FGM ($p = 0.5$) core with an imposed moisture content on the top and bottom surfaces. The figures show the volume fraction of the ceramic phase, the shear modulus, and the moisture content profiles for a thick and a thin structure through their thickness. (**a**) Volume fraction of the ceramic phase V_c. (**b**) Shear modulus μ. (**c**) Moisture content profile of the $R_\alpha/h = 2$ shell. (**d**) Moisture content profile of the $R_\alpha/h = 100$ shell.

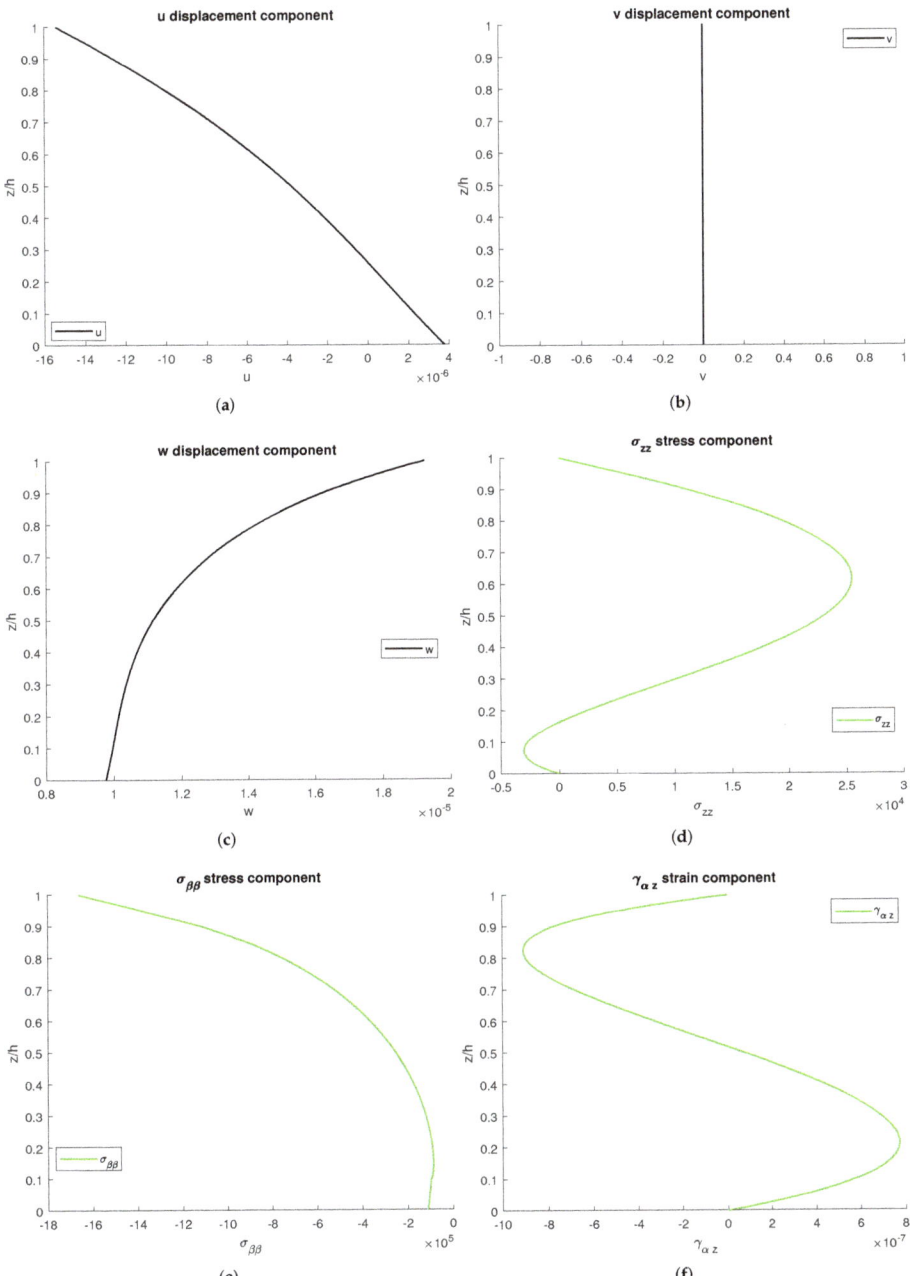

Figure 8. Third benchmark: cylindrical sandwich shell panel featuring an FGM ($p = 0.5$) core with an imposed moisture content on the top and bottom surfaces. The results are calculated for a thick ($R_\alpha/h = 4$) structure via the 3D($\mathcal{M}_c$,3D) model. (**a**) Amplitude of u displacement component. (**b**) Amplitude of v displacement component. (**c**) Amplitude of w displacement component. (**d**) Amplitude of σ_{zz} stress component. (**e**) Amplitude of $\sigma_{\beta\beta}$ stress component. (**f**) Amplitude of $\gamma_{\alpha z}$ strain component.

Table 7. Third benchmark: cylindrical sandwich shell panel with an FGM core subject to an external moisture content applied to the top and bottom surfaces. The results of all the 3D models consider $N = 3$ and $NL = 300$.

R_α/h	2	4	10	50	100
	$\mathcal{M}$ at $(\alpha = a/2, \beta = b/2; \tilde{z} = h/2)$ [−]				
3D($\mathcal{M}_a$)	0.5000	0.5000	0.5000	0.5000	0.5000
3D($\mathcal{M}_c$, 1D)	0.2087	0.2087	0.2087	0.2087	0.2087
3D($\mathcal{M}_c$, 3D)	0.1014	0.1672	0.2008	0.2083	0.2086
	u at $(\alpha = a/2, \beta = 0; \tilde{z} = 3h/4)$ [10^{-3} mm]				
3D($\mathcal{M}_a$)	−16.26	−15.23	−11.79	11.86	41.40
3D($\mathcal{M}_c$, 1D)	−10.53	−9.736	−6.559	15.20	42.34
3D($\mathcal{M}_c$, 3D)	−7.942	−8.844	−6.416	15.20	42.34
	w at $(\alpha = a/2, \beta = b/2; \tilde{z} = h/2)$ [10^{-5} mm]				
3D($\mathcal{M}_a$)	0.5350	1.184	3.310	17.49	35.22
3D($\mathcal{M}_c$, 1D)	0.5000	1.152	3.125	16.16	32.44
3D($\mathcal{M}_c$, 3D)	0.4660	1.116	3.102	16.16	32.44
	σ_{zz} at $(\alpha = a/2, \beta = b/2; \tilde{z} = h/4)$ [kPa]				
3D($\mathcal{M}_a$)	−108.2	−7.415	0.3735	0.2473	0.1354
3D($\mathcal{M}_c$, 1D)	0.6797	5.001	−1.110	−0.5867	−0.3135
3D($\mathcal{M}_c$, 3D)	30.91	6.099	−1.162	−0.5875	−0.3136
	$\sigma_{\beta\beta}$ at $(\alpha = 0, \beta = 0; \tilde{z} = 0)$ [kPa]				
3D($\mathcal{M}_a$)	298.9	142.1	87.89	81.26	81.57
3D($\mathcal{M}_c$, 1D)	−3.571	−92.06	−116.7	−116.9	−116.3
3D($\mathcal{M}_c$, 3D)	−77.09	−112.7	−120.0	−117.0	−116.3
	$\gamma_{\alpha z}$ at $(\alpha = a/2, \beta = 0; \tilde{z} = h/3)$ [10^{-8}]				
3D($\mathcal{M}_a$)	−191.6	−51.36	−11.08	−1.956	−0.9857
3D($\mathcal{M}_c$, 1D)	0.9706	46.64	26.65	5.584	2.790
3D($\mathcal{M}_c$, 3D)	81.83	61.98	27.77	5.593	2.791

The fourth and last benchmark proposes a sandwich spherical shell panel, which embeds an FGM core and features simply supported edges. The lamination scheme is analogous to that discussed in the third benchmark: the bottom skin is metallic, and the top ceramic. Then, the volume fraction of the ceramic phase evolves inside the core through the thickness direction following an exponential law with $p = 0.5$ as chosen coefficients. The hygrometric and elastic properties of the sandwich skin are the same proposed in the previous benchmark and assessments for the metallic and ceramic phases, respectively; those of the core follow the volume fraction law. The exponential trend of the volume fraction V_c vs. the non-dimensional thickness coordinate $\tilde{z}/h$ can be seen in Figure 9a; for completeness, the evolution of the moisture expansion coefficient η through the thickness direction is also given in Figure 10. The spherical shell panel is the only structure among those studied in which both the radii of curvature are non-infinite; furthermore, they take the same value, which equals $R_\alpha = R_\beta = 10$ m. Furthermore, the dimensions of the reference mid-surface are the same in α and β directions as both are a function of the radii of curvature; it holds $a = \frac{\pi}{3} R_\alpha$ and $b = \frac{\pi}{3} R_\beta$. Those dimensions are fixed; however, a wide range of thinner/thicker shells is considered by choosing different thickness ratios: R_α/h ranges from 2 to 100. The amplitude of the moisture content is imposed on the top and the bottom surfaces; it equals $\mathcal{M}_t = 1.0\%$ and $\mathcal{M}_b = 0.0\%$, respectively. As discussed, the external fields are required to have a harmonic form in order for the problem to be exactly solved; $m = 2$ and $n = 2$ are the half-wave numbers considered in this last case study. Table 8 and Figures 9 and 10 summarize an extract of the main results. This fourth benchmark also reports three different sets of results: the elastic model is the same (prefix 3D), but the moisture content profile follows the different approaches. This analysis highlights the distinctions between the three methods. The 3D one is the only model in which no assumptions are made concerning the three-dimensionality of the problem as the moisture content amplitude derives from Fick's law of diffusion. The results coming from 3D($\mathcal{M}_a$) are always wrong because the moisture content evaluation is inaccurate. Considerable differences are present between the calculated and assumed fields. The moisture content profiles of Figure 9c,d once again demonstrate that the 1D and 3D moisture fields get closer in thin structures; despite the thickness, they always differ from the assumed profile, which completely disregards the physics of the problem. This reflects on the results in terms of displacements, strains, and stresses: the differences are high, and 3D($\mathcal{M}_a$) does not provide a reasonable estimate. 3D($\mathcal{M}_c$, 1D) provides acceptable results, but only when the shell is sufficiently thin. As in the previous cases, three displacement components, two stresses, and a strain are shown in their entirety along with the thickness direction. Figure 10 further qualifies the correct introduction of the continuity conditions, elastic/hygrometric properties grading, and mechanical boundary conditions. The transverse stresses $\sigma_{\beta z}$ and σ_{zz} satisfy the external mechanical boundary conditions: they equal 0 at both the top and the bottom surfaces as no external load acts on them. All the quantities are continuous throughout the thickness; this qualifies the division into fictitious layers: they are thin enough to describe the mechanical properties evolution with continuity.

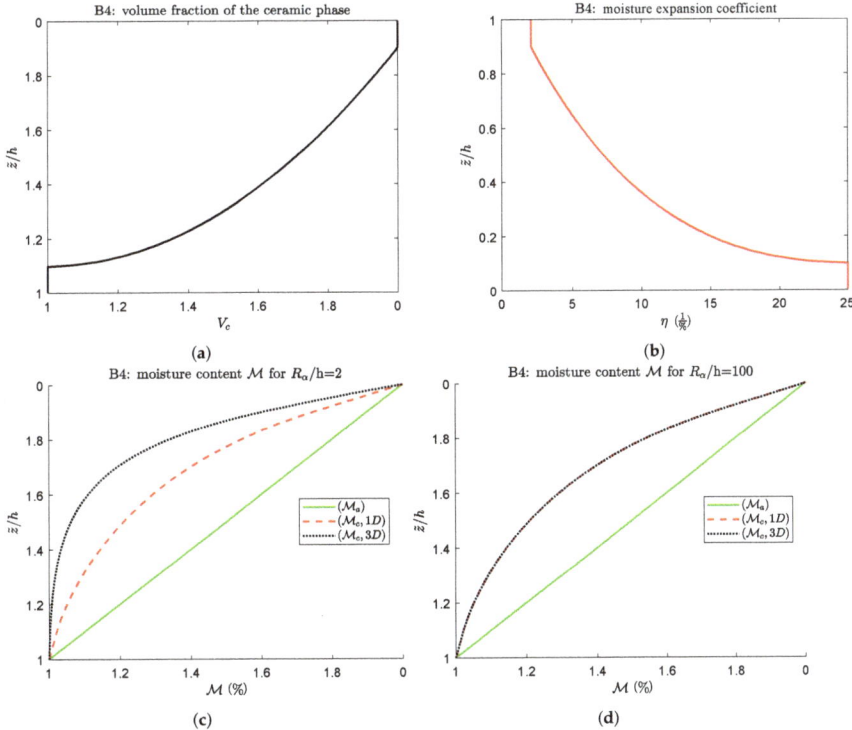

Figure 9. Fourth benchmark: spherical sandwich shell panel featuring a FGM ($p = 0.5$) core with an imposed moisture content on the top and bottom surfaces. The figures show the volume fraction of the ceramic phase, the moisture expansion coefficient, and the moisture content profiles for a thick and a thin structure through their thickness. (**a**) Volume fraction of the ceramic phase V_c. (**b**) Moisture expansion coefficient η. (**c**) Moisture content profile of the $R_\alpha/h = 2$ shell. (**d**) Moisture content profile of the $R_\alpha/h = 100$ shell.

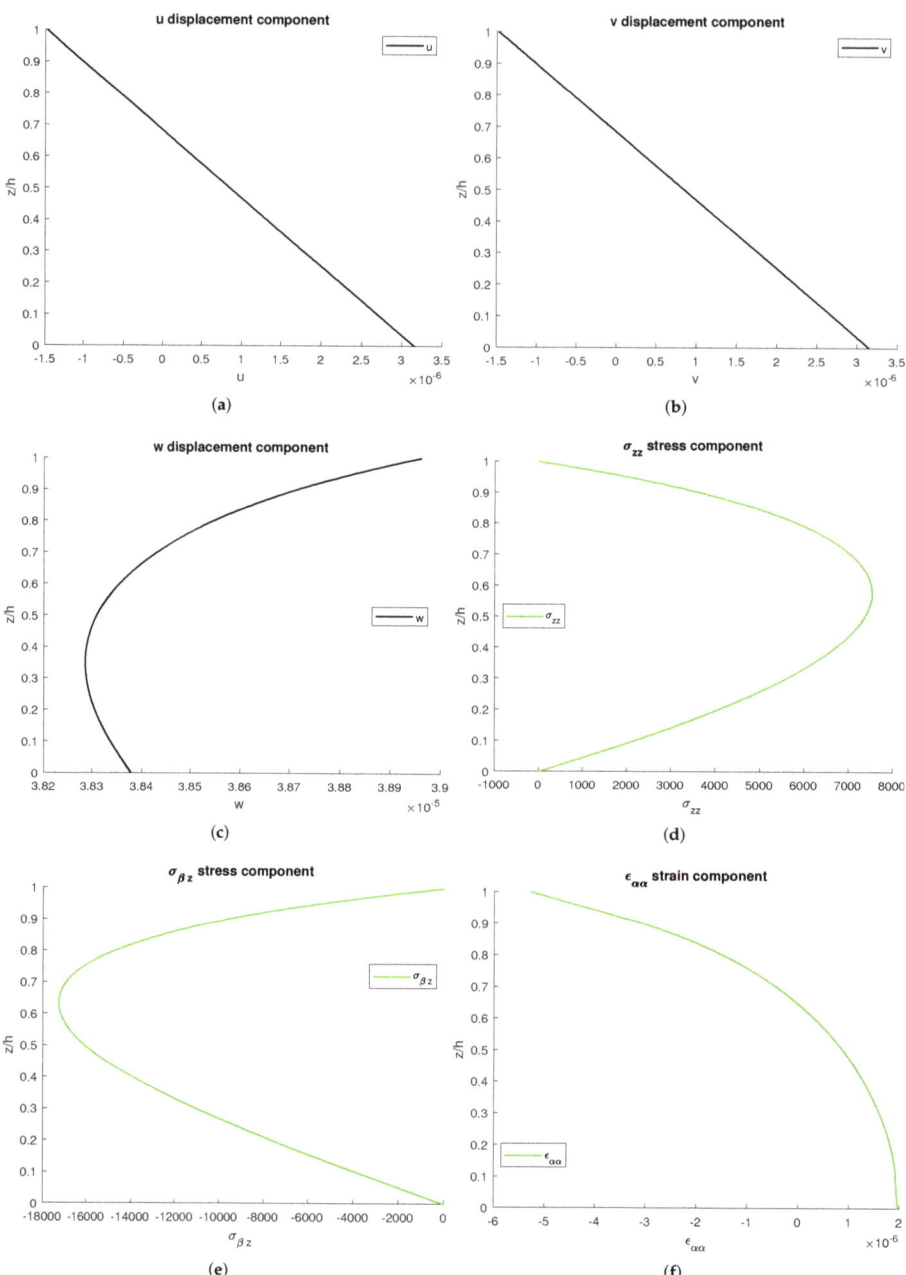

Figure 10. Fourth benchmark: spherical sandwich shell panel featuring a FGM ($p = 0.5$) core with an imposed moisture content on the top and bottom surfaces. The results are calculated for a thin ($R_\alpha/h = 50$) structure via the 3D($\mathcal{M}_c$, 3D) model. (**a**) Amplitude of u displacement component. (**b**) Amplitude of v displacement component. (**c**) Amplitude of w displacement component. (**d**) Amplitude of σ_{zz} stress component. (**e**) Amplitude of $\sigma_{\beta z}$ stress component. (**f**) Amplitude of $\epsilon_{\alpha\alpha}$ strain component.

Table 8. Fourth benchmark, sandwich spherical shell panel with an FGM core subject to an external moisture content applied to the top and bottom surfaces. The results of all the 3D models consider $N = 3$ and $NL = 300$.

R_α/h	2	4	10	50	100
$\mathcal{M}$ at $(\alpha = a/2, \beta = b/2; \tilde{z} = h/2)$ [−]					
3D($\mathcal{M}_a$)	0.5000	0.5000	0.5000	0.5000	0.5000
3D($\mathcal{M}_c$, 1D)	0.2087	0.2087	0.2087	0.2087	0.2087
3D($\mathcal{M}_c$, 3D)	0.0612	0.1376	0.1933	0.2080	0.2085
v at $(\alpha = a/2, \beta = 0; \tilde{z} = h/3)$ [10^{-3} mm]					
3D($\mathcal{M}_a$)	−3.894	−2.822	−1.076	1.741	1.182
3D($\mathcal{M}_c$, 1D)	−1.004	−0.628	0.4564	1.626	1.003
3D($\mathcal{M}_c$, 3D)	0.2033	−0.1980	0.5091	1.625	1.003
w at $(\alpha = a/2, \beta = b/2; \tilde{z} = h/2)$ [10^{-3} mm]					
3D($\mathcal{M}_a$)	4.294	9.003	24.12	59.21	58.93
3D($\mathcal{M}_c$, 1D)	3.366	7.349	18.73	38.36	36.02
3D($\mathcal{M}_c$, 3D)	2.910	6.711	18.31	38.31	36.00
σ_{zz} at $(\alpha = a/2, \beta = b/2; \tilde{z} = h/3)$ [kPa]					
3D($\mathcal{M}_a$)	−337.9	−31.28	−0.2497	6.395	4.422
3D($\mathcal{M}_c$, 1D)	−32.33	35.10	10.25	5.996	3.714
3D($\mathcal{M}_c$, 3D)	85.74	45.62	10.43	5.991	3.713
$\sigma_{\beta z}$ at $(\alpha = a/2, \beta = 0; \tilde{z} = 2h/3)$ [kPa]					
3D($\mathcal{M}_a$)	240.7	125.3	61.39	−8.396	−9.287
3D($\mathcal{M}_c$, 1D)	83.57	25.04	7.627	−17.21	−12.08
3D($\mathcal{M}_c$, 3D)	13.41	3.374	5.225	−17.22	−12.08
$\epsilon_{\alpha\alpha}$ at $(\alpha = a/2, \beta = b/2; \tilde{z} = h)$ [10^{-6}]					
3D($\mathcal{M}_a$)	0.5547	−0.8651	−1.008	−2.249	−3.381
3D($\mathcal{M}_c$, 1D)	−1.988	−3.376	−3.732	−5.260	−6.114
3D($\mathcal{M}_c$, 3D)	−3.912	−4.182	−3.909	−5.268	−6.116

5. Conclusions

The authors proposed a closed-form 3D shell solution that handles the hygro-elastic stress analysis of plates, cylinders, cylindrical shells, and spherical shells while embedding Functionally Graded Material (FGM) layers. First, the author imposed the external moisture content on the top and the bottom surfaces. The moisture conditions act in steady state as an external load; calculating the moisture content profile is a prerequisite. The authors showed that three approaches might be used to determine the moisture content profile along the thickness direction and coupled them with a consolidated elastic solution. The results demonstrated the importance of a correct moisture content profile evaluation in the thickness direction. The 3D Fick's law of diffusion is the only way to obtain exact results when the structures embed FG layers; it is also necessary when the structures are sufficiently thick. On the other hand, the 1D Fick's law of diffusion comes closer to it only when structures are thin; as a rule of thumb, the results of the two models are almost coinciding only from a thickness ratio of 50. The problem relies on a set of differential equations in the thickness direction. The authors demonstrated that the exponential matrix method is a reliable way to solve it, provided that the structures are divided into a sufficiently high number of mathematical layers. This layer-wise approach is critical to get a reliable description of the material properties grading; as a rule of thumb, 300 mathematical layers always deliver the correct results. This achievement is general and does not depend on the geometry, FGM law, and lamination sequence/scheme.

Author Contributions: Data curation, R.T.; Formal analysis, S.B. and R.T.; Methodology, S.B.; Software, S.B.; Supervision, S.B.; Validation, R.T.; Writing—original draft, R.T. All authors have read and agreed to the published version of the manuscript.

Funding: This research received no external funding.

Conflicts of Interest: The authors declare no conflict of interest.

References

1. Zenkour, A.M. Hygro-thermo-mechanical effects on FGM plates resting on elastic foundations. *Compos. Struct.* **2010**, *93*, 234–238. [CrossRef]
2. Tabrez, S.; Mitra, M.; Gopalakrishnan, S. Modeling of degraded composite beam due to moisture absorption for wave based detection. *CMES Comput. Model. Eng. Sci.* **2007**, *22*, 77–90.
3. Swaminathan, K.; Sangeetha, D.M. Thermal analysis of FGM plates: A critical review of various modeling techniques and solution methods. *Compos. Struct.* **2017**, *160*, 43–60. [CrossRef]
4. Allahyarzadeh, M.; Aliofkhazraei, M.; Rouhaghdam, A.S.; Torabinejad, V. Gradient electrodeposition of Ni-Cu-W (alumina) nanocomposite coating. *Mater. Des.* **2016**, *107*, 74–81. [CrossRef]
5. Rezaiee-Pajand, M.; Sobhani, E.; Masoodi, A.R. Free vibration analysis of functionally graded hybrid matrix/fiber nanocomposite conical shells using multiscale method. *Aerosp. Sci. Technol.* **2020**, *105*, 105998. [CrossRef]
6. Sobhani, E.; Masoodi, A.R. Natural frequency responses of hybrid polymer/carbon fiber/FG-GNP nanocomposites paraboloidal and hyperboloidal shells based on multiscale approaches. *Aerosp. Sci. Technol.* **2021**, *119*, 107111. [CrossRef]
7. Sobhani, E.; Masoodi, A.R.; Ahmadi-Pari, A.R. Vibration of FG-CNT and FG-GNP sandwich composite coupled Conical-Cylindrical-Conical shell. *Compos. Struct.* **2021**, *273*, 114281. [CrossRef]
8. Rezaiee-Pajand, M.; Sobhani, E.; Masoodi, A.R. Semi-analytical vibrational analysis of functionally graded carbon nanotubes coupled conical-conical shells. *Thin-Walled Struct.* **2021**, *159*, 107272. [CrossRef]
9. Sobhani, E.; Moradi-Dastjerdi, R.; Behdinan, K.; Masoodi, A.R.; Ahmadi-Pari, A.R. Multifunctional trace of various reinforcements on vibrations of three-phase nanocomposite combined hemispherical-cylindrical shells. *Compos. Struct.* **2022**, *279*, 114798. [CrossRef]
10. Rezaiee-Pajand, M.; Masoodi, A.R.; Mokhtari, M. Static analysis of functionally graded non-prismatic sandwich beams. *Adv. Comput. Des.* **2018**, *3*, 165–190.
11. Rezaiee-Pajand, M.; Rajabzadeh-Safaei, N.; Masoodi, A.R. An efficient curved beam element for thermo-mechanical nonlinear analysis of functionally graded porous beams. *Structures* **2020**, *28*, 1035–1049. [CrossRef]
12. Rezaiee-Pajand, M.; Masoodi, A.R.; Arabi, E. Geometrically nonlinear analysis of FG doubly-curved and hyperbolical shells via laminated by new element. *Steel Compos. Struct.* **2018**, *28*, 389–401.
13. Rezaiee-Pajand, M.; Arabi, E.; Masoodi, A.R. Nonlinear analysis of FG-sandwich plates and shells. *Aerosp. Sci. Technol.* **2019**, *87*, 178–189. [CrossRef]
14. Rezaiee-Pajand, M.; Masoodi, A.R. Analyzing FG shells with large deformations and finite rotations. *World J. Eng.* **2019**, *5*, 636–647. [CrossRef]
15. Das, N.C.; Das, S.N.; Das, B. Eigenvalue approach to thermoelasticity. *J. Therm. Stress.* **1983**, *6*, 35–43. [CrossRef]
16. Altay, G.A.; Dokmeci, M.C. Fundamental variational equations of discontinuous thermopiezoelectric fields. *Int. J. Eng. Sci.* **1996**, *34*, 769–782. [CrossRef]
17. Altay, G.A.; Dokmeci, M.C. Some variational principles for linear coupled thermoelasticity. *Int. J. Solids Struct.* **1996**, *33*, 3937–3948. [CrossRef]
18. Kosnski, W.; Frischmuth, K. Thermomechanical coupled waves in a nonlinear medium. *Wave Motion* **2001**, *34*, 131–141. [CrossRef]
19. Altay, G.A.; Dokmeci, M.C. Coupled thermoelastic shell equations with second sound for high-frequency vibrations of temperature-dependent materials. *Int. J. Solids Struct.* **2001**, *38*, 2737–2768. [CrossRef]
20. Cannarozzi, A.A.; Ubertini, F. A mixed variational method for linear coupled thermoelastic analysis. *Int. J. Solids Struct.* **2001**, *38*, 717–739. [CrossRef]
21. Brischetto, S.; Torre, R. 3D shell model for the thermo-mechanical analysis of FGM structures via imposed and calculated temperature profiles. *Aerosp. Sci. Technol.* **2019**, *85*, 125–149. [CrossRef]
22. Soldatos, K.P.; Ye, J. Axisymmetric static and dynamic analysis of laminated hollow cylinders composed of monoclinic elastic layers. *J. Sound Vib.* **1995**, *184*, 245–259. [CrossRef]
23. Messina, A. Three Dimensional Free Vibration Analysis of Cross-Ply Laminated Plates through 2D and Exact Models. In Proceedings of the 3rd International Conference on Integrity, Reliability and Failure, Porto, Portugal, 20–24 July 2009; pp. 20–24.
24. Fan, J.; Zhang, J. Analytical solutions for thick, doubly curved, laminated shells. *J. Eng. Mech.* **1992**, *118*, 1338–1356. [CrossRef]
25. Brischetto, S. Exact three-dimensional static analysis of single- and multi-layered plates and shells. *Compos. Part B* **2017**, *119*, 230–252. [CrossRef]
26. Brischetto, S. A closed-form 3D shell solution for multilayered structures subjected to different load combinations. *Aerosp. Sci. Technol.* **2017**, *70*, 29–46. [CrossRef]

27. Brischetto, S.; Torre, R. Convergence investigation for the exponential matrix and mathematical layers in the static analysis of multilayered composite structures. *J. Compos. Sci.* **2017**, *1*, 19. [CrossRef]
28. Brischetto, S.; Torre, R. A 3D layer-wise model for the correct imposition of transverse shear/normal load conditions in FGM shells. *Int. J. Mech. Sci.* **2018**, *136*, 50–66. [CrossRef]
29. Brischetto, S. A general exact elastic shell solution for bending analysis of functionally graded structures. *Compos. Struct.* **2017**, *175*, 70–85. [CrossRef]
30. Brischetto, S. Curvature approximation effects in the free vibration analysis of functionally graded shells. *Int. J. Appl. Mech.* **2016**, *8*, 1650079. [CrossRef]
31. Ozisik, M.N. *Heat Conduction*; John Wiley & Sons, Inc.: New York, NY, USA, 1993.
32. Povstenko, Y. *Fractional Thermoelasticity*; Springer: Berlin/Heidelberg, Germany, 2015.
33. Moon, P.; Spencer, D.E. *Field Theory Handbook. Including Coordinate Systems, Differential Equations and Their Solutions*; Springer: Berlin/Heidelberg, Germany, 1988.
34. Mikhailov, M.D.; Ozisik, M.N. *Unified Analysis and Solutions of Heat and Mass Diffusion*; Dover Publications Inc.: New York, NY, USA, 1984
35. Tungikar, V.; Rao, B.K.M. Three dimensional exact solution of thermal stresses in rectangular composite laminates. *Compos. Struct.* **1994**, *27*, 419–439. [CrossRef]
36. Laoufi, I.; Ameur, M.; Zidi, M.; Abbes, A.B.E.; Boushala, A.A. Mechanical and hygrothermal behaviour of functionally graded plates using a hyperbolic shear deformation theory. *Steel Compos. Struct.* **2016**, *20*, 889–911. [CrossRef]
37. Inala, R. Influence of Hygrothermal Environment and FG Material on Natural Frequency and Parametric Instability of Plates. *Mech. Adv. Compos. Struct.* **2020**, *7*, 89–101.
38. Dai, T.; Dai, H.; Li, J.; He, Q. Hygrothermal mechanical behavior of a FG circular plate with variable thickness. *Chin. J. Theor. Appl. Mech.* **2019**, *51*, 512–523.
39. Zenkour, A.M.; Radwan, A.F. Bending and buckling analysis of FGM plates resting on elastic foundations in hygrothermal environment. *Arch. Civ. Mech. Eng.* **2020**, *20*, 234–238. [CrossRef]
40. Boukhelf, F. Hygro-thermo-mechanical bending analysis of FGM plates using a new HSDT. *Smart Struct. Syst.* **2018**, *21*, 75–97.
41. Zidi, M.; Tounisi, A.; Houari, M.S.A.; Bedia, E.A.; Beg, O.A. Bending analysis of FGM plates under hygro-thermo-mechanical loading using a four variable refined plate theory. *Aerosp. Sci. Technol.* **2014**, *34*, 512–523. [CrossRef]
42. Akbas, S.D. Hygro-thermal post-buckling analysis of a functionally graded beam. *Coupled Syst. Mech. Int. J.* **2019**, *8*, 459–471.
43. Chiba, R.; Sugano, Y. Transient hygrothermoelastic analysis of layered plates with one-dimensional temperature and moisture variations through the thickness. *Compos. Struct.* **2011**, *93*, 2260–2268. [CrossRef]
44. Gigliotti, M.; Jacquemin, F.; Molimard, J.; Vautrin, A. Modelling and experimental characterisation of hygrothermoelastic stress in polymer matrix composites. *Macromol. Symp.* **2007**, *257*, 199–210. [CrossRef]
45. Gigliotti, M.; Jacquemin, F.; Vautrin, A. Assessment of approximate models to evaluate transient and cyclical hygrothermoelastic stress in composite plates. *Int. J. Solids Struct.* **2007**, *44*, 733–759. [CrossRef]
46. Khalil, M.; Bakhiet, E.; El-Zoghby, A. Optimum design of laminated composites subjected to hygrothermal residual stresses. *Proc. Inst. Mech. Eng. Part L J. Mater. Des. Appl.* **2001**, *215*, 175–186. [CrossRef]
47. Kollár, L.P.; Patterson, J.M. Composite cylindrical segments subjected to hygrothermal and mechanical loads. *Int. J. Solids Struct.* **1993**, *30*, 2525–2545. [CrossRef]
48. Shen, H.-S. The effects of hygrothermal conditions on the postbuckling of shear deformable laminated cylindrical shells. *Int. J. Solids Struct.* **2001**, *38*, 6357–6380. [CrossRef]
49. Lee, S.Y.; Chou, C.J.; Jang, J.L.; Lin, J.S. Hygrothermal effects on the linear and nonlinear analysis of symmetric angle-ply laminated plates. *Compos. Struct.* **1992**, *21*, 41–48. [CrossRef]
50. Hufenbach, W.; Kroll, L. Stress analysis of notched anisotropic finite plates under mechanical and hygrothermal loads. *Arch. Appl. Mech.* **1999**, *69*, 145–159. [CrossRef]
51. Khoshbakht, M.; Lin, M.W.; Berman, J.B. Analysis of moisture-induced stresses in an FRP composites reinforced masonry structure. *Finite Elem. Anal. Des.* **2006**, *42*, 414–429. [CrossRef]
52. Kundu, C.K.; Han, J.-H. Vibration characteristics and snapping behaviour of hygro-thermo-elastic composite doubly curved shells. *Compos. Struct.* **2009**, *91*, 306–317. [CrossRef]
53. Kundu, C.K.; Han, J.-H. Nonlinear buckling analysis of hygrothermoelastic composite shell panels using finite element method. *Compos. Part B* **2009**, *40*, 313–328. [CrossRef]
54. Marques, S.P.C.; Creus, G.J. Geometrically nonlinear finite element analysis of viscoelastic composite materials under mechanical and hygrothermal loads. *Comput. Struct.* **1994**, *53*, 449–456. [CrossRef]
55. Naidu, N.V.S.; Sinha, P.K. Nonlinear finite element analysis of laminated composite shells in hygrothermal environments. *Compos. Struct.* **2005**, *69*, 387–395. [CrossRef]
56. Patel, B.P.; Ganapathi, M.; Makhecha, D.P. Hygrothermal effects on the structural behaviour of thick composite laminates using higher-order theory. *Compos. Struct.* **2002**, *56*, 25–34. [CrossRef]
57. Sereir, Z.; Adda-Bedia, E.A.; Tounsi, A. Effect of temperature on the hygrothermal behaviour of unidirectional laminated plates with asymmetrical environmental conditions. *Compos. Struct.* **2006**, *72*, 383–392. [CrossRef]

58. Sereir, Z.; Tounsi, A.; Adda-Bedia, E.A. Effect of the cyclic environmental conditions on the hygrothermal behavior of the symmetric hybrid composites. *Mech. Adv. Mater. Struct.* **2006**, *13*, 237–248. [CrossRef]
59. Sereir, Z.; Adda-Bedia, E.A.; Boualem, N. The evolution of transverse stresses in hybrid composites under hygrothermal loading. *Mater. Des.* **2001**, *32*, 3120–3126. [CrossRef]
60. Ram, K.S.S.; Sinha, P.K. Hygrothermal effects on the bending characteristics of laminated composite plates. *Comput. Struct.* **1991**, *40*, 1009–1015. [CrossRef]
61. Ram, K.S.S.; Sinha, P.K. Hygrothermal bending of laminated composite plates with a cutout. *Comput. Struct.* **1992**, *43*, 1105–1115. [CrossRef]
62. Benkhedda, A.; Tounsi, A.; Adda-Bedia, E.A. Effect of temperature and humidity on transient hygrothermal stresses during moisture desorption in laminated composite plates. *Compos. Struct.* **2008**, *82*, 629–635. [CrossRef]
63. Lo, S.H.; Zhen, W.; Cheung, Y.K.; Wanji, C. Hygrothermal effects on multilayered composite plates using a refined higher order theory. *Compos. Struct.* **2010**, *92*, 633–646. [CrossRef]
64. Parhi, P.K.; Bhattacharyya, S.K.; Sinha, P.K. Hygrothermal effects on the dynamic behavior of multiple delaminated composite plates and shells. *J. Sound Vib.* **2001**, *248*, 195–214. [CrossRef]
65. Raja, S.; Sinha, P.K.; Prathap, G.; Dwarakanathan, D. Influence of active stiffening on dynamic behaviour of piezo-hygro-thermo-elastic composite plates and shells. *J. Sound Vib.* **2004**, *278*, 257–283. [CrossRef]
66. Ghosh, A. Hygrothermal effects on the initiation and propagation of damage in composite shells. *Aircr. Eng. Aerosp. Technol. Int. J.* **2008**, *4*, 386–399. [CrossRef]
67. Szekeres, A. Analogy between heat and moisture thermo-hygro-mechanical tailoring of composites by taking into account the second sound phenomenon. *Comput. Struct.* **2000**, *76*, 145–152. [CrossRef]
68. Tay, A.A.O.; Goh, K.Y. A Study of Delamination Growth in the Die-Attach Layer of Plastic IC Packages under Hygrothermal Loading During Solder Reflow. In Proceedings of the 49th Electronic Components and Technology Conference, San Diego, CA, USA, 1–4 June 1999; pp. 694–701.
69. Tay, A.A.O.; Goh, K.Y. A study of delamination growth in the die-attach layer of plastic IC packages under hygrothermal loading during solder reflow. *IEEE Trans. Device Mater. Reliab.* **2003**, *3*, 144–151. [CrossRef]
70. Leissa, A.W. *Vibration of Shells*; NASA SP-288: Washington, DC, USA, 1973.
71. Boyce, W.E.; DiPrima, R.C. *Elementary Differential Equations and Boundary Value Problems*; John Wiley & Sons Ltd.: New York, NY, USA, 2001.
72. Open Document. Systems of Differential Equations. Available online: http://www.math.utah.edu/gustafso/ (accessed on 30 May 2013).
73. Reddy, J.N.; Cheng, Z.-Q. Three-dimensional thermomechanical deformations of functionally graded rectangular plates. *Eur. J. Mech.-A/Solids* **2001**, *20*, 841–855. [CrossRef]
74. Cinefra, M.; Carrera, E.; Brischetto, S.; Belouettar, S. Thermo-mechanical analysis of functionally graded shells. *J. Therm. Stress.* **2010**, *33*, 942–963. [CrossRef]

MDPI
St. Alban-Anlage 66
4052 Basel
Switzerland
Tel. +41 61 683 77 34
Fax +41 61 302 89 18
www.mdpi.com

Applied Sciences Editorial Office
E-mail: applsci@mdpi.com
www.mdpi.com/journal/applsci